SCANDALOUS ERROR

Scandalous Error

Calendar Reform and Calendrical Astronomy in Medieval Europe

C. PHILIPP E. NOTHAFT

OXFORD
UNIVERSITY PRESS

OXFORD
UNIVERSITY PRESS

Great Clarendon Street, Oxford, OX2 6DP,
United Kingdom

Oxford University Press is a department of the University of Oxford. It furthers the University's objective of excellence in research, scholarship, and education by publishing worldwide. Oxford is a registered trade mark of Oxford University Press in the UK and in certain other countries

First Edition published in 2018

Published in the United States of America by Oxford University Press
198 Madison Avenue, New York, NY 10016, United States of America

British Library Cataloguing in Publication Data
Data available

Library of Congress Control Number: 2017956294

ISBN 978–0–19–879955–9

In memory of Fritz Saaby Pedersen
(1945–2016)

Acknowledgements

A sizeable number of institutions and individuals have supported my work on this book over a period of close to four years. For intellectual support, my greatest debt of gratitude is owed to Charles Burnett, Leofranc Holford-Strevens, David Juste, and Immo Warntjes, all four of whom have had a shaping influence on the genesis of this book. I thank them for their advice and occasional criticism, and for reading through earlier drafts. On the institutional side of things, I remain immensely grateful to the administration and staff of the Warburg Institute and to the Warden and Fellows of All Souls College, who confirmed their respective reputations as patrons of abstruse scholarship by offering me a place to work (and live). I also thank the numerous libraries and research institutions—too many to mention—who have granted me access to their holdings or furnished me with reproductions of their manuscripts. Above all, my thanks are due to the library staff of the Monumenta Germaniae Historica (Munich), in particular to Arno Mentzel-Reuters and Tonja Müller-Tamke, for providing swift and repeated access to their microfilm holdings. Many further individuals deserve at least a brief mention for their positive influence on this project, whether as correspondents, editors, conference hosts, mentors, or conversation partners. I thank, in alphabetic order: Matthew Champion, Michał Choptiany, Karen Desmond, Sten Ebbesen, Mordechai Feingold, Giles Gasper, Anthony Grafton, Robert Halleux, Matthieu Husson, Rob Iliffe, Alfred Lohr, Magnus Löfflmann, Edouard Mehl, Dáibhí Ó Cróinín, James Palmer, Fritz Pedersen (†), Olivier de Solan, Sacha Stern, Noel Swerdlow, Faith Wallis, and Helmut Zedelmaier.

Contents

List of Illustrations

List of Tables

Preface

For most of the Middle Ages, European Christians were aware that the calendrical cycles that regulated their civil and religious life were out of tune with the pertinent celestial phenomena. When and how did they first learn about this problem? How did it manifest itself in medieval literature and learned culture? How did it interact with the wider history of astronomy in this period? And why did it take centuries, until the Gregorian reform of 1582, for a viable solution to emerge? My interest in these and other questions relating to the pre-history of the Gregorian calendar dates back to earlier research on the history of historical and technical chronology, which became the core subject of my monograph *Dating the Passion* (= Nothaft 2012). The Christian 'calendar problem' itself was to re-emerge as one of the themes in my second book, on *Medieval Latin Christian Texts on the Jewish Calendar* (= Nothaft 2014d), which brought with it the opportunity to study relevant texts that had received little or no attention in previous scholarship. My decision to devote a comprehensive study to the long medieval history of calendar reform, from Julius Caesar to Johannes Regiomontanus and beyond, provided a perfect excuse to spend the academic years 2013/14 and 2014/15 at the Warburg Institute, University of London, which kindly supported my research by offering me one of its coveted Frances A. Yates long-term fellowships and by granting me access to its matchless open-stack library. Since October 2015 I have had the privilege of continuing my work as a postdoctoral fellow at All Souls College, Oxford, whose Warden and Fellows have provided crucial institutional and financial support during the final stages of research and writing. Both institutions allowed me to work at my own pace and to pre-digest some of the material uncovered over the past four years in the form of articles and critical editions. In the case of Walcher of Malvern and his astronomical-computistical writings, one of these editions has taken the shape of a separate book (= Nothaft 2017e), the introductory study to which gave me the chance to tell a small part of the relevant story in advance, thus relieving the present volume of some of the gorier detail (as far as the eleventh and twelfth centuries are concerned).

Even with the opportunity to hive off certain discussions, the intent to cover the Middle Ages in their entirety has imposed on me the need to make a number of sacrifices, some great and some small, in order to keep the resulting book from reaching unwieldy proportions. Sections and topics originally planned for inclusion have been omitted in the course of successive revisions. Some areas of European astronomy that may have been valuable for providing context or comparison, for example the Jewish astronomical scenes on the Iberian peninsula and in the South of France, will barely receive any attention. Others, such as the calendar-reform discussion that took place in the Byzantine world, will occupy much less space than originally planned. More significant still are the compromises I had to make in the area of documentation. The bibliography for this book, which contained *c.*1,500 entries in earlier drafts, has been decreased by roughly 40 per cent for the present publication. On numerous occasions, references to literature I would have otherwise deemed important had to be excised after the fact, with the result that my claims will sometimes appear less richly documented than initially planned or hoped for. (My failures to cite particular titles should not always be interpreted as signs of ignorance.) For the same reasons, I was generally reluctant to highlight any disagreements between my views and those expressed by other scholars. Such disagreements, whether on specific pieces of data or larger questions of interpretation, are rather frequent, but to spell them all out would have steered this book too far off course. My text will hopefully still contain enough hints for readers to infer where I stand on individual issues.

In order to keep the footnotes as concise as possible, I have opted for an author–date system for all books and articles printed after 1800. For books printed before this threshold, full references will be given in the footnotes, but the titles will not be listed separately in the bibliography. Source editions that belong to a larger series are cited according to their standard abbreviations, as shown in the list of Abbreviations. Unless stated otherwise, all quotations from Latin and other foreign-language sources appear in my own English translation. The original text will be added in the footnotes only if the immediate source is a manuscript and hence still unpublished. In all other cases, readers are kindly referred to the available editions. Since most of the calendrical terminology and other conventions used in this book will be explained in the course of the text, or can be deemed self-explanatory, I decided to forgo including a glossary. Readers who are nevertheless interested in using such a glossary will find excellent examples in the *Oxford Companion to the Year* (= Blackburn and Holford-Strevens 1999: 879–83) and in Immo Warntjes's edition of the *Munich Computus* (= Warntjes 2010c: 341–53).

The mathematics used in this book will rarely exceed primary-school level and should be accessible to readers from any background. One parameter of great utility in discussing the history of calendrical astronomy is the size of the error inherent in a given calendrical cycle, which can be stated as the number of years it will take for the cycle to have gained, or lost, one day in relation to the phenomenon it is supposed to track. Following a convention introduced by John David North (1934–2008; see North 1983: 78), I shall distinguish M, the 'slippage' of the lunar cycle relative to the new and full moons, from J, the 'drift' of the Julian calendar relative to the vernal equinox. Accordingly, '$J = 130$ years' will denote the idea that the vernal equinox drifts through the calendar at a rate of one day in 130 years. Likewise, '$M = 308\frac{1}{2}$ years' will indicate that the error in the lunar cycle increases by one day in $308\frac{1}{2}$ years. To facilitate comparison between various modern and historical values, I shall often cite astronomical parameters in both sexagesimal and decimal notation. Thus, the same value for the mean synodic month may appear side by side as (a) 29;31,50d, (b) 29d 12;44h, and (c) 29.530555d. In (a) and (b), the semicolon marks the start of a sequence of sexagesimal fractions, with the individual sexagesimal places being separated by regular commas, such that '29;31,50d' is meant to be equivalent to $29 + \frac{31}{60} + \frac{50}{3600}$ days. Decimal fractions are usually only given to the sixth place, without rounding. Where the main text or the footnotes refer to the results of calculations based on ancient and medieval astronomical tables, these calculations have mostly been simulated with the help of Raymond Mercier's computer program *Deviations*, which is available for download at <http://www.raymondm.co.uk>. For modern astronomical data and calendrical conversions, I have generally used the program *Kairos 4.0*, available at the same website. Additional data on historical eclipses have been taken from the eclipse catalogues hosted by NASA (<https://eclipse.gsfc.nasa.gov/eclipse.html>). For equinox and solstice times, I have relied on the data provided by the IMCCE (<http://www.imcce.fr/fr/grandpublic/temps/saisons.html>). All websites referenced here and elsewhere in the book were last accessed on 7 September 2017.

Oxford,
September 2017

Abbreviations

AHES	*Archive for History of Exact Sciences*
AIHS	*Archives internationales d'histoire des sciences*
BAV	Biblioteca Apostolica Vaticana
BL	British Library
BML	Biblioteca Medicea Laurenziana
BnF	Bibliothèque nationale de France
BodLib	Bodleian Library
BSB	Bayerische Staatsbibliothek
CCCM	*Corpus Christianorum: Continuatio Mediaevalis*
CCSL	*Corpus Christianorum: Series Latina*
CSEL	*Corpus Scriptorum Ecclesiasticorum Latinorum*
JHA	*Journal for the History of Astronomy*
MGH	Monumenta Germaniae Historica
Epp	*Epistolae*
SS	*Scriptores (in folio)*
NKCR	Národní knihovna České republiky
ÖNB	Österreichische Nationalbibliothek
PG	*Patrologiae cursus completus, series Graeca*
PL	*Patrologiae cursus completus, series Latina*
UFB	Universitäts- und Forschungsbibliothek
ZPE	*Zeitschrift für Papyrologie und Epigraphik*

Introduction

The story of the Western calendar is as old as the city of Rome, whose founder and first king, Romulus, became the first Latin writer to deal with the subject of time reckoning. He is known to have devised a calendar composed of ten months, although the rationale behind this number is no longer perfectly clear. Maybe he considered this the maximum time of a pregnancy, or he chose ten months because it was the period during which Roman widows were required to abstain from marriage after their husband's death. Romulus's successor was Numa Pompilius, a valiant warrior who rendered the Roman year more adequate by adding two months, January and February, to its end and thereby created the twelve-month-long year we are still familiar with. Yet the resulting lunar year of 354 days was still too short to keep the months aligned with the seasons. Julius Caesar, an accomplished astronomer, improved upon Numa's calendar by adding 11 days and 6 hours and, by doing so, inaugurated the solar calendar the Church later came to use for its divine services. Afterwards a Roman abbot named Dionysius put together certain tables for the calculation of Easter, which are still much in use, as well as a widely read compilation on the subject. Efforts to supplement his work were later made by the Venerable Bede and an English abbot named Helperic, who were in turn fiercely criticized by Gerland, a man born in the borderlands between France and Germany. The Church, however, judged Gerland's work too verbose and confusing, and instead continued to adhere to the writings of Dionysius, Bede, and Helperic. Many other authors have followed in their wake, with some conserving ancient lore and others introducing new inventions, such as techniques to calculate the dates of the year on the joints and fingers of one's hand. The demand for textbooks on calendrical reckoning has never since stopped...

* * *

The historical summary just offered does not stem from any modern introduction to the Western calendar—indeed, some of its elements would no longer be recognized as accurate—but freely paraphrases the words of a learned man of the late thirteenth century, who was variously called, or called himself, John of Brunswick, John of Saxony, John the German, or John from the Beautiful Stream (*Iohannes de Pulchro Rivo*).[1] The extant details of his life and career are extremely sparse. Born

[1] See the introduction to John of Brunswick, *Sententia compoti nove compilationis*, as found in MS Vatican City, BAV, lat. 3112, fol. 29v, and quoted in Nothaft 2014c: 232–3. Contrary to the information this passage provides, Romulus and Numa Pompilius played no historically verifiable role in the evolution of the Roman calendar, Dionysius was not an abbot, and Helperic hailed from modern-day France rather than England.

in Brunswick at an unknown date, John appears to have spent some time as a student or master at the University of Paris, where in 1289 he started to write his most popular work, a brief *Compotus manualis* that survives in at least twenty-eight copies. By 1297 he was back in Goslar in his native Saxony, where he compiled a *Compotus novus* as well as a detailed commentary on the same work, completed on 1 April 1298.[2] In the parlance of his day John of Brunswick was a *compotista*, someone who practiced the art of *compotus*—the thirteenth-century spelling of *computus*. On the surface, *computus* was there simply to teach aspiring members of the clergy how to calculate the date of Easter and the other mobile feast days of the liturgical year. In practice, however, it often made them grapple with the problems of time reckoning on a much broader scale, as *computus* texts (or *computi*) were capable of drawing knowledge from a diverse array of disciplines: astronomy and mathematics always occupied front and centre, but the flanks were habitually lined by history, etymology, and theology, sometimes even by astrology, cosmography, poetry, or medicine.[3]

John of Brunswick was hardly the first *compotista* to begin a work on the medieval Church calendar with an outline of its history since Roman times. Similarly worded introductions to the discipline can indeed be encountered in various other manuscripts produced during his lifetime and, just like John's text, they convey the impression that high-medieval practitioners of the art of *computus* considered themselves to be standing on the shoulders of giants.[4] Over many centuries, the calendar and the method to keep it in order had been first established and then perpetuated by a long succession of qualified, sometimes saintly, men. By continuing this tradition and refining it generation after generation, scholars such as John of Brunswick ensured that the liturgical cycles that governed the life of the Church remained firm and undisturbed, guarded by a conscientious and well-trained clergy. An adage his contemporaries were fond of attributing to St Augustine claimed that 'priests are required to know the *computus*, otherwise they are hardly worthy of the name "priest"'.[5] Another quote of this calibre, likewise attributed to the bishop of Hippo, listed 'grammar, music, canon law, and *computus*' as the four things 'necessary in the House of the Lord'.[6] Later medieval authors fortified

[2] See Nothaft 2014c; 2014d: 570–611, for the known details of John's life and work.

[3] No adequate large-scale survey of the history of *computus* exists, but some orientation is provided by Borst 1990; 1998; Wallis 1990b; Lejbowicz 1992, and the literature cited in n. 8 of Ch. 2.1.

[4] Some further examples are cited in nn. 126 and 127 of Ch. 2.3.

[5] William Durand (d. 1296), *Rationale divinorum officiorum* 8.1.1 (CCCM 140B: 131). The pseudo-Augustinian quote is inspired by a passage in the *Decretum magistri Gratiani*, pars I, dist. 38, c. 5 (ed. Friedberg 1879–81: i. 141–2).

[6] See e.g. the opening lines to the *Compotus magistri Gordiani* (s. XIII$^{2/2}$), MS Lüneburg, Ratsbücherei, Ms Miscell. D 4° 46, fols. 25r–42r, at fol. 25ra: 'Ut testatur Augustinus quatuor in domo Domini sunt neccessaria, scilicet grammatica, musica, ius canonicum et compotus.' A more detailed version appears at the start of the *Computus pro rudium intelligentiori noticie capacitate breviter compilatus* (s. XIV?), MS Oxford, BodLib, Canon. Misc. 561, fols. 95r–106v, at fol. 95r: 'In ecclesia militante secundum beatum Augustinum quatuor sunt neccessaria, videlicet canon divinus, per quam vita regatur, grammatica, per quam dictionum accentus et significata noscantur, musica, per quam laus divina cantetur, et computus, per quem pasca ceteraque solempnia sacre fidei cultoribus elucescant.' See also the *Compotus ecclesiasticus secundum Augustinum* (1234), MS Glasgow, University Library,

this commonplace by citing the warning example of the clergy of Bremen, whose ignorance of the *computus* had led them to break the Lenten fast and sing the introit for Easter Sunday (*Resurrexi*) midway into Lent. According to a certain Master Jacob writing in 1436, this act of negligence had been enshrined in a mocking rhyme: 'The asses from Bremen sang *Resurrexi* | while the people of God sang *Oculi mei*' (*Bremenses azini cantaverunt 'Resurrexi' | dum populus Dei cantavit 'Oculi mei'*).[7]

Yet what if God's people, too, sang at the wrong time? For John of Brunswick, as for many other *compotistae* in his day, this was more than just a thought experiment. The truth, which no pious appeal to St Augustine could successfully conceal, was that cracks had long begun to appear in the edifice of ecclesiastical time reckoning and that these cracks continued to widen as time progressed. Throughout the Middle Ages, the calendar of the Roman Church was characterized by the interplay of two cycles. One was a solar cycle of 28 years, which represented the annually changing correlation of the days of the week and the dates of the Julian calendar. The other was a lunar cycle of 19 years, which the Church used to predict the annually changing dates of the new and full moons. Its most important purpose was to identify the date of the *terminus paschalis* or Easter full moon, which served as the key point of reference for Easter Sunday and all the other mobile feast days of the liturgical year, from Shrovetide to Pentecost and Corpus Christi. Defined as the first Sunday after the first full moon to fall on or after the vernal equinox, which was fixed by convention on 21 March, Easter Sunday depended on both these cycles and hence on the motions of both celestial luminaries, Sun and Moon. The problem was that neither of the calendrical devices the Church used to track these motions was entirely successful at its respective task. By the end of the thirteenth century most of John of Brunswick's learned contemporaries would have been well aware that the canonical equinox on 21 March was no more than an ecclesiastical and legal fiction, as both observation and astronomical calculation revealed the true equinox to reside on 13 March or an adjacent date. The straightforward reason for this was that the Julian calendar overestimated the length of the solar year by *c.*11 minutes, causing the vernal equinox to recede from its current position at a rate of close to one day every 130 years ($J \approx 130$ years).[8] The lunar

Hunter 467, fols. 21r–44r, at fol. 21r. The earliest known rendition (with *historia* replacing *musica*) appears in the eighth-century Irish treatise *De ratione conputandi* (ed. Walsh and Ó Cróinín 1988: 117).

[7] *Computus Magistri Jacobi* §5 (ed. Gumbert-Hepp 1987: 50). *Oculi mei semper ad Dominum* is the introit sung on the third Sunday of Lent. See Nothaft 2015c: 107, and Silagiová 2013: 304, who cites further sources for this trope.

[8] Most modern studies attribute to the Julian calendar an error of $J \approx 128$ years, which is based on the length of the mean tropical year, currently measured as *c.*365.24219d. The mean tropical year corresponds to the time it takes for the Sun's mean longitude to increase from 0° to 360°, which differs from the interval between two vernal equinoxes due to apsidal precession. Since pre-modern reform discussions focused on the shifting date of the vernal equinox, an astronomical analysis must be based on the latter parameter, i.e. the 'vernal equinox year', as was recognized by Steel 2000: 184–6, 380–3; Meeus 2002: 357–66; Bien 2007: 45. A current estimate for the vernal equinox year at epoch AD 2000 would be 365.242375d (Meeus 2002: 362), which implies $J = 131.1475\ldots$ years. The value has increased from 365.242138d at the start of the Christian era, but this does not take into account the simultaneous increase of the length of the solar day, which is a result of the Earth's rotational

cycle, for its part, had been losing touch with the visible lunar phases at a slower rate of approximately one day for every 300 years since the early fourth century, with the result that the 'ecclesiastical Moon' now trailed behind the astronomical one at a distance of roughly three days.

If the astronomical criteria were taken seriously, as they usually were, this two-fold failure wrought the disturbing consequence that Easter was often celebrated in the wrong week, sometimes even in the wrong month, and the frequency of this happening could be expected to increase with every century of inactivity. And there was more: not only were the mobile feast days being announced in violation of the ancient rules, but the calendrical drift of the equinoxes and solstices meant that the fixed feasts were beginning to shed their accustomed seasonal context. Long before John of Brunswick wrote about the problem in his *Compotus novus* (1297), other computists had begun to complain that 'winterly feasts', such as Christmas, 'will become summerly, unless this error [of the Julian calendar] is corrected'.[9] In 1276, an anonymous author warned his readers that the growing discrepancy between the calendar and the actual course of the Sun was bound to scandalize the general population in the long run. The only hope was a thorough reform of the Julian calendar, to be decreed by the pope, which was going to excise enough days from the year to restore all feast days to their ancient and legitimate locations in the solar year.[10] By the beginning of the fourteenth century, the awareness that something was seriously amiss about the Roman calendar was no longer reserved exclusively to trained computists. A vivid sign of its widespread currency appears in Dante's *Divine Comedy*, which has Beatrice prophesy the coming of a new leader of humanity 'before January be all unwintered, because of the hundredth part that is neglected below' (*Paradiso* 27.142–3: *Ma prima che gennaio tutto si sverni | Per la centesma ch'è laggiù negletta*). As Dante's early commentator Jacopo della Lana and his countless successors understood only too well, this was a reference to the overestimation of the length of the year by the author of the Julian calendar, who had made the calendrical year one one-hundredth of a day longer than it should have been, thus causing the months to drift through the seasons, and vice versa.[11] Two hundred years after Dante, the chasm between the calendar and astronomical reality was still unbridged, explaining passages like the following at the start of François Rabelais's *Pantagruel*:

deceleration (caused by tidal friction). Such long-term effects, which also affect the length of the synodic lunar month, will be ignored for the purposes of this book, whose focus is on medieval parameters.

[9] The passage appears in Bono of Lucca's *Computus lunaris*, written in 1254 (ed. Arrighi 1991: 8).

[10] See the treatise in MS Florence, Biblioteca Riccardiana, 885, fols. 215r–223v, which will be discussed in Ch. 4.3.

[11] Jacopo della Lana's commentary (1324–8) was first edited by Scarabelli 1866: 415–17. It can now be consulted, together with dozens of other medieval and modern commentaries, in the database of the *Dartmouth Dante Project*, <https://dante.dartmouth.edu>. Note that Lana read the 'unwintering' as a drift of the end of January towards autumn rather than of the vernal equinox towards January. See Andriani 1971, and the extensive survey of literature on *Par.* 27.142–3 in Albertini 1981: 19–38. The revisionist interpretation in Kay 2003 rests on a mistaken conflation of the calendrical drift (responsible for the changing alignment between calendar months and seasons) and the precession of the equinoxes relative to the fixed stars.

> In that year the Kalends were fixed by the Greek date-books, the month of March was outside Lent, and mid-August fell in May. In the month of October, I believe, or perhaps in September—if I am not mistaken, and I want to take particular care not to be—came the week so famous in our annals, that is called the Week of Three Thursdays.[12]

Once again audiences would have had no trouble relating this topsy-turvy scenery to real life. In Rabelais's day, the true vernal equinox oscillated between 10 and 11 March, while the new moons regularly appeared four days in advance of when the Church's breviaries and *computus* tables proclaimed their arrival. The fourteenth and fifteenth centuries had seen several attempts to spur the ecclesiastical hierarchies into action, culminating in an ambitious, if politically stunted, reform attempt at the Council of Basel in the years 1434 to 1440. Astronomers and prelates justified such interventionism by portraying the manifest defects of the ecclesiastical calendar as a major source of grief, one which provoked acrid ridicule from Jews and other unbelievers and threatened to undermine the authority of the clergy in the eyes of the laypeople. And yet the 'scandalous error and erroneous scandal' the eminent Cardinal Pierre d'Ailly had decried in 1411 in his calendrical *Exhortatio* still continued to trouble the Church one-and-a half centuries later,[13] as neither the Fifth Lateran Council (1512–17) nor the Council of Trent (1545–63) managed to bring the widely debated reform to any kind of conclusion. It was thanks in no small part to Ugo Buoncompagni, who donned the papal tiara as Gregory XIII (1572–85), that the scandal did not persist for another generation. Ten years into his papacy, at the age of 80, Gregory issued the bull *Inter gravissimas* (24 February 1582), which has become famous as the birth document of what is still the official civil calendar in most countries on Earth. The most incisive measure Gregory's calendar bull prescribed was the removal of ten days from the year 1582, a bold intervention of a kind not seen since the days of Julius Caesar. In order to keep the calendar from falling back into bad old habits, three out of every four centennial years were to go without a bissextile day, that is, without the leap day traditionally inserted in February. As a result of this modified leap-year rule, there was no 29 February in the years AD 1700, 1800, and 1900, whereas AD 2000 retained its leap day. Less familiar to modern audiences is the other prong of the Gregorian calendar reform, the introduction of a system of lunar 'epacts', which made it possible to retain a cyclical lunar calendar for dating Easter without rapidly losing contact to the actual new and full moons.[14]

The view that a calendar should adhere as closely as possible to the astronomical phenomena it is inspired by—the (tropical) solar year, the (synodic) lunar month,

[12] François Rabelais, *Pantagruel*, ch. 1 (trans. Cohen 1955: 171).

[13] Pierre d'Ailly, *Exhortatio ad concilium generale super kalendarii correctione* 1 (ed. Solan 2016: 204). This text will be discussed in Ch. 7.1.

[14] The literature on the Gregorian calendar is vast, but much useful orientation is provided by the proceedings volume published by Coyne, Hoskin, and Pedersen 1983. Other noteworthy treatments include Kaltenbrunner 1880a; Schmid 1882–4; Schubring 1883; Ginzel 1906–14: iii. 257–79; Wijk 1932; Dutka 1988; Maiello 1989; Steinmetz 2011: 72–214; Armogathe 2017. Additional literature will be cited in Ch. 8.

or indeed both—seems an obvious one to take from the perspective of a modern society heavily invested in high-precision measurement and the technologies that go with it. It is not difficult to trace such attitudes back to the seventeenth and eighteenth centuries, when the Catholic Church's continuing interest in guaranteeing the accuracy of its new calendar led to the employment of astronomers such as Giovanni Domenico Cassini (1625–1712) and Francesco Bianchini (1662–1729), who oversaw the installation of meridian lines (*meridianae*) in churches such as San Petronio in Bologna and Santa Maria degli Angeli in Rome, turning them into Europe's foremost solar observatories.[15] The meridian line, as John Heilbron has shown in his landmark study *The Sun in the Church* (1999), makes for a particularly impressive manifestation of the ways in which the demands of religious life and ecclesiastical administration, on the one hand, and scientific research, on the other, could undergo a fruitful symbiosis during the early modern era. Yet this common focus on the period of the so-called Scientific Revolution makes it easy to forget that the use of devices such as meridian lines had been discussed in writing since at least the eleventh century, when monastic authors such as William of Hirsau puzzled over the question why the equinoxes and solstices were no longer found on the dates the ancients had assigned to them.[16] In a similar vein, the pioneers in discussing how to restore the Roman calendar in a scientifically accurate manner are found not in Renaissance Rome or Reformation Wittenberg, but in the monasteries, schools, and universities of high medieval Europe. It was medieval thinkers who first discovered that the calendar was in error; it was they who first sought out ways to remove it. Historians who ignore this rich medieval background are liable to misjudge the nature of the improved calendar that emerged from it, as when George Sarton, one of the key twentieth-century figures in the study of pre-modern science, dismissed the Gregorian reform by claiming that it 'involved only simple arithmetic and common sense'.[17] To argue this way is effectively to negate centuries of intense debates surrounding the astronomical, legal, theological, and practical components of the calendar problem, debates that were in turn intimately connected to the wider trajectory of change, transformation, and progress that characterizes the intellectual and scientific history of the Middle Ages.

The study of past attempts to reform the calendar may not be everyone's idea of an exciting pastime, and some will no doubt treat this book as further validation of Noel Swerdlow's unvarnished dictum that 'a scholar of sense and taste will readily turn to other labours rather than cultivate this barren field'.[18] Yet the days when historians interested in calendars and their cultural ramifications felt required to excuse themselves for the marginality of their subject are, thankfully, long gone. Some credit for this change in attitude must be given to the momentous media echo generated two decades ago by the approaching Y2K (i.e. the year 2000),

[15] Schiavo 1993; Heilbron 1999.
[16] On the meridian line described by William of Hirsau, see Nothaft 2015a: 1095–9, and Ch. 1.5.
[17] Sarton 1957: 70.
[18] Swerdlow 1974: 49 (n. 2). The context here is the calendar-reform literature of the sixteenth and seventeenth centuries.

which raised public appreciation for the global cultural significance of time reckoning and inspired historians to dig deeper into its various aspects, whether they be material, intellectual, socio-political, or scientific. The shelf of literature, even if limited to titles produced during the past twenty-five years, has grown to such lengths that no attempt can be made here to survey it to any greater extent.[19] It is worth noting, however, that the year 2000 was not the first time in modern history that an anniversary brought the Western calendar back into the limelight. An earlier occasion of this sort was the third centenary of the Gregorian calendar reform celebrated in 1882, the expectation of which galvanized Catholic historians in Italy, where the reform originated, but also in the former Holy Roman Empire north of the Alps, where confessional fragmentation had turned its implementation after 1582 into a source of severe social and political conflict. Among the German-speaking scholars who showed an especially keen interest in this complex of events during the run-up to 1882 was Ferdinand Kaltenbrunner (1851–1902), whose triptych of studies published by the Austrian Imperial Academy of Sciences has never ceased to be cited since it first appeared in the years 1876–81. Kaltenbrunner began his excursions into the topic with a 120-page 'pre-history' (*Die Vorgeschichte der Gregorianischen Kalenderreform*), which traced awareness of the calendar problem and the various attempts to solve it from the early Middle Ages to the Council of Trent. Its sequels, dealing with the quarrels the Gregorian reform generated in the academic and political realm (*Die Polemik über die Gregorianische Kalenderreform*) and with the proceedings that led to the reform itself (*Beiträge zur Geschichte der Gregorianischen Kalenderreform*), continued the story into the seventeenth century.[20]

All three studies contain insights of lasting value, but it would be wrong to imply that no progress has been made in the nearly 150 years since Kaltenbrunner first began to look for new sources. His discussion of historical antecedents, for instance, received a major extension in 1896, when Demetrio Marzi published a 300-page monograph that still remains the standard work on the attempts to reform the calendar in the context of the Fifth Lateran Council (1512–17).[21] Even more work has been done on the genesis of the Gregorian calendar reform and its piecemeal introduction in various parts of Europe, starting with some important

[19] A useful bibliography, specifically for works on medieval *computus* and time reckoning, was created by Faith Wallis, 'Calendars and Time (Christian)', in *Oxford Bibliographies*, doi: 10.1093/obo/9780195396584-0130. Readers are also directed to the judicious overview of calendrical and computistical studies from the early modern period to the twentieth century in Borst 1998: 100–68. See, moreover, the cultural histories by Maiello 1994 and Rüpke 2006, and the multi-disciplinary essays assembled by Hubert 2000; Le Goff, Lefort, and Mane 2002; Geerlings 2002; Herzog 2002; Hameter, Niederkorn-Bruck, and Scheutz 2005. For examples of illustrated 'calendar histories', see Milano 2000, and Naudin 2001. Among the many trade books that were thrown on the 'millennium' market, one may note the 'histories of calendar reform' by Duncan 1998 and Steel 2000, neither of which reaches an adequate standard of accuracy or soundness of judgement. See, moreover, Declercq 2000, for an accessible history of the Christian 'Anno Domini' era occasioned by Y2K.

[20] Kaltenbrunner 1876; 1877; 1880a. See also Kaltenbrunner 1880b, on the infamous Augsburg *Kalenderstreit* (discussed most recently in Koller 2014: 216–45).

[21] Marzi 1896. See also the update in Marzi 1906.

publications by Kaltenbrunner's contemporaries Felix Stieve and Josef Schmid.[22] Two recent books worthy of special mention are Dirk Steinmetz's *Die Gregorianische Kalenderreform* (2011) and Edith Koller's *Strittige Zeiten* (2014), which both use previously unknown source material to shed light on the Gregorian calendar's reception between 1582 and 1700, including the logistics of its implementation and the scholarly debates and polemics that surrounded it. In the case of Steinmetz's wide-ranging study, this includes an up-to-date account of the commission work that preceded the Gregorian reform, while Koller places a special focus on the political discussions that led Germany's Protestant estates to introduce an Improved Calendar (*Verbesserter Kalender*) in 1700.[23]

Kaltenbrunner's work on the medieval pre-history, or *Vorgeschichte*, has not been revised at nearly the same pace, and it can be said with some justification that it has remained the foundational text for anyone interested in the deeper historical antecedents to the Gregorian reform. Inevitably, this means that more recent accounts have left much to be desired.[24] While it would be absurd to fault Kaltenbrunner, who produced this work in the 1870s, for failing modern methodological standards, his ability to trace the history of the calendar problem through the centuries was limited by his writing at a time when the study of medieval science was still in its infancy (or rather, at a pre-natal stage) and a serious scholarly interest in the history of the *computus* had yet to develop. Although he must be credited with identifying many of the key sources that shed light on the medieval debate, Kaltenbrunner was only able to access a fraction of the material relevant to his topic, being confined for the most part to the manuscripts and books available to him in Vienna and at the Vatican Library. Owing to these limitations, his *Vorgeschichte* located the beginnings of a medieval reform debate in the thirteenth century and hence at least one century after its legitimate starting point.[25] Most of the lacunae in Kaltenbrunner's work have been carried over into more recent accounts of the pre-history he set out to write, with the result that important episodes and sources from the period before 1500 are continually left out or fail to receive the attention they deserve. Whoever attempts to fill these various gaps will

[22] Stieve 1879 (which is a review essay on Kaltenbrunner's first two articles); 1880; Schmid 1882–4.

[23] See Steinmetz 2011 and Koller 2014, which both include extensive references to previous literature. Not included in these bibliographies are Delatour 1999; Milz 2001; McNutt 2006. On sources from sixteenth-century Spain, see Carabias Torres 2012; Carabias Torres and Gómez Alfonso 2014, which must both be used with caution (cf. Nothaft 2013b).

[24] Accounts of the medieval pre-history of the Gregorian reform that are largely, or entirely, based on Kaltenbrunner include Schubring 1883: 520–1; Rühl 1897: 221–3; Ginzel 1906–14: iii. 252–7; Hagen 1914: 42–7; Guilland 1926: 282–3; Wijk 1932: 16–21; O'Connell 1975: 189–92; Albertini 1981: 12–19; Gack-Scheiding 1995: 3–46; Aufgebauer 2006; Herold 2011: 23–5; Steinmetz 2011: 46–71. Some greater independence is displayed in Duhem 1913–59: iii. 280–1, 321–2, iv. 43–60, 174–82, and in North 1983, who acknowledges his overall debt to Kaltenbrunner (p. 108 n. 1), but also adds one new source (pp. 91–4). The dissertation by Welborn 1932, which was devoted to the thirteenth century, remained without new insights. A creative, but unreliable survey covering the late Middle Ages and early modern period is Baumgartner 2003.

[25] After a short look at Bede, Kaltenbrunner begins his account with a certain Master Conrad who supposedly wrote in 1200. In actual fact, Conrad was a late fourteenth-century commentator on the popular *Massa compoti* of Alexander of Villedieu. See n. 44 in Ch. 4.1. For an account of the twelfth century, which is absent from Kaltenbrunner's work, see Ch. 3.2–4.

have to keep a close eye on the ways in which the history of the ecclesiastical calendar and its attempted reforms coheres with the wider history of mathematical astronomy, whose twists and turns determined the unfolding of the medieval reform debate in several crucial ways. The literature since Kaltenbrunner has shown a tendency to minimize this technical aspect of the story and little effort has so far gone into discussing the numerical parameters, calculations, and observations that were involved in analysing calendrical errors. The fact remains, however, that the history of calendar reform in medieval Europe cannot be separated from the history of *calendrical astronomy*, which may be defined as the history of those components of mathematical astronomy that were directly relevant to the construction of an accurate calendar.

To be sure, none of this is to suggest that medieval practices and discourses of time reckoning are in themselves a neglected field. An area of research that has experienced a minor boom in recent years is the study and transmission of *computus* in the early Middle Ages, from the pioneering textbooks written by Irish monks in the seventh century to the standardization the discipline went through during the reign of Charlemagne. Scholars such as Arno Borst, Dáibhí Ó Cróinín, Dan McCarthy, and, in more recent years, Immo Warntjes and Alfred Lohr have revolutionized this field with new discoveries and textual editions of high technical merit. What has been missing so far are attempts to connect the excellent work available for the early Middle Ages to those relatively understudied later centuries when the ecclesiastical calendar became the object of intense discussion at the curia of Pope Clement VI (1344/5) and at the major Church councils of the 1400s. Some of the historical fog has cleared very recently thanks to Olivier de Solan's volume of editions pertaining to the calendar reform attempts at the Councils of Constance and Basel,[26] but the need for a bigger-picture account persists.

The goal of this book can hence be stated in very simple terms. It is to tell a story of how the need to regulate civil and ecclesiastical life by means of a calendar posed a technical-astronomical problem to medieval society and how its scholars, astronomers, and churchmen reacted to the challenge at hand. In order to attain a full grasp of this story, it is essential that we go back to the origins of the calendar these medieval writers worried about. That this calendar had historical roots in ancient Rome was a point well understood throughout the period, as was the fact that at least its solar component was the product of a radical reform decreed by Julius Caesar several decades before the birth of Christ. Chapter 1 starts things off by providing a succinct account of the historical background to Caesar's reform in 46/45 BC and the astronomical problems that were a part of its legacy. It proceeds to argue that the survival of Macrobius's *Saturnalia* and the development of the *computus* as a proto-scientific textbook genre in seventh-century Ireland were both instrumental in preserving knowledge of the Julian calendar's Roman-pagan roots. The final two sections show how the existence of diverging traditions for the dates of the equinoxes and solstices in this Julian calendar created an important context

[26] See Solan 2016, to be supplemented by Nothaft 2015i; 2017d.

for the practice of solar astronomy in early medieval Europe, which included the use of observational methods.

In a manner complementary to the first chapter, Chapter 2 deals with the early history of the lunar calendar, which in the Christian tradition was firmly tied up with the calculation of the date of Easter. It begins by tracing the development of the 19-year lunar cycle in late antique Alexandria and its subsequent spread to various parts of Western Europe, where its clashes with alternative forms of Easter reckoning led to fierce controversies, most notably in seventh- and eighth-century Britain and Ireland. The discussion then turns towards the gradual emergence of a standardized ecclesiastical calendar during the early medieval period, which involved the introduction of the so-called 'Golden Number'. While this calendar was still being constructed, the astronomical handicaps of the underlying 19-year cycle had already begun to cause discrepancies between the calendrically predicted and the observable new and full moons. One significant reaction to this problem was the development of a completely new approach to lunar reckoning known as *computus naturalis*, whose practitioners attempted to predict the precise time of the conjunction of Sun and Moon by extracting the length of the mean synodic month from the 19-year cycle and by reckoning forward from observed eclipses. Chapter 2 concludes with a closer look at this little-known episode in the history of medieval astronomy, focusing in particular on the innovations introduced in the eleventh century by the computists Hermann of Reichenau and Gerland.

The invention of the *computus naturalis*, together with the measurements of equinoxes and solstices discussed in Chapter 1, was an important precondition for the emergence of a critical awareness that the existing ecclesiastical calendar was astronomically incorrect and in need of rectification. Yet a full understanding of the problem's technical dimensions depended on a body of knowledge that reached the Latin West only in the twelfth century thanks to a flood of translations of scientific texts from Arabic, Greek, and Hebrew. Chapter 3 relies on previously untapped sources to paint a detailed picture of how the so-called 'Renaissance of the Twelfth Century' reshaped computistical literature and gave rise to a new debate surrounding the merits of the ecclesiastical calendar, which pitted reformers against traditionalists. My account of this twelfth-century debate seeks to close a major historiographical gap that has been left by earlier treatments going back to Kaltenbrunner's *Vorgeschichte*, while stressing the foundational importance that this intellectually dynamic period possessed for the development of calendrical astronomy more generally speaking. Major points of my discussion include the introduction of astronomical tables as instruments for analysing calendrical error, the discovery of the lunar calendars used by Muslims and Jews as potential alternatives to the Church's own style of reckoning, and the adoption of a new understanding of the nature of the solar year, in particular the influential theory of the 'access and recess of the eighth sphere' and its prediction of a variable tropical year.

Chapter 4 continues this story into the thirteenth century, when discussions pertaining to calendrical astronomy found an additional and permanent venue in Europe's newly founded universities. Among the hallmarks of this period is the way the astronomical problems involved in calendrical reckoning were approached in two different classes of *computus* text. The 'vulgar', 'usual', or 'ecclesiastical' *computus*, which was pedagogically minded and primarily devoted to the calendar's traditional rules, was complemented by the technically refined 'philosophical' *computus*, whose texts merged standard computistical lore with new modules of knowledge taken from Arabic and Greek sources. Well-known writers on the philosophical side of *computus*, whose names feature prominently in this book's account of the thirteenth century, include John of Sacrobosco, Robert Grosseteste, and Campanus of Novara, as well as Roger Bacon, who directed a passionate reform appeal to the address of Pope Clement IV (1265–8). Other parts of Chapter 4 are devoted to authors who are rarely or never mentioned in this context, among them Alexander Neckam, Robert Holcot, Giles of Lessines, and an obscure group of Franciscan scholars active in the 1270s to 1290s, who followed Roger Bacon in relying on the Jewish calendar for part of their analysis. Yet more evidence can be gleaned from an anonymous treatise of 1276, which contains the earliest fully developed proposal to restore the Julian calendar to its appropriate state by the drastic means of suppressing a series of days—no fewer than 16 in the case at hand.

The main purpose of Chapter 5 is to identify three important late medieval developments in the area of computational astronomy and to show how each of these put the discussion and analysis of the calendar problem on a new technical footing. The first development to be outlined is the evolution of 'enhanced' lunar calendars, almanacs, and other related works in tabular form, which became a central feature of late medieval astronomical culture. Their dissemination in a great variety of forms and different languages guaranteed that the problems of the ecclesiastical calendar reached an increasingly wide audience in the course of the fourteenth and fifteenth centuries. Next in line are some new departures in observational and mathematical astronomy, roughly datable to the 1290s, which led authors such as William of Saint-Cloud and Walter Odington to reject the established theory of 'access and recess' and reconsider key questions such as the length and long-term variability of the solar year. Above all, however, the calendar debate of this later period was shaped by the introduction of the Alfonsine Tables in 1320s Paris, which laid the basis for the gradual formation of a European-wide standard in computational astronomy. For the next two-and-a-half centuries, a widespread and increasingly habitual reliance on the parameters implicit in these tables would coexist uneasily with a whole conglomerate of worries as to their validity. This holds true especially for the precession model included in the Alfonsine Tables, whose specific features posed some tricky interpretative challenges to its late medieval users.

Chapters 6 and 7 take the matter of calendar reform out of the realm of the universities and into the wider ecclesiastical-political arena of the fourteenth and fifteenth centuries, focusing in particular on the three most serious efforts that were made within the late medieval Church to solve the calendar problem by legislative means. In Chapter 6, the main focus is on the activities at the Avignonese court of Pope Clement VI, who in 1344/5 invited the astronomers Jean des Murs and Firmin de Beauval to assist him in a planned reform of the Golden Number. Next to the various calendrical-astronomical writings attributed to Jean des Murs, my account in this chapter pays close attention to a recently discovered *Expositio kalendarii novi* that the monk Johannes de Termis completed in 1345, presumably at the behest of Clement VI. It also takes a brief look at the parallel discussions that took place in the Byzantine East, where the prospect of a calendar reform was first debated by Nicephorus Gregoras (*c.*1295–*c.*1360) and Barlaam of Seminara (*c.*1290–1348). Moving on from the fourteenth century, the first two parts of Chapter 7 concentrate on (1) the reform initiatives spearheaded in 1411–17 by Cardinal Pierre d'Ailly at the Councils of Rome and Constance, and (2) the repeated efforts toward new calendrical legislation made at the Council of Basel in the years 1434–4, which saw the matter debated by a specially created commission or task force. For each of these clusters of events, the pertinent sections of Chapter 7 briefly introduce the main protagonists and their political-biographical contexts before making an attempt to reconstruct the course and outcome of the discussions they took part in. The third and final part continues the story into the second half of the fifteenth century, which involves a concise look at the role print technology played in the dissemination of calendrical and astronomical knowledge. Special attention is here due to the activities of the astronomer Johannes Regiomontanus, himself the founder of a short-lived scientific printing house, whose premature death in 1476 prevented him from assisting Pope Sixtus IV in preparing a new reform of the ecclesiastical calendar.

A guiding question for the two chapters just summarized is how the modalities of the reform debate changed once the original technical-astronomical problem was expanded to include issues of a legal, liturgical, and logistical nature. While the political circumstances that caused some medieval reform initiatives to founder—as when the Council of Basel saw its efforts hamstrung by a protracted conflict with the papacy—are relatively easy to identify and readily understood, previous accounts of the Gregorian calendar's pre-history have paid remarkable little attention to the practical concerns that informed and, in many cases, inhibited these early reform projects. To restore the ecclesiastical calendar to its ancient, legally prescribed state was easier said than done, given that nearly all of the feasible plans would have involved one or the other of two controversial steps: (1) to remove several days from the year, or (2) to alter the accepted calendrical boundaries of the mobile feast days and hence the shape of the liturgical year. Option (1) was considered delicate or ill-advised for legal and political reasons, while option (2) was bound to cause a host of new problems, not the least of which was the invalidation or necessary alteration of countless books that served the Latin clergy across Europe. The prospect of having to change or scrap mountains of written material was no less daunting in the case of the calendar itself, as some of the envisioned modes of correction would have necessitated the introduction of elaborate new tables. Proposals that avoided such drastic measures were to an extent available, but only at the price of compromising the reform's outcome in terms of astronomical soundness. Although the surviving sources tend to be silent on practical issues and often no more than allude to the dilemmas the protagonists faced when thinking about the problems at hand, Chapters 6 and 7 will make an effort to bring some of these aspects closer to the surface and thereby explain why medieval calendar reform initiatives, in spite of some promising beginnings, never came to full fruition.

Chapter 8 concludes the book with a bird's-eye view of the one hundred years of debate that followed upon Regiomontanus's death and culminated in the Gregorian Reform of 1582. Its much quicker narrative pace compared to the previous two chapters is in part due to my study's declared focus on the medieval period, but it also reflects the ample volume of already existing literature devoted to the sixteenth century, which made it possible to limit my account to the major developments that contributed to the ultimately successful resolution of the calendar problem. The chapter's focus accordingly lies on the debates that raged at the time of the Fifth Lateran Council (1512–17) as well as during the 1570s and early 1580s, when a commission of experts convened by Pope Gregory XIII came to favour an intricate scheme for a cyclical, yet astronomically accurate and freely adjustable calendar, the core elements of which had been designed by the Calabrian physician Luigi Lilio. Although some attention will be paid to the extent to which Copernican heliocentric astronomy may have influenced—or was itself influenced by—the discussion, one of the key points of this chapter, and indeed the rest of the book, is to stress the fundamental continuity that existed between the *computus* of medieval Europe and the calendar reform achieved in 1582. While the Gregorian reform found some compelling solutions to the technical problems medieval authors had uncovered, it did not invent these solutions out of whole cloth. Instead, it rested to a large extent on premises, concepts, and insights that had been formulated during the preceding four centuries.

That the Gregorian solar and lunar calendars have managed to persist in their original form for another four centuries and more since their introduction, in spite of numerous attempts to replace them with more accurate or 'rational' solutions, may simply reflect the inertia inherent in cultural-religious artefacts of this kind. It may also, however, tell us something about the quality of their technical underpinnings, which are deeply rooted in the late Middle Ages, a period when the reckoning of time lay remarkably close to the heart of Europe's intellectual culture and a defect in the reckoning of Easter could attain the status of a 'scandal' great enough to occupy great minds.

1

The Julian Calendar and the Problem of the Equinoxes in the Early Middle Ages

1.1. MACROBIUS ON THE HISTORY OF THE ROMAN CALENDAR

Medieval writers on time reckoning were no different from modern ones in possessing a clear understanding that the calendar they were using traced its origins back to ancient Rome, where the prevailing custom was not to count the days in sequential fashion, but to number them backwards from three fixed points: the Kalends (on the 1st of each month), the Nones (on the 5th or 7th), and the Ides (on the 13th or 15th). By the middle of the first century of our common era, the exposed role of the Kalends (*kalendae*) as the day that marked the month's beginning had given rise to the term *kalendarium*, although its initial use was not reserved for the calendar itself, which the Romans preferred to call *fasti*. Instead, *kalendaria* were written registers of loans, debts, and interests—documents concerned with deadlines, the localization of which depended on the existence of some system of counting days and months. It was only in the early Middle Ages, more specifically in the ninth century, that the word *kalendarium* became firmly attached to textual representations of the counting system itself. These representations usually came in the form of lists of 365 days grouped into twelve months, from January to December. The first line of each month customarily featured a brief sequence of letters in enlarged and sometimes rubricated script: *KAL* or *KL* for *kalendae* (as in Figure 1.1).[1]

The most common etymology for the word *kalendae* is its descent from the verb *calare*, a loan word from Greek (καλειν) whose meaning 'to assemble' or 'call together' reveals to us that the Roman calendar originally depended on the observation of the lunar phases, as did nearly all calendars in the ancient world before *c.*500 BC.[2] At these initial stages of development, the Kalends would have marked the appearance of the thin crescent of the new moon in the evening sky, ready to be spotted and announced to the people's assembly by a priest or magistrate. In a similar vein, the Ides, later fixed to the 13th or 15th inclusive day from the Kalends, would have marked the approximate middle of the month and hence the full moon. The close dependence of the Roman calendar on the course of the Moon was weakened early

[1] On the history of the word 'calendar' and the textual genres related to it, see Borst 1998: 30–99.

[2] See the survey of ancient calendars in Stern 2012. For the predominance of lunar calendars on a global scale, see also Nilsson 1920.

on by giving it a fixed structure, the details of which can still be gleaned from the remaining fragments of the *Fasti Antiates Maiores*, a painted wall calendar from Antium datable to the 60s or 50s BC. While the time between two sightings of the new moon normally takes either 29 or 30 complete days, the Roman Republican calendar had four months of 31 days—March, July, October, and May—which shared the distinction of having the Nones and Ides fall two days later than usual, on the 7th and 15th day of the month. The remaining months, except for 28-day-long February, lasted 29 days, the sum of it all being a quasi-lunar year of 355 days.[3]

One striking feature of the *Fasti Antiates Maiores* is the numbering of the days according to a recurrent sequence of letters from A to H, which represent the 8-day nundinal cycle or 'market week'. If one knew the nundinal letter for a particular year, the *fasti* made it possible to predict at a glance when the year's market days were going to fall.[4] This simple, but efficient system could also be applied to the 7-day week, whether in the guise of the planetary week starting on Saturday or the Judaeo-Christian version starting on Sunday. In 321 Emperor Constantine enacted a law that enshrined the *dies solis* as a day free of legal and commercial activities and thereby laid the groundwork for the full-scale adoption of the Judaeo-Christian week in Rome. As with the nundinal cycle, it was possible to track this week in the Roman calendar via a letter sequence, from A to G, which in the medieval tradition became known as the 'ferial letters' (*litterae feriales*). Once this system was in place, it became possible to distinguish years according to their dominical letter (*littera dominicalis*), which was the ferial letter that picked out all the Sundays in the year in question.[5] Nundinal and ferial letters make a joint appearance in the so-called *Chronograph of 354*, a late Roman compilation of chronological lists and texts headed by a magnificently illustrated calendar that was the work of the calligrapher Furius Dionysius Filocalus. The calendar, which now only survives in early modern copies and drawings, listed the imperial anniversaries and religious holidays of mid-fourth century Rome together with unlucky days (the *dies Aegyptiaci*) and the Sun's entries into the twelve signs of the zodiac. With its characteristic layout of parallel columns for ferial and lunar letters, the calendar of 354 prefigures a whole genre of liturgical-astronomical *kalendaria* that was to rise to full prominence only during the Carolingian period.[6]

By the time Filocalus produced his now-famous *Chronograph*, the calendar of the Roman Republic had long become a relic of the past, the origins of which were the object of lively antiquarian debate. Looking back from the second quarter of the fifth century, the praetorian prefect Macrobius Ambrosius Theodosius displayed his vast and esoteric knowledge of Roman antiquities by putting it in the

[3] The *Fasti Antiates Maiores* are edited in Degrassi 1963: 1–28. On the Republican calendar, see Michels 1967; Samuel 1972: 158–70; Brind'Amour 1983: 28–9, 125–223; Hannah 2005: 98–112; Rüpke 2011: 6–8, 23–43; Stern 2012: 205–11.

[4] Blackburn and Holford-Strevens 1999: 674; Michels 1967: 27–8; Bennett 2003b.

[5] Rüpke 2011: 160–9. On the dominical letter, see Ginzel 1906–14: iii. 125–33; Borst 1998: 402–5.

[6] For the most recent edition and analysis of the *Chronograph of 354*, see Divjak and Wischmeyer 2014 (with references to earlier literature). The calendar was previously edited in Degrassi 1963: 237–62. On the early medieval *kalendaria*, see Ch. 1.4.

mouths of a group of fourth-century aristocrats, who gathered on the evenings of 16–19 December for a series of symposia in celebration of the *Saturnalia*. In elucidating the background of the feast that brought them together (and from which Macrobius's work takes its name), one of the dramatis personae, the distinguished Vettius Agorius Praetextatus, bursts into a lengthy monologue on the history of the Roman calendar (*Sat.* 1.12–16).[7] In the early days, he explains, the priests used to observe the evening sky for the appearance of the new moon. Once the sighting had been made, the high priest would summon the people to the Capitol and announce the number of days until the Nones of the month, 'repeating the word *calo* five times if the Nones fell on the fifth, seven times if it fell on the seventh. The verb *calo* is Greek, meaning "I call", and it was decided that the first of the days "called" should be named the Kalends' (1.15.10–11).

Praetextatus's detailed description of an observation-based calendar appears to contradict what he had previously said about the calendar of Romulus, who was credited with the institution of a badly truncated year of only 10 months and 304 days, running from March to December (1.12.3, 38–39). Four of these 10 months—March, May, Quintilis (later renamed into July), and October—were thought to have comprised 31 days, the others 30 days (1.12.3). In all likelihood, this curious 'Romulean' calendar was no more than an antiquarian invention, founded on the fact that the Roman month names suggested a numbering that originally began with March and ended with December (literally 'the Tenth').[8] Macrobius's Praetextatus explains the meaning of these ten original names at great length (*Sat.* 1.12.5–37), before turning to January and February, which were supposedly added by Romulus's successor Numa Pompilius to yield a proper lunar year of 354 days (1.13.1–7). To these Numa decided to join a 355th day, 'paying honour to the mystery of the odd number that nature revealed even before Pythagoras' (1.13.5). The same reverence made him avoid months of 30 days, giving each an odd length of either 29 or 31 days, with the sole exception of February. In order to keep the year aligned with the annual course of the Sun, it was necessary to intercalate (from *inter* + *calare* = 'call between') additional months or days, which bridged the gap between the lunar year and the solar year. Praetextatus presented another unlikely antiquarian construct when he claimed that Numa chose for this purpose the *octaëteris*, a Greek 8-year lunisolar cycle that added three intercalary months of 30 days over the course of 8 years. In the Roman case, these 90 days were split up into alternating intercalations of 22 and 23 days, which were always inserted in the month of February—not at its end, however, but after the 23rd day, on which the Romans celebrated the Terminalia dedicated to the god who protected boundaries and their makers (1.13.8–15). Moreover, Numa's decision to assign to the common year 355 instead of 354 days created the need for

[7] In what follows, all quotations from Macrobius's *Saturnalia* will be taken from the English translation in Kaster 2011. On the work's historical context, see Cameron 2011: 231–72, who suggests 382 as the year in which the dialogue is set.

[8] Rüpke 2011: 23.

an extra adjustment every 24 years, when the number of days intercalated in a cycle was reduced from 90 to 66 (1.13.13).[9]

In spite of all precautions to keep the calendar properly aligned with the seasons, there came 'a time when intercalation was entirely neglected out of superstition, while the influence of the priests, who wanted the year to be longer or shorter to suit the tax farmers, saw to it that the number of days in the year was now increased, now decreased, and under the cover of a scrupulous precision the opportunity for confusion increased' (1.14.1). Later Republican sources confirm the impression that intercalation could be influenced by private interests and political ambition, with the consequence that the public had little way of knowing in advance when the next intercalation was going to occur. Cicero, who wanted to end his year-long proconsulship in Cilicia (51 BC) as soon as possible, asked a friend in Rome to exert his political influence to prevent the year from being prolonged, but his letters show us that he was unable tell the outcome even as February was drawing to a close.[10] When Julius Caesar, in his capacity as *pontifex maximus* and dictator, reformed the calendar by edict in 46/45 BC, he abolished the Republican intercalation procedure and thus effectively took the power to manipulate time away from both the priesthood and the Roman Senate. In its place he put a fixed calendar that counted 365 days per year and needed no more than the insertion of a single day once every four years to keep functioning. In adopting this extremely simple yet very efficient principle, Caesar's calendar accelerated a general trend in the development of calendars in the ancient world between 500 BC and AD 300, which, as Sacha Stern has shown, went from flexible and empirically determined schemes towards calculated and fixed ones, and from lunar calendars to solar ones.[11] The so-called Julian reform was not only a visible manifestation of Caesar's newly acquired political power, which was unprecedented in Roman history, but it led to the introduction of a calendar system that was ideally suited to administrative tasks and general communication within a Mediterranean empire that covered increasingly vast stretches of land and sea. It was a watershed moment in the history of Rome, one that happened to coincide almost exactly with the end of the Republic and the rise of the Principate.[12]

The precise steps involved in the Julian reform are described in greatest detail by Praetextatus in Macrobius's *Saturnalia* (1.14.2–12), although other sources, such as Censorinus's third-century treatise *De die natali* (20.8–12), offer some valuable additional information.[13] In order to rectify the errors that had crept into the Roman calendar through faulty intercalation, the year 46 BC received an extraordinary intercalation of 90 days (the ordinary 23-day intercalation after February and two

[9] Macrobius is the only Roman source to mention this correction. For modern reconstructions of Republican intercalation methods, see Michels 1967: 146–60; Chantraine 1976: 116–18; Warrior 1991; 1992; Rüpke 2011: 68–86; Bennett 2005.

[10] Cicero, *Ad Atticum* 5.9.2; 5.13.3; 5.21.14; 6.1.12. See Rüpke 2011: 70.

[11] Stern 2012: 425–30, and *passim*.

[12] On the political dimension of Caesar's reform, see Stern 2012: 216–27.

[13] For modern accounts, see Malitz 1987; Bayer 2002; Blackburn and Holford-Strevens 1999: 670–1; Hannah 2005: 112–30; Feeney 2007: 151–60, 193–201; Rüpke 2011: 109–21; Stern 2012: 211–14.

further intercalations totalling 67 days between November and December), extending it to a total of 445 days. To the 355 days of the common year, Caesar permanently added ten more days, distributing them among the months that previously had only taken 29 days. February retained its deficient length of 28 days, but instead of intercalary months it now only received an additional day every fourth year. The corresponding instruction to count the sixth day before the Kalends of March (24 February) twice gave rise to the name *dies bissextilis* or 'bissextile day', which became the ordinary medieval term for the Julian leap day.[14] Only four months retained their old length in Caesar's new calendar: March, May, Quintilis (July), and October, which had already been 31 days in length. They were now joined by January, Sextilis (August), and December, all of which grew by two days, but retained the accustomed way of counting the Nones and Ides. This conservative approach appears to have reflected religious scruples, as Rome's dictator did not want to 'interfere with the declaration of any religious festival' (1.14.9).[15]

Julius Caesar was assassinated only a year after his reform had been implemented, on the 'Ides of March' (15 March 44 BC), which thus became a day of bad luck in his own calendar. Thanks to the political and military success subsequently enjoyed by his adoptive son Octavian Augustus, the Julian calendar, like much else of Caesar's legacy, remained firmly in place. The only real disturbance in this new order was caused by the priests to whom the calendar had initially been entrusted. According to Praetextatus, 'they were supposed to insert the day that made good the four quarters at the completion of every fourth year, before the fifth began', but instead of following this rule 'they were intercalating not at the end of the fourth year, but at the beginning' (1.41.13). In other words: the leap day was inserted in intervals of three, rather than four years, causing the average year to grow longer than intended. In 8 BC Augustus intervened and had the leap-day intercalation suppressed for a period of $3 \times 4 = 12$ years, upon which 'he ordered that the whole system be inscribed on a bronze tablet and so be preserved forever' (1.14.15). Rome's Senate commemorated this second reform by renaming the month of Sextilis after Augustus, thereby repeating an honour that had previously been reserved for the calendar's founder, Julius Caesar, whose name is immortalized in July, the former Quintilis (1.12.34–5). For more than one-and-a-half millennia, between AD 4 and 1582, the Julian calendar continued to run its course without any disruption.[16]

1.2. CALENDRICAL IDEAL AND ASTRONOMICAL REALITY

Macrobius's Praetextatus repeatedly assures us that Julius Caesar was careful to maintain the integrity of sacred days (1.14.8–13), but this can hardly conceal the

[14] Ginzel 1906–14: ii. 277–9; Blackburn and Holford-Strevens 1999: 678–9.

[15] For an explanation of this passage, see Rüpke 2011: 113–15.

[16] For a more nuanced account of what happened to the Roman calendar during the first decades after the Julian reform, see Brind'Amour 1983: 11–25; Bennett 2003a; Hannah 2005: 116–22; Stern 2012: 214–16.

fact that the technical framework of the new imperial calendar was a radical departure from Roman tradition. In deciding to count 365 days per common year, Caesar severed all notional ties the Republican calendar still had to the course of the Moon and instead established a strictly solar count, the historical precedents to which could not be found in ancient Italy, but in Pharaonic Egypt. Owing to the paramount importance of the annual inundation of the Nile to the region's subsistence, the Egyptian civil calendar since the third millennium BC followed not the course of the Moon, but the agricultural seasons of 'inundation', 'emergence', and 'low water'. Each of these three seasons was represented by four months of 30 days, to which a surplus of five 'epagomenal' days was added at the end of each year, bringing the total number of days up to 365. While the resulting simplicity and complete uniformity have earned the Egyptian calendar the predicate of 'the only intelligent calendar which ever existed in human history',[17] the system did not offer an accurate representation of the true length of a solar year, which on any definition was about a quarter-day longer than 365 days. As a result, both the seasons and the rising dates of important stars drifted in relation to the Egyptian months in a cycle that lasted close to $4 \times 365 = 1,460$ solar years or 1,461 calendar years. That this problem did not cause the Egyptians a great deal of grief can be seen from the fact that the first recorded attempt to rectify the situation only dates from the third century BC, by which time Egypt had come under Macedonian rule. In the year 238 BC, King Ptolemy III Euergetes ordered his subjects to celebrate an annual public festival in honour of himself and his wife Berenike, which was to coincide with the heliacal rising of the star Sirius. In order to retain this sidereal event on the same date in the Egyptian calendar, Ptolemy demanded that an additional sixth epagomenal day be added to the calendar once every four years.[18]

It is not unlikely that the calendar reform included in the famous 'decree of Canopus' of 238 BC, which remained politically ineffective as far as the intercalation was concerned (the Egyptian calendar months kept 'wandering' until the Roman conquest), was a response to certain influences from mainland Greece, where astronomers had begun to investigate the precise length of the solar year and its individual seasons. The earliest preserved date in this scientific timeline belongs to two Athenians, Meton and Euctemon, who allegedly found that the summer solstice of 432 BC fell on 13 Skirophorion, which would have been 27 or 28 June in the proleptic Julian calendar.[19] Meton is also credited with constructing a 19-year lunisolar cycle, presumably for the purpose of correlating dates in a *parapegma*, which recorded rising and setting times of stars and weather phenomena, with dates in the lunar calendar ordinarily used in Athens. With its equation of 6,940 days with 19 solar years and 235 lunar months, this 'Metonic' cycle implied

[17] Neugebauer 1957: 81. For a recent discussion of ancient Egyptian calendars, with references to older literature, see Stern 2012: 125–66.

[18] Stern 2012: 137–42.

[19] For discussions of the date and the problematic evidence for this alleged observation, see Bowen and Goldstein 1988: 64–77; Depuydt 1996; Jones 2005: 22–3; Lehoux 2007: 88–90.

an average solar year of $365\frac{5}{19}$ days or *c.*365d 6;18,56h.[20] In the fourth century BC a man named Callippus modified this scheme by omitting one day from every fourth iteration of the cycle. The revised 'Callippic' cycle began in 330 BC and equated 76 solar years with 27,759 days, which implied an average solar year of $365\frac{1}{4}$ days—precisely the length presupposed in the reforms of Ptolemy III (238 BC) and Julius Caesar (46/45 BC).[21] A new chapter in Greek solar astronomy was opened up in the late third century BC by Apollonius of Perga, who showed how one could describe the motion of the Sun around the Earth by using an eccentric circular orbit or an equivalent epicycle, which made it possible to explain the observed asymmetry in the duration of the seasons. Apollonius's approach was adopted in the following century by the Rhodian astronomer Hipparchus, who used the differing intervals between the equinoxes and solstices to establish the eccentricity of the Sun's orbit as $\frac{1}{24}$ of its radius and to determine the precise ecliptical location of the apogee, that is, the point at which the Sun is farthest removed from the earth. The result was a simple, but highly effective solar model capable of yielding predictions that lay within the boundaries of naked-eye observation, provided the eccentricity and apogee remained fixed (which they did not) and the parameter for the Sun's angular motion was correctly determined.[22]

Our main source for these ancient findings is the *Almagest*, a monumental handbook of mathematical astronomy composed in the mid-second century AD by the Alexandrian astrologer Ptolemy, who incorporated the rudiments of Hipparchus's solar theory and supported them with his own observations. According to Ptolemy (*Almagest* 3.1), Hipparchus had come to the conclusion that the solar year, here defined as the period of the Sun's return to the same equinox or solstice, was shorter than $365\frac{1}{4}$ days by a certain fraction of time, which he variously gave as $\frac{0.5}{145}$ or $\frac{1}{300}$ of a day. Comparing Hipparchus's observations of equinoxes and solstices with his own, Ptolemy concluded that $\frac{1}{300}$ was indeed the correct difference, leading to a solar year of 365;14,48d = 365.246666d = 365d 5;55,12h. Modern calculations show this to be an overestimation by *c.*$6\frac{1}{2}$ minutes, as the interval between two vernal equinoxes was only about 365.242138d ≈ 365d 5;48,41h near the beginning of the Christian era.[23] Ptolemy's mistake was bound to affect his measurements of stellar longitudes, which he used to confirm what is now regarded as Hipparchus's most significant discovery: that the equinoxes and solstices change their positions relative to the fixed stars over time. According to Ptolemy, this shift in alignment was due to a slow motion of the sphere of the fixed stars about the poles of the ecliptic, which progressed at a rate of very nearly 1° per century, with a full revolution in 36,000 years. One consequence of this phenomenon, which is

[20] On the Metonic cycle, which had Babylonian precedents, see Bowen and Goldstein 1988: 41–58. On Greek *parapegmata*, see Rehm 1941; 1949; Waerden 1984; Evans 1998: 190–204; Taub 2003: 20–37; Lehoux 2007.

[21] Samuel 1972: 42–9; Lehoux 2007: 93–4; Jones 2000: 141–58.

[22] On Hipparchus's methods and results, see Jones 1991; Maeyama 1998; Duke 2008.

[23] According to Meeus 2002: 362. There is a long-standing suspicion that Ptolemy tinkered with his observational data to support the incorrect Hipparchan value for the tropical year. See Britton 1992: 12–47; Thurston 2002; Gingerich 2002; Jones 2005: 18–27.

commonly known as precession, was the need to distinguish two types of solar year: the tropical year, which ancient and medieval astronomers usually defined as the period of the Sun's return to the true vernal equinox, and the sidereal year, which measured the Sun's revolution with respect to a particular fixed star.[24] Ptolemy never gave an explicit estimate for the latter, but his values for the tropical year and rate of precession jointly imply a sidereal year of approximately $365\frac{1}{4}+\frac{1}{144}$ days $= 365;15,25$d $= 365$d $6;10$h. The fact that this is a fairly good value compared to the modern estimate (365.256363d ≈ 365d 6;9,9,45h) reveals the complementary nature of Ptolemy's overestimation of the tropical year and his underestimation of the speed of precession, for which a more accurate value would have been 1° in $71\frac{1}{2}$ years.

While the system laid out in the *Almagest* obviously came too late to have any kind of influence on the Julian calendar reform, Hipparchus's solar theory must have been widely known among specialists in Julius Caesar's day. Its outlines are recounted at some length in the *Introduction to the Phenomena* written in the first century BC by the astronomer Geminus, although the latter never got more specific as to the length of the solar year than saying that 'the day is very nearly $1/365\frac{1}{4}$ part of the annual period'.[25] Given Caesar's connections with Alexandria, the main centre for applied mathematics in the ancient Mediterranean, it would appear likely that the Julian leap-year rule was not the result of ignorance, but a deliberate effort to keep the calendar simple and uniform.[26] Another possibility is that $365\frac{1}{4}$ days were regarded as a passable compromise between the tropical and sidereal flavours of the solar year, since it was manifest that the Julian year length found its place somewhere between the two.[27] If the Julian year was instead meant to represent the actual tropical year, and hence the astronomical cycle of the four seasons, Hipparchus's $365\frac{1}{4}-\frac{1}{300}$ days would have clearly made for a more adequate estimate and one that would have been easy to implement by erasing a leap day every 300 years. No willingness to go the extra mile seems to have existed in Caesar's camp, notwithstanding Pliny's (*Naturalis historia* 18.212) claim that his adviser Sosigenes, an author of three studies on the subject, 'did not shy away from raising doubts by correcting himself' (*non cessavit tamen addubitare ipse semet corrigendo*). One cannot exclude that the experts in Caesar's entourage were weary of more fine-grained solutions because of existing doubts about Hipparchus's value or because they thought the length of the tropical year varied over time. According to Ptolemy, such variability had already been the object of discussion in Hipparchus's lost work

[24] On Hipparchus's and Ptolemy's discovery of precession, see Pannekoek 1955; Swerdlow 1979–80: 300–6; Goldstein and Bowen 1991: 111–14; Evans 1998: 259–62.

[25] Geminus, *Isagoge* 1.8 (trans. Evans and Berggren 2006: 114).

[26] Sources attesting to the Egyptian-Alexandrian origin of Caesar's reform include Cassius Dio (43.26.2), Macrobius (*Sat.* 1.14.3, 16.39), Appian (*Civil War* 2.154), and Lucan (*Civil Wars* 10.172–218). See also Stern 2012: 211–12, 214.

[27] Such an attitude is attested in the recently published Greek astronomical papyrus P. Fouad inv. 267 A, which appears to date from the later second or early third century. It treats $365\frac{1}{4}$ days as a mean, regular, or uniform (ὁμαλός) length of the year, distinguishing it both from the tropical and the sidereal year. See the edition in Fournet and Tihon 2014, the update in Fournet and Tihon 2016, and the additional discussion in Jones 2016.

On the Displacement of the Solstitial and Equinoctial Points, in which he demonstrated the phenomenon of precession. Ptolemy dismissed these variations, claiming that Hipparchus had not really accepted them himself, but merely reported such observational evidence out of his 'love for truth'.[28] Yet Ptolemy's predecessors in the first century BC, in particular those who advised Caesar in his reform, may have read Hipparchus's evidence differently, concluding that $365\frac{1}{4}$ days was a good enough approximation to the long-term average.

1.3. CAESAR'S REFORM IN IRISH MONASTERIES

Whatever its deficits or merits, the Julian calendar became a permanent fixture. Backed by the irresistible force of Caesar's dictatorship and the Principate that it inspired, it managed to persist beyond the fall of the Roman Empire and hence marked 'a victory in the realm of culture more lasting than any Roman victory on land or sea'.[29] Much of this long-term victory must no doubt be credited to the influence of the Christian Church, which used the Julian calendar as the framework for its expanding liturgical year and in doing so carried it to places where Rome had never even attained a foothold in political terms. Not only did the Julian calendar itself survive the dramatic upheavals brought about by the Barbarian invasions and the collapse of the Roman Empire in the West, but so did the source that to this day best elucidates its history and significance: Macrobius and his *Saturnalia*. While the extant manuscript transmission for the complete *Saturnalia* does not set in before the ninth century, a stand-alone excerpt of Book I, chapters 12–15, containing the bulk of Praetextatus's monologue on the Roman calendar, already circulated on the British Isles two centuries earlier. Known as the *Disputatio Chori et Praetextati* (*Horus* being mentioned in the title for his brief intervention in *Sat.* 1.15.1–3), this excerpt first appears as a fellow-traveller on board a seventh-century Irish anthology of late antique tracts, tables, and especially letters, which were all dedicated to the problem of finding the date of Easter.[30] The spread of these texts in early medieval Ireland and the controversies they generated gave rise to a new tradition of computistical textbook writing, which was to have a definitive influence on the way calendrical reckoning and basic astronomy were taught and practised in Western Europe. Even when the Carolingian educational reforms of the ninth century sparked a revival of the ancient seven liberal arts (*septem artes liberales*), the study of *computus* continued to be the principal area of scientific

[28] Ptolemy, *Almagest* 3.1 (trans. Toomer 1998: 136). On Ptolemy's motivations, see Jones 2005: 25.

[29] See Grafton 1993: 233, who here paraphrases the opinion of the Renaissance philologist Joseph Justus Scaliger (1540–1604).

[30] See Arweiler 2000 and Ó Cróinín 2008: 263, who emphasizes the Irish background of the known manuscripts, as does Warntjes 2010c: lxvii (n. 171). A critical edition of the *Disputatio Chori et Praetextati* by Leofranc Holford-Strevens is currently in press. For the medieval manuscript transmission of the *Saturnalia*, see Kaster 2010: 3–27. On the 'Sirmond group' of computistical manuscripts, to which the *Disputatio* is connected, see Jones 1937; 1943b: 75–7, 105–12; Ó Cróinín 1983; Springsfeld 2002: 68–80; Wallis 1999: lxxii–lxxix; Graff 2010; Warntjes 2010a; 2010c: xxii (n. 37), xxviii–xxix (n. 55); 2015: 41, 52–3.

activity in Western Christendom, a status it only slowly ceded with the advent of Graeco-Arabic astronomy in the eleventh and twelfth centuries.[31]

The discipline's Irish roots are best represented by the so-called *Munich Computus*, a work compiled in 718 or 719 by an anonymous scholar located in the south of Ireland, who could already draw on a substantial body of earlier Irish *computus* texts.[32] Another anonymous Irish treatise, entitled *De ratione conputandi secundum solem et lunam*, was apparently written a few years after the *Munich Computus*, between 719 and 727.[33] Both authors made extensive use of the Macrobian excerpts in the *Disputatio Chori et Praetextati*, which enabled them not only to explain the etymologies behind the Roman months and their Kalends, Nones, and Ides, but to trace the evolution of the Roman calendar in some detail from Romulus's ten-month year, to which Numa added January and February, to the reform decreed by Julius Caesar.[34] Their rendering of Macrobius's information was not always pitch-perfect, as when the *Munich Computus* claimed that the Julian leap day had already been discovered 'in nature' by 'Gignus', king of the Sabinians (here classified as a 'Greek' tribe), without telling his readers anything about Caesar's role in its institution.[35] The Julian reform received greater attention in *De ratione conputandi*, whose author interpreted Praetextatus's statement that Numa added a day to January 'in honour of the odd number' (*Sat.* 1.13.5) as a reference to the leap day, which led him to qualify King Numa's action as 'stupid', presumably because he believed that February would have been a more appropriate location.[36] He was nevertheless aware that Numa's calendar remained in place only until Gaius Caesar reordered the year to contain 365 days 'and a quarter, from which the *bissextus* is produced in the fourth year. And this ordination will stay with the Latins until judgement day—and it is the one we use [today].'[37]

The wording creates a contrast between the 'Latins' and the group the computist and his readers belonged to, reminding us that the calendar described in this treatise was a foreign artefact that bore no direct relation to the indigenous traditions of

[31] The Irish computistical textbook tradition and its significance for the history of medieval science are explored in Warntjes 2010c; 2011a; 2013b, 2016a; Mc Carthy 2011.

[32] The text of the *Munich Computus* is edited and translated in Warntjes 2010c, who also covers the early Irish computistical tradition at large.

[33] The text of *De ratione conputandi* is edited in Walsh and Ó Cróinín 1988: 115–213. See, in addition, Ó Cróinín 1982a; Borst 1998: 180–1; Holford-Strevens 2008: 198–201. The date is established in Warntjes 2010c: lv, cxci–cci.

[34] *Munich Computus* 12–28, 41 (ed. Warntjes 2010c: 44–81, 120–3); *De ratione conputandi* 26–44, 49–57 (ed. Walsh and Ó Cróinín 1988: 134–53, 159–69).

[35] *Munich Computus* 41 (ed. Warntjes 2010c: 122). See also ibid. 53 (p. 206), where Gignus appears as the first person to have divided the solar year into 12 months. The source for this passage was Isidore of Seville, *De natura rerum* 4.5 (ed. Fontaine 1960: 189, 341–42), in the course of whose seventh-century insular transmission the name *Sancus* was corrupted into *Cignus*, based on a confusion with *Cingius*, a Roman grammarian mentioned in the *Disputatio Chori et Praetextati* (Macrobius, *Sat.* 1.12.2). See also *De ratione conputandi* 28 (ed. Walsh and Ó Cróinín 1988: 136), where the named is rendered as *Gingius*. The corruption *Rucingus* appears in the only known copy of the eighth-century text edited by Borst as *Dial. Langob.* 13A (ed. Borst 2006: 442), which is heavily reliant on the Irish textbook tradition. On the latter work, see Warntjes 2010c: clxxiv–clxxix.

[36] *De ratione conputandi* 28 (ed. Walsh and Ó Cróinín 1988: 138).

[37] Ibid. (ed. Walsh and Ó Cróinín 1988: 139).

Ireland.[38] Unlike the late antique Church Fathers, who could casually mention liturgical and historical dates in their writings while knowing that the framework of the Julian calendar would be taken for granted by their listeners and readers, Irish computists often felt the need to introduce their audience to the Roman way of counting days or the exotic month names attached to the Julian calendar. One way to clarify this material, both to readers and to students in the classroom, was by the occasional use of glosses and words in the vernacular, which are particularly noticeable in the earliest preserved Irish textbook, the *Computus Einsidlensis* (*c.*700). There, the Roman designations of the first thirteen days of January (*kl.*, *IIII non.*, *III non.*, etc.) are juxtaposed with ordinal numbers in Old Irish (*oin*, *ailiu*, *tres*, etc.).[39] Two decades later, the author behind the *Munich Computus*, which likewise retains some instances of Irish vocabulary, confronted his readers with the question 'why the names of months which originated among the heathen were adopted', and surmised that this might have facilitated the conversion of these heathens to the true faith.[40] Another computistical treatise, written a few years later in the Germanic East of the Frankish Empire, but heavily influenced by Irish sources, gave a more mundane and fatalistic answer: old habits die hard, especially where they have become ensconced in a nation's language.[41]

Yet the Irish authors of computistical handbooks and their continental successors did more than just comment on what they had copied down from sources such as Macrobius. They also brought new thoughts to bear on the technical structure of the Julian calendar, the most thought-provoking element of which was the intercalation of a day every four years. The *Munich Computus*, for instance, highlighted its importance by inviting readers to reflect on the consequences of not intercalating the bissextile day. After 360 years 'three months would correspond to three different months', as 90 natural days remained unaccounted for, such that 'winter would be called by the name of spring'. Nevertheless, one could also imagine the opposite scenario of leap days being inserted without any correspondence to nature. In this case, too, 360 years would 'change the denomination of three months into three different months' (unbeknownst to the Irish author, it was a version of the latter scenario that was actually slowly unfolding in his day).[42] Those mathematically inclined could try and find how much each day and month contributed to the bissextile increment of one-quarter day per year. The Munich Computist solved the problem with bravura: 365 days divided by 12 yielded 30 days and 10 hours as the average length of the Julian month. For each of these months, the *bissextus* added 6 hours divided by 12 or exactly one half-hour, which in Irish computistical terminology could be expressed as 20 'moments'

[38] McCluskey 1998: 25–8.

[39] Warntjes 2010c: cci; 2013a: 170–5; 2013b: 53–4; Bisagni and Warntjes 2008: 96–102. The discovery of the *Computus Einsidlensis* was first announced Warntjes 2005–6. On the use of the vernacular in early insular computistics, see also Bisagni and Warntjes 2007; Warntjes 2009.

[40] *Munich Computus* 24 (trans. Warntjes 2010c: 69).

[41] *Quaest. Austr.* 2.1C (ed. Borst 2006: 478). On this text, see Warntjes 2010c: clxix–clxxi.

[42] *Munich Computus* 41 (trans. Warntjes 2010c: 133).

(40 moments = 1 hour). Reckoned per day, the corresponding increment was almost exactly $20 \div 30 = \frac{2}{3}$ of a moment.[43]

Another salient question went as follows: where to insert this surplus once it had accrued to a full day? From Macrobius (*Sat.* 1.14.6) it was known that the Romans had chosen to double the sixth day before the Kalends of March (24 February), such that the first of these two days became known as the *bissextus*; but the Iberian bishop Isidore of Seville (d. 636) had sent a distracting signal by suggesting that the calculation of the *bissextus* started with the sixth day before the Nones of March (2 March). Irish computists added to these a third option: if the world had been created on the twelfth day before the Kalends of April (21 March), as was widely assumed, it was reasonable to base the reckoning of the year on this date and to count it twice whenever the bissextile increment demanded it.[44] Macrobius's stance on the matter was treated as the only viable option by the Northumbrian monk Bede of Jarrow, whose *De ratione temporum* (725) was to become the most influential representative of the computistical textbook tradition that had sprung up in Ireland in the previous century.[45] The impact of Bede's works on cosmology and time reckoning (which also include the much briefer tracts *De temporibus* and *De natura rerum*) on the theory and practice of *computus* was such that it has taken modern scholars a long time even to recognize the Irish models and influences behind them. Still, the manuscript transmission of Bede's sources leaves no doubt that nearly all of the material he worked with, including the *Disputatio Chori et Praetextati*, reached his scriptorium via Ireland.[46] Basing himself on the *Disputatio*, he declared that the bissextile day was intercalated at the sixth day before the Kalends of March as a sign of devotion to the god Terminus, whose feast fell on the preceding 23 February. 'For a long time', Bede explained, the Romans

> inserted many days after this [feast] had been duly celebrated. However as their expertise increased over the centuries, they inserted one day every four years. Alas, what wretches those fools were, who did not know Him whose kingdom will have no end [...] but believed that Terminus should be honoured with divine worship, and sacrifices and rites offered to him.[47]

[43] Ibid. (ed. Warntjes 2010c: 136). Other early medieval calculations and various related doctrines concerning the bissextile increment, which often involved the peculiar notion of a 12-hour intercalary day, are documented in Smyth 2017.

[44] Isidore of Seville, *Etymologiae* 6.17.27 (ed. Chaparro Gómez 2012: 107); Isidore of Seville, *De natura rerum* 6.7 (ed. Fontaine 1960: 197); *Munich Computus* 41 (ed. Warntjes 2010c: 120, 130); *De ratione conputandi* 35, 52, 57 (ed. Walsh and Ó Cróinín 1988: 146–7, 163–4, 168–9). See, in addition, Borst 1998: 640–6; Warntjes 2010c: cxliii–cxliv, cxcviii; Smyth 2017: 231–4, and the summary of the eighth-century Iro-Frankish treatise *De bissexto* in Springsfeld 2002: 203–14. Further discussions engendered by the Julian bissextile day until the thirteenth century are documented by Schulz 1945, with a particular focus on legal sources.

[45] Bede's computistical works were edited by Charles W. Jones in Jones 1943b, and again in CCSL 123B and 123C. English translations and commentaries are available in Wallis 1999; Kendall and Wallis 2010. For a French commentary on *De temporum ratione*, see Pillonel-Wyrsch 2004. See now also the brief introduction in Wallis 2010: 119–25.

[46] This was first demonstrated by Jones 1937, and subsequently confirmed by Ó Cróinín 1983; 2003: 201–12. See also Wallis 1999: lxxii–lxxix; Warntjes 2010c: xlvii–xlviii (nn. 109–10).

[47] Bede, *De temporum ratione* 40 (trans. Wallis 1999: 109–10).

Bede's many authoritative statements on the history of the Roman calendar and the meaning of its vocabulary were to make sure that the pagan origins of the Julian year and the reform that had given rise to it were never forgotten as long as it remained in use in the Latin West.[48]

1.4. MEDIEVAL CALENDARS AND CONFLICTING EQUINOXES

The calendar's centrality to early medieval learning is on impressive display in the hundreds of manuscripts containing computistical texts and tables that have come down to us from the period between 500 and 1000. In the vast majority of cases, these were copied in the ninth or tenth century on the territory once covered by the Frankish Empire, where the long reign of Charles the Great brought in its wake a revival of learning on all fronts as well as an explosion in book production.[49] As one of the outcomes of this 'Carolingian Renaissance', computists began to expand the scope of their codices by branching out into the rudiments of classical cosmology made available to them by the handful of sources that had survived the collapse of the Roman Empire in the West. From the fifth-century pen of Macrobius, who had already taught Irish and Anglo-Saxon computists the history of the Roman calendar, came the *Commentarii* on Cicero's *Somnium Scipionis* as a rich mine of astronomical information, some of it in diagram form.[50] To this could be added the extended synopsis of astronomy contained in Martianus Capella's *De nuptiis Philologiae et Mercurii*, which became the foundational text for the revival of the seven liberal arts in Carolingian centres of learning.[51] More subdued, but still significant, was the reception of the astronomical sections in Calcidius's commentary on the *Timaeus*, the only work by Plato available at the time.[52] One noteworthy feature of the texts just mentioned is their occasional endorsement of non-Ptolemaic cosmological ideas, which thus managed to enjoy a second lease of life during the ninth to twelfth centuries. Rather than having the two inferior planets, Venus and Mercury, circle the Earth in spheres below the Sun, Martianus Capella equipped them with circumsolar orbits, which are duly reflected in a number of medieval diagrams and glosses.[53]

[48] See Bede, *De temporibus* 6 (CCSL 123C: 589); Bede, *De temporum ratione* 12–13 (CCSL 123B: 319–27); Jones 1943b: 348–9; Pillonel-Wyrsch 2004: 140–52. For examples of the influence of these passages, see Helperic, *Liber de computo* 9 (PL 137: 28B–9A); Gerland, *Computus* 1.4, 1.12, 1.19–20, 1.36 (ed. Lohr 2013: 101–3, 122–5, 137–9, 168–9); Borst 1998: 461, 600; 2001: 423–8, 1636–7; Mosimann 1994: 356–77.

[49] On the production and transmission of computistical texts and manuscripts during the early Middle Ages, see Borst 2006: 131–89, 205–317; García Avilés 2001; Palmer 2011a; Dobcheva 2013.

[50] More than 200 medieval manuscripts are listed in Eastwood 1994a. See further Eastwood 1994b; 2007: 31–94; Caiazzo 2002: 27–85.

[51] Leonardi 1959–60; Eastwood 2007: 179–311. On the tenth-century reception, see Mayr-Harting 2007: 195–229.

[52] Eastwood 1999; 2007: 313–72; Müller 2008: 29–92.

[53] Eastwood and Graßhoff 2004; Eastwood 2007: 238–59; 1997: 245–7; 2000; 2001; 2002: 268–82; 2011: 200–7.

From a computistical vantage point, the classical Roman text most heavily in demand was Pliny the Elder's *Naturalis historia*, especially the astronomical and meteorological sections in Books II and XVIII, which had already been quoted profusely in the Venerable Bede's works on nature and time reckoning, *De natura rerum* and *De temporum ratione*.[54] Pliny's data also played a visible role in the evolution of the early medieval *kalendarium*, which came to differ from late antique manifestations in the liberal way in which it combined liturgy, *computus*, and astronomy into one seamless package. By the ninth century, individual calendar pages were typically headed by entries that indicated the lengths of the given Julian month and its corresponding lunation (i.e. the synodic lunar month) as well as the values of the 'regulars' that were used to calculate weekdays and lunar ages (see Ch. 2.2). An 'interval zone' occupying the left flank of each page counted the days of the month in vertical sequence, to which could be joined various columns of letters for tracking the course of the Moon or the day of the week. The remaining space was used to endow the individual dates with significance by assigning to them saints, martyrs, and ecclesiastical feasts as well as references to biblical events. A second class of such entries, oftentimes placed slightly further to the right than the first, provided additional information of computistical, astronomical, meteorological, or seasonal relevance (see Figure 1.1).[55]

An indispensable marker in any *kalendarium* of this sort was the date of the vernal equinox, which played a key role in regulating the date of Easter. Statements highlighting the importance of the 'rule of the equinox', which kept the *terminus paschalis* or Easter full moon from falling below this date, were widespread in the computistical literature of the day and clearly showed that this appeal to the vernal equinox was understood to be more than just a calendrical convention. A prominent example is Bede's admonition to his readers that 'the Pasch is not to be sacrificed before the equinox and before the darkness has been overcome',[56] which conveyed how the equilibrium between day and night reached at the time of the equinox played into the symbolic and typological content of the Easter celebration. According to Bede's influential exegesis of the fourth day of creation (Genesis 1:14–19), God placed the newly created Sun at the vernal point and the Moon in opposition to it, making it rise as a full moon at sunset. In doing so, the Creator effectively foreshadowed the calendrical rule the Church was required to obey in celebrating Easter:

> For the only Paschal rule to observe is that the spring equinox be completed, with a full Moon following. But if the full Moon precedes the equinox by a single day, it is

[54] Borst 1994: 98–176; Eastwood 1986; 1987; 2007: 95–178. On Pliny's medieval afterlife, see also Chibnall 1975; Reeve 2012.

[55] See the monumental edition by Borst 2001, and the detailed treatments in Borst 1994: 126–38; 1998; 2004; 2006: 36–51, which depict the characteristic layout of the *kalendarium* as the result of a conscious 'calendar reform' during the reign of Charles the Great. Alternative views, with a greater emphasis on Bede's role, have been proposed by Meyvaert 2002 and Bullough 2003. See also Bullough 2004: 287–93; Starostine 2006.

[56] Bede, *De temporum ratione* 30 (trans. Wallis 1999: 89). See also Bede, *De temporum ratione* 61 (CCSL 123B: 451).

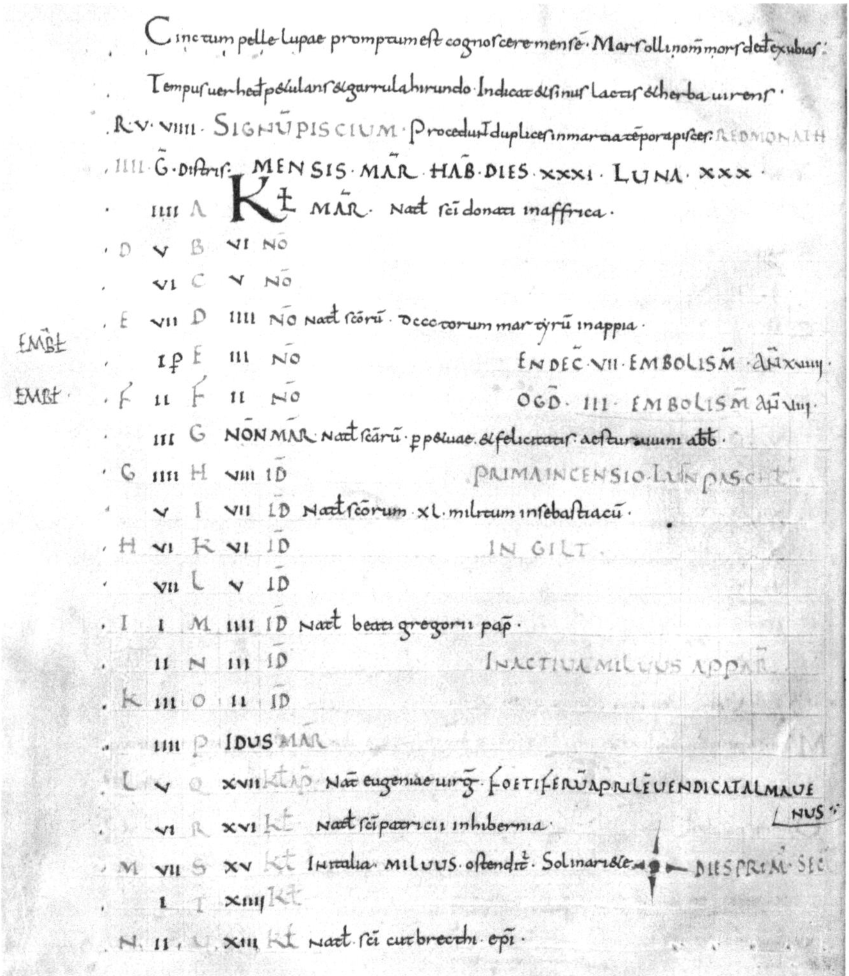

Figure 1.1. Carolingian *kalendarium*, page for March, Prüm Abbey, *c.*840. Berlin, Staatsbibliothek, Phillipps 1869, fol. 2v.

Source: Staatsbibliothek zu Berlin—Preußischer Kulturbesitz, Handschriftenabteilung

considered to be the Moon of the last month, and not the first. [...] This is for the sake of a certain symbolism, because the created Sun which lights up all the stars signifies the *true* and eternal *light which lighteth every man that cometh into the world* [John 1:9], while the Moon and stars, which shine, not with their own light (as they say), but with an adventitious light borrowed from the Sun, suggest the body of the Church as a whole, and each individual saint. These, capable of being illumined but not of illuminating, know how to accept the gift of heavenly grace but not how to give it. And in the celebration of the supreme solemnity, it was necessary that Christ precede the Church, which cannot shine save through Him. For if anyone were to argue that the full Moon can come before the equinox, he would be stating either that Holy Church existed in its perfection before the Saviour came in the flesh, or that one of the faithful, before the bestowing of His grace, can have something of the supernal light.[57]

When it came to identifying the date itself, computists who, like Bede himself, followed the reckoning of the Alexandrian Church were compelled to put the vernal equinox on 21 March, which in Roman parlance was the twelfth day before the Kalends of April.[58] Victor of Capua, a sixth-century bishop and advocate of the Alexandrian style of Easter reckoning, was more precise in his determination when he pointed out that the equinox by necessity occupied not just one date, but two adjacent ones, 21 and 22 March. This was due to the discrepancy between the Julian common year of 365 days and the solar year of $365\frac{1}{4}$ days, whose surplus of six hours moved the actual moment when the Sun crossed the vernal point forward with each year until the insertion of a bissextile day in the fourth year restored everything to order.[59] A certain amount of confusion on this point is betrayed by early medieval *computus* treatises that simply assign the vernal equinox 'according to the Greeks' to 22 March, without also including the canonical 21 March.[60] The latter is very clearly identified as the 'Greek' equinox in the *Munich Computus* of 718/19, whose author used its calendrical position on the twelfth day before the Kalends of April to assign analogous dates to the other three turning points of the year, which came to fall on 20 June (summer solstice), 20 September (autumn equinox), and 21 December (winter solstice).[61] As he correctly observed, the resulting lengths of the seasons were 90 days for winter, 91 days for spring, 92 days for summer, and 92 days for autumn.[62] A tendency to assign equal lengths to the four seasons can be detected in the writings of Bede, who repeatedly claimed that the Sun remains in each zodiacal sign for 30 days and $10\frac{1}{2}$ hours, which is $\frac{1}{12}$ of the average Julian year of $365\frac{1}{4}$ days.[63] In line with this position, Bede's letter to the

[57] Bede, *De temporum ratione* 6 (trans. Wallis 1999: 25–6). See the commentaries on this chapter in Jones 1943b: 37–9; Wallis 1999: 273–5; Pillonel-Wyrsch 2004: 77–88. Bede repeats the same point in *De temporum ratione* 50 (CCSL 123B: 436–7).

[58] Late antique sources on Easter reckoning that explicitly mention the equinox as falling on 21 March include St Ambrose's letter to the Emilian bishops, written 387 (*Ep. extra collectionem* 13.16; ed. CSEL 82.3: 230); the *Prologus Theophili* of 385/95 (ed. Krusch 1880: 223), Proterius's letter to Pope Leo of 455 (ed. Krusch 1880: 276–7), the *Prologus Cyrilli* (ed. Krusch 1880: 339), and Dionysius Exiguus's letter to Petronius of 525 (ed. Krusch 1938: 65). The Alexandrian mode of Easter reckoning will be discussed in Ch. 2.1.

[59] The pertinent fragments from Victor of Capua's lost *Libellus de cyclo paschali*, as found in MS Paris, BnF, lat. 12309, fol. 88r–v, are printed in Pitra 1852: 297. See also Bede, *Epistola ad Wicthedum* 1–8 (CCSL 123C: 635–40); *Comp. Col.* 3.4 (ed. Borst 2006: 903).

[60] See e.g. *Dial. Neustr.* 29 (ed. Borst 2006: 416); *Dial. Langob.* 18 (ed. Borst 2006: 448–9); *Cap. comp.* 5 (ed. Borst 2006: 1043); *Bobbio Computus* 45 (PL 129: 1297).

[61] *Munich Computus* 11, 36, 38 (ed. Warntjes 2010c: 42, 98, 102–3, with further references in the *apparatus*). See also *De ratione conputandi* 47 (ed. Walsh and Ó Cróinín 1988: 157).

[62] *Munich Computus* 38 (ed. Warntjes 2010c: 104).

[63] Bede, *De natura rerum* 17 (CCSL 123A: 209–10); Bede, *De temporum ratione* 2, 11, 16, 39 (CCSL 123B: 275, 316, 334, 402); Alcuin, *Epistola* 148 (MGH Epp 4: 238–9). A duration of 91 days and $7\frac{1}{2}$ hours per season is deduced in the *Annalis libellus* of 793 (ed. Springsfeld 2002: 363–4; Borst 2006: 749). See also Rabanus Maurus, *De computo* 28, 34, 40, 53 (CCCM 44: 232, 244, 252, 265).

presbyter Wicthed (725/31) locates the autumn equinox on 19 September, which is 182 days or roughly half a Julian year after the spring equinox.[64]

Carolingian scholars who gathered and excerpted the surviving bits and pieces of Greek astronomy were in a position to know better. Martianus Capella used the Earth's eccentric position relative to the circular path of the Sun to explain why it took the Sun different lengths of time to pass through the signs of Gemini and Sagittarius (32 vs. 28 days) and why the winter half of the year lasted only 180 days and was hence considerably shorter than the summer half.[65] Seizing on Capella's account, a ninth-century glossator writing in Laon was able to warn his readers that Bede's 30 days and $10\frac{1}{2}$ hours per sign were no more than an average.[66] In a similar vein, excerpts from Pliny as well as diagrams based on them kept a consciousness alive that the circular orbits of the seven planets, the Sun included, had apogees and therefore were eccentric with regard to the centre of the universe.[67] Other excerpts and diagrams, derived from Calcidius's commentary on the *Timaeus*, underscored this point even further by showing how the spring equinox was separated from the summer solstice by exactly $94\frac{1}{2}$ days, the summer solstice from the autumn equinox by $92\frac{1}{2}$ days, the autumn equinox from the winter solstice by $88\frac{1}{8}$ days, and the winter solstice from the spring equinox by $90\frac{1}{8}$ days.[68] The very same intervals, which had already characterized the solar theories of Hipparchus and Ptolemy, were also present in Pliny's *Naturalis historia*, Book XVIII of which was dedicated to the theoretical knowledge necessary to run a farm.[69] An extended part of this book was taken up by an astro-meteorological register or *parapegma*, which recorded the rising of stars and the occurrence of certain weather phenomena in the course of a year based on Caesar's newly established solar calendar. Pliny in fact claimed to have derived some of the dates in his own *parapegma* from none other than Julius Caesar himself, which seems to suggest that an agricultural calendar of this kind had already accompanied the original reform of 46/45 BC.[70] Modern calculation indicates that at the time of this reform the two equinoxes would have fallen on 23 March and 25 September, but these are dates in the proleptic Julian calendar, which may differ from the actual dates in the Roman civil calendar before AD 4, owing to disturbances in the intercalation cycle and the subsequent realignment of the calendar during the reign of Augustus (see Ch. 1.1).[71] M. Terentius Varro, writing in 37 BC, placed the two equinoxes on 24 March and 26 September (*De re rustica* 1.28), but Pliny's *parapegma* simplified

[64] Bede, *Epistola ad Wicthedum* 6 (CCSL 123C: 637).

[65] Eastwood 2007: 275–88; 2011: 207–14; McCluskey 1998: 162–3; 2011: 227–9.

[66] See the gloss on Bede, *De natura rerum* 17 (ed. CCSL 123A: 208–10, esp. p. 209). See also the gloss on Bede, *De temporum ratione* 30 (ed. CCSL 123B: 372), and Eastwood 2007: 339–40; 2011: 209; McCluskey 2011: 233–4.

[67] Eastwood and Graßhoff 2004: 30–5; Eastwood 2007: 109–18.

[68] Eastwood and Graßhoff 2004: 73–82; Eastwood 2002: 282–90; 2007: 320–3, 333–54; McCluskey 1998: 120.

[69] Pliny, *Naturalis historia* 18.220–1. This passage was copied into the computistical encyclopedia of 818, which is also known as the 'Three-Book-Computus'. See *Lib. calc.* 107B (ed. Borst 2006: 1444).

[70] Pliny's *parapegma* is reproduced in English translation in Lehoux 2007: 253–61. See also Rehm 1941: 43–121; 1949: 1356–8; Malitz 1987: 121–4; Pedersen (Olaf) 1986: 170–4.

[71] Bennett 2003a: 231.

things by locating all four cardinal points of the solar year on the eighth day before the Kalends of the appropriate month (25 March, 24 June, 24 September, and 25 December), which disagreed with the previously cited seasonal lengths according to Hipparchus and Ptolemy.[72] Whatever its precise origins, Pliny's schematic placement soon became a firmly entrenched part of the Roman calendrical tradition,[73] which eventually spilled over into the works of the Latin Church Fathers. Examples of the latter include Ambrose and Augustine, but also more obscure sources such as the third-century treatise *De pascha computus*.[74] Isidore of Seville, even though he used a version of the Alexandrian Easter cycle, only ever mentioned Pliny's dates on the eighth day before the Kalends when writing about the calendar in his famous *Etymologiae* and his treatise *De natura rerum*.[75] The authority of this 'Latin' dating tradition was also strongly supported by the liturgical calendar, where, according to the *Munich Computus*:

> 25 December marks the birth of Christ and the increase of daylight; 25 March (marks) his incarnation in the womb of the virgin and the equinox; 24 June (marks) the birth of John the Baptist, because (daylight) is full on that date; 24 September marks the conception of John the Baptist and the equinox.[76]

For those who followed the 'Greeks' in assuming that the equinoxes and solstices occurred four days earlier (on the twelfth rather than the eighth day before the Kalends), this ancient association of John's and Jesus's commemorative dates with the turning points of the solar year could appear illogical, but few were as open in their criticism as Heriger, abbot of Lobbes (d. 1007), who in *c.*995 concluded that his Church committed a persistent, if forgivable, error when celebrating Christmas on 25 rather than 21 December.[77] More generally speaking, it was precisely its connection to the liturgical year and the story of Christ's incarnation that allowed the 'Latin' equinoctial date on 25 March to survive in the computistical literature despite the universal acceptance that the 'Greek' mode of Easter reckoning enjoyed by end of the eighth century. In practice, this survival manifested itself in an

[72] Pliny, *Naturalis historia* 18.221, 246, 256, 312. The implied seasonal lengths are 90 (winter), 91 (spring), 92 (summer), and 92 days (autumn). On Varro's dates, see Brind'Amour 1983: 15–21, who argues that they may go back to Julius Caesar himself.

[73] See, for instance, Columella, *De re rustica* 9.14.1, 11–12, and the fifth-century *Fasti* of Polemius Silvius, ed. in Degrassi 1963: 263–76; Lehoux 2007: 309–33 (with English translation). Solstices on 25 December and 24 June are marked on a mid-third century portable Roman sundial now at the Museum of the History of Science, Oxford (inv. 51358). See Stebbins 1958.

[74] Ambrose of Milan, *De Noe* 17.60 (CSEL 32.1: 456); Augustine of Hippo, *Sermones* 287.3, 288.5 (PL 38: 1302, 1308); pseudo-Cyprian, *De pascha computus* 3–4 (CSEL 3.3: 251).

[75] Isidore of Seville, *Etymologiae* 5.34 (ed. Yarza Urquiola and Santos 2013: 113–15); Isidore of Seville; *De natura rerum* 8.1 (ed. Fontaine 1960: 205). See also pseudo-Martin of Braga, *De pascha* 5 (ed. Barlow 1970: 273), the *Tractatus Adthanasii* 4 (ed. Krusch 1880: 332), the spurious *Acts of the Council of Caesarea* 3 (ed. Krusch 1880: 310), and the *Disputatio Morini* (ed. Cordoliani 1945–6: 30–4; Graff 2010: 141–2).

[76] *Munich Computus* 38 (trans. Warntjes 2010c: 107). The same dates are already listed in a seventh-century addition to the *Argumenta* of Dionysius Exiguus (ed. Krusch 1938: 79–80), which first appears in the so-called *Computus Digbaeanus* of 675. See Warntjes 2010b: 92–4.

[77] Heriger of Lobbes, *Epistola ad quemdam Hugonem monachum* (PL 139: 1132–4; Cordoliani 1949a: 483–4). See Borst 1994: 211–12; Verbist 2010: 15–33; Nothaft 2012: 101–2.

all-inclusive approach towards the existing traditions, as calendrical scribes typically annotated both the *equinoctium secundum Graecos* and the *equinoctium secundum Latinos* on their respective dates (and similarly for the solstices).[78] With the hindsight afforded by today's knowledge, the existence of two different conventions for dating the equinoxes and solstices is easily explained as a historical outcome of the Julian calendar's overestimation of the length of the tropical year, which made the calendar incapable of retaining the Sun's turning points on the same dates for more than 130 years. The 'Latin' equinox on 25 March must hence be understood as a relic of the time of Caesar's reform, when the vernal equinox resided close to 24 March, whereas the Alexandrian 'rule of the equinox' came into use during the fourth century AD, by which time the astronomically correct date had already moved up to 20 March. For early medieval onlookers the situation was not nearly as obvious. Since none of the available sources recorded any credible estimate for the length of the year aside from the Julian $365\frac{1}{4}$ days, they were naturally inclined to assume that Caesar's leap-day intercalation was an exact reflection of the time it took the Sun to traverse the zodiac and return to the vernal point. Even those confident of their calendar's accuracy, however, could still be puzzled by the conflicting, indeed incompatible, traditions as to when this vernal point was reached. If Latins and Greeks differed by four days with respect to the date of the equinox, the obvious inference was that at least one party was mistaken in their assumption—but which one?

The confusion surrounding this question is especially palpable in two Irish *computus* treatises of the first half of the eighth century, neither of which took a clear stance on the matter apart from suggesting answers such as 'the Greeks are closer to sunrise than the Latins, (and) so they can more easily investigate the course of the Sun'.[79] Some of this caution may have been due to the existence of a separate tradition of Easter reckoning within the Irish-Celtic Church, which for the longest time had not only followed its own Easter cycle, known as the *latercus* (see Ch. 2.1), but had observed 25 March instead of 21 March as the equinoctial limit for Easter Sunday. Vehement opposition against this practice came from the Northumbrian twin monastery of Monkwearmouth-Jarrow, which during the first decades of the eighth century was home to the Venerable Bede and his abbot Ceolfrith (d. 716). When Nechtan, king of the neighbouring Picts, wrote to Ceolfrith requesting his advice on the calculation of Easter, the abbot of Jarrow answered with an appeal to

[78] Borst 2001: 28–9, 684 (17 Mar.), 694 (20 Mar.), 697 (21 Mar.), 710 (24 Mar.), 713 (25 Mar.), 719 (26 Mar.), 1001 (20 June), 1004 (21 June), 1013 (24 June), 1306 (19 Sept.), 1309 (20 Sept.), 1312 (21 Sept.), 1317 (23 Sept.), 1320 (24 Sept.), 1324 (25 Sept.), 1599 (19 Dec.), 1604 (21 Dec.), 1613 (24 Dec.), 1615 (25 Dec.). See further Borst 1998: 195–6, 264, 275, 287, 297, 436–7, 567, 640–53, 672. See also Wormald 1934: 167, 172, 186, 189, 192, 195, 220, 262; Wildhagen 1961: 90, 94–5; Karasawa 2015: 28–9, 181. According to McCluskey 2006: 143–8, the 'Roman' equinoctial dates are frequently reflected in the orientation of medieval English village churches. See also McCluskey 2004.

[79] *Munich Computus* 38 (trans. Warntjes 2010c: 107). In spite of this statement, the same text (ch. 31, ed. Warntjes 2010c: 90) later defines the 'solstitial year' as one that extends from 25 December to 24 June. See also the remarks on equinoxes and solstices in *Dial. Langob.* 18 (ed. Borst 2006: 449), which was an Irish treatise according to Warntjes 2010c: clxxiv–clxxix.

'the opinion of all eastern nations and especially the Egyptians, who took the palm from all other learned men in calculations'. According to this Egyptian doctrine, the vernal equinox 'usually falls on the twenty-first of March, as we can also prove by inspecting a sundial'.[80] The same tantalizing hint, according to which the controversy between the two dates could be solved by observational means, is found repeated in Bede's treatise *De temporum ratione*, which boldly claimed that the veracity of 21 March was borne out not only by the authority of the Greek Fathers, but also by *horologica consideratio*—investigation by means of a sundial.[81]

How far such a remark can be taken to reflect actual observational practice seems impossible to tell from a distance of 1,300 years, but there are some intriguing signs that the basic idea of 'horological proof' inspired later authors to widen the role of empirical considerations in their computistical writings.[82] The key example appears in a popular *computus* textbook that the monk and teacher Helperic of Auxerre and Grandval composed around AD 900. Like most Carolingian computists before him, Helperic was heavily reliant on Bede's *De temporum ratione*, whose contents he sought to summarize and simplify for the sake of his students. This plan did not keep him from occasionally introducing new material, which included a vivid and detailed depiction of an observational method that his former teachers at the famous monastic school of Saint-Germain d'Auxerre (Burgundy) had allegedly used to adjudicate between the Latin and Greek dates for the equinoxes and solstices. To discover the truth, Helperic recommended the use of an upstairs room (*coenaculum*) facing due east. The idea was to create a small aperture in the eastern wall through which the light of the rising Sun could fall against the opposing western wall, allowing observers to track the direction of movement (*ascensus et descensus*) of the Sun's rising point over the course of the year.[83] Although this setup was not entirely unrealistic, much of Helperic's account was clearly didactic in nature, as seen from his expectation that the Sun would enter the sign of Capricorn each year on 18 December and then reach the winter solstice exactly three days later. This was standard textbook knowledge in the ninth century, but remained far removed from the astronomical situation in AD 900, when the Sun already crossed the equinoctial points on 15 March and 18 September and reached the solstices on 16 June and 16 December. While there is no need completely to deny that solar observations were carried out at Auxerre in the ninth century, or at Jarrow in the eighth, it does not seem that they ever were intended to demonstrate anything that was not already known. Later works such as Honorius Augustodunensis's widespread *Imago Mundi* (*c.*1110/39) still continued to use the 'horological proof' as a rhetorical device to

[80] Bede, *Historia ecclesiastica* 5.21 (ed. and trans. Colgrave and Mynors 1969: 542–3). On the background, see now Ohashi 2015.

[81] Bede, *De temporum ratione* 30 (CCSL 123B: 374–6). See the commentaries in Jones 1943b: 365–6; Wallis 1999: 312–15; Pillonel-Wyrsch 2004: 231–6.

[82] The issue is discussed in greater detail in Nothaft 2015a.

[83] Helperic, *Liber de computo* 31 (PL 137: 40D–1B). On the author, text, and manuscript transmission, see now Jullien 2010: 421–9.

buttress the Greek equinoctial and solstitial dates, regardless of the fact that the actual dates had shifted even further.[84]

1.5. NEW IMPULSES: ASTROLABES AND 'MODERN' DATES

Helperic's textbook still survives in more than eighty copies, whose scribes carried the idea of observing equinoxes and solstices into numerous continental libraries and even across the English Channel. The way to its overseas reception was paved at the end of the tenth century by Abbo of Fleury (*c*.945–1004), who spent two years (985–7) as a visiting teacher at Ramsey Abbey in Cambridgeshire. As a dedicated scholar of the *computus*, Abbo introduced his Anglo-Saxon students to texts and tables that had previously been unavailable in the British Isles, including Helperic's *Liber de computo*, whose 'Abbonian' recension, produced in 978, became by far the most popular version of the text.[85] Among the monks who learned the art of *computus* directly from Abbo was Byrhtferth of Ramsey (d. *c*.1020), author of a massive *Manual* (*Enchiridion*) in Latin and Old English, in which the Alexandrian or 'Greek' dates for the equinoxes and solstices were effectively treated as a fixed part of God's created order and hence as timelessly valid.[86] Byrhtferth's attitude toward the Julian year was similar to that of his contemporary Ælfric of Eynsham (d. *c*.1010), who in his vernacular rewriting of Bede's computistical works critiqued a long-standing belief according to which the Julian bissextile day and its annual increment originated as a result of the miracle of the Sun mentioned in the Book of Joshua (10:12–14). Ælfric protested that the leap-year increment had been steadily accumulating at the usual rate of one quarter-day per year since the very beginning of time, notwithstanding the fact that the Roman habit of inserting the bissextile day in February was a later invention.[87] As for Abbo of Fleury himself, he had dealt with the course of the Sun in a brief treatise, *De ratione spere* (978), in which he offered a simple method of computing the exact day and hour

[84] See the edition in Flint 1982: 108, 113–14. The 'Greek' dates were also still used in twelfth-century texts such as William of Conches, *Dragmaticon Philosophiae* 4.12 (CCCM 152: 113), or the *Liber Hermetis Mercurii Triplicis de VI rerum principiis* (ed. Silverstein 1955: 301–2).

[85] On the transmission, see Traube 1893: 131–8; Van de Vyver 1935: 147–9; Cordoliani 1943: 62–3; McGurk 1974. One early manuscript of Abbo's computistical anthology (s. X/XI) contains three chapters excerpted from Helperic's *computus*, including the one on observing solstices. See MS Bern, Burgerbibliothek, 250, fol. 27v, and the remarks in Cordoliani 1958: 135–45; Wiesenbach 1991: 122–3.

[86] Byrhtferth, *Enchiridion* 1.1 (ed. Baker and Lapidge 1995: 8). A descendant of the computistical manuscript Byrhtferth assembled and commented upon in his *Manual* is preserved in MS Oxford, St John's College, 17 (s. XII$^{in.}$). It includes computistical material Abbo had brought to Ramsey in the 970s. See Baker 1981; Baker and Lapidge 1995: xxv–xxvi, lii–lx, 373–427; Hart 1970; 2003: 309–56; McCluskey 1998: 153–7; Wallis 1985; and Faith Wallis's website *The Calendar and the Cloister* (<http://digital.library.mcgill.ca/ms-17/index.htm>).

[87] Ælfric, *De temporibus anni* 7 (ed. and trans. Blake 2009: 88–91). The connection between Joshua's solar miracles and the bissextile day was still affirmed in the mid-twelfth-century computistical miscellany contained in MS Cambridge, Trinity College, O.7.41, fol. 25v. See on the background Pesic 2009; Smyth 2017: 258–63.

at which the Sun will enter each zodiacal sign. Following Pliny and Bede, he fixed the starting point of the Sun's annual course on 18 March and took its stay inside each sign to last precisely 30 days and $10\frac{1}{2}$ hours.[88] In good computistical fashion, Abbo thus completely ignored the non-uniformity of the Sun's motion brought about by its eccentric path, which was described at length in Calcidius's *Timaeus* commentary, a work he was to study intensively during his later years.[89]

Calcidius was only one of several rare astronomical sources available in the library of the French monastery of Fleury, whose monks elected Abbo as their abbot after his return from England. One of the most exotic relics of ancient learning to which Abbo and his brethren had access was a sixth-century translation of Ptolemy's *Handy Tables*, known as the *Preceptum canonis Ptolomei*, of which only the tables for the Sun and Moon have been preserved. The accompanying canons (explanations and operating instructions) were partly drawn from Theon of Alexandria's fourth-century *Little Commentary* on Ptolemy's tables, but ended up so suffused with transliterations from the Greek that it is doubtful whether the work was ever of much practical use to its Latin readers.[90] In contrast to the Ptolemaic *Preceptum*, which apparently had slumbered in monastic libraries for centuries before the earliest extant manuscript was produced at Fleury around the year 1000,[91] other texts that were being copied and read in Abbo's circles had been completely unavailable only a generation earlier. This holds true in particular for a group of brief astronomical tracts that had originated during the second half of the tenth century in Catalonia, which at the time provided a crucial cultural link between Christian Europe and the advanced astronomical learning of Islamic al-Andalus. The bulk of this knowledge transfer consisted in a family of widely diffused astrological texts known as the *Alchandreana*.[92] In addition, there was a loosely connected corpus of works describing the construction and use of instruments such as the solid sphere, the quadrant, and, most important of all, the planispheric astrolabe.[93] Among the new bits of technical information that came in the astrolabe's wake was a schematic table for tracking the course of the Sun, which began to circulate soon after AD 1000 as part of the treatise *De utilitatibus astrolabii* (Figure 1.2). It shows the entry of the Sun into all twelve signs on the 16th day of the respective month, thereby implying solstitial and equinoctial dates on

[88] *De ratione spere* is edited in Thomson 1985: 130–1. See also Thomson 1988, for an edition of two short appendices to this text, and the remarks in Van de Vyver 1935: 140–7; Eastwood 1997: 251–3; McCluskey 1998: 152–3; 2011: 234–6; Juste 2004: 195–200.

[89] Calcidian planetary diagrams and corresponding excerpts appear as part of Abbo of Fleury's computistical anthology in MS Berlin, Staatsbibliothek, Phill. 1833, fols. 36v–38r, 39v. See Eastwood 1999: 178–86, 195–9; 2002: 282–7; Caiazzo 2006: 17–20, 33–8.

[90] Pingree 1990; 1997; 2000; McCluskey 1998: 114–16. On the importance of Fleury as a centre for astronomical studies, see Blume, Haffner, and Metzger 2012: 83–102.

[91] This manuscript is now MS London, BL, Harley 2506, fols. 55v–69r. See Pingree 1997: 7–9.

[92] Van de Vyver 1936; Juste 2007.

[93] Key sources are printed in Millás Vallicrosa 1931: 271–335. On the transmission of this material, see most recently Burnett 1998; Poulle 1998: 77–84; Samsó 2000; 2004a; Jacquemard, Desbordes, and Hairie 2007; Juste 2016. The role of non-written forms of communication in the early diffusion of astrolabe knowledge is emphasized in Borrelli 2008: 79–157.

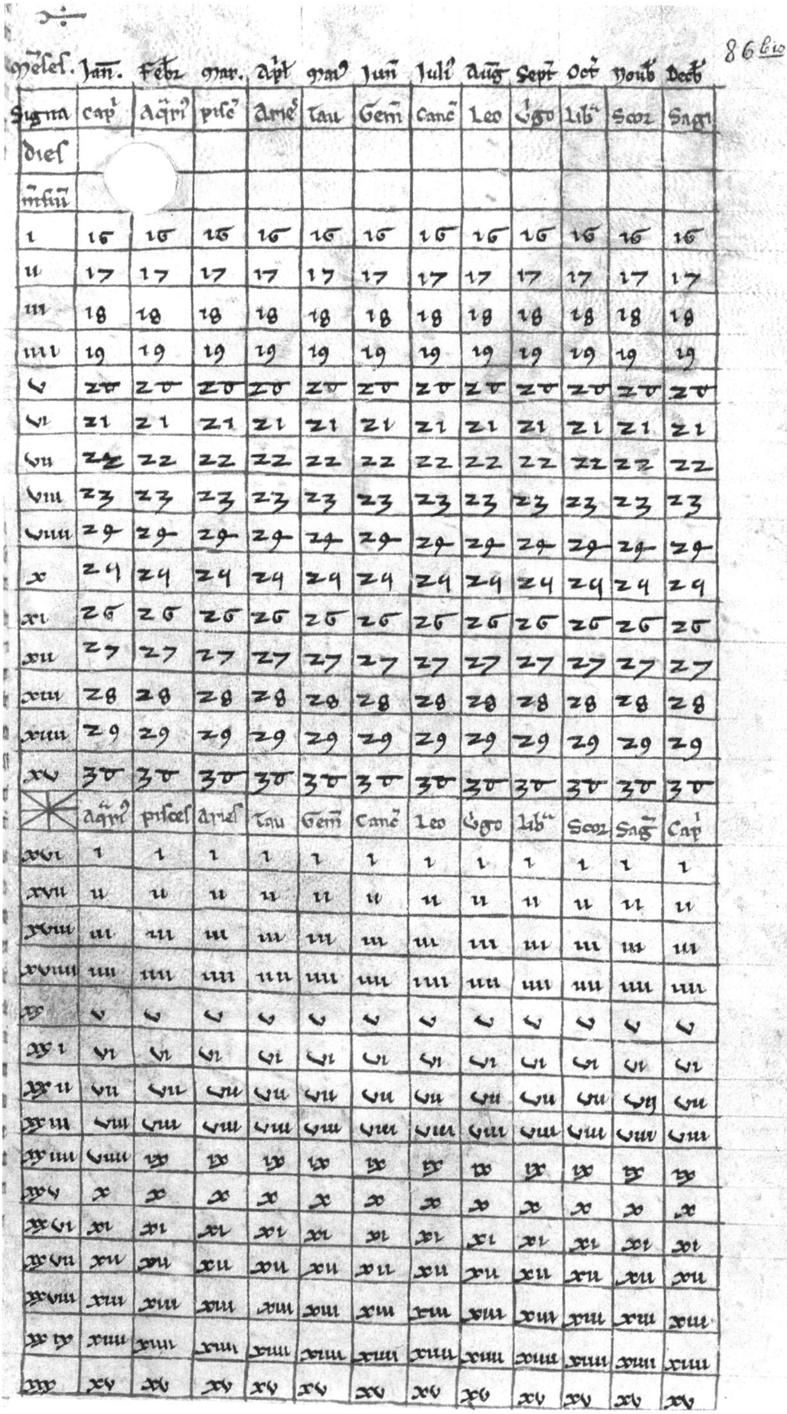

86 bis

Meses.	Jan.	Febr.	Mar.	Apr.	Mai.	Jun.	Jul.	Aug.	Sept.	Oct.	Nov.	Dec.
Signa	Cap.	Aqri	Pisc	Arie	Tau	Gem	Canc	Leo	Vgo	Lib	Scor	Sag
Dies												
mensium												
i	16	16	16	16	16	16	16	16	16	16	16	16
ii	17	17	17	17	17	17	17	17	17	17	17	17
iii	18	18	18	18	18	18	18	18	18	18	18	18
iiii	19	19	19	19	19	19	19	19	19	19	19	19
v	20	20	20	20	20	20	20	20	20	20	20	20
vi	21	21	21	21	21	21	21	21	21	21	21	21
vii	22	22	22	22	22	22	22	22	22	22	22	22
viii	23	23	23	23	23	23	23	23	23	23	23	23
viiii	24	24	24	24	24	24	24	24	24	24	24	24
x	25	25	25	25	25	25	25	25	25	25	25	25
xi	26	26	26	26	26	26	26	26	26	26	26	26
xii	27	27	27	27	27	27	27	27	27	27	27	27
xiii	28	28	28	28	28	28	28	28	28	28	28	28
xiiii	29	29	29	29	29	29	29	29	29	29	29	29
xv	30	30	30	30	30	30	30	30	30	30	30	30
	Aqri	Pisces	Aries	Tau	Gem	Canc	Leo	Vgo	Lib	Scor	Sag	Cap
xvi	i	i	i	i	i	i	i	i	i	i	i	i
xvii	ii	ii	ii	ii	ii	ii	ii	ii	ii	ii	ii	ii
xviii	iii	iii	iii	iii	iii	iii	iii	iii	iii	iii	iii	iii
xviiii	iiii	iiii	iiii	iiii	iiii	iiii	iiii	iiii	iiii	iiii	iiii	iiii
xx	v	v	v	v	v	v	v	v	v	v	v	v
xxi	vi	vi	vi	vi	vi	vi	vi	vi	vi	vi	vi	vi
xxii	vii	vii	vii	vii	vii	vii	vii	vii	vii	vii	vii	vii
xxiii	viii	viii	viii	viii	viii	viii	viii	viii	viii	viii	viii	viii
xxiiii	viiii	ix	ix	ix	ix	ix	ix	ix	ix	ix	ix	ix
xxv	x	x	x	x	x	x	x	x	x	x	x	x
xxvi	xi	xi	xi	xi	xi	xi	xi	xi	xi	xi	xi	xi
xxvii	xii	xii	xii	xii	xii	xii	xii	xii	xii	xii	xii	xii
xxviii	xiii	xiii	xiii	xiii	xiii	xiii	xiii	xiii	xiii	xiii	xiii	xiii
xxix	xiiii	xiiii	xiiii	xiiii	xiiii	xiiii	xiiii	xiiii	xiiii	xiiii	xiiii	xiiii
xxx	xv	xv	xv	xv	xv	xv	xv	xv	xv	xv	xv	xv

Figure 1.2. *De utilitatibus astrolabii*, table for the Sun's sign entries. Paris, Bibliothèque nationale de France, latin 16208, fol. 86bisr.

Source: BnF

16 March, 16 June, 16 September, and 16 December.[94] In recording these dates, *De utilitatibus* became the first textual provider of an alternative to the Latin and Greek dates that had populated Western *kalendaria* up to this point. A noteworthy attempt to confirm the 'new' dates by empirical means was made around 1055 by the monastic reformer William of Hirsau (*c.*1031–91), who spent his early years, until 1069, as a monk at the monastery of St Emmeram in Regensburg. William's scientific pursuits at St Emmeram's Abbey are briefly summarized by the chronicler Bernold of Constance (*c.*1054–100), who calls to mind the 'many monuments of his natural talent', among them the invention of a 'natural sundial modelled on the celestial hemisphere' as well as the ability to investigate astronomical parameters 'by means of reliable experiments'. As examples of such investigations Bernold mentions the 'natural solstices or equinoxes' as well as the degree of inclination of the axis of the world, which he, following William's own parlance, refers to as the *status mundi*.[95]

Bernold's words receive full support from William of Hirsau's treatise *De astronomia*, the opening section of which survives in a manuscript from St Emmeram. In this opening section William confesses how his ignorance of the true dates of the equinoxes and solstices used to impede his astronomical pursuits, in particular those relating to 'the theory of the sphere, the construction of sundials, and the subtleties of the astrolabe'.[96] In order to dispel this sort of ignorance, he proposes a detailed observational setup that can be seen as constituting a considerable improvement upon the approach Helperic had described more than one-and-a-half centuries earlier. Not only did William's method involve a switch from marking the Sun's rising points to measuring its altitude at noon on a meridian line, which was capable of producing much less ambiguous results, but he completely dropped the idea of observing the solstitial point directly. Attempts to do so were indeed greatly impeded by the Sun's very slow change in declination as it approached the turning points, which was less than 1° for two weeks on either side of the solstice. William's adequate reaction to this problem was to create a detailed record of the Sun's intermediate positions, which allowed him to use linear interpolation and the principle of equal altitudes to infer retrospectively the approximate time at which the Sun had reached the solstice.[97] A meridian line drawn in the left margin of the only preserved manuscript gives a vivid idea of the north–south path the beam cast by the Sun at noon was going to travel in the course of a year (see Figure 1.3).

[94] The table is edited in Bubnov 1899: 125. That the 'first part' of Aries and Libra was supposed to correspond to the equinoxes is confirmed in ch. 8.3 (p. 132). The authorship and textual history of *De utilitatibus astrolabii* has been a matter of debate. See Bergmann 1985: 131–68, and the overview in Borrelli 2008: 83–7. A very effective critique of Bergmann and Lindgren, who ascribed the treatise to Gerbert of Aurillac, is provided in Gautier Dalché 2013: 261–75.

[95] Bernold of Constance, *Chronicon* a.d. 1091 (ed. MGH SS 5: 451). William's scientific pursuits are discussed in more detail in Bultot 1978; Wiesenbach 1991; Jacquemard 2015.

[96] MS Munich, BSB, Clm 14689, fol. 86v: 'Igitur a solstitiis et equinoctiis ordo adinventionum nostrarum sumit exordium quia, horum adhuc naturalem terminum ignorans, multa, fateor, et in spericis rationibus et in horologicis institutionibus et in astrolabicis subtilitatibus et in illa ad quam modo tendimus mundani status diversitate, frustra laborabam.'

[97] Nothaft 2015a: 1096–8.

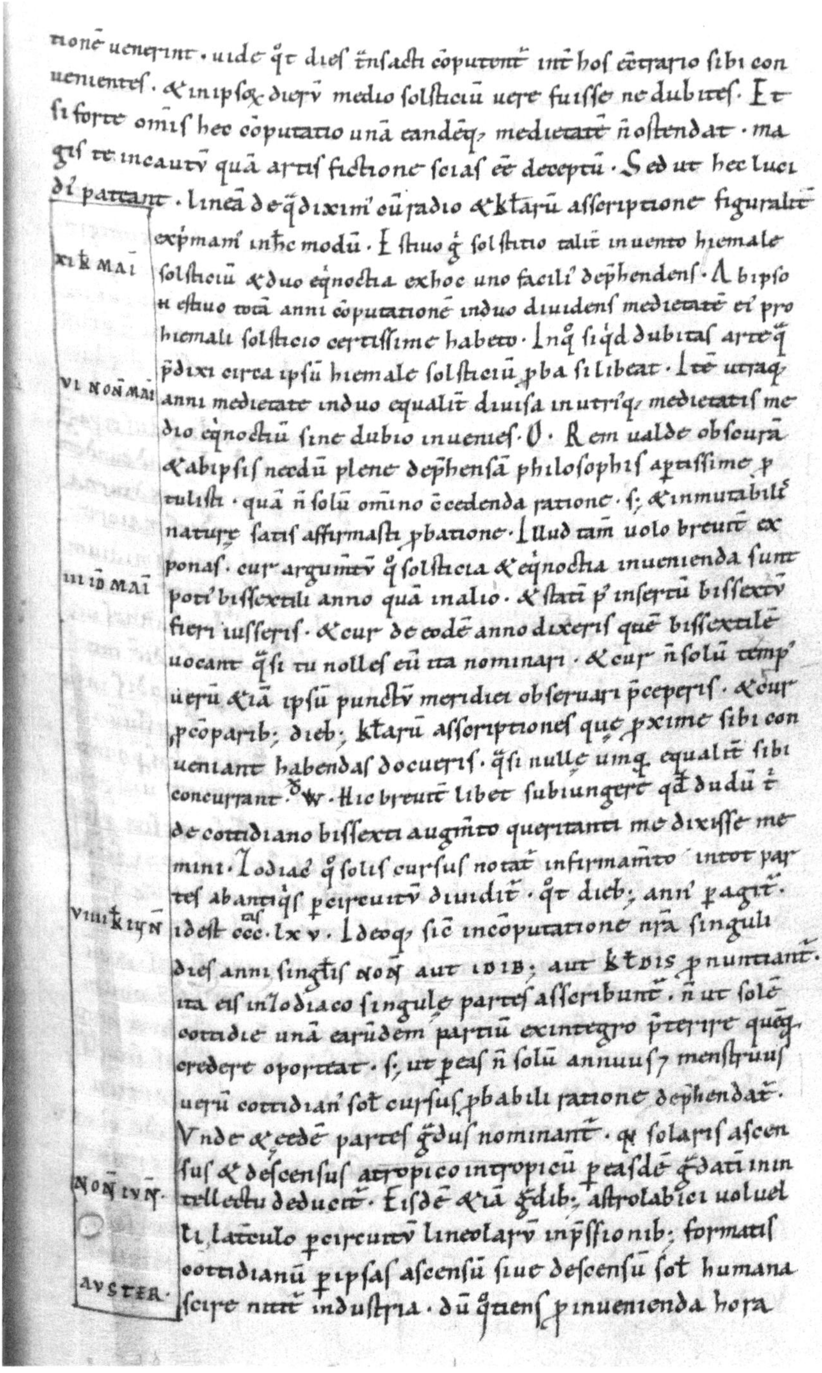

Figure 1.3. Meridian line in William of Hirsau's *Astronomica*. Munich, Bayerische Staatsbibliothek, Clm 14689, fol. 87r.

Source: Bayerische Staatsbibliothek

Although William himself does not explicitly say so, the examples of 'equivalent days' (i.e. days of equal solar noon altitude) his text provides make it clear that he believed the summer solstice to fall on 16 June, which happens to conform to the astronomical situation around 1050–60, but also to the information relayed by *De utilitatibus astrolabii*, a copy of which appears in the same codex as William's *De astronomia*.[98] If he followed this source closely, he would have had reason to place all four cardinal points on the 16th day of the respective month, although some of his remarks, according to which the winter solstice occurs half a year after the summer solstice while the two equinoxes are located exactly midway between these points, might instead be taken to imply that he placed the autumn equinox and winter solstice on 15 September and 15 December.[99] A preference for the exact dates seen in *De utilitatibus* is signalled in a *kalendarium* drawn up in 1074 by William's admirer Bernold of Constance, who retained the usual two sets of dates according to the 'Greeks' (on the 12th day before the Kalends) and 'Latins' (on the 8th day before the Kalends), but in addition entered the equinoxes and solstices 'according to the moderns' (*equinoctium/solstitium modernorum*), placing each on the 16th day of the month in question.[100]

This departure from tradition may have been too radical to catch on immediately. In a treatise on astrolabe construction (*De mensura astrolabii*) written in the 1030s or 1040s by the southern German monk Hermann of Reichenau, the vernal point is moved back to 18 March, a date Hermann associates with the belief of the 'moderns' (*quod moderni primo gradui Arietis, id est XV kl. Aprilis, ascribendum credunt*).[101] It would appear that Hermann here relied on the traditional date for the Sun's entry into the sign of Aries, which Pliny and numerous later authors such as Bede and Helperic assigned to the 15th day before the Kalends of April.[102] By altering the date in *De utilitatibus* by two days and labelling it the 'modern' variant, Hermann may have sought to maintain at least some contact to authoritative traditions. A similar tendency was exhibited at the end of the eleventh century by

[98] MS Munich, BSB, Clm 14689, fol. 75v.

[99] Ibid., fol. 87v. 'Modern' equinoxes and solstices on 16 March, 16 June, 15 September, and 15 December are indeed recorded in the calendar at the start of MS Lambach, Stiftsbibliothek, Cml CIII, fols. 2r, 3v, 5r, 6v (s. XIII^{in}). See Zinner 1988: 63–4. An identical set of dates for the Sun's sign-entries *secundum modernos* appears in a twelfth-century Latin version of the so-called *Calendar of Córdoba* (originally written in Arabic in the 960s). See the edition in Martínez Gázquez and Samsó 1981: 36, 51, 63, 76, and the discussions of this calendar in Borst 1994: 203–8; 1998: 344–50; Viladrich 1996; Vernet and Samsó 1996: 250–1; McCluskey 1998: 166–71; Martínez Gázquez 2011.

[100] See Bernold's autograph in MS Munich, BSB, Clm 432, fols. 2v, 4r, 5v, 7r (ed. Kuithan and Wollasch 1984: 500–1, 508, 514–15, 521–2). It is worth noting that the Sun's entry into Aries is here preponed from the 18 March (15th day before the Kalends) to 16 March (17th day before the Kalends), whereas the entries into Cancer, Libra, and Capricorn remain on the 15th day before the Kalends (17 June, 17 September, and 18 December). On Bernold's calendar, see Wollasch 1983; Wiesenbach 1986: 336; Borst 2001: 318–20.

[101] MS Munich, BSB, Clm 14836, fol. 23r. See also the edition in Drecker 1931: 211.

[102] This is the conclusion drawn by Zinner 1988: 63; McCluskey 1998: 179. For the tradition of placing the sign entries on the 15th before the Kalends, see e.g. Pliny, *Naturalis historia* 2.81; 18.221; Bede, *De natura rerum* 17 (CCSL 123A: 209–10); Bede, *De temporum ratione* 6, 16 (CCSL 123B: 294, 332–7); Helperic, *Liber de computo* 2 (PL 137: 23D–25A); Borst 1998: 263, 274, 286, 296, 433–6, 653–60; Nothaft 2017e: 254–61.

the astronomer Walcher of Malvern, who, in spite of his familiarity with the astrolabe and the text of *De utilitatibus*, preferred to infer the dates of the Sun's sign entries from his own tables, which he based on a scheme popularized by Abbo of Fleury. In Walcher's tables, the summer solstice oscillated between 17 and 18 June, whereas the winter solstice could fall as early as 16 December. The vernal equinox, by contrast, was fixed on 18 March.[103]

Further attempts to resolve a confusing proliferation of dates by observational means may be reflected in a text first attested in manuscripts from around 1100, which describes how to find the dates of the equinoxes and solstices by paying attention to the shadow cast by a gnomon at sunrise. The required method is here described only in the vaguest terms, which fits the fact that the writer made no choice between the three received sets of dates, here claimed to correspond to the 15th, 12th, and 8th day before the Kalends.[104] The same range of options crops up in the *Liber decennalis* (1092) by the chronicler Sigebert of Gembloux, who pits a student against his master in a drawn-out discussion of computistical and chronological problems. A prominent theme in Sigebert's dialogue are the limitations of man's ability to scrutinize celestial cycles, which manifest themselves in the way 'philosophers from so many nations have for so many ages worked assiduously on the course of Sun and Moon [...] and yet nothing, I believe, has until now been brought to a full conclusion'.[105] The teacher lends confirmation to these words, uttered by his student, by invoking the dates of the equinoxes, which:

> the ancients put on the 8th before the Kalends; those who came after them (*iuniores*), since they were endowed with greater perspicacity (*perspicatiores*), moved them to the 12th before the Kalends; the moderns, being even more perspicacious, used the motion of shadows to note with their eyes and hands the ascent and descent of the Sun at the solstice and thought it more correct to assign the Sun's tropics of Cancer and Capricorn to the 15th day before the Kalends.[106]

At first glance, the teacher's statement may seem to depict these modifications of doctrine as instances of scientific progress, especially given his allusion to Priscian's famous formula *quanto iuniores, tanto perspicatiores*,[107] which suggested that the findings of later generations trump those of their ancestors. If this was the case, Sigebert's contemporaries had every reason to be critical of the Easter *computus* with its assumed vernal equinox on 21 March, which failed to receive support from the discoveries of the 'moderns'. Sigebert's master remained strikingly silent on this crucial point, as he limited his remarks to the two solstices in the signs of Cancer

[103] See ch. 5.1–3 of Walcher of Malvern's *De lunationibus*, as edited, translated, and provided with a commentary in Nothaft 2017e: 168–75, 251–61.

[104] Millás Vallicrosa 1931: 326. The text is attested in MSS Oxford, Corpus Christi College, 283, fol. 109v; Vatican City, BAV, Barb. lat. 92, fol. 38v, and twice in MS London, BL, Royal 15.B.IX, fols. 74v and 76v. The latter codex contains a modified version of the solar table in *De utilitatibus astrolabii* (fol. 55v), which places the twelve sign entries on the 18th instead of the 16th of the month.

[105] Sigebert of Gembloux, *Liber decennalis* 2.65 (ed. Wiesenbach 1986: 252). On Sigebert's *Liber decennalis*, see Borst 1998: 334–5; Verbist 2002a; 2010: 173–237.

[106] Sigebert of Gembloux, *Liber decennalis* 2.65 (ed. Wiesenbach 1986: 253).

[107] Priscian, *Institutiones Grammaticae* (ed. Hertz 1955: 1).

and Capricorn. Far from drawing any unwelcome consequences for the reckoning of Easter, he relativized the current variety of opinions by quoting the fifth-century grammarian Servius, according to whom ancient astrologers used to follow divergent conventions for the date of the Sun's entry into the zodiacal signs.[108] He concluded the discussion on a sceptical note, declaring that 'a thing vexed by such variety has obviously not yet been solidified by truth', while reminding his student that so-called wise men have often enough failed to behold more than the shadow of the thing they tried to pursue.[109]

Sigebert of Gembloux's *Liber decennalis* gives us a rare glimpse of the impasse that eleventh-century computists experienced in relation to the date of the vernal equinox. Over the course of ten centuries, its gradual shift in the Julian calendar had created a cacophony of diverging calendrical traditions, which could be variously assigned to 'ancient', 'modern', 'Greek', or 'Latin' schools of thought. While newly fashioned observational instruments and procedures, like the one espoused by William of Hirsau, had the potential of casting serious doubt on the dates marked in conventional *kalendaria*, Sigebert and his contemporaries were still one step removed from the conclusion that would have best explained why ancient and moderns had managed to observe the equinox on different dates: an error in the Julian calendar's estimation of the length of the solar year. The newly imported corpus of astrolabic texts had helped Latin scholars refine their existing doctrines to a considerable degree, but another push beyond the tipping point was needed if a reform of the Julian calendar was to become a conceivable, and openly discussable, option. As we shall see in Chapter 3, this push was soon going to be provided by the rapid influx of texts translated from Arabic in the course of the twelfth century, which changed the face of Western astronomy forever. Before we can examine how the Julian year fared under the pressure exerted by these texts, it will first be necessary to trace the parallel fate of the ecclesiastical lunar calendar during the early Middle Ages, which by the end of the eleventh century had undergone scrutiny in its own, and more far-reaching, way.

[108] See Sigebert of Gembloux, *Liber decennalis* 2.65 (ed. Wiesenbach 1986: 253), and Servius's commentary on Virgil's *Georgics* 1.205 (ed. Thilo 1878: 179).

[109] Sigebert of Gembloux, *Liber decennalis* 2.66 (ed. Wiesenbach 1986: 253).

2

The Ecclesiastical Lunar Calendar and Its Critics, 300–1100

2.1. EASTER: CYCLES AND CONTROVERSIES

When medieval computists looked back at the calendar reform of 46/45 BC, some of them credited Julius Caesar with more than just the institution of a solar year. Those who learned about the structure of the liturgical year from William Durand's massively influential *Rationale divinorum officiorum* (1286) could take from it the claim that the Roman dictator diligently marked the dates of the new moon in his calendar with the 'Golden Number', a device so named 'by comparison, because just as gold surpasses all other metals, so this number excels all other means of lunar reckoning'.[1] While the specific trope of Caesar as inventor of the Golden Number was a product of the thirteenth century (see Ch. 2.2), the more general idea that he was somehow invested in lunar computations can already be encountered in an obscure Irish *computus* tract from the middle of the eighth century (AD 754), which offered a detailed discussion of the natural length of a lunar month.[2] Against the standard custom of the computists, who counted the age of the Moon by alternating 'full' months of 30 days and 'hollow' months of 29 days, the anonymous author insisted that the lunar month possessed a uniform length of slightly more than $29\frac{1}{2}$ days, which entailed that the Moon's 'kindling' (*incensio*) and 'extinction' (*extinctio*) did not always occur at the same time of the day. As for the habit of assigning to each month a full number of days, the author thought it could be traced back to C. Julius Caesar and his trusted scribe Flavianus (mentioned in Macrobius, *Sat.* 1.14.2), 'who did this for the sake of simplicity, so that uneducated men could follow the Moon's course more easily'.[3]

[1] William Durand, *Rationale divinorum officiorum* 8.11.1 (CCCM 140B: 165). A nearly identical passage already appears in the *Computus lunaris* of Bono of Lucca, written in 1254. See the edition in Arrighi 1991: 23.

[2] The text appears in MS Paris, BnF, lat. 6400B, fols. 274r–284r. It was first discussed by Warntjes 2013–14; 2016b: 297–9, who argues for an Irish origin. See also the additional remarks at n. 93 of this chapter.

[3] MS Paris, BnF, lat. 6400B, fol. 275r: 'Gaius Cesar et Flavianus scriba hoc fecerunt et hoc propter facilitatem peractum est, ut ineruditi homines facilius lune cursum intenderent.' Ibid., fol. 275v: 'Dicimus quod XXVIIII dies et dimidium die lunae cursus secundum naturam habet et a Gaio Caesare et Flaviano scriba per artem alia luna XXX[ma] alia XX[a]VIIII[a] conpossita est.' Medieval claims about Caesar's lunar calendar were probably founded on Macrobius, *Sat.* 1.14.13: 'Sic annum civilem Caesar habitis ad lunam dimensionibus constitutum edicto palam posito publicavit.' Modern editions replace *ad lunam*, as found in the manuscripts, with the conjectural emendation *ad limam*. See Bayer 2002: 59 (n. 41).

There is a sense in which the alternative history presented in these medieval texts negated one of the most significant facts about the Julian reform, which is the way it completely divorced Roman calendrical reckoning from the phases of the Moon and in doing so broke with an astronomical principle that most ancient calendars had respected until this point in time. In Asia Minor and the Levant, the advent of the Julian calendar gave rise to a confusing multitude of regional variants, which adapted the basic structure of the 365-day year with a 4-year intercalation cycle to existing local traditions, characterized by diverging month names and beginnings of the year.[4] In the case of Egypt, which was annexed by the Romans in 30 BC, this adaptation of the Julian calendar amounted to little more than the addition of a sixth epagomenal day in every fourth year—a modification already envisaged in the 'decree of Canopus' of 238 BC (see Ch. 1.2). Just like the traditional calendar of ancient Egypt, the revised 'Alexandrian calendar' counted 30 days for each month and began the year on 1 Thoth, which was now perpetually fixed to 29/30 August relative to the Julian year. While the Egyptians were thus able to hold on to their calendrical customs, the spread of the Julian calendar and its Eastern variants severely pushed back the use of lunar calendars in many regions of the empire, leading to a situation where the evidence for the majority of pre-Julian calendars on Roman territory has been lost entirely. The situation was different for the Jewish and Samaritan communities in Palestine and the Diaspora, who retained their lunar calendars, as did the Greek city states of the East.[5] It was different, moreover, for the newly emerging Christian communities, for whom the phases of the Moon were to maintain a deep religious and liturgical significance.

The use of a lunar calendar by ancient Christians found its ultimate justification in the New Testament accounts of the Last Supper, the crucifixion, and the subsequent resurrection of Jesus Christ, which used Passover—with its imagery of the sacrificial lamb—as a narrative and symbolic backdrop. The force of this imagery was such that the early Church saw itself compelled to continue using the date of Passover as its temporal point of departure for the commemoration of said events. The original Christian *Pascha* was celebrated in strict analogy to Passover, on the evening of 14/15 Nisan, whereas a later custom, which was perhaps only fully developed in the second century, tied the termination of the paschal fast to the first Sunday after Passover.[6] This emphasis on Sunday, the day of the week on which Jesus rose from the dead, highlighted the resurrection as the feast's central theme, whereas the Jewish calendar date 14/15 Nisan naturally tended to put a greater emphasis on the Last Supper and crucifixion.[7] During these early stages of its history, the date of Easter was hence little more than a Christian variant of Passover, which still depended on the lunar calendar used by the local Jewish community. While some churches would continue this practice of relying on the Jews for

[4] Stern 2012: 259–94; Samuel 1972: 171–88.

[5] Stern 2012: 299–353.

[6] On the early history of Easter, see Gerlach 1998; Leonhard 2006, with references to further literature. On the standardization of Easter reckoning in late antiquity, see Stern 2012: 380–424.

[7] There is a well-known discrepancy between John and the synoptic Gospels with regard to the precise Jewish date of the crucifixion (14 or 15 Nisan), which has spawned a vast and intricate literature. For orientation, see Nothaft 2012; Saulnier 2012.

several centuries, the larger Christian centres of the Roman Empire soon developed a willingness to emancipate the observance of Easter from its Jewish roots and to look for independent methods of reckoning. From the third century onwards there are clear signs that the bishops of Rome and Alexandria announced the annual Easter date on the basis of specially designed lunar cycles. The purpose of these cycles was to simulate the occurrence of the lunation of Passover and predict the corresponding paschal full moon long before the onset of the Lenten fast, as well as to locate all pertinent dates in the Julian solar calendar, which had become the frame of reference for anniversaries in the liturgical year, including the date of Jesus's nativity (Christmas).[8]

In attaching a lunar calendar to the established Roman calendar year, Christians were helped by the existence of an ongoing Latin tradition of lunar dating, which survived the Julian reform and is still manifested in inscriptions.[9] There was, in addition, a long-standing Babylonian and Greek expertise in creating theoretical lunisolar cycles. The most successful of these cycles, known to the Greeks quite simply as the '19-year period' or *enneakaidekaëteris*, equated 235 lunations to 19 solar years. This equation, which among the Greeks was already used by Meton of Athens in the fifth century BC (see Ch. 1.2), can be understood as an attempt to find the correct ratio between the length of a solar year and the length of a lunation. In the case of the Christian Easter cycle, the ratio in question is best conceived of as the quotient of the average Julian year ($365\frac{1}{4}$ days) and the mean synodic month, presently measured at *c.*29.5306d, from which it follows that there are $365.25 \div 29.5306 = 12.3685\ldots$ lunations per year. From a modern mathematical point of view, the obvious next step would be to convert the remainder of 0.3685... into continued fractions and then proceed to extract convergents.[10] One of the better approximations in the case at hand would be $\frac{7}{19} = [0;2,1,2,2] = 0.3684\ldots$, suggesting that a satisfactory solution can be achieved in a relatively short span of years if 7 out of 19 lunar years are made to contain 13 rather than 12 months, in which case the total number of months contained in 19 years will be $12 \times 19 + 7 = 235$.

Our sources contain no signs that late antique Christians used such refined methods of analysis when approaching the problem of lunar computation, which were at any rate unnecessary to arrive at a satisfactory answer. Armed with the basic insight that the solar year of *c.*365 days exceeds the lunar year of *c.*354 days ($= 12 \times 29\frac{1}{2}$ days) by 11 days, they could have easily come to recognize that this excess approaches three complete lunations at the end of 8 years, since $8 \times 11\text{d} = 88\text{d}$,

[8] The best and most up-to-date survey on the topic of Easter reckoning in late antiquity is Mosshammer 2008: 40–316, although older contributions, such as Mac Carthy 1901: xxx–clxxvii, and Schwartz 1905, remain useful. Specifically for the Latin West, the most detailed introduction is offered by Jones 1943b: 6–122. The lunar calendars that were constructed out of Easter cycles are surveyed in Holford-Strevens 2008. See, in addition, Rühl 1897: 107–69; Schmid 1907; Ginzel 1906–14: iii. 210–51; Wijk 1936: 3–45; Harrison 1979; Pedersen (Olaf) 1983; McCluskey 1998: 80–4; Wallis 1999: xxxiv–lxiii; Declercq 2000: 49–95; 2002; Lejbowicz 2006, and the essays collected in Warntjes and Ó Cróinín 2010; 2011.

[9] Stern 2012: 313–30; Holford-Strevens 2008: 206–8.

[10] See Lejbowicz 2006: 19–21, 23–4 (n. 56), whose insistence on this approach in the context of late antique Easter reckoning must be rejected as anachronistic.

which is equal to two 'hollow' months of 29 days and one 'full' month of 30 days. By intercalating these three months over a period of 8 years it was possible to obtain a simple lunar cycle, the so-called *octaëteris*, but only at the cost of lowering the average lunation in this cycle to $< 29\frac{1}{2}$ days. This solution could not stand, since experience taught that the interval between two sightings of the new-moon crescent was 30 days somewhat more often than it was 29 days, showing the average lunation to be $> 29\frac{1}{2}$ days. One possible reaction was to increase the number of days from $8\times 354+88=2{,}920$ to $8\times 354+90=2{,}922$ in order to reach the exact length of 8 Julian years $(2{,}922=8\times 365\frac{1}{4})$.[11] Another was to add three more years to the cycle to arrive at a surplus of $11\times 11=121$ days, to be distributed over four additional months. Even if all these months are counted 'full', an excess of 1 day will remain, since $121=4\times 30+1$. It takes no immense mathematical ingenuity, however, to realize that if the 8- and 11-year periods are combined, their respective 'errors' will cancel out in a favourable way, such that $19\times 11=209$ days amount to six 'full' months of 30 days and one 'hollow' month of 29 days. In more formal terms, these various cycles can be conceived of as attempts to satisfy the two-part relation $29\frac{1}{2}\times q<11\times p<30\times q$, where p is the number of years and q represents the number of intercalary or 'embolismic' months.[12] The 8-year intercalation cycle ($q=3$; $p=8$) fails at the first part $(29\frac{1}{2}\times 3>11\times 8)$, while the 11-year cycle ($q=4$; $p=11$) cannot satisfy the second one $(11\times 11>30\times 4)$. The 19-year cycle, by contrast, offers the least integral solution for the problem, with $p=19$ and $q=7$.

The introduction of the *enneakaidekaëteris* into Christian Easter reckoning can be traced back to the Alexandrian scholar Anatolius, the later bishop of Laodicea, who is believed to have designed an Easter cycle of this type at some point in the third quarter of the third century, perhaps in the 270s. Most of the details relating to Anatolius's original design and the initial steps in the development of the Alexandrian 19-year cycle remain unclear or uncertain, to the extent that scholars have put forward radically diverging views on the matter.[13] For present purposes, it will suffice to take a closer look at the Alexandrian 19-year cycle in its 'classical' form, which appears to have developed out of the original Anatolian cycle during the fourth century, and which later became normative for nearly the entire Christian world. The key parameter that explains the system as a whole is the so-called 'epact', which corresponds to the age of the Moon on some reference date in the Julian calendar. In the medieval interpretation of the Alexandrian system, the reference date was taken to be 22 March, although in the original version it would have been the last day of the Egyptian year, which began on 1 Thoth = 29/30 August. The starting point of the 19-year cycle coincided with AD 284/5, the first year of

[11] Easter cycles based on this period are attested for the third century. See Lejbowicz 2006: 13–18; Mosshammer 2008: 109–29; Holford-Strevens 2008: 167–72, and the references listed in Warntjes 2010c: xxxv (n. 75).

[12] See on this point Goldstein 1966; Bowen and Goldstein 1988: 42–3.

[13] See e.g. Mac Carthy 1901: xlv–xlix; Schwartz 1905: 15–19; Strobel 1984: 24–32 (with references to further literature); Mc Carthy and Breen 2003; Mosshammer 2008: 130–203; Zuidhoek 2017, and the caveats expressed in Stern 2012: 402–4.

the widely used regnal era of the Emperor Diocletian. In this year 1 Thoth was assumed to also be the 1st day of a lunar month, meaning that the final day of the preceding year had the lunar age 30, which corresponds to an epact of 0.[14] For the calculation of Easter, the key parameter is the 14th day of the Moon of the first lunar month in spring, which serves as the *terminus paschalis* or boundary date for Easter Sunday. In year 1, the epact of 0 shows 22 March to be the 30th and final day of a lunation, which implies that 23 March is the start of the next lunation. The 14th day of this lunation, 5 April, is defined as the paschal *terminus* for year 1. In year 2, the *terminus* drops from 5 April to 25 March due to the 11-day discrepancy between a common lunar year of 354 days and a common Julian year of 365 days. The insertion of an embolismic month of 30 days between the paschal lunations of years 2 and 3 causes the *terminus* to ascend by $30 - 11 = 19$ days, from 25 March to 13 April. Further intercalations are implemented for years 6, 8, 11, 14, 17, and 19 of the cycle, ensuring that the date of the *terminus* never drops below the threshold of 21 March, or 26 Phamenoth, which the Church of Alexandria adopted as the date of the vernal equinox. The earliest possible date of Easter Sunday is reached in year 16 of this cycle, in which the *terminus* always falls on the equinoctial boundary of 21 March. If this date coincides with a Saturday, the following 22 March will be the date of Easter Sunday. The other extreme is reached in year 8, which has the *terminus* on 18 April. If this *terminus* is itself a Sunday, the feast has to be postponed by a week, to 25 April.

In practice, a 19-year cycle of this kind will contain either 4 or 5 bissextile days, the average being $4\frac{3}{4}$. Whenever a bissextile intercalation occurs in the course of these 19 years, one day must be added to the tally of lunar days contained in the cycle. This addition, however, will not interfere with the calculation of the epact, which rises by 11 (modulo 30), regardless of whether the Julian year is common or bissextile. Together with the seven embolismic months of 30 days, the average number of Julian intercalations in a 19-year cycle will amount to $7 \times 30 + 4\frac{3}{4} = 214\frac{3}{4}$ days, yielding a total sum of $19 \times 354 + 214\frac{3}{4} = 6{,}940\frac{3}{4}$ lunar days per cycle. This result is one day in excess compared to the target of 19 Julian years, which on average only contain $6{,}939\frac{3}{4}$ days. In order to restore the balance, it is necessary to drop one day from the lunar count, which is most easily achieved by curtailing a 'full' 30-day month and making it 'hollow', preferably in the final year of the cycle. Medieval computists referred to this intervention as the 'leap of the Moon' or *saltus lunae*, because the Moon could be understood to jump over or skip one day in its path. If the *saltus* is placed near the end of year 19, which thus becomes an embolismic year of 383 rather than 384 days, the epact-difference between the final year and the first year of the cycle will grow from 11 to 12. As seen from Table 2.1, this effectively means that the date of the paschal *terminus* drops from 17 to 5 April, while the value of the epact increases from 18 to $30 = 0$.

Although there is a good deal of uncertainty as to when exactly the familiar standard version of the Alexandrian 19-year cycle was introduced, it is clear that it

[14] The Alexandrian epact is discussed in more detail in Mosshammer 2008: 75–80.

Table 2.1. The structure of the Alexandrian 19-year cycle

Year	Epact	Paschal new moon	Paschal *terminus*	Character of year
1	0	23 March	5 April	common
2	11	12 March	25 March	common
3	22	31 March	13 April	embolismic
4	3	20 March	2 April	common
5	14	9 March	22 March	common
6	25	28 March	10 April	embolismic
7	6	17 March	30 March	common
8	17	5 April	18 April	embolismic
9	28	25 March	7 April	common
10	9	14 March	27 March	common
11	20	2 April	15 April	embolismic
12	1	22 March	4 April	common
13	12	11 March	24 March	common
14	23	30 March	12 April	embolismic
15	4	19 March	1 April	common
16	15	8 March	21 March	common
17	26	27 March	9 April	embolismic
18	7	16 March	29 March	common
19	18	4 April	17 April	embolismic

was in place by the beginning of the fifth century, when the Alexandrian monk Annianus wrote a world chronicle that is still attested in some fragments transmitted in the ninth-century work of George Syncellus. It appears that Annianus's chronological account of world history was based on a 532-year Easter period, which started in the year of the creation of the world (5492 BC) and whose underlying 19-year cycle was already identical to the 'classical' form known from later sources.[15] The 532-year period was based on the fact that the date of Easter Sunday depends not just on the Moon, but also on the day of the week, which keeps shifting in relation to the Julian calendar. A common year of $7 \times 52 + 1 = 365$ days contains 52 complete weeks plus one additional day, whereas the bissextile intercalation in every fourth year increases this surplus to two days. This effectively means that the correlations between weekdays and dates in the Julian calendar recur in sequence only after $7 \times 4 = 28$ years, the so-called *cyclus solaris* or 'solar cycle'. The cyclical recurrence of all possible dates of Easter Sunday is determined by the least common multiple between this 28-year solar cycle and whatever lunar cycle is being used. In the case of the 19-year cycle, the answer is $19 \times 28 = 532$ years.

While Annianus's chronicle shows us that Alexandrian Christians had become aware of this 532-year period by the start of the fifth century,[16] it may have been

15 Mosshammer 2008: 198–202; Nothaft 2012: 58–65.

16 A tenth-century Byzantine *florilegium* makes the doubtful claim that the 532-year cycle was already known to Anatolius of Laodicea in the late third century. See Fernández 2011: 198.

deemed too large for practical use, explaining why Alexandrian Easter tables were generally based on a 95-year period consisting of only five 19-year cycles. A period of the latter type contains an average of $34{,}698\frac{3}{4}$ days, which is only one quarter-day short of a multiple of 7, meaning that most Easter dates will stay exactly as they are from one repetition of the 95-year table to the next, the exceptions being confined to bissextile years. An even more convenient solution, however, was to find a lunar cycle whose period was also a multiple of the solar cycle. The shortest period of this type lasts $3 \times 28 = 84$ years, for which the total epactal increment is $11 \times 84 = 924$ days. This is only six days short of the number of days contained in 31 'full' 30-day lunar months and hence fulfils the mathematical requirement for an adequate 'Julian' lunar cycle, since $29\frac{1}{2} \times 31 < 11 \times 84 < 30 \times 31$. Use of this 84-year cycle became customary in the Roman Church in the late third or early fourth century, which is why one version of it is also known as the *Supputatio Romana*.[17]

The *Supputatio*'s deviations from the Alexandrian 19-year cycle were heightened by the Roman habit of celebrating Easter no later than 21 April, which was to ensure that the Lenten fast did not interfere with the celebration of the *Parilia*, the city of Rome's birthday festival. In contrast to the Alexandrian way of letting Easter fall on the 15th to 21st day of the paschal lunation, in emulation of the seven days of the biblical Feast of Unleavened Bread (15–21 Nisan), the Roman bishops maintained that Easter Sunday could not be celebrated before the 16th day of the Moon, or *luna* 16, which represented the date of the resurrection implied in the Gospel of John (where Jesus died on 14 Nisan, hence *luna* 14). Rome's readiness to make concessions reached its limits in the 440s, when Pope Leo I staged an unsuccessful protest against the Alexandrian decision to celebrate on 23 April in the upcoming year 444. Instead of supporting the pope's stance, one of his advisers, Bishop Paschasinus of Lilybaeum, informed him that only the Alexandrian 19-year cycle accurately followed the 'legal calculation of the Hebrews'.[18] Seven years later, upon discovering that the Alexandrians intended to fix Easter of 455 on 24 April, Leo decided to bring the matter before the Emperor Marcian, which resulted in the pope being schooled on the topic of Easter reckoning by the Alexandrian patriarch Proterius, who insisted that his Easter rules were good enough for the entire Church, Rome included.[19] In order to forestall future embarrassments of this kind, the curia commissioned the construction of a new Easter table from the mathematician Victorius of Aquitaine, whose *Cursus paschalis* ended up mixing Alexandrian and Roman principles and comprised a cycle of 532 years.[20] Its success was greatest in Merovingian Gaul, whose bishops officially

[17] On the *Supputatio Romana*, see Mosshammer 2008: 204–44; Holford-Strevens 2008: 173–8. References to further literature are found in Warntjes 2010c: xxxvi (n. 76).

[18] Paschasinus of Lilybaeum, *Epistola ad Leonem papam* 1 (ed. Krusch 1880: 248).

[19] The relevant sources are edited in Krusch 1880: 251–78. For their background, see Krusch 1880: 129–38; Jones 1943b: 55–61; Holford-Strevens 2011: 6–11.

[20] The relevant texts are edited in Krusch 1938: 16–52. The best technical analysis is provided by Schwartz 1905: 72–80. See also Holford-Strevens 2008: 193–6. For references to further literature, see Warntjes 2010c: xxxviii (n. 82).

adopted the *laterculus Victorii* at the council of Orléans in AD 541 and held on to its use for the next two centuries.[21] Rome was less consistent in its endorsement of the Victorian Easter table, allowing for an alternative to be produced by the monk Dionysius Exiguus, who worked for the Holy See as a translator of Greek texts and is otherwise best known for his Latin editions of papal decretals and the canons of ecclesiastical councils.[22]

As Dionysius told the otherwise unknown bishop Petronius in a letter datable to 525, his 95-year Easter table, which covered the years 532 to 626, was simply the continuation of an earlier table by the Alexandrian patriarch Cyril, who had tabulated Easter data from 437 to 531.[23] Dionysius decided to alter this table in one important respect, which was to exchange the Alexandrian habit of numbering the years according to the regnal era of Diocletian with a count 'from the incarnation of our Lord', thereby spawning the *Annus Domini* era still in use today.[24] Another contribution to the history of *computus* can be discerned in Dionysius's insistence that his Easter table represented 'the absolute and true opinion of the venerable 318 bishops, who convened in Nicaea, a city in Bithynia, to combat the madness of Arius' and used the opportunity to decide upon a cycle of 19 'firm and immovable' Easter full moons.[25] While the idea that it was the Council of Nicaea (325) that had formally sanctioned the 19-year cycle and equinoctial limit on 21 March was already propagated by St Ambrose in the fourth century,[26] it is mainly thanks to Dionysius Exiguus and his followers that medieval computists came to regard the Alexandrian Easter principles as invested with the full authority of the First Ecumenical Council.[27] Dionysius did relatively little to elucidate the arithmetical structure of this supposedly Nicene 19-year cycle, the finer details of which appear to have escaped him,[28] but he still rendered a great service to later users by attaching to his letter a series of computistical algorithms that made it possible to calculate Easter dates without the aid of a table and also to compose continuations for the current 95-year table, which was to expire in 626. The further transmission and development of these reckoning formulae, also known as *argumenta*, became an

[21] *Concilium Aurelianense* § 1 (CCSL 148A: 132). On the fate of the Victorian table in Gaul and elsewhere, see Krusch 1884: 122–41; Schmid 1907: 38–107; Jones 1934: 412–21; 1943b: 65–6; Warntjes 2010c: xxxix–xli, lxxxiii–lxxxix, clvi–clviii; 2015.

[22] Dionysius's writings on *computus* are edited in PL 67: 483–520, which reproduces an earlier edition and study by Wilhelm Jan (Wittenberg, 1718), and in Krusch 1938: 59–86. The best discussion is in Mosshammer 2008: 3–9, 57–106. References to further literature are found in Warntjes 2010c: xxxix (n. 85).

[23] Dionysius Exiguus, *Epistola ad Petronium* (ed. Krusch 1938: 64). The Cyrillan table had already been circulating in a Latin version by the end of the fifth century, as seen from the pseudepigraphic 'preface' discussed in Mosshammer 2013.

[24] The origins of this era, which may well pre-date Dionysius's use of it, are a contentious matter. See most recently Mosshammer 2008: 339–437; 2013: 75–6.

[25] Dionysius Exiguus, *Epistola ad Petronium* (ed. Krusch 1938: 63). Part of this passage was reproduced in a letter written in 627 by the Spanish monk Leo to the archdeacon Sesuldus (ed. Krusch 1880: 301). See also *Dial. Neustr.* 15C (ed. Borst 2006: 396); *Epist. Rat.* 1 (ed. Borst 2006: 1027).

[26] Ambrose of Milan, *Ep. extra collectionem* 13.6 (CSEL 82.3: 230). See Lejbowicz 2010.

[27] Krusch 1884: 106–7; Schmid 1905: 100–10; Jones 1943a: 206–9; Mosshammer 2008: 51–2, 64–5.

[28] See Warntjes 2010c: xlv–xlvi (nn. 103, 106); 2013b: 50–1, 66, on the limits of Dionysius's computistical abilities.

important aspect of the slow spread of the Dionysiac Easter table through various parts of the former Western Empire, in the course of which the corresponding *argumenta* were repeatedly updated, augmented, or modified. This process accelerated once the Dionysiac reckoning reached the British Isles, where substantially expanded formularies were produced in 675 and again in 688/9, after which the *argumenta* were further popularized in the works of the Venerable Bede.[29] One version of these early insular formularies, known as the *Computus Cottonianus* (689), served the Anglo-Saxon monk Willibrord on his mission to the Frisians, which began in 690. He can thus be credited with having introduced Dionysius's *argumenta* to the lands under Frankish rule, where the dissemination and refinement of computistical algorithms flourished during the eighth and ninth centuries.[30]

Before Dionysius's table could fully establish itself in Willibrord's native Britain it first had to prevail in a protracted struggle against two other systems of Easter reckoning that had arrived there previously.[31] The older of these was the 84-year *latercus* devised by Sulpicius Severus (*c.*360–*c.*420/5), whose use in Ireland may go back to the first half of the fifth century. The strong authority that Sulpicius's system continued to enjoyed in the *regiones Scottorum* (as well as among the Britons and Picts) eventually brought the Irish out of step with the Merovingian episcopacy in Gaul, who in 541 switched to dating Easter in accordance with Victorius of Aquitaine. Conflicts inevitably arose once the so-called Columbanian mission led to the foundation of Irish influenced monasteries in the Frankish and Lombard kingdoms. In a justificatory letter written to Pope Gregory the Great around 600, St Columbanus, the figurehead of this Irish mission to the continent, made it clear that his party saw no reason whatsoever to abandon the *latercus* in favour of the Victorian method, boldly declaring that the latter 'has not been accepted by our teachers' or by 'the mathematicians most skilled in reckoning chronology', who had found Victorius's calculation worthy of 'ridicule or indulgence' rather than any kind of reverence.[32] This resolute stance had considerably weakened by 628/9, when the clergy at home in Ireland received a concerned letter from Pope Honorius I urging them to demonstrate their orthodoxy by conforming to Roman practice.[33] After an Irish delegation to Rome had brought home first-hand evidence of the Easter reckoning followed in the capital city of Latin Christianity, the clergy in the south of the island, in 632 (or perhaps 643), decided to adopt what turned out

[29] A detailed account is provided in Warntjes 2010b. See also Cordoliani 1960–1, pt. 2: 173–80; Ó Cróinín 2008. For the presence of *argumenta* in Bede's works, see *De temporibus* 14 (CCSL 123C: 598–9); *De ratione temporum* 47, 49, 52, 54, 57, 58 (CCSL 123B: 429, 434–5, 441, 443, 446–7).

[30] Warntjes 2011b; 2016a: 166–8. An indispensable earlier study on the spread of the Dionysiac reckoning in the West is Krusch 1884.

[31] The latest and best assessment of the insular Easter controversy is Warntjes 2015, which offers extensive references to the older literature. Other recent accounts include McCluskey 1998: 87–93; Ohashi 1999; Blackburn and Holford-Strevens 1999: 794–6; Richter 1999: 204–16; Charles-Edwards 2000: 391–415; Corning 2006.

[32] Columbanus, *Epistula* 1.4 (ed. Walker 1957: 7). On the background, see Corning 2006: 19–64; Leso 2013.

[33] Bede, *Historia ecclesiastica* 2.19 (ed. Colgrave and Mynors 1969: 198).

to be the Victorian table, while the monasteries further to the north continued to cling to their 84-year cycle.[34] In neighbouring Britain, the existence of diverging Easter reckonings eventually became the cause of discord at the Northumbrian court of King Oswiu (642–70), who had been educated by Irish monks and followed the Celtic practice of celebrating Easter on *luna* 14–20, as calculated by the *latercus*. Upon marrying his second wife, the Kentish princess Eanflæd, Oswiu was soon confronted with the fact that she and her entourage maintained the Victorian Easter calculation that had been introduced into Kent by the Augustinian mission sent from Rome in 596/7. More often than not, the diverging cycles caused king and queen to celebrate Easter a week or even month apart, with the result that 'the king might seek to pay the conjugal debt while the queen was observing Lenten abstinence'.[35] These marital disturbances were put to an end in 664, when a synod convoked by Oswiu moved to abolish the Irish reckoning in favour of the Roman one. From Bede's vivid account of the negotiations it appears that the driving force behind this decision was Wilfrid of Ripon, the later bishop of York, who had learned the rules of Easter reckoning in Rome. Significantly, this education had caused him to champion the Alexandrian 19-year cycle in its Dionysiac form, which indicates that the Victorian alternative, though still popular on the British Isles, could no longer claim papal support.[36]

When it comes to Dionysius's eventual victory over his Irish and Victorian contenders, there can be little doubt that the reasons underpinning it were mainly theological and doctrinal in nature. Seventh-century opponents of the 84-year *latercus*, which allowed Easter to fall on *luna* 14, repeatedly accused its Irish users of being 'Quartodecimans', thereby linking them to the late antique heresy of celebrating Easter on 14 Nisan along with the Jews without regard for the day of the week.[37] Heresy was also invoked when the Northumbrians Ceolfrid and Bede asserted that putting the paschal full moon before the equinox, as Victorius did in years where Easter fell on 22 March and *luna* 16, was tantamount to Pelagianism, that is, agreeing 'with those who trust that they can be saved without the grace of Christ'.[38] At the Synod of 664, Bishop Wilfrid allegedly managed to convince

[34] The main source for these events is a letter sent by Cummian to Ségéne, abbot of Iona (623–52), and a certain Béccán, which is edited and translated in Walsh and Ó Cróinín 1988: 57–97. The possible dates of the delegation are discussed in Warntjes 2015: 41–9.

[35] See Holford-Strevens 2010: 150, who offers an illuminating analysis of the conflict.

[36] The two main accounts of the Synod of 664, which is also known as the Synod of Whitby, are Bede, *Historia ecclesiastica* 3.25 (ed. Colgrave and Mynors 1969: 294–308), and the *Vita Sancti Wilfrithi* 5, 10 (ed. Colgrave 1927: 12, 20–2). See Jones 1943b: 103–4; Ohashi 1999: 156–76; Wallis 1999: lxi–lxiii; Corning 2006: 114–29; North 2007; Dailey 2015, and the literature cited in Warntjes 2010c: xl (n. 87).

[37] On the theology behind different lunar limits and the exegetical arguments used in the insular Easter controversy, see Cullen 2007: 109–251; Charles-Edwards 2000: 396–405, 411–15. On the charge of Quartodecimanism, see also Corning 2006: 114–20, 127–8, 132–3, 137–8, 170–2. On early medieval strategies to establish the right method of Easter reckoning, see also McCluskey 2003: 202–5.

[38] Bede, *Historia ecclesiastica* 5.21 (ed. Colgrave and Mynors 1969: 544). See also Bede, *De temporum ratione* 6. The Pelagian heresy is already denounced in a letter sent by pope-elect John IV to the Irish clergy in 640, possibly in connection with the Irish or Victorian Easter reckoning. Bede, *Historia ecclesiastica* 2.19 (ed. Colgrave and Mynors 1969: 198–202); Harrison 1984; Ó Cróinín 1985; Ohashi 1999: 115–46; Corning 2006: 89–92, 131, 160; Thacker 2014: 52–5.

Table 2.2. The *latercus* Easter full-moon dates compared with the corresponding Victorian and Dionysiac dates and the approximate times of true opposition (AD 655–64)

AD	Latercus *luna* 14	Victorian *luna* 14	Dionysiac *luna* 14	Full moon (true opposition)
655	23 March	26 March	27 March	27 March, 8 p.m.
656	11 April	14 April	15 April	14 April, 5 p.m.
657	31 March	3 April	4 April	4 April, 7 a.m.
658	18 April	21 April	22 April	23 April, 8 a.m.
659	8 April	11 April	12 April	13 April, 0 a.m.
660	28 March	31 March	1 April	1 April, 1 p.m.
661	15 April	18 April	19 April	20 April, 8 a.m.
662	4 April	9 April	10 April	9 April, 11 a.m.
663	24 April	26 April	27 April	28 April, 2 a.m.
664	12 Apr	16 April	17 April	16 April, 6 a.m.

Note: In AD 658, 661, and 663, the Roman Easter fell a month before the Celtic Easter, such that the Dionysiac and Victorian full-moon dates displayed in the table belong to the month *after* the Easter lunation. See the comparison of Easter dates and epacts in AD 643–64 in Holford-Strevens 2010: 146–8, and the more extensive table for AD 590–773 in Corning 2006: 183–90.

King Oswiu by linking his 'Roman' reckoning to Saint Peter, who in matters of apostolic authority trumped any saints the Irish might harness in their defence.[39] Irish defectors to the Alexandrian way of reckoning could buttress their position by pointing to the Latin version of an epistle once written by Cyril of Alexandria to the Council of Carthage, which had been augmented by a computistical section at the beginning of the seventh century. Adding to his borrowings from Dionysius Exiguus, pseudo-Cyril imagined that the 19 Easter full moons of the Alexandrian cycle had been revealed by an angel to the Egyptian saint Pachomius and that their reliability had been proven by the waxing and waning of the magical light cast by a 'selenite stone' in Persia.[40]

While there is no direct evidence that different cycles were ever assessed according to their astronomical merit, it is still worth observing that the comparison between predicted and observed phases of the Moon would have given Dionysius (and, to a lesser extent, Victorius) a fairly clear advantage over the 84-year *latercus* used by the Irish. This is readily seen from Table 2.2, which shows that the Irish paschal full moons (*luna* 14) of the seventh century ran ahead of the true opposition of Sun and Moon (calculated for Greenwich Mean Time), and hence of the exact time of the full moon, by four, sometimes by five, days. By contrast, Dionysius's Alexandrian dates for *luna* 14 managed to stay within one day's difference from the astronomical event. Of the ten cases listed in Table 2.2, three (AD 655, 657, 660) show that the true opposition would have fallen within the 24-hour period designated

[39] Bede, *Historia ecclesiastica* 3.25 (ed. Colgrave and Mynors 1969: 306); Eddius Stephanus, *Vita Sancti Wilfridi* 10 (ed. Colgrave 1927: 22). See Corning 2006: 124–8, 173–4; Pelteret 2011.

[40] *Epistola Cyrilli* 5 (ed. Krusch 1880: 346–48). See Jones 1943a: 201–3, 207; 1943b: 93–7; Cordoliani 1945–6: 24–8; Harrison 1982: 315–16; Walsh and Ó Cróinín 1988: 38–40; Wallis 1999: lx–lxi; Warntjes 2010c: lxiv–lxv (n. 164), with references to further literature.

by Dionysius's table as *luna* 14. In three other cases (AD 656, 662, 664), the beginning of *luna* 14, if counted from the previous sunset, would have coincided with the rising of the Moon at its fullest state of illumination. Only in the four remaining cases was Dionysius's *luna* 14 a complete day ahead of the full moon. Since Victorius of Aquitaine had displaced the *saltus lunae* of the Alexandrian cycle, his lunes often fell one day earlier than Dionysius's dates, as is the case with all dates listed in Table 2.2. The advantage hence tended to lie with his Dionysiac contender. Whether anyone in the seventh century tried to gauge the accuracy of the three cycles in this way, we simply do not know.[41]

2.2. THE EMERGENCE OF AN ECCLESIASTICAL LUNAR CALENDAR

For all its apparent advantages over other contenders, the Dionysiac Easter table also posed an array of difficult questions to those who converted to it during the sixth to eighth centuries. Since Dionysius Exiguus offered relatively few details with regard to the computational system that generated the data on display in his tables, early medieval scholars writing in his wake could only wonder about issues such as the correct location of the *saltus lunae* and the seven embolismic months that were needed to ensure that 235 lunar months equalled 19 Julian years. In their attempts to shed light on these issues, insular computists began to collect and systematize the existing knowledge on calendars and Easter reckoning and soon supplemented this body of knowledge with new ideas, paving the way for the further development of *computus* as a learned discipline.[42] As a result of these labours of seventh- and eighth-century Irish and Anglo-Saxon monks, the medieval Church eventually received a standardized lunar calendar, where the correlation of days in the Julian calendar and the respective age of the Moon was fixed for the entire 19-year cycle. Part and parcel of this development was the invention of new algorithms and reckoning aids, which made it possible to obtain vital data for the calculation of mobile feast days without recourse to tables. A basic example already in use in the seventh century were the twelve so-called 'solar' regulars, which could be added to the annual value of the 'concurrent', a number equivalent to the weekday on 24 March, to yield the initial weekday of each Julian month.[43] Another computational shortcut of this kind was provided by the 'ferial' or 'paschal' regulars, which could be combined with the concurrent of a given year to produce the ferial number of the Easter full moon. These regulars gained

[41] What *is* known is that Irish computists produced comparative tables for the Dionysiac and Victorian epacts and paschal full moons, as documented in Warntjes 2010c: clii–clviii, 319–29.

[42] On the innovations introduced by Irish computists, see Holford-Strevens 2008: 198–201; Warntjes 2010c: xliv–li, lxix–lxxiii, cxxxiii–clii, clix–clxviii, cxcv–cc, 256–77; Warntjes 2013b: 51–5; 2016a.

[43] The solar regulars make their first appearance in the *Computus Cottonianus* of 689, which can be traced back to the Anglo-Saxon monastery Rath Melsigi in Ireland. See MS London, BL, Cotton Caligula A.XV, fol. 75v, and Warntjes 2011b: 190–1. See also Bede, *De temporum ratione* 21 (CCSL 123B: 349–50); Borst 2001: 1640–3; *Lect. comp.* 1.3, 2.2 (ed. Borst 2006: 546–7, 566–9).

Table 2.3. The monthly lunar regulars for the age of the Moon (Roman and Egyptian) and the solar regulars for the day of the week

	Jan.	Feb.	Mar.	Apr.	May	June	July	Aug.	Sept.	Oct.	Nov.	Dec.
Roman	9	10	9	10	11	12	13	14	16	16	18	18
Egyptian	9	10	9	10	11	12	13	14	5	5	7	7
Solar	3	6	5	1	3	6	1	4	7	2	5	7

immense popularity in the form of 19 mnemonic verses (*Nonae Aprilis norunt quinos*, etc.),[44] which, following a passage in the seventh-century *Epistola Cyrilli*, were widely reported to be the revelation of an angel, either to St Pachomius or to the bishops of the Nicene Council.[45] To further assist in the determination of mobile feast days, later computists added to the 19 paschal regulars the *claves terminorum*, a sequence of 19 calendrical 'boundary keys' whose addition to different reference dates (*sedes clavium*) made it possible to find the calendrical *termini* for five important Sundays of the liturgical year: Septuagesima, Quadragesima, Easter, Rogation, and Pentecost.[46]

When it came to the age of the Moon, an essential device arose in the form of the twelve 'lunar regulars' (Table 2.3), the earliest description of which appears in Bede's *De temporum ratione*.[47] In a manner analogous to the solar regulars, the lunar regulars could be added to the annual epact to indicate the lunar age at the beginning of each Julian month. In working with these regulars, medieval computists could choose between a 'Roman' variant, where the lunar year began on 1 January, and the manner of the 'Egyptians'. In the Egyptian system, which became the more influential of the two, the epacts changed in September, giving rise to a different set of regulars for the final four months of the year.[48] With the reckoning aid provided by the lunar regulars it became relatively easy to extrapolate to a complete lunar calendar for all 19 years of the cycle. Attempts to display such a lunar calendar in a simple and user-friendly format led to the development of the *litterae punctatae*, which could be inserted into any *kalendarium* as a separate column that assigned to each day of the Julian year its position in the Moon's synodic cycle (Table 2.4).[49] In their most common form, the *litterae* were a recurring sequence of 59 dates, which covered the 30 days of a 'full' lunation followed by the 29 days of a 'hollow' lunation. Of these 59 lunar days, the first 20 were standardly represented

[44] Strecker 1923: 670–1. See also *Lect. comp.* 6.2 (ed. Borst 2006: 616–19); *Lib. ann.* 31b–32 (ed. Borst 2006: 724–8); *Lib. calc.* 7A (ed. Borst 2006: 1391); Irtenkauf 1957: 10–12; Ó Cróinín 1982b; Borst 1998: 495–60.

[45] On the evolution and popularity of this trope, see Jones 1943a.

[46] The *claves* make one of their first appearances in MS Paris, BnF, lat. 9432, fol. 3r (Amiens, s. IX^{ex}/X^{in}). See Ginzel 1906–14: iii. 147–8; Borst 1998: 426.

[47] Bede, *De temporum ratione* 20 (CCSL 123B: 346–9). See Jones 1943b: 354–7; Wallis 1999: 293–6; Pillonel-Wyrsch 2004: 180–5.

[48] Sickel 1861: 186–90; Ginzel 1906–14: iii. 141–8; Springsfeld 2002: 137–8; *Lect. comp.* 1.4, 2.3 (ed. Borst 2006: 547–8, 569–71). See also Helperic, *Liber de computo* 13 (PL 137: 29D–30A).

[49] Sickel 1861; Rühl 1897: 137–8; Ginzel 1906–14, iii. 138–40; Lindhagen 1914: 11–14, 22–6; Wijk 1936: 25–31; Jones 1939: 65–8, 108–9; 1943b: 358–9; Cordoliani 1955: 300–10; 1958: 138–45; 1960–1, pt. 1: 120–8; Wallis 1990a: 53–7; Borst 1998: 405–11, 485–7; 2001: 30–1, 411–16; Germann 2006: 96–9.

Table 2.4. The *litterae punctatae*

Date	Letter	Date	Letter
1 January	*A*	1 July	*E*
20 January	*V*	1 August	*Q.*
21 January	*A.*	1 September	*H*
9 February	*V.*	1 October	*S.*
10 February	*.A*	1 November	*K*
28 February	*.T*	1 December	*V.*
1 March	*A*	2 December	*.A*
1 April	*M.*	20 December	*.T*
1 May	*C*	21 December	*A*
1 June	*O.*	31 December	*L*

Source: Sickel 1861: 159.

by the letters *A* to *V*, the second 20 by *A.* to *V.* (with a dot to the right of each letter), and the final 19 by *.A* to *.T* (with a dot to the left). For each year of the 19-year cycle, a different pair of letters could be assumed to pick out the new-moon dates of the year's full and hollow lunations.

One significant problem with this system was that it ran smoothly only in the common years of the lunar cycle, which came with a regular sequence of full and hollow months. The order was bound to be disturbed in embolismic years and Julian bissextile years, both of which contained at least two 30-day months in a row. Disturbances of this kind even affected calculations based on the 19 lunar regulars, which yielded incorrect results for certain months in years 8, 11, and 19 of the cycle, as described in some detail by the Venerable Bede.[50] That this failure only occurred in three years rather than all seven intercalary years was due to a clever placement of the embolismic months, which Bede attributed to the 'Romans', but which is more likely to have been invented by Irish followers of the Dionysiac table.[51] The Irish and Anglo-Saxon investigations into the Alexandrian 19-year cycle were to have a formative influence on the practice of *computus* in the Frankish kingdom, which gradually progressed from Victorian into Dionysiac usage in the course of the eighth century.[52] While many of the earlier Frankish productions carry the unmistakable stamp of the Irish textbook tradition,[53] the ninth century was to feel the dominance of Bede's *De temporum ratione*, which is

[50] Bede, *De temporum ratione* 20 (CCSL 123B: 347–8). See also Wijk 1936: 123–5; Blackburn and Holford-Strevens 1999: 813–14; Springsfeld 2002: 131–2.

[51] Bede, *De temporum ratione* 45 (CCSL 123B: 422). The placement is explained in greater detail in the eighth-century text *De ratione embolismorum* (CCSL 123C: 685–9). See Sickel 1861: 174–80; Ginzel 1906–14: iii. 136–7; Borst 1998: 427–9; Holford-Strevens 2008: 200–4; Warntjes 2010c: xlviii–xlix (n. 111), lxx–lxxi, clxv–clxvi, 268–9.

[52] The first 'Dionysiac' manual written in Francia dates from 737 and is edited as *Dial. Neustr.* in Borst 2006: 375–423. See Krusch 1910; Warntjes 2010c: lxxiv, ciii–civ, clxxii–clxxiv. On the introduction of Dionysiac reckoning in Francia, see Krusch 1884: 129–31, 136–41; Declercq 2000: 160–4, 179–88; Palmer 2011b; Warntjes 2011b.

[53] On early Frankish computistics and its various influences, see Warntjes 2010c: clix–cxcil; 2011a: 16–21; 2012. Many of the relevant texts are edited in Borst 2006.

believed to survive in more than eighty complete or fragmentary copies made before the year 900.[54] Bede's astounding success in continental scriptoria was fostered at least in part by the intellectual presence of Alcuin of York (*c.*735–804), who had been invited by Charlemagne to organize educational matters at his court in Aachen. Prior to his journey to the continent in 782, Alcuin had spent several decades at the cathedral school of York, where he had received his education under archbishop Ecgbert (732–66) before ascending to the school's headship under Ecgbert's successor Ælberht (766–80). Both Ecgbert and Ælberht had been students of Bede, making Alcuin part of an intellectual lineage that reached from early eighth-century Northumbria to Carolingian Aachen.[55]

The decades subsequent to Alcuin's arrival in Francia saw the production of elaborate computistical 'encyclopediae', whose creators sought to synthesize the available techniques and background information in a comprehensive and didactically serviceable manner. In the case of the imposing Seven-Book-Computus of 809 and the even more voluminous encyclopedia of 818, which came divided into 115 chapters, the range of relevant topics was expanded to include substantial sections on astronomy, metrology (i.e. the study of weights and measurement), and cosmology, thus transforming the *computus* into a general repository of natural knowledge.[56] On the purely computistical side of things, one of the most significant developments of this period was the creation of a full-scale lunar cycle calendar, which mapped out the lunar age on every single day contained in a 19-year cycle. Since the bissextile February had to be charted separately from the common 28-day version, the resulting series of tables recorded no fewer than 7,486 individual lunar days.[57] An impressive glimpse of the degree of refinement that Latin lunar computistics had reached by the end of the tenth century is offered by the manuscripts associated with Abbo of Fleury (*c.*945–1004), whose full-page tables and diagrams display every conceivable combination of lunar, solar, and ferial data.[58] Among his various creations was a 'perpetual calendar' that attached the complete 19-year lunar cycle calendar to a Julian *kalendarium* featuring not one, but five different systems of synodic lunar letters (see Figure 2.1). Each of these came with its own

[54] On the manuscripts of *De temporum ratione*, see Jones 1943b: 140–61; Stevens 1995: 54–8; Wallis 1999: lxxxv–xcvii; Germann 2006: 46–8. See also Hart 2003: 219–307, on the so-called 'Byrhtferth glosses'. On Bede's continental reception more generally, see Contreni 2005; Westgard 2010; Palmer 2014: 150–9.

[55] Alcuin's work as a computist and his role in Carolingian computistical studies is studied in great depth in Springsfeld 2002. See also Borst 1994: 121–6; Bullough 2004: 287–93, 359–61.

[56] The texts are introduced and edited in Borst 2006: 1054–334, 1367–451. They have been discussed to varying extents in Borst 1993; 1994: 138–74; 1998: 231–44; 2006: 83–95, 144–50; Springsfeld 2002: 91–127; Germann 2006: 29–31, 88–106; Warntjes 2010c: xxv–xxvii (nn. 45–7); Dobcheva 2013: 221–8. See now also the art-historical study of Ramírez-Weaver 2017, which focuses on the astronomical illustrations in the Seven-Book-Computus.

[57] The earliest preserved copy, which is at the root of all later versions, appears in the original manuscript of the computistical encyclopedia of 818, which is believed to have been created under the auspices of Archbishop Arn of Salzburg. See MS Munich, BSB, Clm 210, fols. 145v–162v. The entire lunar cycle calendar was edited by Borst 2001: 1647–727. A modern tabular representation is included in Springsfeld 2002: 402–8.

[58] On the 'tabular turn' in medieval computistics, see Germann 2006: 94–106. On Abbo's computistical works, see Engelen 1993; Obrist 2006; 2008; Germann 2006: 106–76; 2008a.

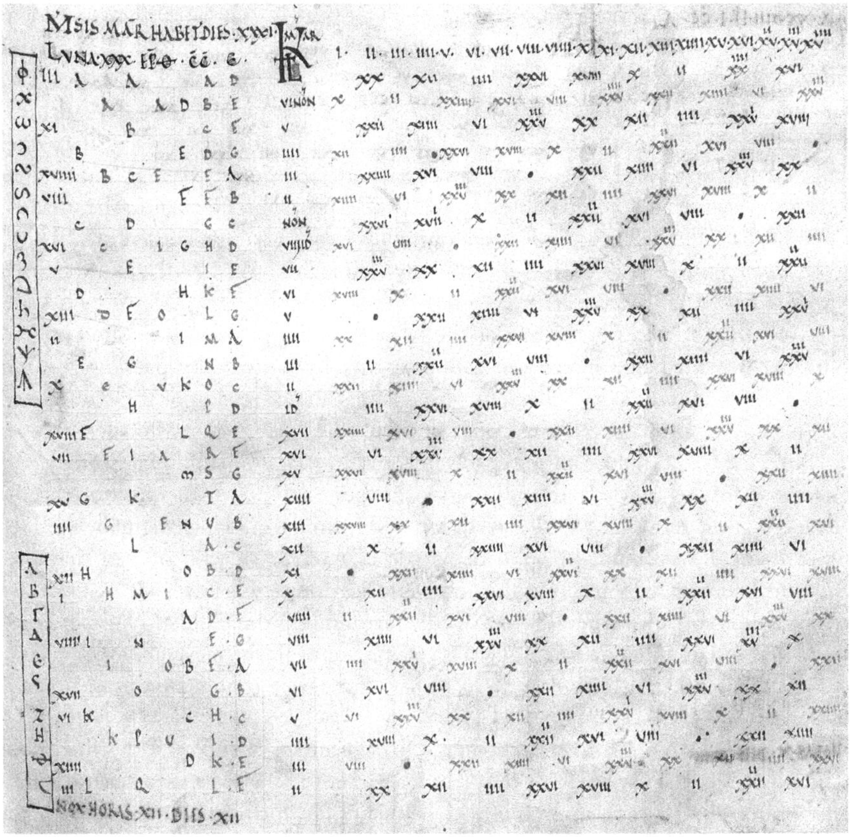

Figure 2.1. Abbo of Fleury's perpetual calendar, page for March, Fleury-sur-Loire, *c.*1000. Berlin, Staatsbibliothek, Phillipps 1833, fol. 24r.
Source: Staatsbibliothek zu Berlin—Preußischer Kulturbesitz, Handschriftenabteilung

auxiliary tables, to which Abbo added his *Ephemerida*, an intricate acrostic poem in tabular form. The latter allowed users to extract the sequence of the lunar letters and the sign of the zodiac in which the Moon will be located on a given date as well as other types of computistical information.[59]

One significant addition to Abbo's expanded calendar was a column to the far left that marked the first day of each lunation in the 19-year cycle by placing the Roman number of the year in question next to the corresponding Julian date. By the end of the twelfth century, this device had almost fully supplanted the much more cumbersome, and less reliable, lunar letters that had been in use since Bedan times.[60] In *c.*1163 one Master William already referred to the numbers that identified the

[59] MS Berlin, Staatsbibliothek, Phill. 1833, fols. 23r–34r. For the *Ephemerida*, see also the edition in Lapidge and Baker 1997: 1–6, 12–21, and Wallis 1990a: 54–7.

[60] On the history of the Golden Number, see Borst 1998: 704–8, and the earlier studies by Wijk 1936: 24–33; Van de Vyver 1947; Lithberg 1953: 66–104; Moreton 1995a: 569–71.

new moons in a conventional *kalendarium* as the *Aureus numerus* or 'Golden Number', so called 'because it is more precious than the other [numbers]. For this one teaches the age of the Moon for every day without any [need for] epacts or regulars and through it one can know the *termini* for any mobile feast day, such as Easter.'[61] In 1171 another computist expanded on this point by writing that 'one calls the number "golden" because it was customary in antiquity to paint it in golden letters, or because it is comparable to gold for the utility it renders unto the institutions of the Church'.[62] Roger of Hereford, who wrote a comprehensive treatise on the *Compotus* in 1176 (see Ch. 3.4), similarly claimed that the Golden Number had received its name 'because it used to be written in gold letters; for it provides an amazing trick to discern the age of the Moon'.[63] The popularity of the Golden Number was furthered in the thirteenth century by writers of didactic *computus* texts such as Alexander of Villedieu, who claimed that a version of these numbers had already accompanied the calendar devised by Julius Caesar, who thereby was able to remedy the 'failure' of the epacts in certain years.[64] Its supposed Roman origins were also acknowledged by Robert Grosseteste (*c.*1220–30), who once again alleged that the term *Aureus numerus* descended from the gold ink originally used to write them down,[65] and by John of Sacrobosco (*c.*1232), who opined that the Romans had originally received this innovation from the Chaldaeans.[66] Curiously enough, this sort of medieval legendarium surrounding the Golden Number continues to exert an influence on modern accounts, some of which even invoke ancient Athens as its original birthplace.[67] In reality, the device cannot be traced further back than the ninth century.[68]

The 235 Golden Numbers that made up the lunar calendar of the medieval Church are on display in Table 2.5. They represent the 235 lunar months of the Alexandrian 19-year cycle, seven of which are embolismic months marked with an

[61] *Summa magistri Willelmi de compoto*, MS Oxford, BodLib, Digby 56, fols. 212ra–217ra, at fol. 215vb: 'Dictus est autem aureus numerus, quia pretiosior est aliis. Ipse enim etatem lune docet singulis diebus absque omni epacta et regulari et per hunc potest sciri terminus cuiuslibet sollempnitatis mobilis, ut pasche etc.'

[62] *Compotus Petri*, MS Paris, Bibliothèque Mazarine, 3642, fol. 34rb (ch. 54): 'Numerus autem aureus vocari solet, quia literis aureis antiquitus depingi solebat vel quia propter utilitatem sui quam ecclesiasticis institutis prebet auro comparabilis est.' For more on this source, see Ch. 3.3.

[63] Roger of Hereford, *Compotus* 1.3 (CCCM 272: 142, ll. 48–50).

[64] Alexander of Villedieu, *Massa compoti* (ed. Wijk 1936: 52). A German translation of this passage was incorporated into the *Mainauer Naturlehre* of *c.*1300, which is uniquely preserved in MS Basel, Universitätsbibliothek, B.VIII.27, fol. 301va (ed. in the 'Beiheft' to Mosimann 1994: 37). See also William Durand, *Rationale divinorum officum* 8.11.1 (CCCM 140B: 165), cited in n. 1, and Gumbert-Hepp 1987: 56–8.

[65] Robert Grosseteste, *Compotus* 9 (ed. Steele 1926: 253, ll. 21–254, l. 2).

[66] See the *Computus* included in John of Sacrobosco, *Libellus de Sphaera*, ed. Philipp Melanchthon (Wittenberg: Clug, 1538), sig. E8r–v. Sacrobosco may have drawn this passage from ch. 17 of the anonymous *Compotus ecclesiasticus* found, for instance, in MS London, BL, Add. 27589, fol. 21rb. See also Giles of Lessines, *Summa de temporibus* III.2.14 (ed. Steele 1926: 125).

[67] See Archer 1941: 4; Eisner 1980: 5; Zemanek 1987: 45; Dutka 1988: 57, and the examples cited in Wijk 1936: 85.

[68] Early appearances in *kalendaria* include MSS Trier, Stadtbibliothek, 2500, fols. 10r–15v (Reims/Laon, *c.*840), and Karlsruhe, Landesbibliothek, Aug. CLXVII, fols. 16v–17v (Laon, *c.*848). See Borst 1994: 174; 2001: 145–6; 228–30.

Table 2.5. The Golden Numbers in the Julian calendar

	Jan.	Feb.	Mar.	Apr.	May	June	July	Aug.	Sept.	Oct.	Nov.	Dec.
1	3		3		11		19	8	16	16		
2		11		11		19	8	16*	5*	5	13*	13,2*
3	11*	19	11		19	8		5		13	2	
4		8		19	8	16	16		13	2		10
5	19		19*	8		5	5	13	2		10	
6	8	16	8*	16	16			2		10		18
7		5		5	5	13	13		10		18	7
8	16		16			2	2	10		18	7	
9	5	13	5	13	13				18	7		15
10		2		2	2	10	10	18	7		15	4
11	13		13					7		15	4	
12	2	10	2	10	10	18	18		15	4		12
13						7	7	15	4		12	1
14	10	18	10	18	18			4		12	1	
15		7		7	7	15	15		12	1		9
16	18		18			4	4	12	1		9	
17	7	15	7	15	15			1		9		17
18		4		4	4	12	12		9		17	6
19	15		15			1	1	9		17	6	
20	4	12	4	12	12				17	6		14
21		1		1	1	9	9	17	6		14	3
22	12		12					6		14	3	
23	1	9	1	9	9	17	17		14	3		11
24						6	6	14	3		11	19
25	9	17	9	17	17			3		11	19	
26		6		6	6	14	14		11	19		8
27	17		17			3	3	11	19		8	
28	6	14	6	14	14			19		8		16
29				3	3	11	11		8		16	5
30	14		14				19	8		16	5	
31	3		3		11					5		13

asterisk (*). The sequence begins with Golden Number *3* on 1 January, which is thereby identified as the seat of a new moon in the third year of the 19-year cycle. The same number appears 30 and 59 days later next to 31 January and the first day of March, which month features an identical pattern of Golden Numbers to January. Of the 365 days of the common Julian year, 131 remain perpetually without a Golden Number and hence without any calendrical new moon. By contrast, 2 December is the only date to receive two Golden Numbers, *13* and *2*, for reasons lucidly explained by Robert Grosseteste in his influential *Compotus* (see Ch. 4.2).[69] An interval of eight years separating two new moons that fall in the same Julian month usually comprises 96 common lunations of $96 \times 29\frac{1}{2} = 2{,}832$ days and 3 embolismic lunations of $3 \times 30 = 90$ days. The resulting $2{,}832 + 90 = 2{,}922$ days are two days in excess of eight common Julian years of $8 \times 365 = 2{,}920$ days, which explains why the Golden Number *1* is always followed two days later by the Golden Number *9*. Things look different if the 8 years in question include the final year of the 19-year cycle, in which case the *saltus lunae* reduces the interval from 2,922 to 2,921 days. The Golden Numbers *13* and *2* normally follow on adjacent dates for this very reason, but an additional factor must be borne in mind in the one case where Golden Number *13* falls on 2 December. Not only will the 8 years that begin here be diminished by the 'leap of the Moon', but the embolismic month that nominally belongs to year 14 has already been inserted on 2 November of year 13. The total number of days contained in the 8-year interval hence reduces to 2 embolismic lunations of $2 \times 30 = 60$ days and 97 common ones, of which the first and last are a 'hollow' lunation of 29 days. The sum total of 2,920 days is the same as 8 common Julian years of 365 days, 'which is why it necessarily happens that with the completion of these eight years the Moon is again one day old on the first day of the ninth year. Hence it is necessary that two new moons are put on the same day.'[70] Aside from this minor point, the only real adjustment the scheme required was caused by the Julian bissextile day. Normally, this bissextile day remained 'invisible' to the calculation, as it was quietly incorporated into the lunar count by adding a 30th day to the lunation that ended in February. In years, however, where the insertion of the lunar *bissextus* preceded its solar equivalent, because the lunation had already ended before 24 February, it became necessary to interpret the day that was marked by the subsequent Golden Number as the final day of the current rather than the first day of the following lunar month.[71]

The gradual standardization of an ecclesiastical lunar calendar, which culminated with the introduction of the Golden Number, unfolded independently from any empirical input or appeals to observational data, as computists were understandably reluctant to challenge a computational structure that they believed had been defined as 'firm and immovable' by the Council of Nicaea, or had been the subject of angelic revelation to St Pachomius. A good indicator of the unsuspecting attitude that still prevailed at the end of the first millennium can be found in the

[69] Robert Grosseteste, *Compotus* 9 (ed. Steele 1926: 254–7).
[70] Ibid. (ed. Steele 1926: 257, ll. 19–21).
[71] This is explained, for instance, in Alexander of Villedieu, *Massa compoti*, ll. 36–8 (ed. Wijk 1936: 53).

works of Abbo of Fleury, who, rather than debating the accuracy of the lunar and solar cycles that made up the *computus*, spoke of them as the 'cycles formed by nature' (*circulos quos natura formavit*), which he believed had been running without interruption since the creation of the world.[72] Abbo used this assumption in a series of attempts to wrestle with a chronological problem that had been uncovered by Bede in chapter 47 of *De temporum ratione*, where he had shown that the Dionysiac dating of Christ's birth to AD 1 (or 1 BC) could not be reconciled with the common patristic assumption that the crucifixion had occurred in a year when Friday coincided with 25 March and *luna* 15. Firm in his conviction that the 'natural order' of the cycles trumped historiographic tradition, Abbo of Fleury came to the conclusion that Jesus's death had taken place not in AD 34, as popularly asserted, but in AD 12, which was the closest year nearby to show the desired combination of data. The same method could be applied to the date of the world's creation and the date of death of Benedict of Nursia, always assuming, of course, that the 19-year and 28-year cycles were faithful representations of the lengths of the lunar month and the solar year and that they had been recurring in perfect uniformity since the beginning of time.[73]

Despite Abbo's conviction that the lunar cycle he used was a representation of nature itself, the truth was that its predictions had grown increasingly less adequate ever since its construction in the fourth century. By equating the 235 lunations contained in the cycle with $19 \times 365\frac{1}{4} = 6{,}939\frac{3}{4}$ days, the Alexandrian system implied a mean lunation of $29.530851\text{d} \approx 29\text{d}\ 12;44,25,32\text{h}$, which exceeds the length of the mean synodic month—currently measured as $c.29.530589\text{d} \approx 29\text{d}\ 12;44,2,53\text{h}$—by approximately 22.65 seconds. If this value were to remain constant over time (which is not quite the case), a lunar calendar based on Alexandrian principles could be expected to amass an error of 1 day close to every $308\frac{1}{2}$ years ($M \approx 308\frac{1}{2}$ years), which is considerably more moderate than the error of $J \approx 130$ years that plagued the Julian calendar. One additional factor that tended to mask the error of the Church's lunar *computus* for the first centuries of its existence was an ambiguity in the way the paschal new moons and *termini* were supposed to relate to the actual cycle of the Moon's phases. At its inception in the early fourth century the Alexandrian cycle had still placed *luna* 1 on or close to the day of conjunction, but nothing prevented later computists from instead associating the start of the lunation with the first sighting of the new-moon crescent, which was usually made one or two full days after conjunction. In a similar vein, it was possible to tie *luna* 14 to the rise of the full moon and hence to the approximate time when the Sun and Moon were in opposition.

From Table 2.6, it can readily be seen that during the second half of the seventh and the first half of the eighth century, when Bede and the Irish computists produced their seminal works, *luna* 14 of the paschal lunation still frequently graced the actual day of the true opposition of Sun and Moon. This holds true, at any rate,

72 This expression is found in Abbo of Fleury's *Epistola prima*, MS Berlin, Staatsbibliothek, Phill. 1833, fol. 56v. On Abbo's attitudes towards his cycles, see also Germann 2008b.

73 Verbist 2006; 2010: 35–84. See also Borst 1984: 385–8; 1998: 712–14.

Table 2.6. New- and full-moon times in the age of Bede, computed for the longitude of Jarrow (1;30° West) in year 16 of the 19-year cycle

AD	True conjunction	True opposition
661	6 March 14:04	21 March 17:46
680	6 March 14:40	21 March 01:06
699	7 March 12:09	21 March 11:20
718	7 March 02:09	21 March 05:29
737	6 March 09:48	21 March 05:12

for year 16 of the 19-year cycle, in which the paschal *terminus* was always supposed to fall on 21 March. The conjunction that preceded this paschal full moon tended to fall on either 6 or 7 March, which suggests that the start of the calendrical lunation on the evening prior to 8 March would have more often than not aligned with the first visibility of the new-moon crescent. For those who believed that *luna* 1 was supposed to mark the day of conjunction, such new-moon sightings on 8 March might have indicated that the 19-year cycle lagged behind its official target. Yet on a reading that associated *luna* 1 with first visibility it could even be the case that the calendrical start of the lunation occurred in advance of when the Moon was actually observable in the evening sky. It is the latter type of case that receives an unlikely amount of attention in Charlemagne's correspondence with Alcuin of York, eight letters among which touch upon astronomical and computistical questions.[74] In 799, the king and his entourage had noticed that the Moon on 18 March seemed too small for the age it should have possessed according to the standard *computus*, which predicted the paschal *luna* 14 for 25 March, making 18 March equivalent to *luna* 7. If the Moon first became visible on *luna* 1, the date of *luna* 7 should have been the occasion of a properly formed half-moon, yet this was not the case in the year in question. Indeed, the true conjunction of this lunation had occurred on 11 March, while the Moon would have become visible only after the sunset that marked the start of 13 March, which was already the beginning of *luna* 2 in the calendrical count. Although the resulting discrepancy between calculated and observed Moon cannot have been dramatic, Charlemagne was no doubt justified in writing to Alcuin that the Moon on 18 March looked smaller than 7 days old and demanding an explanation.[75] In responding to the king's query, Alcuin took recourse to two arguments that, oddly enough, had originally been devised to deal with the opposite scenario, where the observed Moon anticipated the calendrical one. They make their first appearance in the 43rd chapter of Bede's *De temporum ratione*, a chapter dedicated to the question 'Why the Moon sometimes appears

[74] On computistics and astronomy in Alcuin's correspondence with Charlemagne, see Lohrmann 1993; Borst 1993: 64–7; 1998: 146–51; Springsfeld 2002: 38–59.

[75] Alcuin, *Epistola* 170 (MGH Epp 4: 279–80). See Springsfeld 2002: 55–7, 168–71; McCluskey 2003: 209–11.

older than its computed age'.[76] In this revealing discussion, Bede links the start of a lunation not to the Moon's visibility, but to the evening that follows upon the moment when 'the Moon is lit up by the Sun'. Since it is indicated that the latter can occur at different times of the day, it is clear that Bede was here thinking of the moment of conjunction. By shifting the start of *luna* 1 to the following evening Bede highlighted the way the calendar tracked the Moon's phases in a discrete rather than a continuous fashion, making the computistical lunar age change only once every 24 hours. In cases where the conjunction preceded the start of *luna* 1 by almost a complete day, it was 'not surprising that as a new Moon, it can appear very conspicuously in the heavens, seeing that it has traversed so many hours'.[77] Bede's other argument to explain premature appearances of the new crescent invoked the *saltus lunae*, which in his estimation was necessitated by the fact that the Moon moved quicker than expected:

> For how could it be that over [the course of] nineteen years it becomes necessary to subtract one from the usual number of lunar days, were it not that a certain acceleration of the lunar orbit gradually brings it into effect over the whole period of the 19-year cycle?[78]

According to Bede's own calculation, which involved dividing the 24 hours of the *saltus* by 19 while counting 5 *puncti* per hour, the Moon anticipated its proper course by the equivalent of 1 hour, 1 *punctus*, and $\frac{1}{19}$ of a *punctus* for each year of the 19-year cycle.[79] This surplus continued to add up until immediately before the implementation of the *saltus* in year 19, by which time the observed Moon had become one day 'larger' than the tabulated one. In discussing these calendrical perturbances, Bede's stated goal was to explain discrepancies of up to two days, which could come about if both factors—increment of the leap of the Moon and the count of lunar ages from sunset—combined forces in the final year of a 19-year cycle. At no point, however, did he argue that such a scenario invalidated the 19-year cycle as it was currently in use. Instead, Bede assured his readers that the Fathers of the Council of Nicaea, who had given this cycle their special blessing, had been perfectly aware of the problem, but recognized such deviations of one or two days for what they were: unavoidable temporary imprecisions in a cycle that was otherwise blessed with 'firm stability' and could 'never waver and never fail'.[80] Bede underscored his point by referring to the various authorities that vouched for the divine sanction the Alexandrian *computus* appeared to enjoy. These ranged from the legend of St Pachomius presented in the *Epistola Cyrilli* to an anecdote about miraculously gushing baptismal fonts recounted in Paschasinus

[76] See Bede, *De temporum ratione* 43 (CCSL 123B: 412–18), and the translation in Wallis 1999: 115–19. For commentaries on this chapter, see Jones 1943b: 378–9; Wallis 1999: 328–32; 2010: 123–4; Springsfeld 2002: 167–8; Pillonel-Wyrsch 2004: 294–300.

[77] Bede, *De temporum ratione* 43 (trans. Wallis 1999: 117).

[78] Ibid. 42 (trans. Wallis 1999: 113).

[79] Ibid. (CCSL 123B: 407–8). On the influence of this passage, see Wiesenbach 1986: 139–49.

[80] Bede, *De temporum ratione* 43 (trans. Wallis 1999: 117). See also Alcuin, *Epistola* 170 (MGH Epp 4: 280).

of Lilybaeum's letter to Pope Leo I.[81] If a man of Pope Leo's stature was eventually won over to the Alexandrian method by this kind of evidence, how could anyone responsibly demur?

> Hence, on this same question, I seek for myself and for my people no better way of acting or speaking than what I know he did, who so far excels us in knowledge, merit and authority—that is, to follow the sure judgement of the most reverend Fathers in those matters which are doubtful and unsure to us. It is not to be thought that we either understand the variation of the lunar course with greater acuity than the ancients, or that we are able more advantageously [than they] to distinguish what path it is preferable to follow in such variations.[82]

A careful reading of his chapter 'Why the Moon sometimes appears older than its computed age' makes it clear that Bede regarded deviations between the observable and the calendrical Moon as mere temporary displacements, brought about by the necessity of tracking the Moon's natural course, which amounted to approximately $29\frac{1}{2}$ days per lunation, by means of an artificial count of complete days, which was interrupted by elements such as the *saltus*, the *bissextus*, and the insertion of 30-day embolismic months. Rather than exposing any hidden flaws in the *computus*, the bottom line of Bede's discussion was that these temporary disturbances were no cause for alarm, as they had already been anticipated by the founders of the 19-year cycle and could even in extreme cases never amount to more than two days. If Bede had any further doubts about the validity of his cycle, he certainly hid them well.[83]

2.3. TOWARDS A *COMPUTUS NATURALIS*

The picture that eighth-century sources paint of the relation between the computed and the observed phases of the Moon turns out to be far from straightforward. When Bede, in AD 725, spoke of up to two days by which the Moon 'sometimes appears older than its computed age', he may have been done so for purely theoretical reasons, based on an understanding that there were two separate features of the 19-year cycle that were liable to create temporary deviations. By contrast, the problem Charlemagne and Alcuin discussed 74 years later almost certainly related to a real observed event, yet in this instance the Moon had been seen smaller and hence 'younger' than the age the *computus* assigned to it. A different and more consistently reliable approach to testing the calendar against observation was offered by eclipses of the Sun or Moon, which followers of the Roman encyclopedic tradition would have been able to associate with the moments of conjunction or opposition and hence with the final or 15th day of a lunar month.[84] An intriguing sign that at least some early medieval observers compared eclipses

[81] Paschasinus of Lilybaeum, *Epistola ad Leonem papam* 3 (ed. Krusch 1880: 249–50).
[82] Bede, *De temporum ratione* 43 (trans. Wallis 1999: 119).
[83] This is worth emphasizing in light of the deeply flawed argument put forward in Moreton 1998.
[84] See the section on early medieval eclipse theory in Nothaft 2017e: 36–42.

with the predictions made by their 19-year cycle appears in a mid-ninth-century copy of an elaborate calendar originally devised in 789 at the abbey of Lorsch.[85] Among the various astronomical annotations that accompany this calendar are records of two solar and four lunar eclipses spotted in the years 760 to 787, to which the annotator in four cases added the calendrical age of the Moon on the date in question. Of particular interest is the lunar eclipse he recorded to have taken place on 2 November 784, 'around the eighth hour of the night', noting that this was the 14th rather than the 15th day of the current lunation. The annotator sought to explain this discrepancy by pointing out that one-half of a lunation lasted 14 days and 20 hours (356 hours) rather than a full number of days, which would have allowed for cases where a conjunction on the final day of the previous lunation was followed by an opposition on *luna* 14.[86] It is not clear if the same principle was meant to apply to the three other lunar eclipses he recorded for 15 February 770, 22 September 777, and 18 March 778. The latter two are expressly stated to have occurred on *luna* 14, but the eclipse of 777 is also indicated to have happened at the beginning of the night, which would have been too early to make the date reconcilable with a conjunction on the final day of the preceding lunation. A lunar age is omitted entirely from the report on the earlier eclipse on 15 February 770, presumably because this was *luna* 13 and hence impossible to explain away. Both the date and the lunar age were suppressed, moreover, in the entry for the solar eclipse of 760, which fell on 15 August and hence on the 28th day of a 29-day lunation—too early to coincide with the conjunction.[87]

The annotator's silence may be telling, but one should be careful in assuming that early medieval observers would have automatically taken discrepant eclipse dates as evidence that the Church's lunar count was in some substantial condition of error. A striking reminder that theory kept being underdetermined, rather than undermined, by data comes from Fulcher of Chartres's chronicle of the First Crusade, in which he identifies the reddish darkening of the Moon on 16 June 1117 as a miracle, precisely because the date in question was the 13th rather than 14th or 15th day of the Moon.[88] Similar claims were made—in contemporary chronicles from Cambrai, Rolduc, and Reichersberg—about a total solar eclipse seen over parts of central Europe and the Levant on Wednesday, 2 August 1133,

[85] The calendar is preserved in the mid-ninth-century MS Berlin, Staatsbibliothek, Phill. 1869, fols. 1r–11v (Prüm Abbey, *c.*840), edited in Borst 1998: 254–98, and described in Borst 2001: 56–9.

[86] MS Berlin, Staatsbibliothek, Phill. 1869, fol. 4r. See the editions in Borst 1998: 268–9; 2001: 834. For this argument to work, the eclipse of 2 November 784 would have had to occur within the final four hours of the day. Since the stated eclipse time is '*VIII hora noctis*', this may indicate that the day is here counted from sunrise. A similar calculation of the interval between conjunction and opposition appears in the computistical encyclopedia of 818, where the hours contained in one lunation are given more plausibly as $29 \times 24 + 12 = 708 = 2 \times 354$. See *Lib. calc.* 113 (ed. Borst 2006: 1449–50).

[87] MS Berlin, Staatsbibliothek, Phill. 1869, fols. 3r, 7v, 8v, edited in Borst 1998: 265, 284, 288; 2001: 735, 1196, 1245, 1342. The lunar eclipse of 778 is dated to 23 March, but the lunar age given (*luna* 14) points to 18 March, which matches the fact that the eclipse happened on 17 March after sunset. This suggests a corruption from *XV. kl. April.* to *X. kl. April.*

[88] Fulcher of Chartres, *Historia Hierosolymitana* 2.69 (ed. Hagenmeyer 1913: 604–5). See also Bartlett 2008: 67, and the further examples provided in Dall'Olmo 1979.

luna 27.[89] An early glimpse of the idea that eclipses might be a useful corrective to the traditional lunar *computus* can nevertheless be caught from a brief text entitled *De accensione lunae*, which circulated as an annex to the Seven-Book-Computus of 809.[90] Inspired, perhaps, by the unusually high number of solar eclipses observed in his own time,[91] the anonymous author pointed out that one could realign the new-moon dates of the 19-year cycle according to the times and dates of such events, which were bound to coincide with the date of conjunction. Eager to optimize the accuracy of new-moon calculations even further, he made the important step of working not just with 'full' and 'hollow' months of 30 or 29 days, as was done in the ecclesiastical calendar, but with a precise mean value, which he took to be '29 complete days of 24 hours each and 14 hours and 9 *ostenta*. And wherever this lunation shall come to its end, be it during the day or during the night, there one will know the beginning of the next one to be.'[92]

This exaggerated value of more than 29d 14h per mean lunation was a far cry from the important breakthrough in exploring the arithmetical fine-structure of the 19-year cycle that computists had already managed to make during the previous century. Evidence of this breakthrough appears in an obscure *computus* treatise of the year 754, in which an anonymous Irish (or, possibly, Breton) author commented on how the duration of a lunar month was not adequately captured by alternating sums of 30 and 29 days. The average implied by such a sequence would have been $29\frac{1}{2}$ days, but the matter was complicated by the insertion of 7 embolismic months, the leap of the Moon, and the $4\frac{3}{4}$ bissextile days that accrued over the course of 19 years. Each of the seven embolisms contained 30 days and hence exceeded the usual lunation by $\frac{1}{2}$ of a day. If these $3\frac{1}{2}$ days were added to the $4\frac{3}{4}$ bissextile days while one day was subtracted for the *saltus*, the remaining sum was $7\frac{1}{4}$ days or 174 hours. In the view of our computist, these 174 hours represented 'fragments of nature' (*nature fragmenta*), which had to be equally distributed over 235 months to obtain the average lunation 'according to nature' (*secundum naturam*). Its length turned out to be $29\frac{1}{2}$ days $+\frac{1}{2}$ of an hour $+9$ moments $+\frac{1}{2}$ of a moment $+\frac{1}{10}$ of a moment $+\frac{4}{47}\times\frac{1}{5}$ of a moment, with one moment or *momentum* being $\frac{1}{40}$ of an hour.[93] One of the few Carolingian texts to dare to ascend to this level of analysis was the treatise *De lunae*, which in spite of its attribution to St Columbanus was probably composed not much earlier than 800.[94] Like his predecessor of 754, pseudo-Columbanus understood that the embolismic months and bissextile days created a surplus of $7\frac{1}{4}$ days, which amounted to 174 hours or 348 semi-hours. Of these 348 semi-hours, one could be added to each of the 235

89 See MGH SS 7: 550; MGH SS 16: 453, 710; MGH SS 17: 454, and the examples cited in Dall'Olmo 1975: 240–3; Nothaft 2015f: 196–8.

90 *Lib. comp.* 4.33a (ed. Borst 2006: 1240–2).

91 See McCluskey 2003: 206–8, and the record of solar eclipses from 760 to 812 in *Lib. comp.* 5.10 (ed. Borst 2006: 1277–9).

92 *Lib. comp.* 4.33a (ed. Borst 2006: 1240). The *ostentum* was mentioned by Bede as being equivalent to $\frac{1}{60}$ of a *momentum* $= \frac{1}{2400}$ of an hour, but later redefined by Rabanus Maurus (*c.*820) as $\frac{1}{60}$ of an hour. See Bede, *De temporum ratione* 3 (CCSL 123B: 278); Rabanus Maurus, *De computo* 12 (CCCM 44: 219).

93 MS Paris, BnF, lat. 6400B, fols. 275v–276r. See n. 2.

94 See Ó Cróinín 1997: 269–70; Borst 1998: 325; Warntjes 2013b: 62–5; 2016b: 299–301.

lunations contained in the 19-year cycle. The lunation's average duration hence grew to $29\frac{1}{2}$ days + $\frac{1}{2}$ of an hour, leaving 113 semi-hours to be dealt with. At this point, the author saw the opportunity for a convenient shortcut. He converted 113 semi-hours into 226 quarter-hours, rounded their number to 235, and repeated the earlier distribution, assigning to each month of the cycle another 10 moments (equivalent to $\frac{1}{4}$ of an hour), but only in approximation, as he made clear by writing that one lunar month contained $29\frac{1}{2}$ days + $\frac{1}{2}$ of an hour + 'almost 10 moments'.[95] Pseudo-Columbanus's calculation of the mean lunation was subsequently inserted into two Carolingian *computus* textbooks, one written in Cologne *c.*805, the other created in 827 at the northern Italian monastery of Bobbio, which had been founded by St Columbanus and his companions in 614.[96]

Another Irish foundation in this area was St Gall Abbey, where the centre for *De saltu lunae*'s modest manuscript transmission can be located. A reference to pseudo-Columbanus's calculation found its way into a set of computistical *Questiones* put together *c.*1015 by Notker Labeo, also known as Notker the German, who was a teacher and translator at St Gall. Notker realized that Columbanus's calculation was accurate, but not quite complete, leading him to quip:

> 'Almost' is what he says, however, and not 'exactly'. If you recoil from the ambiguity, then go on and examine these minutiae in order to find a precise number for their quantity and express the complete [value], and know yourself smarter than Columbanus.[97]

The text of the *Questiones* soon travelled north to the island monastery of Reichenau in Lake Constance, where a monk named Hermann, who was tied to his chair by physical disability, took Notker's comment on pseudo-Columbanus as an encouragement to rethink the mathematical problem in his own terms. The outcome of these reflections is detailed in Hermann of Reichenau's brief epistolary treatise *De mense lunari*, written *c.*1040–2 and addressed to an unidentified friend named Herrand. Picking up where pseudo-Columbanus had left off, Hermann took the $7\frac{1}{4}$ days that exceeded $235 \times 29\frac{1}{2}$ days and converted them into moments, arriving at a sum of $7\frac{1}{4} \times 24 \times 40 = 6{,}960$ moments. Dividing this sum by 235, he obtained 29 moments per lunation with a remainder of 145. In order to deal with this remainder, Hermann took recourse to the definition of an 'atom' as $\frac{1}{564}$ of a moment or $\frac{1}{22560}$ of an hour, which had been available since the eighth century. The 145 remaining moments were therefore equivalent to $145 \times 564 = 81{,}870$ atoms, which happened to be evenly divisible by 235, yielding an addition of 348 atoms to each and every lunation of the cycle.[98] Inspired by this result, Hermann repeated his calculation in a full-scale treatise on the *computus*, coyly titled

95 Pseudo-Columbanus, *De saltu lunae* (ed. Walker 1957: 212–14).

96 MSS Cologne, Erzbischöfliche Diözesan- und Dombibliothek, 83^{II}, fol. 47v (= *Comp. Col.* 3.7B, ed. Borst 2006: 906); Milan, Biblioteca Ambrosiana, H 150 inf., fol. 19v (ed. PL 129: 1298). On the relationship between these texts, see Warntjes 2013b: 63, 70–2; 2016b: 301–2.

97 Notker Labeo, *Computus* 8 (ed. King and Tax 1996: 327). An Old High German version of the work was discovered by Kruse 2003. See, in addition, Borst 1984: 415–17; 1998: 324–5, 698; Warntjes 2016b: 303.

98 Borst 1984: 476. See Germann 2006: 185–99. On Hermann's computistical oeuvre and its significance, see the useful survey in Warntjes 2016b, which will soon be followed by a new edition of the corpus with extensive introduction.

Abbreviatio compoti cuiusdam idiotae (1042),[99] in which he avoided the large numbers generated by the atom by instead working with *puncti* and *particulae*, such that 5 *puncti* = 1 hour and 47 *particulae* = 1 *punctus*, or 235 *particulae* = 1 hour. As a result, each lunation received 29 days, 12 hours, 3 *puncti*, and 33 *particulae*.[100]

Hermann's true originality as a computist, however, consisted not so much in his pushing Columbanus's calculation to its final conclusion as in applying this conclusion to a problem that had become more frequent and more noticeable since Bede first addressed it in chapter 43 of *De temporum ratione*. Bede's insight according to which the 19-year cycle allowed for months and years when the new moon appeared one, sometimes even two days earlier than predicted, was now regularly confirmed by observation, creating embarrassing situations where 'one would think it absurd to follow the rules in claiming that the Moon has not yet appeared when everyone, even the peasants, can clearly see it new'.[101] With this frank admission Hermann implicitly contradicted his source, Notker Labeo, who in his *Quaestiones* insinuated that one only needed to step outside on the evening of the *saltus lunae* to find the new moon appear exactly as predicted in the tables, quite as if the 'leap of the Moon' was not just a calendrical convention, but had its equivalent in nature.[102] Now that Hermann had improved upon Notker's account by extracting the precise length of the mean lunation implied by the 19-year cycle, it was a tempting further step to use this value for the creation of a revised lunar *computus*, which accurately tracked the beginning and end point of each lunation. The guiding assumption of such an approach was the absolute uniformity (*aequalitas*) of the Moon's motion, which had merely been concealed by the ecclesiastical calendar's habit of counting 29 days for some lunar months, but 30 days for others. That the 'decree of nature' (*constitutio naturae*) allotted to all months the exact same length was a fact that, in Hermann's eyes, 'no one, not even the demented, may call into question'.[103] On his way towards a table that faithfully obeyed this 'decree of nature' Hermann began with the realization that 12 lunar months, according to his newly derived value, contained 354 days, 9 hours, 4 *puncti*, and 20 *particulae*, which differed from a Julian year by 10 days, 21 hours, and 27 *particulae*, almost 3 hours less than the habitual 11-day increment of epacts. This called for a revised table of epacts, which reflected the precise change in the age of the Moon, down to the hour's points and particles, from one year to the next (see Table 2.7a).[104] Naturally, these more fine-grained epacts could only assist in the calculation of lunar ages if at least one new or full moon had already been timed to a comparable degree of precision. In trying to obtain such a starting point,

[99] The text is printed in Germann 2006: 314–40. For discussion, see Cordoliani 1963: 168–80; Borst 1984: 427–31; 1998: 329–31, 584–5; Germann 2006: 199–219.

[100] Hermann of Reichenau, *Abbreviatio* 32–5 (ed. Germann 2006: 330–1).

[101] Ibid. 25 (ed. Germann 2006: 326).

[102] Notker Labeo, *Computus* 8 (ed. King and Tax 1996: 327). Compare this with Hermann, *Abbreviatio* 36 (ed. Germann 2006: 332), who reminded his readers that there was no *saltus* or embolism in nature.

[103] Hermann of Reichenau, *Abbreviatio* 27 (ed. Germann 2006: 327). Hermann's concept of *aequalitas* is analysed at length in Germann 2006: 232–78, 302–4.

[104] Hermann of Reichenau, *Abbreviatio* 36–8 (ed. Germann 2006: 331–4).

Hermann deemed it best to go back to the hypothetical beginning of the solar and lunar cycles at the creation of the world. In the eighth century, Bede had popularized the notion that the fourth day of creation, on which God suspended the Sun and the Moon in the sky, had been on 21 March, such that the equinox of spring coincided with a full moon.[105] If the term 'full moon' was taken as literally as possible, this implied that the Moon had been exactly at the mid-point of a lunation (14 days, 18 hours, 1 *punctus*, and 40 *particulae*) when it rose in the evening that marked the beginning of the next day. To the resulting hypothetical lunar age on 22 March, Hermann added 10 more days to get to the beginning of the following Julian month, 1 April, which thus started with a lunar age of 24 days, 18 hours, 1 *punctus*, and 40 *particulae*. He now had a calendrical fixed point from which to derive the lunar age on the day of the Kalends of each of the 228 Julian months contained in a 19-year cycle. Besides attaching to his treatise a complete list of these 228 lunar ages, which started with 1 April of the year of the creation (see Figure 2.2), Hermann also drew up a table of 'reformed' lunar regulars, which could be added to the epact of a given year of the 19-year cycle to yield identical results (Table 2.7b).[106]

Even with these improvements in place, however, the precision of Hermann's lunar tables was merely an artificial one as long as his work continued to assume the traditional placements of the new and full moons, as implied by the old 19-year cycle. In order truly to improve upon the predictions this cycle made, one had to be prepared to relocate the starting point for the entire sequence of calendrical lunations, preferably by aligning it to the time and date of a solar eclipse, as had already been suggested in the early ninth century by the author of *De accensione lunae*. Hermann's ability to learn from failure and further refine his method is on impressive display in his *Prognostica* on eclipse prediction, a treatise he wrote at some point between 1049 and his death in 1054.[107] In the initial sections of this text Hermann extended his approach of calculating lunar ages 'according to nature' from the synodic month to the sidereal month and hence to the time the Moon needed to perambulate the entire zodiac. In extracting the length of this period from the 19-year cycle, Hermann assumed that 19 Julian years were equal to 254 sidereal months, which forced him to introduce a completely new unit of time: the *portiuncula*, which he defined as $\frac{1}{127}$ of an hour. One sidereal month hence lasted 27 days, 7 hours, and 92 *portiunculae*.[108] If the Moon moved through the zodiac at uniform speed, it became possible to state the precise time it needed to wander from one of its nodes—one of the two points of intersection between the Moon's path and the ecliptic—to another. According to Hermann's theory this was precisely one-half of a sidereal month, or 13 days, $15\frac{1}{2}$ hours, and 46 *portiunculae*. If

[105] Bede, *De temporum ratione* 6 (CCSL 123B: 290–92).

[106] Hermann of Reichenau, *Abbreviatio* 39–47 (ed. Germann 2006: 334–40). The values of the regulars correspond to the age of the Moon at the start of the Julian months in year 16 of the table shown in Fig. 2.2.

[107] The text is printed in Germann 2006: 341–50. For discussion, see Cordoliani 1963: 180–6; Borst 1984: 436–43; 1994: 217–18; 1998: 332–3; Germann 2006: 219–32.

[108] Hermann of Reichenau, *Prognostica* 2 (ed. Germann 2006: 341–3).

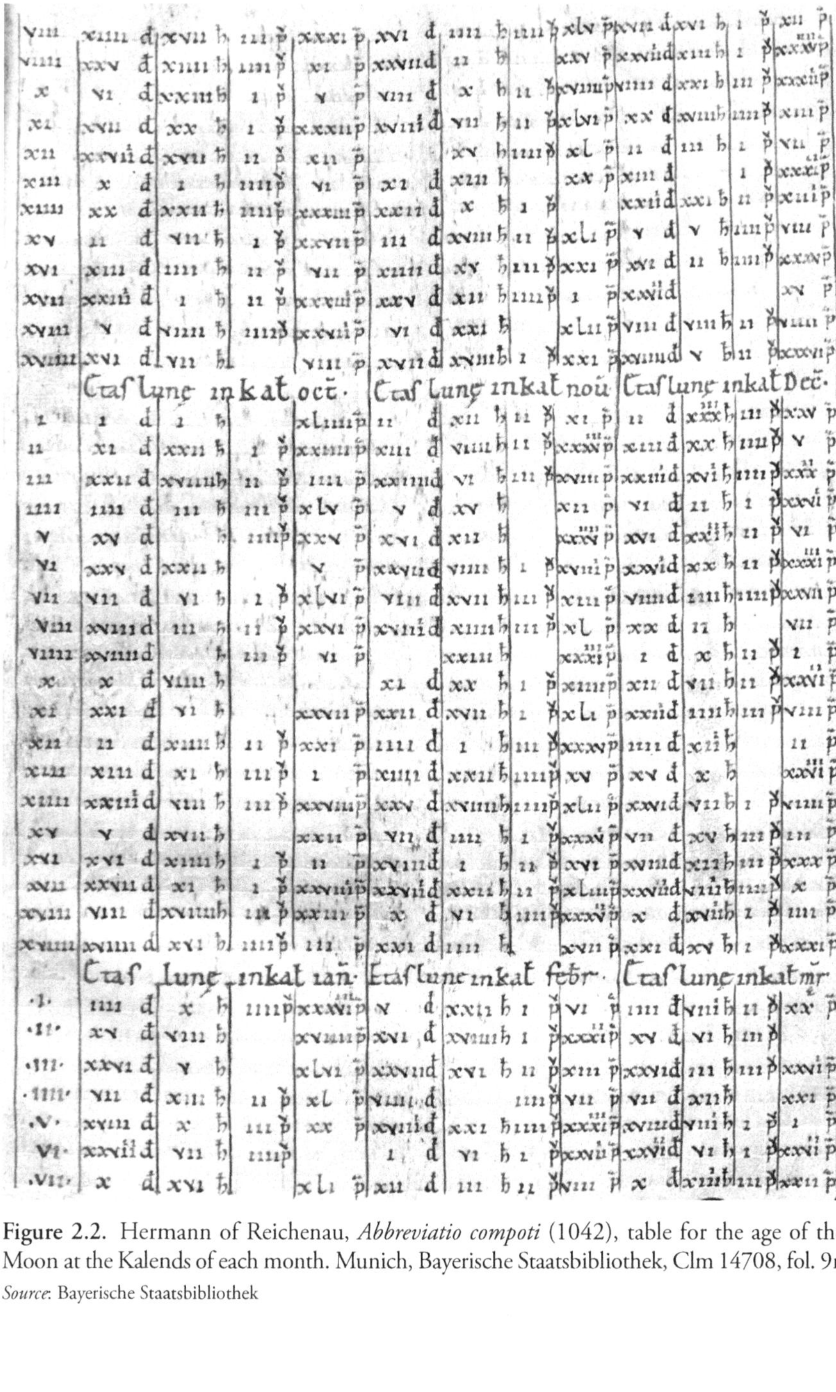

Figure 2.2. Hermann of Reichenau, *Abbreviatio compoti* (1042), table for the age of the Moon at the Kalends of each month. Munich, Bayerische Staatsbibliothek, Clm 14708, fol. 9r.

Source: Bayerische Staatsbibliothek

Table 2.7a. The revised lunar epacts in Hermann of Reichenau's *Abbreviatio compoti* compared with the traditional epacts

Year	Days	Hours	Puncti	Particulae	Traditional epacts
2	10	21	0	27	11
3	21	18	1	7	22
4	3	2	3	1	3
5	13	23	3	28	14
6	24	20	4	8	25
7	6	5	4	2	6
8	17	2	1	29	17
9	27	23	2	9	28
10	9	7	4	3	9
11	20	4	4	30	20
12	1	13	1	24	1
13	12	10	2	4	12
14	23	7	2	31	23
15	4	15	4	25	4
16	15	13	0	5	15
17	26	10	0	32	26
18	7	18	2	26	7
19	18	15	3	6	18
1	29	12	3	33	0

Table 2.7b. The revised lunar regulars in Hermann of Reichenau's *Abbreviatio compoti* compared with the traditional regulars

Month	Days	Hours	*Puncti*	*Particulae*	Traditional regulars
Jan.	9	2	4	17	9
Feb.	10	14	0	31	10
Mar.	9	1	1	45	9
Apr.	10	18	3	12	10
May	11	5	4	26	11
June	12	17	0	40	12
July	13	4	2	7	13
Aug.	14	15	3	21	14
Sept.	16	2	4	35	16
Oct.	16	14	1	2	17
Nov.	18	1	2	16	18
Dec.	18	12	3	30	18

a 19-year cycle contained 235 synodic months as well as 254 sidereal months, it was inevitable that one of the $2 \times 254 = 508$ nodal crossings would coincide with the conjunction of Sun and Moon and hence become the occasion of a solar eclipse. The very same principle applied to the opposition, which produced a lunar eclipse whenever the Moon was located at one of the nodes.[109] One of Hermann's

[109] Ibid. 3–4 (ed. Germann 2006: 343–4).

innovations in this respect was the introduction of a new epact count for what may be termed the 'ecliptical lunar age', which grew at an annual rate of 10 days, $1\frac{1}{2}$ hours, and $10\frac{1}{2}$ *portiunculae*. In a way analogous to the standard lunar calendar, these epacts could be combined with ecliptical regulars to find the ecliptical lunar age at the beginning of each of the 12 months of the Julian calendar.[110]

Once combined with a new set of synodic lunar epacts and regulars of the type already introduced in the *Abbreviatio*, Hermann's ecliptical parameters provided the basis for what promised to be a functioning eclipse theory. Whenever a completed ecliptical period of 13 days, $15\frac{1}{2}$ hours, and 46 *portiunculae* was found to coincide with a completed synodic month, a solar eclipse should have been the outcome. Where it coincided with exactly one-half of a synodic month, the time was ripe for a lunar eclipse. Hermann made an important step towards greater accuracy by calibrating his synodic regulars with the help a solar eclipse witnessed on 29 June 1033, which the chronicle of Reichenau Abbey accurately recorded for the sixth hour of the day.[111] Yet he failed to do the same for his ecliptical lunar regulars, whose prediction of the Moon's position in the zodiac disagreed by more than a day with what was implied by the eclipse of 1033. This resulted in some serious discrepancies between Herman's calculations and the observable phenomena, for instance when the Moon underwent an eclipse on 16 August 1049, 'around the third hour of the night', even though his tables claimed it was still the equivalent of 2 days, $7\frac{1}{2}$ hours, and $49\frac{1}{2}$ *portiunculae* removed from the ecliptic.[112] Hermann's text insinuated that this interval was small enough (and the Earth's shadow large enough) to allow for a partial eclipse in the case at hand, but even he had to admit that the results his theory yielded were far from perfect. In a brief epilogue to his *Prognostica* Hermann freely confessed to the possibility that the 'uniformity and congruity in the courses of Sun and Moon', on which his own calculations relied, might be 'not as pure as has been transmitted by previous writers'. He closed by leaving it to 'more careful investigators of these things' to inquire into the true state of affairs. 'Nevertheless, I have no regrets about the work I have undertaken, by which I was at least able to find something close to the natural course.'[113]

The eight copies of Hermann's highly original *Prognostica* that are known to survive tend to treat the text as a companion piece to his more popular *Abbreviatio*, for which there is a score of witnesses. A number of these only transmit the first part of the *Abbreviatio*, which outlined the traditional doctrines of the *computus*, without including the ground-breaking discussion of the length of the lunar month that follows in the second part. That these two parts could be regarded as different works representing separate branches of lunar computation is shown very clearly in a thirteenth-century copy once kept at the Abbey of Saint-Victor near Paris, which

[110] Ibid. 5–8 (ed. Germann 2006: 344–6).

[111] Ibid. 12 (ed. Germann 2006: 349); *Chronicon Suevicum universale* a.d. 1033 (MGH SS 13: 71). Cf. Hermann of Reichenau, *Chronicon* a.d. 1033 (MGH SS 5: 121), where the seventh hour is mentioned. The eclipse in question was annular and reached its time of greatest eclipse at 11:38 GMT. See Warntjes 2016b: 292.

[112] Hermann of Reichenau, *Prognostica* 13 (ed. Germann 2006: 349–50).

[113] Ibid. 14 (ed. Germann 2006: 350).

merges the *Abbreviatio* and *Prognostica* into a seamless whole. Here the first 24 chapters of the first treatise are designated as *compotus usualis* to distinguish them from the following *compotus naturalis*.[114] This conceptual distinction had already been established in the course of the twelfth century, when Latin scholars began to regard the traditional epacts, regulars, and Golden Numbers as manifestations of only one kind of *computus*, the *computus usualis* or *vulgaris*, which was taught as standard at monasteries, cathedral schools, and—from the thirteenth century onwards—at universities. To this basic form of the discipline they added a more advanced and speculative pursuit, the *computus naturalis*, which gave its users greater insight into the arithmetical structure of natural cycles. In contrast to the *computus usualis*, which only worked with approximations based on complete numbers of days, the *computus naturalis* distilled the exact length of the lunar month from the 19-year cycle by using minute divisions of the hour. This not only allowed for more accurate predictions, but gave expression to the assumed uniformity and regularity (*aequalitas*) of nature, where a single numerical value could apply to each and every lunation.[115]

One of the first scholars fully to capitalize on the distinction between 'usual' and 'natural' *computus* tentatively suggested by Hermann of Reichenau's works was a man named Gerland, who seems to have received his mathematical training at one of the cathedral schools in Lotharingia.[116] To both medieval computists and modern scholars, the best-known element of Gerland's voluminous *Computus* is its revision of the date of Christ's Passion, which followed similar efforts made at the end of the tenth century by the learned abbots Heriger of Lobbes and Abbo of Fleury. Like Heriger, Gerland placed the crucifixion of Jesus on Friday, 23 March in AD 42, which was *luna* 15 according to the standard 19-year cycle, leading him to conclude that the year of nativity had to be moved to AD 8.[117] Gerland defended this correction in the first part of his work, which he appears to have completed no later than 1081. At some point between 1081 and 1093 he decided to add a second book, dedicated to the *computus naturalis*, in which he abandoned the well-worn tracks of the 19-year cycle and thus implicitly undermined his own criticism of the incarnation era. His starting point was a calculation already presented in Book I, which used duodecimal fractions to divide by 235 the $7\frac{1}{4}$ days by which the 19-year cycle surpassed $235 \times 29\frac{1}{2}$ days, arriving at a surplus of $\frac{2}{3}$ of an hour $+2\frac{11}{12}$ moments $+\frac{1}{48}$ of a moment $+7\frac{1}{4}$ atoms per lunation.[118] In Book II Gerland converted these fractions into the much handier 29 moments and 348 atoms first envisaged in Hermann's letter to Herrand. One of his many tables showed how it took 19 years for this surplus to grow to 6,960

[114] MS Paris, BnF, lat. 14960, fol. 260v. See Cordoliani 1963: 171; Warntjes 2016b: 304–16.

[115] For the twofold definition of *computus*, see e.g. Cunestabulus, *Compotus* 1 (CCCM 272: 61–2, ll. 2–8); Roger of Hereford, *Compotus* I.1 (CCCM 272: 132–3, ll. 15–37). These two works, written in England in 1175 and 1176, will be discussed in Ch. 3.3 and 3.4.

[116] For all the pertinent information, see the edition of Gerland's works by Lohr 2013.

[117] Gerland, *Computus* 1.24–26 (ed. Lohr 2013: 143–59). See also the partial edition in Cordoliani 1949a: 484–7. For further discussion, see Verbist 2010: 147–71.

[118] Gerland, *Computus* 1.11 (ed. Lohr 2013: 115–22).

moments (6,815 moments + 81,780 atoms), which was the same as 174 hours or $7\frac{1}{4}$ days.[119]

As in the case of Hermann of Reichenau and his *Prognostica*, Gerland hoped that his painstaking calculations, which must have involved countless hours of sweating over multiplication tables and abacus boards, were going to put him in a position to predict the occurrence of eclipses based on the 'natural' lengths of the sidereal and synodic months. His main innovation in this area was to insist that the two nodes of the lunar orbit were located at the start of Libra and Aries, thereby limiting eclipses of the Sun and Moon to the times of the autumn and spring equinoxes.[120] Needless to say, this approach was no more successful at predicting eclipses than Hermann's attempt had been. Not only did the equation of 235 synodic months and 254 sidereal months with 19 Julian years yield incorrect average durations, but Gerland's theory failed to take account of the difference between the sidereal and the draconitic month, which was caused by a slow rotation of the lunar nodes. While Gerland may have been willing to dismiss the eclipses of 1033 and 1049 mentioned in Hermann's *Prognostica*, he could not ignore the widely observed annular solar eclipse of 23 September 1093, which occurred neither on the date nor in the year predicted by his theory. Without openly commenting on the contradiction this involved, Gerland decided to use the recalcitrant eclipse as the basis for a revised lunar calendar, the explication of which took up the final chapter of Book II. To the eclipse date of 23 September, the calendar assigned the letter A and the numeral XV, indicating that a conjunction had occurred during the 15th hour from the previous sunset in what was now the first year of Gerland's revamped lunar cycle. This starting point lay 2 days and *c.*9 hours earlier than the corresponding new moon in the regular *computus* (26 September) and thus made for a significant improvement compared to the traditional lunar cycle calendar. All further conjunctions were obtained by counting forward from this starting point with a rounded interval that increased from 29d 12h to 29d 13h whenever the additional fractions contained in a natural lunation demanded it. In addition to replacing the conventional lunar letters and Golden Numbers with more accurate indicators, Gerland extended the number of columns that flanked each calendar page from one to four, which enabled him to represent 940 consecutive lunations spread out over $4 \times 19 = 76$ years. Since 76 Julian years comprise 27,759 days without any fractional surplus, this step also allowed him to solve a problem that had still plagued earlier tables based on Hermann's works, where the bissextile increment of the Julian calendar caused a time-shift of six hours between consecutive iterations of the 19-year cycle. In Gerland's scheme, the conjunction on 23 September 1093 at the 15th hour was followed 19 years later, in 1112, by a conjunction on the same day at the 9th hour. It was preceded, moreover, by two cycles with corresponding conjunctions on 24 September, 3h (1055), and 23 September, 21h (1074).[121]

119 Ibid. 2.2 (ed. Lohr 2013: 172–7).

120 Ibid. 2.14–16 (ed. Lohr 2013: 204–16).

121 Ibid. 2.17 (ed. Lohr 2013: 216–32). On Gerland's calendar, as found in MS Paris, BnF, lat. 15118, fols. 52v–57v, see Borst 2001: 265–7.

Like Hermann of Reichenau, Gerland seems to have been principally interested in constructing a 'natural' system of lunar computation for speculative reasons and for the purpose of eclipse calculation.[122] At no point did he use his insights to attack the validity of the 19-year cycle the Church used to determine the date of Easter, which he had still presupposed in Book I of his *Computus*, but which his own calculations now showed to be off-track by more than two days. Neither did he display any critical awareness of the excess in the Julian year that caused the equinoxes and solstices to migrate towards earlier dates. In Gerland's scheme of things all four cardinal points remained fixed on the 12th day before the Kalends of January, April, July, and October, where they had already been placed by authorities such as Bede and Helperic. Regardless of these limitations, the *Computus Gerlandi* was undoubtedly a work ripe with ingenuity and novelty. It was to receive a wide dissemination during the twelfth and thirteenth centuries, making its author a much-cited, much-respected, but also much-criticized name in the subsequent computistical literature.[123] One of the first authors explicitly to mention his name was Philippe de Thaon, author of an Anglo-Norman calendar poem in more than 3,500 lines (*Li compuz*, *c.*1120), who referred to the Lotharingian computist as 'Le vaillant clerc Gerlant'.[124] Roger Bacon later went as far as calling him 'a famous man whom all computists and astronomers follow in computistical matters'.[125] Twelfth- and thirteenth-century writers on the *computus* often put him in one line with Dionysius Exiguus, Bede, and Helperic, but also expressed misgivings about the rash way in which he opposed the doctrines of the Church and those who wrote before him.[126] One of the more telling passages of this kind appears at the

[122] He calls the information in Book II 'compiled from nature's womb' (*de naturae gremio compilatum*). See the prologue to Gerland, *Computus*, bk. II (ed. Lohr 2013: 170).

[123] The transmission is discussed in Lohr 2013: 14–15, 28–58, who describes 37 manuscripts. Numerous copies contain annexes that supplement, modify, or improve the material found in the main body of the text. Nineteen such addenda are edited, translated, and commented upon in Lohr 2013: 233–63, 356–83, 437–50. Some chapters from the *Computus Gerlandi* were integrated into the *Compotus Petri* of 1171, MS Paris, Bibliothèque Mazarine, 3642, fols. 36va–38va, 39rb–41vb (chs. 56–9, 62–4). See also Caiazzo 2012: 30–2, 35–6, who edits excerpts from a mid-twelfth-century commentary on Gerland's *Computus*.

[124] See the edition in Short 1984: 39 (l. 2952). Further references to Gerland are found on pp. 7 (l. 211), 18 (l. 1314), 19 (l. 1386), 28 (l. 2079), 29 (l. 2130), 32 (l. 2360), 33 (l. 2495), 39 (l. 2910), 41 (ll. 3109, 3119), 42 (l. 3208), 44 (ll. 3326, 3340). See also p. 33 (ll. 2476–91), where the length of the lunar month is stated to be 29 days, 12 hours, 29 moments, and 348 atoms. Another reference to Gerland in a poetic context is Alexander Neckam, *Suppletio defectuum* 2.951–2 (CCCM 221: 150), on whom see Ch. 4.1.

[125] Roger Bacon, *Opus maius*, pars IV (ed. Bridges 1897: 205). See also Bacon, *Opus tertium* 57 (ed. Brewer 1859: 223), where Gerland is referred to as 'omnium computistarum doctissimus'.

[126] MS London, BL, Add. 40744, fol. 14va (s. XII): 'Inter ecclesiasticos autem viros Dionisius, Beda, Theophilus Alexandrie episcopus, Helpericus abbas, postremo autem Gerlandus postissimi fuerunt. Gerlandus autem subtilibus rationum calculis innixus in multis longo usui ecclesie contradixit. Unde ipsum pro minimo habet ecclesia. Simplices enim fidei, non perversores querit ecclesia.' MS London, BL, Royal 2.D.VI, fol. 141r: 'Inter catholicos Dionisius, abbas Romanus, prius in hac scientia viguit, qui eam celeberrime edocens tabulas ad inveniendos terminos depinxit. Deinde Elpericus, abbas Anglicus, huic scientie multa adiecit. Postmodum Beda de hac arte tractatum composuit. Ad ultimum Gerlandus, Latoringie oriundus, de hac scientia multa disseruit, qui tres predictos viros catholicos in multis nimis audacter asseruit oberrasse. Cuius doctrinam omittens ecclesia, tum quia tribus catholicis viris ita audacter contradixit, tum quia eius doctrina nimis difficilis erat et

start of an anonymous thirteenth-century commentary on John of Sacrobosco's widely copied *Computus ecclesiasticus* or *De anni ratione*, where Gerland is described as a man from the border region between France and Germany who 'boldly and fiercely gnawed away at the writings' of his predecessors, 'claiming that they all speak the untruth, and yet he himself fell into error just like the others'.[127]

Gerland's *Computus* enjoyed an especially vivid reception in twelfth-century England, which spawned an exceptionally high amount of original work in the areas of *computus usualis* and *computus naturalis*. Hints at an ongoing controversy between adherents to both 'schools' appear in the preface to a comprehensive *Compotus* in five books written in 1176 by Roger of Hereford, which will be discussed in more detail in Chapter 3.4. Roger, who was happy to cover both traditions, used Gerland as his main source when he dedicated the fourth book of his work to the 'natural' *computus*.[128] The centrepiece of Book IV was a calendar with four columns for the date and hour of the new moon that were lifted directly from the *Computus Gerlandi* (see Figure 2.3). Roger openly acknowledged this debt and explained to readers the principles and assumptions that went into the calendar's construction. While he maintained Gerland's doctrine that the change of epacts and beginning of the lunar year had to be sought in September, Roger's own *kalendarium* took its beginning from 22 March, the earliest possible date for Easter Sunday. Another update of sorts was the inclusion of a column for the traditional Golden Number, here designated as the *cyclus vulgaris*, which made it easier to gauge the extent to which the traditional *computus* had begun to lag behind the 'natural' one.[129] Roger's version of Gerland's lunar tables enjoyed a remarkable revival during the thirteenth and fourteenth centuries, when it was copied dozens of times in the more conventional format of a *kalendarium* starting on 1 January (see Figure 2.4).[130] In this particular format, the calendar was sometimes attributed to Robert Grosseteste, bishop of Lincoln from 1235 to 1253 (hence the occasional

obscura, potius tres predictos viros in observatione huius artis, huius scientiae sequitur.' This is taken from the introduction to a thirteenth-century *computus* (fols. 141r–148r), starting: 'Antequam principale propositum aggrediamur...' Similar examples appear in MSS Glasgow, University Library, Hunter 467, fol. 21v; London, BL, Cotton Titus D.VII, fol. 14r; London, BL, Cotton Vitellius A.XII, fols. 98r, 102r; Paris, BnF, lat. 2020, fol. 198ra, and in Giles of Lessines, *Summa de temporibus* II.1, MS Bologna, Biblioteca Universitaria, 1845, fol. 20vb (= Steele 1926: 190). See also Wright 1847: 72–5; Lohr 2013: 17 (n. 42).

127 MS Paris, BnF, lat. 7422, fols. 17ra–29va, at fol. 17rb–va: 'Novissimus autem omnium istorum Garlandus in confinio Gallie et Germanie oriundus scripta eorum satis audacter et acriter remordens falsitatis redarguit et erroris, dicens quod omnes mentiti sunt et tamen ipse una cum aliis mentitus est.' Further copies of this commentary appear in MSS London, BL, Harley 1, fols. 64ra–69va; Munich, BSB, Clm 10273, fols. 153r–160v (incomplete). Grabmann 1936: 527, reproduces the relevant passage in the latter codex. See also the similar remark on Gerland in MS Oxford, BodLib, Digby 193, fol. 27rb.

128 Roger of Hereford, *Compotus* IV.20–2 (CCCM 272: 185–212).

129 Ibid. IV.21 (CCCM 272: 198–209). See the description of Roger's calendar in Borst 2001: 299–300. The addition of Golden Numbers also characterizes the version of Gerland's *kalendarium* found in MS London, BL, Cotton Vespasian A.IX, fols. 62v–68r. The accompanying text (fols. 68v–70v), dated to 1131, was edited as 'App. R' and discussed in Lohr 2013: 255–9, 444–5.

130 The calendar and its canons were edited in Lindhagen 1916: 15–41. Another version of the canons appears in Pedersen 1983–4: 365–6. Manuscripts are listed in Thomson 1940: 106–7, who expands on Baur 1912: 65*–66* (n. 5). See also Moreton 1995a: 580–1. Later offshoots include the *Kalendarium optimum* found in MS Erfurt, UFB, CA 4° 345, fols. 10r–15v.

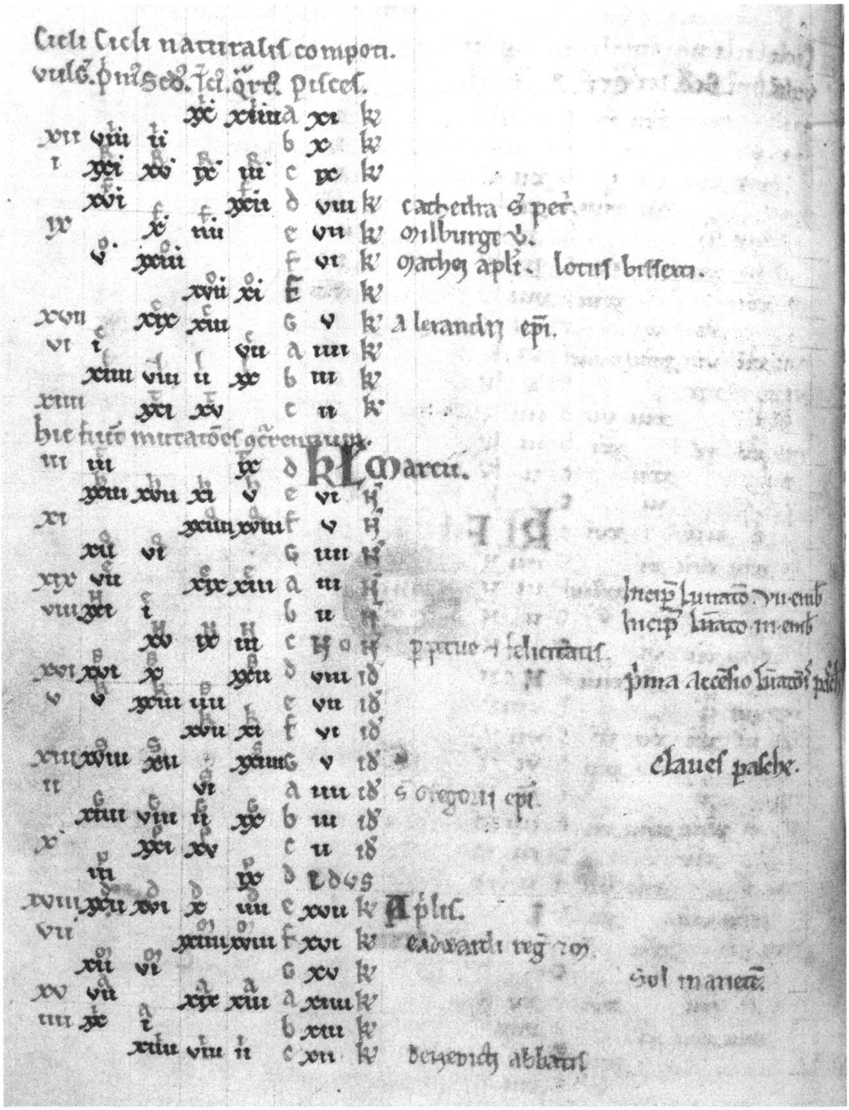

Figure 2.3. Roger of Hereford, *Compotus* (1176), calendar with new-moon times, page for February/March. Oxford, Bodleian Library, Digby 40, fol. 43v.

title *Kalendarium Lincolniensis* or *Lincolniense*), although its true origins were evidently still known to one English scribe who made a sumptuous copy of it soon after 1273.[131] No such awareness speaks from the various small extensions that were later added to the canon that accompanies the calendar in most copies. In one

[131] MS Cambridge, St John's College, F.25, fol. 40r: '. . . ad dilucidationem dicti kalendarii quedam philosophie dicta de compoto magistri Rogeri Herefordensis, qui dictum kalendarium ad instantiam

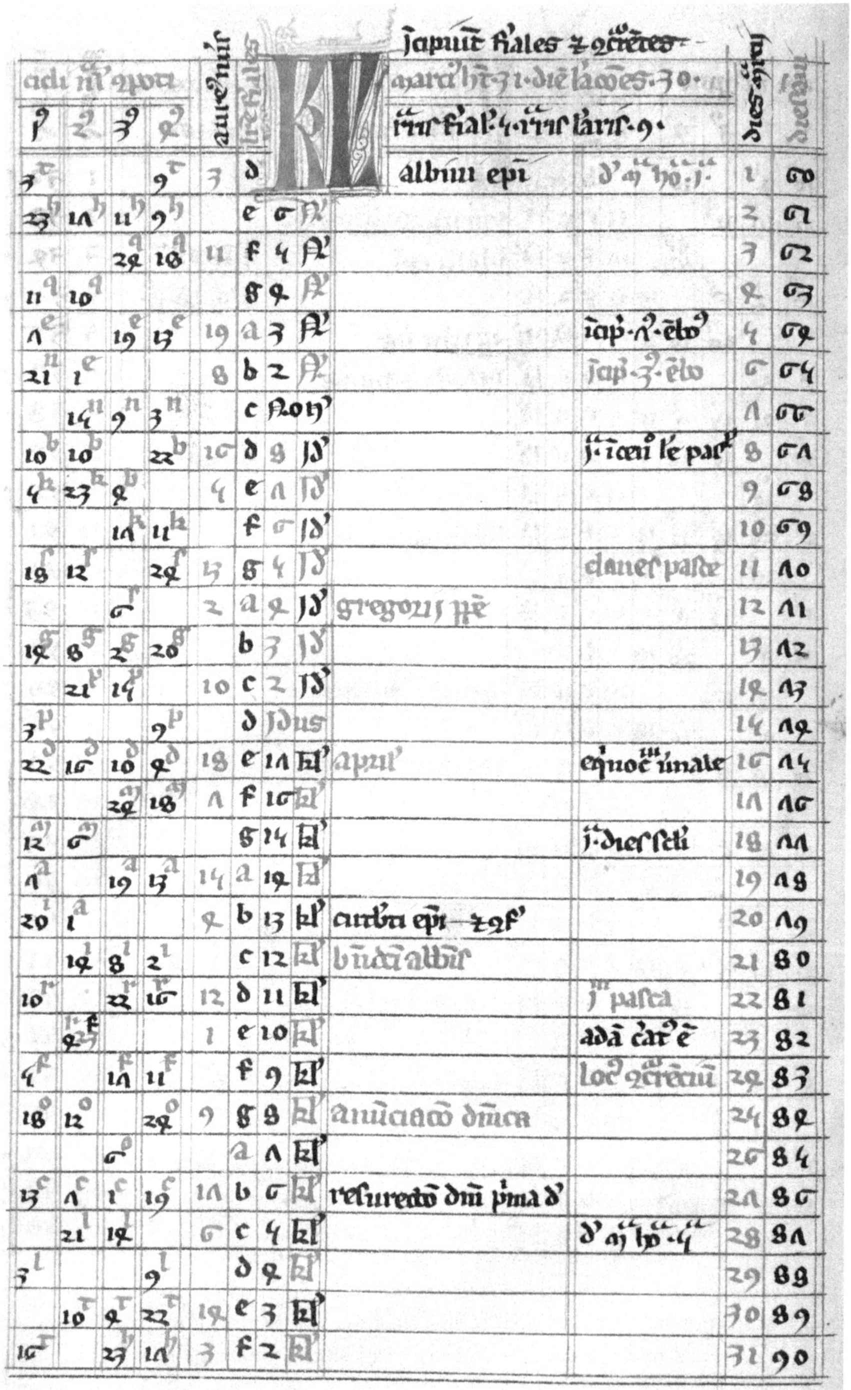

Figure 2.4. *Kalendarium* of pseudo-Robert Grosseteste, page for March. Paris, Bibliothèque nationale de France, latin 7421, fol. 5v.

Source: BnF

of these extensions, it is even suggested that the new-moon times shown in the calendar's four principal columns might be based on the meridian of Paris.[132]

The 'natural' lunar calendar first invented by Gerland and later popularized in the form of the *Kalendarium* of pseudo-Grosseteste presented a particularly advanced way of improving the predictions yielded by the Alexandrian 19-year cycle without giving up the cyclical framework itself. What it could not do was remove the fundamental problem inherent in this cycle, namely its failure adequately to keep track of the lunar phases for longer periods of time. Users interested in maintaining a reliable lunar calendar for astrological and medical purposes were understandably impatient with these shortcomings and eventually abandoned the *computus naturalis* in favour of the more accurate parameters Latin computists had received from Arabic astronomical sources in the course of the twelfth century. The end result of this development was that the formal layout of four columns covering 4×19 years was maintained, but the arithmetical structure of the 19-year cycle replaced with more exact (and exacting) astronomical calculations, which no longer presupposed a repetition of all new-moon dates after 76 years. The evolution of Latin lunar tables during the later Middle Ages thus constitutes a calendar reform 'from below', which took place in countless astronomical and astrological manuscripts produced during the late thirteenth, fourteenth, and fifteenth centuries, albeit without any practical consequences for the calculation of Easter. Before we can pick up this thread again in Chapter 5, it is necessary first to address the introduction of new astronomical sources into Latin Europe and the critiques of the existing calendar they engendered in the course of the twelfth and thirteenth centuries. The following Chapters 3 and 4 will reconstruct the unfolding of this discussion in the necessary detail.

domini Gileberti Londonensis episcopi composuit anno ab incarnatione domini M° C° LXX° VI°, excerpere curavi.'

132 MS London, BL, Harley 3735, fol. 6rb (s. XIII/XIV): 'Et notandum quod tam dies quam hore primacionis lune in 4 ciclis huius kalendarii sunt secundum mediam coniunctionem luminarium super meridiem Parisius, ut credo.' The same passage appears in MSS Basel, Universitätsbibliothek, O.II.7, fol. 43r; Escorial, Real Biblioteca de San Lorenzo, O-II-10, fol. 4r; Cambridge, University Library, Ii.3.3, fol. 4r.

3

Calendrical Astronomy in the Twelfth Century

3.1. WALCHER OF MALVERN AND THE TIPPING POINT OF LATIN ASTRONOMY

By the end of the eleventh century, the innovative efforts of Hermann of Reichenau and his successor Gerland to 'naturalize' the *computus* had pushed the potential of the Alexandrian 19-year cycle to its very limits and in doing so had erected a conceptual bridge between *computus* as a form of calendrical arithmetic and mathematical astronomy more properly speaking. The *computus naturalis* movement reached an intellectual peak around 1092, the year when Walcher of Malvern (d. 1135), a Lotharingian émigré living in England, used his astrolabe to determine the mid-point of a lunar eclipse he had spotted before the break of dawn on 18 October. His observation was good enough to make the conjunction tables that resulted from it the most accurate tool for lunar computation that had ever been produced by a medieval Latin scholar up to that point in time.[1] Even though Walcher seems to have worked independently of his contemporary Gerland, there are some striking parallels between their respective approaches to constructing a 'natural' *computus*, suggesting a shared intellectual milieu in their native Lorraine.[2] Similar to Gerland, who created a 76-year lunar cycle with a start in AD 1055/56, Walcher centred his work on lunar computation (*De lunationibus*) on a set of tables for the 76 years from 1036 to 1111. Instead of inserting his data into the format of a calendar, however, he preferred to present the 940 conjunction times as a consecutive list ordered by individual years, which in one of the preserved manuscripts takes up no fewer than 19 pages (4 years per page).[3] Rather than trying to fashion the 19-year cycle into a tool for the prediction of eclipses, as had been done without much success by Gerland and their shared predecessor Hermann of Reichenau, Walcher focused much of his efforts on developing an improved method of tracking the courses of Sun and Moon through the zodiacal signs. The coordinate system needed for this purpose was conveniently provided by the Julian year, which Walcher ended up dividing into $365\frac{1}{4}$ zodiacal 'parts'. Using the cumbersome 'Roman' or duodecimal fractions known to the abacists of his day, he inferred that

[1] See the edition, translation, and commentary on the first four chapters of Walcher of Malvern's *De lunationibus* in Nothaft 2017e: 90–169, 219–51.

[2] See Nothaft 2017e: 15–18, 27–31, 221–4, 266–8, and *passim*.

[3] MS Cambridge, St John's College, I.15, pp. 307–25.

the Moon was endowed with an approximate daily motion of $13 + \frac{1}{3} + \frac{1}{36} + \frac{1}{144}$ zodiacal parts as well as an hourly motion of $\frac{1}{2} + \frac{1}{24} + \frac{1}{72} + \frac{1}{864} + \frac{1}{3456}$ of one part.[4]

According to Walcher's own testimony, his arithmetical *tour de force* had been inspired at least in part by the idea that the 'the change in the Moon's waxing and waning' had effects on the human body and therefore was relevant to the practice of medicine.[5] Given this practical orientation of his project, predictive accuracy was obviously paramount, which may explain why Walcher continued to check the 'kindlings' of the Moon listed in his tables against the conjunctions and oppositions revealed by observed eclipses. Slight imprecisions had already become noticeable in 1093, when the widely recorded solar eclipse of 23 September occurred during the day even though Walcher's tables announced a conjunction for the fourth hour of the preceding night. When a total eclipse of the Moon was witnessed close to midnight on 11 January 1107, his tables suggested an opposition of Sun and Moon that fell a quarter of a day earlier, at the end of 10 January. An even greater error was encountered a year later when the Moon underwent a partial eclipse on the evening that heralded the beginning of 1 January 1108, whereas the tables had led Walcher to expect the corresponding opposition 16 hours before this event. What made this second lunar eclipse especially troubling was the way it failed to agree even with the previous one on 11 January 1107, which occurred during the fifth hour of the night. Based on a mean lunation of approximately 29 days and $12\frac{3}{4}$ hours, the subsequent lunar eclipse twelve months later should have occurred after 354d 9h and hence 10 hours before it was actually observed. As Walcher put it, the Moon had illegitimately 'usurped' 10 additional hours during the intervening year, since on a uniform reckoning these hours should have been shared between the previous and subsequent years.[6]

The empirical findings Walcher of Malvern discussed at the end of his work clashed with one of the theoretical pillars of the eleventh-century *computus naturalis*: the idea of kinematic uniformity, or *aequalitas*, which effectively negated the existence of irregularities in the apparent orbital motions of Sun and Moon and hence the need to differentiate between mean and true motions or longitudes. By 1108, Walcher had been pushed towards the conclusion that this uniformity was an illusion and the Moon's motion was instead characterized by inconstancy and variability (*varietas*). Yet this variability was a phenomenon his existing theoretical toolkit, as provided by the Latin computistical tradition, proved incapable of explaining or modelling in any useful way. The impasse Walcher thus found himself in lasted until 1120, the approximate year he made the acquaintance of Petrus Alfonsi, a Jewish-Aragonese astronomer and natural philosopher who had converted to Christianity in 1106. From the evidence provided by a treatise *De Dracone* from Walcher's pen, it can be inferred that Alfonsi acted as his teacher in astronomical matters, familiarizing him with concepts that had long been known

[4] See Walcher of Malvern, *De lunationibus* 5 (ed. and trans. Nothaft 2017e: 168–89), and the commentary in Nothaft 2017e: 251–8.

[5] Walcher of Malvern, *De lunationibus* 2 (ed. and trans. Nothaft 2017e: 94–5).

[6] Ibid. 6 (ed. and trans. Nothaft 2017e: 188–91).

to scholars in the Islamic parts of the Iberian peninsula but were hardly available in the Latin West.[7] Thanks to the introductory course he received from his Hebrew 'Master Petrus', Walcher became the first medieval Latin author properly to expound the notion that the Moon's orbital nodes—here referred to, in oriental fashion, as the 'Head and Tail of the Dragon'—performed their own motion along the zodiac in a direction contrary to that of the Sun and Moon. The possibility of predicting eclipses hence crucially depended not just on the ecliptical longitudes of the luminaries, but also on that of the nodes, which Walcher now expressed using the sexagesimal division of the zodiac commonly found in Greek and Arabic sources. For the Moon, this meant that its motion could be stated as 13;10,24,52°/d, while the lunar nodes travelled 0;3,10,50°/d in the opposite direction, completing a full revolution in slightly more than 18 years and 7 months. Other elements of lunar and solar astronomy, such as eccentricities, anomalies, and the corresponding equations, were briefly hinted at in *De Dracone*, but not explained in any greater detail, as Master Petrus had left his books 'beyond the sea', presumably in neighbouring France. Referring back to his own treatise on the 'natural kindling' of the Moon, Walcher pointed out that the Moon performed not just one but three types of motion ('mean', 'greater', and 'lesser'), ignorance of which was the reason why his lunar tables had repeatedly failed correctly to predict the time of conjunction or opposition.[8]

Although the astronomical knowledge presented in *De Dracone* remained rudimentary and functionally inadequate, Walcher's intellectual encounter with Petrus Alfonsi can be appreciated as a key moment in the history of European astronomy. It transformed Walcher, who was then the prior of Great Malvern Priory (Worcestershire), from a devotee of an indigenous type of Latin astronomy, represented by the *computus naturalis*, into the harbinger of a completely new phase in Western scientific thought, brought about by the rapid influx of texts translated from Arabic as well as from Greek and Hebrew. At the root of this intellectual sea-change were cities on the Iberian peninsula that had recently been reconquered by Castilian and Aragonese forces. By the middle of the twelfth century, libraries in formerly Islamic centres of learning such as Toledo had begun to attract scholars from abroad who were eager to expand their knowledge of subjects such as astronomy, astrology, mathematics, magic, and medicine and also to make the requisite Arabic texts available to a wider Christian audience.[9] Of the Toledan translators active in this period, the most prolific by far was Gerard of Cremona (d. 1187), whose Arabic-to-Latin rendering of Ptolemy's *Almagest* became the

[7] See the introduction, edition, translation, and commentary devoted to *De Dracone* in Nothaft 2017e: 46–55, 194–217, 271–97.

[8] Walcher of Malvern, *De Dracone* 2.2, 4 (ed. and trans. Nothaft 2017e: 199–8, 204–5).

[9] The literature on the twelfth-century translators and the transmission of their texts is vast. For orientation, see the work of Charles Burnett, including Burnett 1978; 1992; 1995; 1997; 2011, and the articles collected in Burnett 2009. See, in addition, Thorndike 1923–58: ii. 66–93; Haskins 1924; Millás Vallicrosa 1963; Hourani 1972; Lindberg 1978; Pedersen (Olaf) 1978: 308–14; Alverny 1982; Beaujouan 1982; Montgomery 2000: 138–85.

standard version of that text until efforts at retranslation were made in the fifteenth century.[10] Next to Ptolemy and other surviving remnants of Greek astronomy, Latin readers were confronted with a much larger mass of textual material that had originated in the Islamic world during the intervening centuries, with ninth-century authors such as al-Battānī, Thābit ibn Qurra, Māshā'allāh, al-Farghānī, and Abū Maʿshar soon emerging as the most widely cited authorities on the science of the stars. The rapid assimilation of their astronomical and astrological wisdom, together with the requisite computational techniques, cannot be understood apart from the existing continuities between 'old' computistics and 'new' astronomy, which at the very least show that the mathematical tracking of celestial motions was a concept already familiar to Latin Christian minds.[11]

With Walcher of Malvern, who sought instruction from Petrus Alfonsi after his own attempts to improve Latin lunar tables had alerted him to the inconstancy of the Moon's course, we have already encountered a key example of how the wide availability of computistical training in the Latin West prior to the twelfth century created a fertile ground for the integration of Graeco-Arabic mathematical astronomy after 1120. Of particular importance in this regard were the newly translated sets of astronomical tables, known in Arabic as *zīj*, which promised their users a much greater predictive success at lunar and solar computations than Easter tables could possibly offer. Following a long-established tradition that went back to Ptolemy's *Almagest* and *Handy Tables*, *zīj*es expressed celestial motions in the form of numerical columns, the entries of which could be combined to yield the longitudes of the Sun, Moon, and five other planets as a function of time. Next to the core tables for mean motions and equations, users could avail themselves of a wealth of auxiliary tables and lists dealing with topics such as calendrical conversions, trigonometric functions, planetary latitudes, visibilities or retrogradations, syzygies (i.e. conjunctions and oppositions), star positions, geographic coordinates, lunar parallax, or the magnitudes and durations of eclipses.[12] The richness of this newly imported technical knowledge was increased even further by the way the practice of mathematical astronomy in the early Abbasid period had been shaped not just by the *Almagest* but by alternative models and parameters from India, which had been introduced via Persia during the second half of the eighth century. Among the chief hallmarks of this 'Hindu' tradition was the use of the background of fixed stars as a reference frame for computing planetary longitudes, which contrasted with the tropical zodiac presupposed by Ptolemy and those working in his wake. Tables that used such a sidereal coordinate system included the *zīj al-Sindhind* by Muḥammad ibn Mūsā al-Khwārizmī (active in the first half of

[10] Kunitzsch 1992; Burnett 2010b. Other translations are discussed in Haskins 1924: 103–10, 157–64, 191–3; Mercier 1987: 114–15; Burnett 1999: 159–63; 2003: 23–36.

[11] A telling example of this conceptual continuity is a commentary on the pseudo-Aristotelian (or pseudo-Ptolemaic) *Iudicia* from *c.*1138, in which the field of astronomy is divided into the branches of *compotus* and *iudicia*. See MS London, BL, Cotton App. VI, fol. 20vb.

[12] For general orientation, see Mercier 1987; King and Samsó 2001; Chabás and Goldstein 2012a; Chabás 2014. On the term *zīj*, see Mercier 2000.

the ninth century), which later was to become the first proper set of astronomical tables to make its way from Muslim Spain into Western Europe. It appears plausible that this transfer was engineered by Walcher's teacher Petrus Alfonsi, to whom some manuscripts attribute an early Latin adaptation of al-Khwārizmī's tables that features an epoch date of 1 October 1116. A second Latin rendering, one much more faithful to the Arabic source, was due to Adelard of Bath, a prolific translator and propagator of Arabic wisdom, whose studies had led him to travel as far as Sicily, perhaps even to Antioch. Adelard's version of the Khwārizmīan *zīj*, entitled *Ezic Elkaurezmi*, follows immediately after Walcher of Malvern's *De Dracone* in a manuscript from Worcester cathedral that also includes Gerland's computistical work.[13]

Al-Khwārizmī's influence on the practice of Latin astronomy was on the whole rather short-lived and remained largely confined to England, as was the case for different set of tables based on the lost tenth-century *zīj* of al-Ṣūfī. This work, which reached Europe from the Crusader States (Antioch), used a tropical frame of reference and is notable for its very accurate estimate of the tropical year of nearly $365\frac{1}{4} - \frac{1}{131}$ days or 365d 5;49d.[14] In a scenario similar to Petrus Alfonsi and the Khwārizmīan *zīj*, the Latin dissemination of al-Ṣūfī's tables is connected to the travelling Jewish scholar Abraham Ibn Ezra (1089–*c.*1161), who supplemented his adaptation for the meridian of Pisa (*c.*1143) with an introductory astronomical treatise, datable to 1154, which only survives in Latin.[15] All of these various tables were eventually rendered obsolete by the wide circulation of the so-called Toledan Tables, which were translated into Latin on at least two separate occasions. Although essentially a hybrid work heavily dependent on the tables of al-Battānī and al-Khwārizmī, the Toledan Tables also contained some new parameters that had been added to the mix in late eleventh-century Toledo by a group of astronomers assembled and financed by the local *Qāḍī* Ṣāʿid al-Andalusī (d. 1070). The most prominent among these Andalusian astronomers was Azarquiel (al-Zarqālī or al-Zarqālluh, d. 1100), to whom the tables and their accompanying canons were frequently attributed in the Latin manuscript tradition, which comprises over 150 extant codices, most of them dating from the thirteenth and fourteenth centuries.[16] Their pervasive influence was further amplified by numerous adaptations to the Christian calendar and local European meridians, a tradition that started

[13] MS Oxford, BodLib, Auct. F.1.9, fols. 99v–159v. The best study of the various versions of al-Khwārizmī's tables is Mercier 1987: 88–102. See also Samsó 1992: 84–93; Vernet and Samsó 1996: 252–6; King and Samsó 2001: 33–5. On Petrus Alfonsi's version, see Haskins 1924: 117–18; Casulleras 1996, and the transcription and analysis of MS Oxford, Corpus Christi College, 283, fols. 114r–145r, in Neugebauer 1962b: 133–234. Adelard of Bath's astronomical works and their possible relation to Petrus Alfonsi are examined in Dickey 1982.

[14] Mercier 1987: 108–12; 2004; Pedersen 2002: 1208–9. The Antiochene connection of this and other translated works is studied by Burnett 1999; 2000; 2003; 2006.

[15] Abraham Ibn Ezra's *Liber de rationibus tabularum* was edited by Millás Vallicrosa 1947. See, moreover, Millás Vallicrosa 1943–50: 289–347; Samsó 2004b: 286–93; 2012.

[16] See the four-volume edition by Pedersen 2002, together with the literature cited in Nothaft 2017e: 64 (n. 224).

in 1141 with Raymond of Marseille's *Liber cursuum planetarum* and the tables for Marseille that formed part of this work.[17]

3.2. ARABS, HEBREWS, AND LATINS IN CALENDRICAL CONFRONTATION

To any computist interested in putting the new astronomy to calendrical use, one of the most attractive parts of a common-or-garden variety *zīj* would have been the tables for mean syzygy, which tracked the monthly times of conjunction and opposition, often together with other parameters useful for eclipse predictions such as the corresponding longitudes of the Sun and Moon, the mean lunar anomaly, and the mean argument of lunar latitude. Tables of this type can be traced back to Ptolemy's *Almagest* (6.3), which tabulated the root-dates for syzygies at intervals of 25 years, as calculated in the old Egyptian calendar (where 1 year = 365 days). In computing these values, Ptolemy used an extremely accurate estimate of the mean synodic month of $29;31,50,8,20\text{d} = 29\text{d}\,12;44,3,20\text{h} = 29.530594\text{d}$, which was only about 0.4 seconds longer than the presently accepted value of 29.530589d. If this value, which traces its origin back to Babylonian sources, is inserted into a 'Callippic' cycle of 76 years = 940 months, the resulting length of $940 \times 29;31,50,8,20 = 27,758;45,30,33,20$ days will be nearly $\frac{1}{4}$ of a day shorter than 76 Julian years of 27,759 days. A rounding to this quarter-day may account for the excessive length of the tropical year accepted by Ptolemy and his predecessor Hipparchus (see Ch. 1.2), since $27,758;45\text{d} \div 76 \approx 365;14,48\text{d}$ or $365\frac{1}{4} - \frac{1}{300}$ days,[18] although a more precise calculation on the basis of the same value would show the tropical year to contain $27,758;45,30,33,20\text{d} \div 76 \approx 365\frac{1}{4} - \frac{1}{314.68}$ days. The latter result also suggests a joint estimate for the errors that affected the solar and lunar components of the ecclesiastical calendar, which on the Ptolemaic value would be $J = M = 314\frac{2}{3}$ years.

The Arabic astronomers who inherited Ptolemy's syzygy tables usually retained their structure and parameters, but they naturally saw fit to replace his Egyptian years with the lunar calendar standardly used in the Islamic world (henceforth 'Arabic calendar').[19] The latter stands out from among most lunar calendars known to history for its refusal to adjust the lunar year to the solar year via intercalation. Since there are no embolisms, each of the twelve months of the Arabic calendar year will constantly drift through the seasons at an approximate rate of 11 days per year. In the standard form of this calendar, as used for civil and religious purposes, individual months began on the evening of the first appearance of the new moon crescent, which had to be determined empirically. The ad hoc

[17] See the edition of Raymond's *Liber cursuum planetarum* in Alverny, Burnett, and Poulle 2009: 126–341. An even earlier witness to the Toledan Tables in Latin may be the aforementioned commentary on the *Iudicia* (see n. 11 in this chapter), where the tables are mentioned in connection with the year 1138. See MS London, BL, Cotton App. VI, fol. 21va.

[18] Swerdlow 1979–80: 292–300; Goldstein 2003; Lehoux 2007: 98–9. See also n. 31 in this chapter.

[19] See e.g. Suter 1914: 183–6; Toomer 1968: 79–81; Pedersen 2002: 1327–40.

nature of this calendar made it unsuited to most astronomical and astrological purposes, which often required one to project into the past or future. Early Islamic astronomers solved the problem by furnishing their calendar with a fixed cycle based on a steady succession of 30-day and 29-day months counted from 1 Muḥarram = 15 July AD 622, the epoch of the Hijrah. In a second step, the average length of the lunar year was increased from 354d to 354;22d = 354d 8;48h by adding a 355th day to 11 out of 30 years. The result was a perpetual 30-year cycle of 10,631 days in which the average length of the synodic month was 10,631d ÷ 360 = 29;31,50d = 29.530555d = 29d 12;44h, only $3\frac{1}{3}$ seconds below the Ptolemaic value. The first documented attempt in the Latin West to study this Arabic calendar cycle and its possible implications for the *computus naturalis* was made by an English computist in the second quarter of the twelfth century, whose text has been preserved in a manuscript linked to Tewkesbury Abbey, Gloucestershire.[20] Taking as his starting point the information found in Adelard of Bath's translation of the *Ezic* of al-Khwārizmī, the author used the newly introduced sexagesimal minutes (*dakaice*) and seconds (*elthenie* or *minutie*) to establish that there was an excess of 0;4,10d separating 19 Julian years from 235 lunations as reckoned by the Arabs, which amounted to more than a day after 285 years. His insight led him to complain that the conventional lunar *computus*, which apparently had been losing track of the Moon since the days of Julius Caesar, now 'totters beyond good measure'. One clear indicator of this predictive failure was that the Moon was frequently spotted full and rising due east on the 13th day of the computed lunar month, while solar eclipses, like the one witnessed over England on 2 August 1133, could even appear on the 27th day, well ahead of the calendrical new moon.[21] His overall critique of the 'Roman' lunar reckoning was a significant departure from earlier authors such as Gerland, to whom he referred as 'the most learned man in this art among the Romans'.[22] Not even Walcher of Malvern, who had played an important role in introducing the new astronomy to neighbouring Worcestershire, had as much as hinted at the possibility that the mean lunation of his lunar tables might differ in any significant way from the value accepted by Arabic astronomers.

What the anonymous English computist did not say was whether the error he had uncovered should be remedied and how one might go about doing this. A pragmatic answer to the latter sort of question emerges from an anthology of computistical notes that once belonged to Reichenbach Abbey, a monastery in the diocese of Regensburg, Bavaria. In the midst of a series of conventional instructions dating from 1143 or shortly thereafter, the scribe inserted a conversion table for Arabic and Christian years valid for AD 1117 to 1408, which according to the accompanying text could be used reliably to determine the 'natural' age of the Moon.[23] Similar ideas are on display in a slim booklet of computistical material

[20] See the study, edition, and translation in Nothaft 2015f.
[21] *Collatio compoti Romani et Arabici* 5 (ed. Nothaft 2015f: 203).
[22] Ibid. 1 (ed. Nothaft 2015f: 201).
[23] MS Vienna, ÖNB, 275, fol. 29v. See Nothaft 2014g: 46–52.

that someone put together in the archdiocese of Salzburg during the 1150s. It begins with the first half of Hermann of Reichenau's *Abbreviatio compoti*, but instead of also reproducing the second half dedicated to the *computus naturalis*, the booklet closes with a series of *argumenta* on how to translate Christian dates and years into the Arabic lunar calendar, whose dates were again meant to represent the age of the Moon 'according to nature' (*secundum naturam*).[24] The scribe's attempt to reinvent Christian lunar computations along these lines culminated in a table of lunar epacts and regulars based not on the conventional 19-year cycle, but on a 30-year cycle modelled after the Arabic calendar and with a starting point in 1155. Creative as this idea may have been, it is not surprising that it failed to find many imitators, as 30 Arabic years constitute a lunar but not a lunisolar cycle. The 30-year cycle therefore offered no real advantage when it came to calculating the new moon within the framework of the Julian calendar.[25]

For all their limitations, the two Bavarian texts offer valuable testimonies to the early incursion of Arabic knowledge into the traditional *computus*, which is further underscored by the way their calendrical tables already made use of Hindu-Arabic numerals in their Western form. The slow but ineluctable spread of these numerals among Latin scribes in the course of the twelfth century was helped to a very great extent by adaptations of a work *On Indian Calculation* by al-Khwārizmī, whose name lurks behind the textual genre *Algorismus* as well as the modern term 'algorithm'.[26] One of the more significant early Latin texts of this sort was the *Liber ysagogarum Alchorismi*, which left traces in both of the southern German manuscripts just discussed.[27] Rather than being a text on pure mathematics, the *Liber ysagogarum* combined chapters on arithmetic and geometry with passages on calendars and astronomy, thus highlighting the close connection between the dissemination of texts on 'Indian' arithmetic and the incipient spread of Arabic astronomical tables in the second quarter of the twelfth century.[28] The adequate handling of *zīj*es typically required mastery both of the new numerals with their decimal place-value system and of the sexagesimal notation used by Arabic astronomers, who divided the day into minutes, seconds, *tertiae*, and so on. The author of the *Liber ysagogarum* traced the use of these sexagesimal units back to the ancient Egyptians, while stressing the increased convenience they offered in comparison with the *puncta*, *momenta*, and *instantia* (= atoms) known to computists.[29]

Although he was primarily focused on the wisdom of the Indians and Arabs, the brief calendrical section in the astronomical Book V of his *Liber ysagogarum* also reserved a few paragraphs for the calendar used by the contemporary 'Hebrews' or Jews, which had been transmitted to Europe from the rabbinic schools of

[24] MS Vienna, ÖNB, 2453, fols. 6r–7r, ed. Nothaft 2014g: 53–5. See also Warntjes 2016b: 313–14.

[25] Nothaft 2014g: 45–6.

[26] See, in general, Allard 1991; Folkerts 2001; Berggren 2002, and the articles assembled in Burnett 2010a.

[27] The *Liber ysagogarum* was edited in full by Dickey 1982: 251(bis)–328. The first three books on arithmetic are found in Allard 1992: 23–61. The fourth book is edited in Burnett 2013.

[28] In MS Paris, BnF, lat. 16208, fols. 67ra–71ra, the text is headed *Liber ysagogarum Alchorismi in artem astronomicam*.

[29] Allard 1992: 24–5. Sexagesimal fractions are the topic of Book II (pp. 38–50).

Babylonia.[30] The computational basis for this calendar, which reached its present form in the ninth or early tenth century, was the *molad* (מולד), which literally denotes the 'place of birth' of the Moon. It is equivalent to the time of the mean conjunction of the Sun and Moon, as reckoned from equinoctial sunset (6 p.m.) of a given day. In the system used by medieval and modern-day Jews, two successive *moladot* are always and everywhere separated by a uniform length of 29 days, 12 hours, and 793 'parts' or *ḥalaqim* (חלקים), where one *ḥeleq* is equivalent to $\frac{1}{1080}$ of an hour or $3\frac{1}{3}$ seconds. This makes the length of the lunation in the Jewish calendar identical to the 29d 12;44,3,20h = 29;31,50,8,20d = 29.530594d that had been passed down from the ancient Babylonians to Ptolemy and onwards from Ptolemy to Arabic astronomers.[31] Indeed, the difference between the Babylonian 29;31,50,8,20d and the 29;31,50d implicit in the Arabic 30-year cycle is equivalent to exactly one Jewish 'part' or *ḥeleq*, meaning that the Arabic lunation can also be expressed as 29d 12h 792p—a value that already crops up in the aforementioned *computus* from Reichenbach Abbey.[32]

Compared to the Arabic calendar, whose 30-year cycle comprised a complete number of days and hence could be used without resorting to any fractions, the Jewish calendar was a considerably more complicated tool. Its *molad*-system entailed that the new moon of Tishri had to be calculated annually down to the hour and *ḥeleq* until it could be decided when the New Year festival (*Rosh Hashanah*) was going to take place. Such fine-grained calculations, which were close in spirit to the *computus naturalis* practised by Hermann of Reichenau and Walcher of Malvern, entailed that the time of conjunction in the Jewish calendar was fully cyclical only after 98,496 years or 5,184 consecutive 19-year cycles, a number that increased to 689,472 years if the day of the week was taken into account. At the same time, however, the Jewish calendar possessed at least one crucial advantage from a Christian perspective, which was its use of the familiar 19-year intercalation cycle to keep the lunar months in line with the solar year. With its average solar year of *c.*$365\frac{1}{4} - \frac{1}{314.68}$ days the Jewish calendar hence allowed for a more intuitive comparison with the ecclesiastical calendar, whereas the beginning of the Arabic lunar year kept drifting through the seasons. This affinity was further strengthened by the fact that both the Christian and the Jewish version of the 19-year cycle employed the same sequence of common and embolismic years, with intercalations taking place in years 3, 6, 8, 11, 14, 17, and 19. The only major difference between them, as Abraham Ibn Ezra had already noted in 1154 in his Latin *Liber de rationibus tabularum*, concerned their respective starting point, which was three years later in the Jewish case. As a result of this divergence, the Church inserted embolismic months in two years (the 8th and 19th) that happened to be common

[30] *Liber ysagogarum Alchorismi*, ed. Dickey 1982: 314–15. The twelfth-century Latin reception of the Jewish calendar is discussed in more detail in Nothaft 2014d: 51–68.

[31] See on this point Swerdlow 1979–80: 307–9; Neugebauer 1989; Stern 2001: 207–10; Depuydt 2002.

[32] MS Vienna, ÖNB, 275, fol. 29v. This usage is probably derived from the *Liber ysagogarum Alchorismi*. See MS Munich, BSB, Clm 13021, fol. 30rb, and the edition in Dickey 1982: 313.

for the Jews (the 5th and 16th), each time resulting in a one-month gap between Easter and Passover.[33]

Among the first Christian writers to address the existence of this gap and its underlying cause was an otherwise unknown computist named Peter, whose *Compotus* of 1171 explained these different starting points with the Hebrew habit of counting all years from the creation of the world, which had not been followed by the Egyptian inventors of the Roman 19-year cycle.[34] Several chapters in the voluminous *Compotus Petri* were recycled only a few years later by a certain Master Cunestabulus, whose own *Compotus*, written in England in 1175, survives in a fire-damaged codex from Salisbury Cathedral, now at the British Library.[35] Even though Cunestabulus's stance on the subject of time-reckoning was on the whole a conservative one (see Ch. 3.4), it is evident that he was heavily invested in the advanced astronomy recently inherited from Greek and Arabic writers, among whom he expressly mentioned Hipparchus, Ptolemy, al-Farghānī, al-Battānī, Thābit ibn Qurra, and Azarquiel.[36] A particularly exotic dart in his quiver of sources was a certain *Samuel Iudeus*, 'who ascended to the master's chair as the fourth after Akiva, who had been president when Titus destroyed the Temple'.[37] This was a reference to Samuel of Nehardea, to whom the Babylonian Talmud (*Eruvin* 56a) attributed a method of calculating the dates of the four *tequfot*, that is, the dates of the equinoxes and solstices in the Jewish calendar.[38] As Cunestabulus was well aware, Samuel's *tequfot* were based on a solar year of $365\frac{1}{4}$ days, in which all four cardinal points were separated by the same interval of 91 days and $7\frac{1}{2}$ hours. Not only did 'Samuel the Jew' endorse the year length of the Julian calendar, but his vernal equinox (*tequfat Nisan*) fluctuated between 25 March at sunset (18:00h) and 26 March at noon (12:00h) over a four-year period, which made it possible to assimilate his doctrine to that of Pliny the Elder and other champions of the 'Latin' tradition.[39] Information of this kind is unlikely to have reached Cunestabulus via any route other than oral transmission, and it is indeed striking to find him appeal to the Jews themselves (*ipsos interrogate!*—'ask them!') as witnesses to the doctrine that the 17th year of each 19-year cycle must be embolismic.[40] Other signs of his acquaintance with the present-day Jewish calendar include his designation of the year of writing as the 4,935th year according to the era of the 'modern Jews',[41]

[33] Abraham Ibn Ezra, *Liber de rationibus tabularum* (ed. Millás Vallicrosa 1947: 99–100).

[34] See the *Compotus Petri* in MS Paris, Bibliothèque Mazarine, 3642, fols. 13r–49v, at fol. 38v (ch. 28).

[35] The main witness to the text is MS London, BL, Cotton Vitellius A.XII, fols. 87ra–97va. For further details, see Alfred Lohr's introduction in CCCM 272: xiii–xix, xxxiv–xl. Cunestabulus's knowledge of the Jewish calendar was previously discussed in Nothaft 2014d: 66–7. See Cunestabulus, *Compotus* 17 (CCCM 272: 79–80), for his discussion of the 19-year cycle.

[36] Cunestabulus's use of sources is examined in Nothaft 2017a. See already Haskins 1924: 87, and Steele 1926: xx.

[37] Cunestabulus, *Compotus* 36 (CCCM 272: 106, ll. 36–7).

[38] On the background, see Ginzel 1906–14: ii. 96–9; Nothaft 2014d: 32–3; Wartenberg 2014: 100–5.

[39] Cunestabulus, *Compotus* 25, 36, 39 (CCCM 272: 89, 106, 110, 118).

[40] Ibid. 24 (CCCM 272: 88, l. 81).

[41] Ibid. 32 (CCCM 272: 102, ll. 32–3). This chapter is based on the *Compotus Petri*, MS Paris, Bibliothèque Mazarine, fol. 22vb (ch. 28), where ad 1171 is equated to 4931 in the Jewish world era.

which is AD 1174/5, or his acknowledgement that the beginnings of the Jewish months could often fall two days after the calculated conjunctions due to the influence of postponement rules.[42] In order to find the first month of the Jewish year (Tishri) one had to look for the lunation that began on a date between 27 August and 27 September according to the ecclesiastical calendar, but it was important to realize that this beginning was by no means identical with the one computed by the Jews:

> For it frequently happens that we proclaim the first day of the Moon when it is the second, or third, or even the fourth according to the Jews. Which is also the reason why some lunation that for us begins on 27 August begins for them either on the same date or earlier.[43]

An extreme case of the type indicated could be witnessed two years after the completion of Cunestabulus's work, in 1177, when the Alexandrian 19-year cycle was in its final year, demanding that the Easter lunation be reckoned from Monday, 4 April. In the Jewish calendar, this was no longer the first but the second month of spring (Iyyar), whose corresponding *molad* or conjunction had fallen not two, not three, but four days earlier, on Thursday, 31 March, 16h 925p after sunset or 10:51h in the morning. For the small elite of Christian computists who, like Peter and Cunestabulus, were actually familiar with the Jewish calendar, such discrepancies could be troubling to observe. Not only did the Jews base their new-moon times on the best astronomical parameter available, enabling them to attain a computational accuracy the ecclesiastical calendar could only dream of, but the diverging starting points of the two 19-year cycles meant that the Church regularly failed to place its Easter in the month of Passover, as celebrated by the Jews, even though this was the feast during which Jesus's death and resurrection had taken place.

One contemporary Christian writer who found this state of affairs utterly unacceptable was Reinher of Paderborn, whose *Compotus emendatus*, written in 1170/1, contains the most outspoken critique of the ecclesiastical calendar and call for its reform to have been preserved from the twelfth century.[44] Little is known about the author himself, whose name is documented between 1154 and 1179 among the members of the cathedral chapter of Paderborn in Westphalia. One of the fifteen extant charters signed by Reinher (in 1177) expressly mentions him as *magister scolarum*, which indicates that he spent some time, probably since the 1160s, as head of Paderborn's cathedral school. In the opinion of the contemporary author of the *Vita* of Bishop Meinwerck of Paderborn (1009–36), this school was home to a flourishing crowd of mathematicians and astronomers,[45] which would appear to resonate well with the presence of *magister* Reinher. His mathematical abilities were certainly considered exceptional by whoever composed the heading in the two earliest extant copies of Reinher's *Compotus*, which characterizes the author

[42] Cunestabulus, *Compotus* 39 (CCCM 272: 117, ll. 79–83; 119, ll. 164–70).
[43] Ibid. 17 (CCCM 272: 79, ll. 33–7).
[44] Reinher of Paderborn's *Compotus emendatus* was recently edited by Alfred Lohr in CCCM 272: 1–55, superseding the earlier editions by Wijk 1951 and by Herold 2011. The latter discusses and edits all the documents pertaining to Reinher's life (see pp. 47–55, 164–99).
[45] *Vita Meinwerci episcopi Patherbrunnensis* 157 [CLX] (ed. Berndt 2009: 188).

as a 'most astute' or 'most perspicacious' reckoner (*perspicacissimus calculator*).[46] Such praise would seem all the more well deserved if Reinher was indeed the inventor of the calendrical conversion tables included in the same work, which employed Hindu-Arabic instead of Roman numerals 'for the sake of economy in writing and calculating'.[47] Reinher made this statement in the brief preface to his treatise, in which he also gave his readers a very precise idea of why he had taken on the task of producing an 'Improved *Computus*' (*Compotus emendatus*):

> Those who attack the Catholic faith rejoice that they manage to find an error in the *computus* employed by the Church. For they believe and affirm that we also err in other things. Eager to put an end to disgrace and mockery, not only on the part of our adversaries, but also of many believers, some of my friends have asked me to ponder how this scandal and this state of confusion might be removed.[48]

Reinher's awareness that the Church was not easily persuaded to abandon long-standing customs and habits made him hesitant at first. 'Yet seeing the error increase more and more, I finally put my mind to it, so I could at least be of use to posterity', once future necessity made a correction inevitable.[49] As he was quick to stress, however, it was not his intention to invent a completely new calendar. What he promised his readers instead was to introduce them to the ancient Hebrew *computus* that had once been accepted by the Fathers of the early Church before it was unwittingly abandoned by their successors. The main culprit in this story was Dionysius Exiguus, who had been a target of computistical criticism ever since Abbo of Fleury first suggested that he had misplaced Jesus's birth by 21 years. According to Reinher of Paderborn, the introduction of Dionysius's 532-year cycle in the sixth century had been part of an effort to end a period of confusion during which different Churches had used different Easter computations after breaking away from the Church of Alexandria.[50] Two problems in particular plagued the Dionysiac cycle, making it impossible for the Latin Church to celebrate Easter in full accordance with the Law of Moses. For one thing, Dionysius Exiguus and his followers had defined the Easter full moon as the first full moon that fell on or after the vernal equinox, even though no such rule existed in the ancient Hebrew Law. To make matters worse, they believed that the equinox would always fall on the same date in the Julian calendar, 21 March, which was evidently false. Their second mistake was to assume that 235 lunations were exactly equal to 19 Julian years and that the new moons could therefore be expected perpetually to return to the same dates in the Roman calendar—an assumption refuted by the observed discrepancy

[46] 'Incipit praefatio magistri Reinheri decani Patherburnensis perspicacissimi calculatoris in compotum emendatum.' This initial line is found in MSS Leiden, Universiteitsbibliotheek, B.P.L. 191E (s. XII$^{4/4}$), fols. 129r–140r; Hannover, Niedersächsische Landesbibliothek, IV 373 (s. XIII^{in}), fols. 3v–10v. For references to further manuscripts and literature, see Alfred Lohr's introduction in CCCM 272: vii–xiii, xxvi–xxxiv.

[47] Reinher of Paderborn, *Compotus emendatus*, praef. (CCCM 272: 3, ll. 20–2).

[48] Ibid. (CCCM 272: 3, ll. 2–8). [49] Ibid. (CCCM 272: 3, ll. 12–14).

[50] Ibid. 1.10 (CCCM 272: 14–15). This chapter seems to be based in part on Bede, *De temporum ratione* 44 (CCSL 123B: 418–19).

in the lunar count, which amounted to two days or more.[51] Reinher's verdict was harsh and unwavering: in abandoning the sophisticated computation once used by the ancient Hebrews and adopting a simpler one, Dionysius and his followers had ended up subverting both their own intentions and the institutions of the Holy Fathers. They should have perceived that:

> Moses, who had been educated in every wisdom of the Chaldaeans and Egyptians, and who was indeed full of the spirit of the Lord, would have made sure to correct the ancient *computus* had he not approved of it, because he understood that the precepts of the Law concerning the observance of feasts cannot be kept unless one also observes reliable rules for their computation.[52]

In order to remedy the defects of the Dionysiac reckoning, Western computists were called upon to relearn the *computus* of the ancient Hebrews, whose roots appeared to reach back to the divine revelation Moses had received at Mount Sinai.[53] As the bulk of the *Compotus emendatus* shows, Reinher identified this ancient Hebrew computation with the present-day Jewish calendar, more specifically with its system of computing the times of the conjunctions or *moladot*, which he was the first Latin author to describe in detail. His description culminated in a table, valid for the years 1171 to 1270, which plotted the lunations of the Hebrew *computus* against the canvas of the Julian calendar, showing the 'Hebrew' age of the Moon at the beginning of each month (see Figure 3.1).[54] Users interested in delving deeper and creating such tables for themselves were offered a set of tables for the conversion between Julian and Jewish dates, which calculated the difference that had accrued between the two calendar systems since the beginning of the Christian era, more specifically since noon of 31 December 1 BC. At the precise moment of this epoch or 'root'-date (*radix*), the Jews already counted 3,760 years, 3 lunations, 15 days, 21 hours, and 952 parts. For each period of 76 years since this date, the Hebrew computation gained an additional 5 hours and 860 parts relative to the Julian calendar.[55] From the worked example Reinher provides, it is evident that he put down these rules not too long after Sunday, 13 September 1170, on whose noon the Hebrew count had reached 4,930 years, 6 hours, and 668 parts. By a lucky coincidence, this was not only the date of the *molad* Tishri in the Jewish calendar, but also the beginning of the year 566 in the Arabic calendar. During the following night, Reinher observed a conjunction of Mars and Jupiter, which lasted until the tenth hour.[56]

In the second part of his treatise, which was aimed at those who planned on becoming 'studious computists',[57] Reinher went on to compare the Hebrew lunar month of 29 days and $12\frac{793}{1080}$ hours with the mean lunation implicit in the

[51] Reinher of Paderborn, *Compotus emendatus* 1.11 (CCCM 272: 16).
[52] Ibid. 1.12 (CCCM 272: 17, ll. 20–9).
[53] Ibid. 1.1, 13 (CCCM 272, 7, ll. 4–14, and pp. 17–18).
[54] Ibid. 1.21 (CCCM 272: 29–33).
[55] Ibid. 1.14–21 (CCCM 272: 18–29).
[56] Ibid. 1.19 (CCCM 272: 26–7). See also ibid. 2.5–6 (CCCM 272: 39–42), where Reinher uses the same date to show that the epoch of the Jewish world era falls exactly 1,800,657 days and 18h 876p earlier, on Monday, 7 October, 5h 204p.
[57] Ibid. 1.23 (CCCM 272: 35, ll. 16–19).

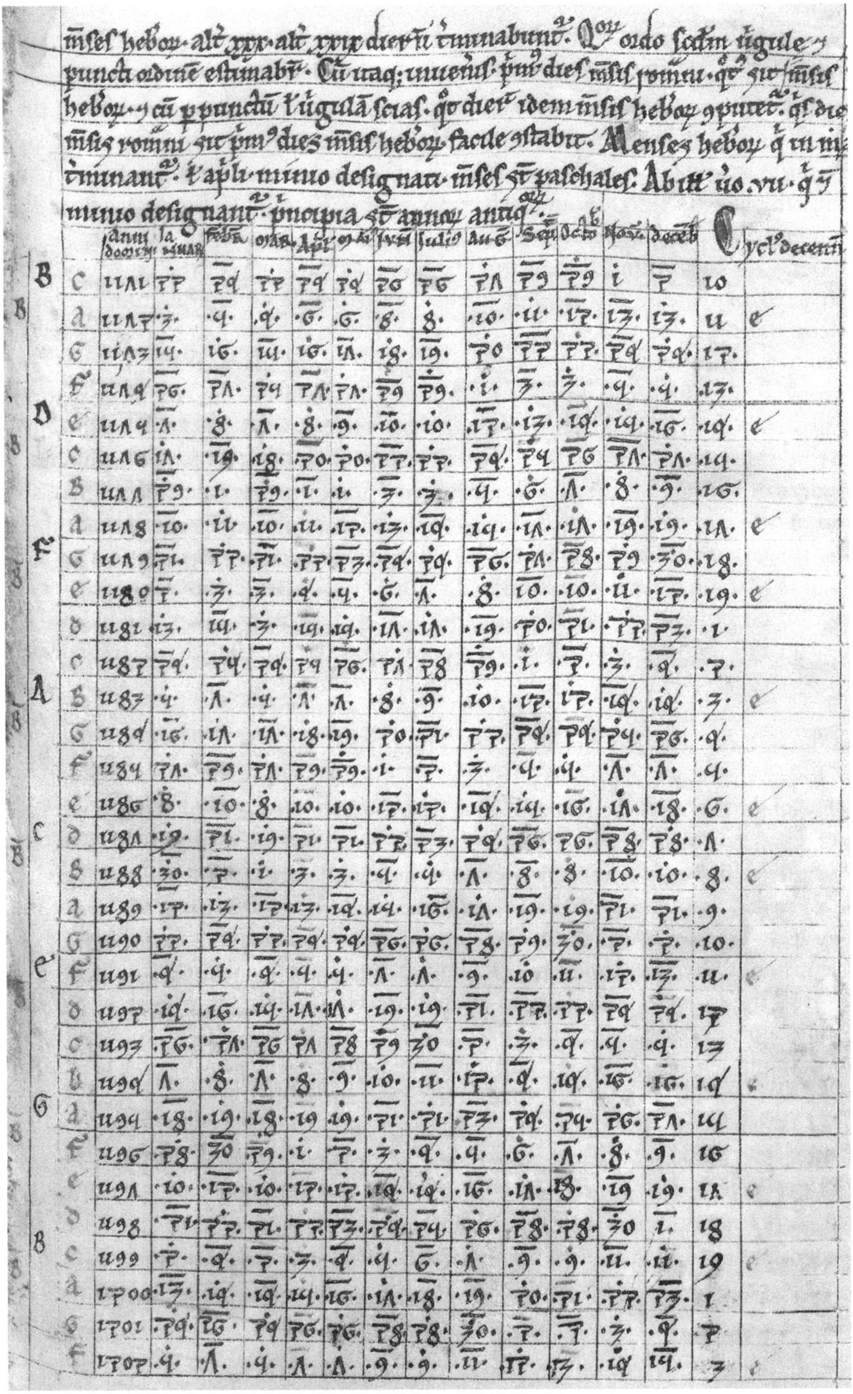

Figure 3.1. Reinher of Paderborn, *Compotus emendatus* (1170/71), table for the 'Hebrew' age of the Moon in the Julian calendar, AD 1171–1202. Leiden, Universiteitsbibliotheek, B.P.L. 191E, fol. 135r.

Source: Universiteitsbibliotheek Leiden, Special Collections

Dionysiac cycle, which could be expressed as 29 days and $12\frac{174}{235}$ hours. Using his skills as a 'most astute reckoner', the Westphalian cathedral canon established that the difference between these values amounted to $\frac{313}{50760}$ of an hour, reaching a full day after a little less than 315 years.[58] This result seemed to be in good agreement with the fact that slightly more than $2 \times 315 = 630$ years had gone by since the beginning of the Dionysiac calculation, whose tabulated Moon currently lagged about two days behind that of the Hebrew calendar. Indeed, the error could be expected to increase further and further, until the new moon according to the Hebrews coincided with a full moon according to the *computus* of Dionysius.[59] Reinher's belief in the technical superiority of the Jewish calendar, coupled with his erroneous views regarding its ancient Mosaic origin, made him confident that the Christian Church could do no better than to part ways with the accustomed method of Easter computation and return to its Hebrew roots. In his effort to underscore this point, the schoolmaster from Paderborn spent the final eight chapters of his treatise on a detailed discussion of the chronological problems surrounding the crucifixion of Jesus, which had already agitated eleventh-century authors such as Marianus Scottus and Gerland.[60] The latter's attempts to reconcile a resurrection on Sunday 25 March in the 34th year of Jesus's life with the lunar data provided by the 19-year cycle had led him to argue that the Church missed the mark by seven years when it counted the 'Years of the Lord' from AD 1. Reinher openly rejected this approach, which anachronistically applied Dionysius's 532-year cycle to dates in the Jewish calendar of Jesus's time. As he demonstrated at the very end of his *Compotus emendatus*, a calculation on the basis of the actual Jewish calendar allowed for Christ's crucifixion to have taken place in AD 34 provided it was dated to Friday, 26 March, which in Reinher's flawed conversion was equivalent to 16 Nisan. Although this was still one day later than the Jewish date indicated in the synoptic Gospels (15 Nisan), a fact that necessitated the reinterpretation of key scriptural passages, Reinher found his result heartening enough to exclaim: 'Behold how according to the *computus* of the Hebrews everything fits together!'[61]

That things were not nearly as simple was shown four years later by Master Cunestabulus, who devoted the final chapter of his treatise to a lengthy refutation of the arguments Marianus Scottus and Gerland had levelled against the validity of the Dionysiac incarnation era. Like Reinher, he agreed that it was unwarranted to assume that the Jews of Christ's time operated with the Alexandrian 19-year cycle, which, for all anybody knew, had only been introduced by Bishop Eusebius of Caesarea in the fourth century.[62] But did this automatically mean that the full moon that accompanied the crucifixion was best computed with the aid of the *molad*-system now in use among European Jews? Cunestabulus expressed strong doubts, which were fuelled by the fact that the contemporary Jewish calendar did

[58] Ibid. 2.1–4 (CCCM 272: 36–8).
[59] Ibid. 2.4 (CCCM 272: 39).
[60] Ibid. 2.8–15 (CCCM 272: 43–55). For the background to this problem, see Verbist 2010.
[61] Reinher of Paderborn, *Compotus emendatus* 2.15 (CCCM 272: 54, ll. 37–40). See Nothaft 2012: 135–46, for further details.
[62] Cunestabulus, *Compotus* 39 (CCCM 272: 116–17, ll. 72–82). On the notion that Eusebius invented the 19-year cycle, see Bede, *De temporum ratione* 44 (CCSL 123B: 418).

not allow for 15 Nisan to fall on a Friday, as should have been the case for the day of Christ's Passion if the Synoptic Gospels were right.[63] Armed with these critical insights, Cunestabulus proceeded to calculate that the Jewish calendar would have placed the paschal *molad* of AD 34 on 9 March, which was a full three days too early if Jesus's death on Friday, 26 March, was supposed to coincide with 15 Nisan. An attractive alternative was opened up by 'the equations of Ptolemy and Hipparchus', which suggested that the crescent new moon only became visible on the evening that marked the start of 11 March. This was still one day short of the targeted result, yet 'who can exclude that the Jews of that time had a *computus* according to which the Moon was called "first" on 12 March . . . and thus only one day later than it was seen?'[64] One reason this scenario seemed acceptable was that Christians in the far West, such as England, were more likely to see the new moon in the evening than those who inhabited the eastern part of the world, owing to the time interval between sunsets at different longitudes. That the Jews in Jerusalem would have failed to spot the new moon on 11 March AD 34 and waited until 12 March to start their paschal month was therefore far from implausible.[65] It is not perfectly clear whether Cunestabulus really meant to say that the Jewish calendar of antiquity was entirely first-visibility based, as argued by the Dominican scholar Giles of Lessines one century later,[66] but Reinher of Paderborn's approach of projecting the present-day *molad* computation of the Jews back into the biblical past would have certainly struck him as unjustified. Neither would he have fallen in line with Reinher's belief that this computation was destined to replace the ailing Dionysiac *computus*, especially given his awareness that the current range of dates for the beginning of the Passover lunation failed to match the paschal limits the Fathers of the Church had defined in antiquity.[67] In 1175, the question of whether the Church should follow the Arabs, the Jews, or none of the above was obviously yet to be decided.

3.3. THE VAGARIES OF THE SOLAR YEAR

Reinher of Paderborn's groundbreaking *Compotus emendatus* presented the reform of the ecclesiastical calendar primarily as a matter of changing the mode of lunar computation back to the forgotten *computus* of the ancient Hebrews, which he claimed had been unduly abandoned in late antiquity. Given these priorities, Reinher refrained from expressing any great misgivings about the solar calendar the Church had inherited from the ancient Romans, which could be maintained

[63] Cunestabulus, *Compotus* 39 (CCCM 272: 117, ll. 79–83).

[64] Ibid. 39 (CCCM 272: 120, ll. 195–200).

[65] See Cunstabulus's remarks, ibid. 38 (CCCM 272: 112–13, ll. 18–27), and ibid. 39 (CCCM 272: 120–1, ll. 200–13).

[66] Giles of Lessines, *Summa de temporibus* II.12, MS Arras, Bibliothèque Municipale, 674 (722), fol. 69v. A different recension of this chapter (here: II.3.8) is found in MS Bologna, Biblioteca Universitaria, 1845, fol. 33ra.

[67] Cunestabulus, *Compotus* 25 (CCCM 272: 89–90, ll. 12–46).

with good conscience once the tables of Dionysius Exiguus were discarded in favour of those included in his own treatise.[68] At the same time, he readily admitted that Dionysius Exiguus and his followers had erred in assuming that the vernal equinox would always remain fixed on the same place in the calendar.[69] In the here and now, this astronomical event was witnessed several days prior to its official date on 21 March, which had the troubling and frequent consequence that the Church located Easter not in the first month of spring but in the second, as was the case whenever the full moon fell between the actual vernal equinox and 21 March. In such years Dionysius and his followers contravened their own precept of always celebrating Easter after the first full moon to fall on or after the equinox.[70] For Reinher, this was all the more reason to abandon the present regime and adopt the *molad*-system of the Jewish calendar, which made it possible to identify the first spring month (Nisan) without ambiguity and without recourse to the Alexandrian rule of the equinox. Since the latter was nowhere mentioned in the Mosaic code for the celebration of Passover, as laid down in the Pentateuch, Reinher felt he could safely ignore it.

For scholars who refused to follow Reinher in his fervent commitment to the Hebrew *computus*—and they were probably in the majority—the evident calendrical recession of the equinoxes and solstices remained a serious problem, which had to be dealt with in one way or another. Thanks to the experiments carried out by the likes of William of Hirsau in the eleventh century, denialism was no longer viable. In 1171, the author of the aforementioned *Compotus Petri* went on the offensive by using Abraham Ibn Ezra's Pisan Tables for a calculation of the approximate date and time of all four cardinal points of the solar year. For the vernal equinox, the result for the current year was 13 March during the fourth hour of the day. This was the third day before the Ides of March, but contrary to what established conventions suggested, the other cardinal points refused to fall on equivalent Roman dates. The summer solstice was found at the third hour of the night that followed the 18th day before the Kalends of July (14 June), while the autumn equinox could be expected for the 16th day before the Kalends of October (16 September), about four hours before daybreak. For the winter solstice, the astronomical calculation yielded the 19th day before the Kalends of January (14 December), 'a little before sunrise'.[71] Elsewhere in the same treatise, Peter alleged that the winter solstice, which had once been observed on 25 December, had now already arrived on the feast day of St Lucy and hence on 13 December, 'as the peasants say and perceive. If this is true, it is necessary that one day collected

[68] See Reinher of Paderborn, *Compotus emendatus* 1.2–9 (CCCM 272: 8–13), which chapters are dedicated to the Julian calendar and its solar cycle.

[69] Ibid. 1.11 (CCCM 272: 16, ll. 10–12).

[70] Ibid. 1.12 (CCCM 272: 16–17, ll. 6–13).

[71] *Compotus Petri*, MS Paris, Bibliothèque Mazarine, 3642, fol. 19ra–b (ch. 17): 'Anno igitur ab incarnatione domini M° centesimo septuagesimo primo invenimus equinoctium vernale III° idus Martii hora diei quarta. Solstitium estivum in tertia hora noctis que sequitur diem XVIII° kl. Iulii; equinoctium autumpnale XVI kl. Octobris fere quatuor horis ante diem; solstitium yemale XIX° kl. Ian. paulo ante solis ortum; sumptis tamen horis secundum meridiem Pise.' The stated times stay within an hour's range from the times predicted by the Pisan Tables (see n. 15 in this chapter), provided the date-epoch is placed at sunset.

over many years be subtracted or added, lest the winter solstice after the course of time end up where there is now the summer solstice or one of the equinoxes.'[72] Astronomical computation and the experiential knowledge of the common man thus conspired to tell the same story, which was not that the ancients had been wrong in assigning the solstice to later dates, but that its location had drifted through the calendar in the course of the centuries. In a memorable passage, one later to be reused by Master Cunestabulus, Peter put the point as follows:

> Nowadays, however, any layman who observes the ascent and descent of a sunray at his window will not doubt that the summer solstice occurs around the 17th before the Kalends of July [15 June] rather than the eighth day [24 June]. But nobody, I think, will be so insane as to opine that the sleep of ignorance crept upon the vigilance of the ancients to have them commit such an obvious error in a matter where even the layman's simplicity can find out something close to the truth. Perish the thought! Instead, it is doubtlessly the case that the computists did not investigate the period the Sun takes to travel through the entire zodiac with full exactitude or perhaps, what is more credible, that they freely elected for themselves this convenient tradition concerning the length of the year, because they understood that, in order to avoid difficulty in computation, it was useful to speak in an approximate way.[73]

That the Julian calendar was less than perfect in reproducing the length of the solar year and that this had been the cause of the observed changes in the equinoctial and solstitial dates was still a relatively novel idea to Christian computists in the twelfth century, but one that had started to receive support from various directions. There were, for instance, Reinher's conversion tables for the Hebrew *computus*, which showed that 76 years in the Jewish calendar were 5h 860p shorter than 76 Julian years or 27,759 days. If the Hebrew 19-year cycle was assumed to be a perfect lunisolar cycle, the conclusion could be drawn that the true solar year stayed roughly $\frac{1}{320}$ of a day below the Julian $365\frac{1}{4}$ days, which is the estimate Abraham Ibn Ezra attributed to the Jews in his *Liber de rationibus tabularum* of 1154. As Ibn Ezra knew, however, things were not so easily settled, given that many Arabic astronomers believed the difference to be closer to $\frac{1}{136}$, $\frac{1}{130}$, or even $\frac{1}{106}$ of a day.[74] Progressively minded computists such as Peter added yet another estimate to the mix when they explained that the regression of the equinoxes since antiquity

[72] Ibid., fol. 14ra (ch. 6): 'Nec tamen ibi modo est, sed in festo sancte Lucie, ut dicunt et sentiunt rustici. Quod si verum est, oportet ut unus dies in multis annis collectus auferatur vel addatur, ne solsticium iemale contingat post curriculum temporis ubi modo est solstitium estivale vel equinoctium.' See also ibid., fol. 19ra (ch. 17): 'Magistri tamen nostri asserunt invenisse solstitium yemale circa festum sancte Lucie.'

[73] Ibid., fol. 18va (ch. 17): 'Nunc autem quilibet laicus ad fenestram suam considerans ascensum et descensum radii solis non dubitat solstitium estivum magis contingere circa XVII° kl. Iulii quam circa VIII° kl. Neminem autem arbitror esse tam insani capitis ut opinetur vigilantia<m> antiquorum ignorancie sompnium obrepsisse ut tam manifeste mentirentur in ea re quam prope verum indagare potest laicalis simplicitas. Absit. Sed procul dubio non ad unguem disquisitum est a compotistis spacium temporis quo sol peragrat totum zodiacum vel fortasse, quod magis credibile est, gratis hanc compendiosam traditionem de anni spatio sibi elegerunt, nam ut devitarent deficultatem computationis intellexerunt magis expedire ut prope verum dicerent.' Most of this passage is repeated verbatim by Cunestabulus, *Compotus* 36 (CCCM 272: 108, ll. 86–96).

[74] Abraham Ibn Ezra, *Liber de rationibus tabularum* (ed. Millás Vallicrosa 1947: 75–6, 100).

was caused by a surplus of $\frac{1}{5}$ of an hour in the average Julian year, which presupposes a solar year of 365d 5;48h or $365\frac{1}{4}-\frac{1}{120}$ days.[75] This was a relatively good value, though not quite as accurate as Peter's parallel claim that Ptolemy had found $365\frac{1}{4}$ days to be the period in which the Sun manages not just one full revolution, but an additional 0;0,27° of the zodiac. The actual source for this parameter was not the *Almagest*, but Ibn Ezra's Pisan Tables based on the *zīj* of al-Ṣūfī, where the mean Sun was allowed 20 full revolutions + 0;9° in 20 Julian years, revealing a tropical year of very nearly $365\frac{1}{4}-\frac{1}{131}$ days or 365d 5;49h.[76] Peter's slightly shorter and computationally handier 365d 5;48h = $365\frac{1}{4}-\frac{1}{120}$ days ($\frac{1}{120}$ of a day being $\frac{1}{5}$ of an hour), which were to play a significant role in subsequent calendrical discussions, may have originated as a mere simplification of the latter value.[77]

The confusion this variety of recorded estimates could arouse was hardly mitigated by the way Greek and Arabic astronomers distinguished between two different definitions of the solar year, tropical and sidereal. A very clear exposition of this fact and its consequences was offered in 1145 by an anonymous Latin work closely connected to the Pisan Tables, which began by stating that Ptolemy and many other wise men had measured the solar year 'according to the Sun's return from one juncture between the circles to the same juncture' (*secundum reditum solis a iunctura circulorum... ad eandem iuncturam*), the junctures being the two equinoctial points. For the ancient Indians, by contrast, the solar year was the period of the Sun's return to the same fixed star, which implied not only a different length of the year, but a different reference frame for plotting planetary longitudes. The solar year of $365\frac{1}{4}$ days standardly reckoned by the Christians conformed to neither of these definitions, but took something like the middle ground by falling halfway between Ptolemy's tropical year of $365\frac{1}{4}-\frac{1}{300}$ days and the Indian sidereal year of $365\frac{1}{4}+\frac{1}{120}$ days.[78]

References to Indian parameters in Latin literature were usually linked to the non-Ptolemaic astronomical tradition that had first arrived in Europe in the guise of al-Khwārizmī's *zīj al-Sindhind*, where the sidereal year was taken to be 365d 6;12,9h or 365;15,30,22,30d ≈ $365.25+\frac{1}{118}$ days.[79] It has already been noted that the tables of al-Khwārizmī were first brought to the attention of Latin scholars by

[75] See *Compotus Petri*, MS Paris, Bibliothèque Mazarine, 3642, fol. 19ra (ch. 17), where Peter ascribes this value to 'magistri nostri', and ibid., fol. 19rb.

[76] Ibid., MS Paris, Bibliothèque Mazarine, 3642, fols. 18va–b, 19rb (ch. 17). See Mercier 2004.

[77] Another early attestation of 365d 5;48h appears in a brief text with the incipit 'Inter omnes recte philosophantes Egipcii', found in MS Parma, Biblioteca Palatina, 718–20, fols. 422v–423v, at fol. 423r–v, where 365d 5;48h and $365\frac{1}{4}-\frac{1}{131}$ are treated as interchangeable. The text identifies the present year as AD 1169 and shares much of its content with an anonymous work on the Pisan Tables in MS Cambridge, Fitzwilliam Museum, McClean 165, fols. 67r–76v, at fol. 68r–v, which was written in 1145 (see n. 78 in this chapter). For attestations of a topical year of 365d 5;48h in thirteenth-century works on the computus, see Ch. 4.1.

[78] MS Cambridge, Fitzwilliam Museum, McClean 165, fol. 67r–v (see n. 77 in this chapter). The author states that 1,144 years have passed before the present year AD (ibid., fol. 70r). The same Indian year length appears in Abraham Ibn Ezra, *Liber de rationibus tabularum* (ed. Millás Vallicrosa 1947: 75).

[79] Neugebauer 1962b: 131; Samsó 2012: 179. See also *Compotus Petri*, MS Paris, Bibliothèque Mazarine, 3642, fol. 19va (ch. 17), where the length of the sidereal year is expressed as $365\frac{1}{4}$ days and the equivalent of a daily revolution of 3;2° or 'paulo plus, scilicet circa quintam partem unius hore'.

Petrus Alfonsi, whose knowledge of this sidereal year is enough to explain why his student Walcher of Malvern was forced to admit that the Sun was too slow to go full circle on the ecliptic in a Julian year of $365\frac{1}{4}$ days.[80] The same basic conclusion urged itself upon users of the Toledan Tables, who would have calculated solar longitudes based on a mean daily motion of 0;59,8,11,28,27°. This implied that it took the Sun $365.2565247\text{d} = 365;15,23,29,20,6,43,12\text{d} \approx 365\frac{1}{4} + \frac{1}{153}$ days to complete a full revolution. Unlike al-Khwārizmī's 'Indian' tables, which maintained an exclusive focus on sidereal motions, the Toledan Tables helped their users convert sidereal longitudes into tropical ones, making them viable tools for all those interested in drawing up horoscopes according to Ptolemaic rules, but also those who worried about the equinoctial limit for the calculation of Easter. A conversion algorithm of this sort had to take into account the precession of the eighth sphere, which caused the vernal equinox and the beginning of the sidereal zodiac to grow apart over time. In the case of the Toledan Tables, this elongation value was obtained from a set of tables for the *motus* of the sphere of the fixed stars that relied on a theory known as the *accessio et recessio* or *accessus et recessus* (from Arabic: *al-iqbāl wa'l-idbār*) of the eighth sphere. A straightforward English equivalent would be to speak of the 'access and recess' or 'toing and froing' of the sphere in question, although modern historians have generally accepted the term 'trepidation'.[81]

The type of precession model that is described by these words originated in a tenth-century Islamic context after the Arabic inheritors of Ptolemaic astronomy had begun to compare the data and constants recorded in the *Almagest* with their own increasingly meticulous observations. One crucial source for these early revisions of Ptolemaic theory is the treatise *On the Solar Year*, which was originally written in the ninth century in the circle of the Banū Mūsā, a group of brothers working in Baghdad during the reign Caliph al-Maʾmūn ibn Harūn (813–33).[82] In *De anno solis*, as the treatise was called in its twelfth-century Latin translation, the Banū Mūsā reacted to the discovery that the apogee of the solar orbit had increased by some 15° in tropical longitude since the days of Ptolemy, who had falsely claimed that the line of apsides remains fixed with the apogee located at 65;30° from the vernal point.[83] Their findings instead suggested that the apogee of the Sun underwent precession together with the remaining planetary apogees and that the rate of this precession over the past centuries had been much quicker than previously

80 Walcher of Malvern, *De Dracone* 3.3 (ed. and trans. Nothaft 2017e: 204–5).

81 Useful orientation on the history of precession and trepidation during the Middle Ages is provided by Dobrzycki 1965/2010; Mercier 1976–7; Ragep 1996. See now also Nothaft 2017c. As far as I have been able to ascertain, the first Latin text to employ the term *trepidatio* to describe the motion of the eighth sphere was Nicolaus Comes de Comitibus's *Tractatus de accessu et recessu motus octave spere*, written in 1450 and extant in MSS Arras, Bibliothèque Municipale, 688, fols. 66r–69r; Florence, BML, Ashburnham 208, pp. 407–17; Vatican City, BAV, lat. 3379, fols. 1r–4v; Vicenza, Biblioteca Civica Bertoliana, 208 (132), fols. 258r–265v.

82 The Latin version, which survives in MS London, BL, Harley 1, fols. 19r–22r (s. XIII$^{1/2}$), was edited in Carmody 1960: 63–79, and translated into English by Neugebauer 1962a: 265–89. The Arabic text and a French translation appear in Morelon 1987: 27–67. On the text's authorship, see pp. xlvi–liii.

83 See the discussion of Ptolemy's findings in Petersen and Schmidt 1967.

proposed. A comparison of present observations of the longitude of the star Regulus (α Leo) with the data transmitted from Hipparchus indicated that the fixed stars moved at a rate of 0;0,49,39° per year or 1° in $72\frac{1}{2}$ years, which was an excellent value in terms of accuracy and far exceeded the slow pace of 1°/100y Ptolemy had assumed centuries earlier.[84] The observational work recorded in *De anno solis* was built upon by astronomers such as al-Battānī, whose *zīj* for the meridian of ar-Raqqah on the Euphrates (*c.*880) became known in the Latin West thanks to translations by Plato of Tivoli (1134/8) and Robert of Ketton (now lost). Among al-Battānī's many accomplishments was a new derivation of the length of the solar year, which at $365;14,26\text{d} = 365\text{d}\ 5\text{h}\ 46\text{m}\ 24\text{s} = 365\frac{1}{4} - \frac{3.4}{360} \approx 365\frac{1}{4} - \frac{1}{106}$ days was 9 minutes and 12 seconds shorter than the $365\frac{1}{4} - \frac{1}{300}$ days known from the *Almagest*. In line with this drastic decrease of the Sun's tropical period, which amounted to an underestimation compared to the modern value, al-Battānī accepted a very swift rate of precession of close to 1° in 66 years.[85]

Today it is widely understood that overt disagreements between Ptolemy's data and his Arabic successors' were primarily due to errors, if not outright manipulations, on Ptolemy's part.[86] For ninth- and tenth-century onlookers in the Islamic world, who had only recently begun to construct a pool of reliable observations, things were not quite as obvious. In the context of Western Islamic astronomy, as practised in al-Andalus and the Maghrib, attempts to reconcile the discrepant data eventually led to the development of a highly original model for the motion of the eighth sphere, which was made available to twelfth-century Latin readers in the treatise *De motu octave spere* falsely attributed to the Sabian astrologer-mathematician Thābit ibn Qurra (d. 901). The work's central premise was that the eighth sphere (or sphere of fixed stars) moves neither steadily nor about the poles of the ecliptic, but in a non-uniform manner, which is the result of two points on a 'mobile' ecliptic, defined as 0° Aries and 0° Libra, being carried on small circles with a radius of 4;18,43° (see Figure 3.2). The centres of these circles correspond to the points of intersection between the celestial equator and a hypothetical mean or 'fixed' ecliptic projected onto the ninth sphere. As a consequence of its circular motion around these points, the position of sidereal 0° Aries will be identical with the vernal point whenever it comes to lie on celestial equator. Its maximum distance from the vernal point is ±10;45° and will be reached after an angular motion of 90° or 270° has been completed. The accompanying tables show 0° Aries to move at an angular motion of 0;5,9,56,8° per Arabic lunar year, implying that it takes 4,181.5144 lunar years or approximately 4,057 Julian years for the eighth sphere to undergo a complete cycle of 'access and recess'. During this period both the rate of precession and the length of the tropical year will undergo a continuous sinusoidal increase from minimum to maximum and back,

[84] For a detailed analysis of the contents of *De anno solis*, see Moesgaard 1974; Morelon 1996: 26–31.

[85] Al-Battānī, *De motu stellarum*, trans. Plato of Tivoli (Nuremberg: Petreius, 1537), fols. 26v–27v, 79r–81r (chs. 27, 51–2); Mercier 1987: 112–14. On the history of al-Battānī's precession parameter, see Kurtik 1990; Mozaffari 2016: 303–7.

[86] See n. 23 in Ch. 1.2.

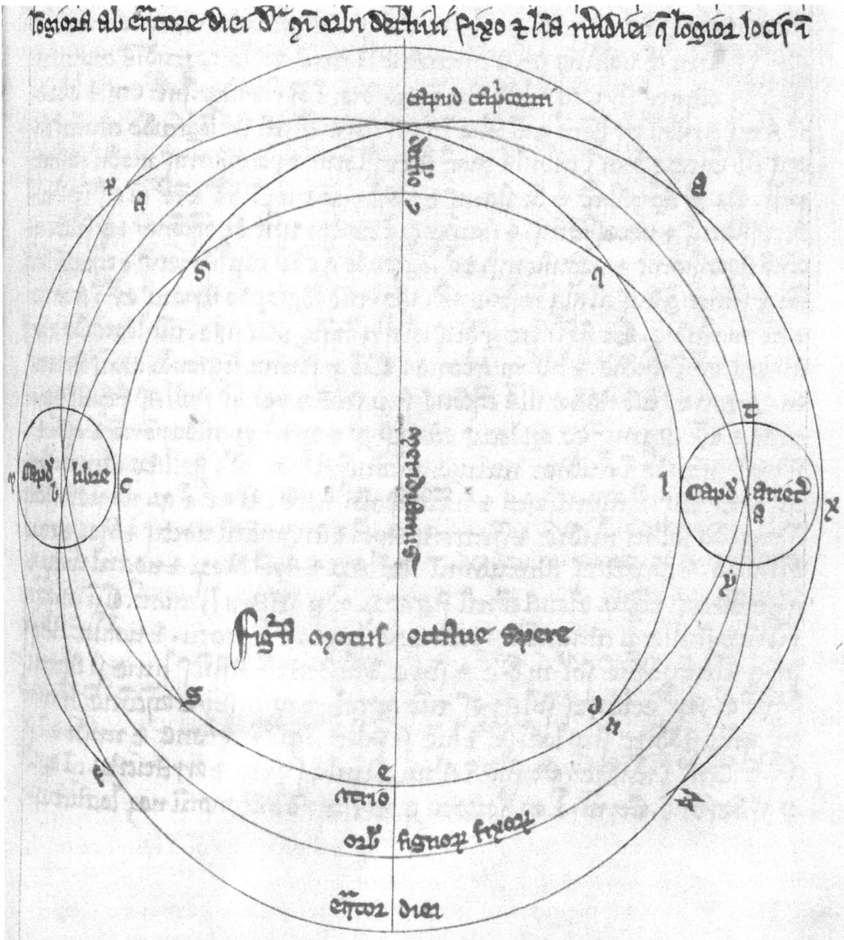

Figure 3.2. *De motu octave spere*, diagram for the access and recess of the eighth sphere. Oxford, Bodleian Library, Bodley 464, fol. 149r.

whereas the sidereal year remains perpetually fixed at what is effectively the average length of the tropical year over the entire cycle. In addition, the theory predicts long-term changes in the obliquity of the ecliptic, which had been given an exaggerated value by Ptolemy.[87]

[87] Editions of *De motu octave spere* appear in Carmody 1960: 102–13 (two versions); Millás Vallicrosa 1945 = 1943–50: 496–509. The text was translated into English by Neugebauer 1962a: 291–9. I discuss the specifics of pseudo-Thābit's model in Nothaft 2017c: 213–15. See also Goldstein 1964–5: 232–8; North 1976: iii. 155–8; Dobrzycki 1965/2010: 20–30; Toomer 1968: 118–22; Mercier 1976–7, pt. 1: 209–20; pt. 2: 62–5; Swerdlow 1973: 446–8; Comes 2001: 300–2; Chabás and Goldstein 2012a: 44–7, and the discussion of the model's zero-point in Goldstein 2011.

Pseudo-Thābit's specific model of access and recess is unlikely to pre-date the construction of the Toledan Tables, which incorporate tables for the motion and 'equation' of the eighth sphere that are identical with those found in *De motu octave spere*. The use of this model made it possible for the Toledan tropical mean Sun to retrace with great exactitude the solar longitudes that resulted from Hipparchus's and Ptolemy's ancient observations as well as from those carried out by al-Battānī in the ninth century. In addition, its results conformed to the solar observations carried out in *c.*1075 by Azarquiel, the putative main author of the Toledan Tables.[88] It is to his name that the model is ascribed in a brief astronomical work written or translated in the mid-twelfth century by John of Spain, which contains one of the earliest independent descriptions of this 'Toledan trepidation model' in Latin.[89] While Azarquiel's authorship is definitely more plausible than Thābit's, it should be stressed that the theory incorporated into the Toledan Tables differs in a number of respects from the three trepidation models presented in Azarquiel's *Treatise on the Fixed Stars*, which is only preserved in a Hebrew translation. They show that the Andalusian astronomer introduced further circles about the poles of the equator, which enabled him to separate the periodic variation in the obliquity of the ecliptic from the access and recess of the fixed stars.[90] Azarquiel made further important contributions to solar theory in his treatise *On the Solar Year*, which ascribed to the Sun's apogee a proper motion of 1° every 279 Julian years that was separate from precession. In addition to this discovery of what is now known as the anomalistic year, Azarquiel created a completely new model for the Sun's circular path around the Earth, the eccentricity of which he believed to vary over time.[91]

Latin Christian astronomers were understandably slow in catching up with these finer developments of Andalusian astronomy, busy as they were trying to understand the new tables, texts, and theories as a whole. A helpful resource in this overall learning process was the *Liber de rationibus tabularum* Abraham Ibn Ezra wrote in 1154 as a supplement to his Pisan Tables, as this text summarized a whole gamut of views and parameters that Greek and Arabic astronomers had put forward over the centuries. When it came to the motion of the solar apogee, Ibn Ezra subscribed to the synchronicity between stellar and apsidal precession proposed in the ninth-century treatise *De anno solis* and dismissed Azarquiel's belief in an independently moving apogee as an illusion caused by observational imprecisions. The imperfections inherent in many instruments and observational procedures also made him sceptical as to the reality of the access and recess of the

[88] Mercier 1987: 106; 1996. See also Ragep 1993: ii. 400–8; Samsó 1994b: 2–5; Comes 2001: 294–6; Pedersen 2002: 54–5.

[89] Edited in Millás Vallicrosa 1936: 469–73. On the author's identity, see Robinson 2007: 64–9.

[90] Millás Vallicrosa 1943–50: 239–343; Goldstein 1964–65: 238–46; Mercier 1996: 327–36; Samsó 1987c; 1992: 227–40; 1994b. A related model was included in the twelfth-century *Muqtabis zīj* by Ibn al-Kammād, which was translated into Latin in 1260. See Chabás and Goldstein 2015: 596–8. On the history of trepidation in al-Andalus and North Africa, see Comes 1996; 2001; 2002; Samsó 1998: 93–6; 2001: 169–74; Díaz-Fajardo 2001.

[91] Toomer 1969; 1987; Samsó 1992: 207–18; Samsó and Millás 1994; Pedersen 2002: 1228–9.

eighth sphere, which he rejected in favour of the tropical reference frame and linear precession adhered to by authorities such as Ptolemy and al-Battānī.[92]

In spite of Ibn Ezra's doubts, the access-and-recess model described in *De motu octave spere* was bound to find widespread acceptance as a result of it being carried across Europe on board the Toledan Tables. It was only a matter of time until a computist capable of understanding the new theory applied it to evaluate the calendrical recession of the equinoxes and solstices. The earliest Latin writer who can be shown to have done precisely that is the aforementioned Peter, who dedicated one of the 70 chapters of his *Compotus* of 1171 to the long-standing question: 'Why the solstices and equinoxes are no longer found where they were once found.'[93] Some of the elements of Peter's answer have already been mentioned above. He explained the observed change in equinoctial and solstitial dates since antiquity with an appeal to the idea that the Julian calendar added up to 12 minutes or $\frac{1}{5}$ of an hour to the actual length of the tropical year, which threatened over time completely to subvert the accustomed seasonal alignment of the Julian months. One straightforward way of dealing with the problem would have been to omit every 30th bissextile day, assuming that $J = 120$ years. Peter was aware of this solution, but felt it necessary to warn his readers against any attempt to implement it, citing the 'not inconsiderable disagreement' that raged between astronomers concerning the length of the solar year:

> For there are some who believe that the solar year must be defined according to the Sun's motion from and return to the same solstice or equinox. Others, however, assert that the solar year must be defined according to the Sun's departure from and return to the same star. These two motions are without doubt different from each other.[94]

The Julian calendar failed on both fronts, as $365\frac{1}{4}$ days exceeded the length of the tropical year, but were not enough to make the Sun return to the same reference star. Another point that needed to be borne in mind was that only the latter period was universally recognized to be a constant. The constancy of the tropical year, by contrast, was thrown into doubt by the observational work carried out by past astronomers. Ptolemy, when comparing the two types of solar year, had found the rate of precession to be 1° per century, whereas al-Battānī, 'who worked with the same instruments as Ptolemy', found the tropical year to be shorter than Ptolemy and concluded that 'the fixed stars in his own time proceed against the firmament

[92] See Abraham Ibn Ezra, *Liber de rationibus tabularum* (ed. Millás Vallicrosa 1947: 75–86), and the discussion of precession in the closely related treatise in MS Cambridge, Fitzwilliam Museum, McClean 165, fols. 67v–68r. On Ibn Ezra's views, see also Samsó 2004b: 288–90; 2012: 177–85, 193–6.

[93] *Compotus Petri*, MS Paris, Bibliothèque Mazarine, 3642, fols. 18va–20ra (ch. 17: 'Quare solstitia et equinoctia non modo inveniantur ubi olim inveniebantur'). This chapter became the basis for Cunestabulus, *Compotus* 36 (CCCM 272: 105–11), for which see Nothaft 2017a: 38–51.

[94] MS Paris, Bibliothèque Mazarine, 3642, fol. 19rb: 'Verumtamen nemo modo consilio temere presumet hunc mutationem facere. Nam inter ipsos astronomos de quantitate anni solaris non modica versatur dissentio. Sunt enim quidam qui opinantur annum solis accipi debere secundum motum solis ab uno solstitiorum sive equinoctiorum et reditum ad illud. Alii autem asserunt annum solis debere accipi secundum recessum solis a quavis stella et reditum ad eandem, qui duo motus proculdubio inter se differunt.'

by one degree every 60 years. Others, however, made different findings in their own times.' Fortunately, an explanation for this diversity had been discovered by Thābit, son of Qurra, 'the greatest philosopher among the Christians', who 'declared that none of the aforementioned authors had been in error. For he said that the fixed stars sometimes move against the firmament (now faster, now slower) and sometimes with the firmament.'[95] Peter made sure his readers understood the consequences of this motion. As long as the stars moved against the firmament in an easterly direction, the Sun, which was carried inside the sphere of the fixed stars, was guaranteed to reach the vernal point sooner, keeping the tropical year shorter than the sidereal year. This had been the situation since before the time of Ptolemy and was going to continue to be case for many more years before the stars reversed direction and started to move with the firmament to the west. Once the sphere of the fixed stars entered this phase of its motion, the length of the tropical year was bound to exceed the length of the sidereal year, which necessarily also meant that it exceeded the Julian year of $365\frac{1}{4}$ days. 'If this is the case, then the winter solstice will after another great number of years return to the eighth before the Kalends of January [25 December] and to the other dates from which it had previously receded. The other solstices and equinoxes will likewise return to the places they had once abandoned.' Peter deftly used this future scenario as a warning against any 'thoughtless despiser of antiquity', who by subtracting a day from the calendar might end up defacing the existing *computus* in a way that:

> might force our descendants after some years not only to make a new addition, but also to compensate for the damage caused by this unlawful removal. For Ptolemy himself asserts that the most ancient astronomers said that the solar year was greater than $365\frac{1}{4}$ days.[96]

That Peter's warning did not go by unheard is clear from the *Compotus* written in 1175 by Cunestabulus, who not only copied parts of Peter's chapter on the equinoxes and solstices, but greatly enriched its argument by adding new data drawn from Ptolemy's *Almagest* and *De anno solis*, which he believed to be another

[95] MS Paris, Bibliotheque Mazarine, 3642, fol. 19va: 'Albategni autem pluribus annis post Ptholomeum cum eisdem instrumentis cum quibus est operatus Ptholomeus deprehendit in tempore suo quantus esset idem motus solis, scilicet a solstitio ad solstitium, et invenit spacium temporis illius motus esse minus quam Ptholomeus invenerat. Invenit etiam in tempore suo singulis LX annis stellas fixas gradum unum procedere contra firmamentum. Alii autem in temporibus suis diversis modis invenerunt. Sed Thebit filius Core, Cristianorum philosophus optimus, neutrum precedentium auctorum fuisse mentitum asseverat. Dixit enim stellas fixas quandoque moveri contra firmamentum, modo cicius, modo tardius, quandoque cum firmamento, similiter nunc cicius, nunc tardius, nunc tantum temere, sed motu circulari.' The false claim that Thābit ibn Qurra was a Christian goes back to Abraham Ibn Ezra, *Liber de rationibus tabularum* (ed. Millás Vallicrosa 1947: 76).

[96] Ibid., fol. 19vb: 'Quod si verum est post multos annos adhuc redibit solstitium hyemale ad VIII° kl. Ian. et ad alias datas a quibus prius recesserat. Aliud quoque solstitium et equinoctia ad loca ante relicta revertentur. Caveat igitur improvidus antiquitatis despector alicuius diei detraxcione usitatam compoti nostri dispositionem ne deterius incidat violare, ne forte cogantur posteri post aliquot annos non solum novam additionem facere, sed etiam huius iniuriose detractionis dampna recompensare. Ipse namque Ptholomeus asserit antiquissimos astronomos dixisse annum solis maiorem esse CCCLXV diebus et quadrante.' The reference is to Ptolemy, *Almagest* 3.1 (trans. Toomer 1998: 139).

work by Thābit. Thanks to his access to these recently translated texts, Cunestabulus was in a position to compare equinoctial dates observed by astronomers over a period of more than 1,200 years, which on balance left no doubt that the Julian calendar had been slightly overestimating the length of the solar year, causing the equinox 'to recede for all eternity, if this shortness of the year were to persevere. But will it really persevere?'[97] Cunestabulus gave the answer by listing various historical estimates of the length of the solar year, which seemed to exhibit a progressive acceleration of the Sun's tropical rate of motion, from the $365\frac{1}{4}+\frac{1}{76}$ days in a year that Ptolemy attributed to the Athenian astronomers Meton and Euctemon to al-Battānī's notoriously short length of $365\frac{1}{4}+\frac{1}{106}$ days.[98] This, of course, was exactly the sort of change more recent astronomers such as Azarquiel had accounted for by hypothesizing a motion of access and recess, which:

> indicates that all of the above-mentioned investigators told the truth in their own time. For it teaches that the solar year sometimes increases and sometimes decreases, and in doing so sometimes adds to the quarter-day, sometimes subtracts from it. If this is true, then the equinox will recede towards the beginning of March as long as the year is shorter than $365\frac{1}{4}$ days and once [the year] adds to the quarter-day, it will return to the dates on which it was found by the ancients.[99]

Far from just writing about access and recess in the same general terms as his predecessor, however, Cunestabulus was in a position to cite precise data, which appear to have come to him as the result of hard arithmetical labour in the quarry of the Toledan Tables. According to his bold prediction, the current calendrical regression of the equinoxes was going to come to an end in the year 1392 of the Christian era, by which time the true vernal equinox would be found to occur on 13 March, shortly after noon, as calculated for the meridian of Jerusalem. After this date, the equinoxes and solstices were going to revert towards their traditional dates for a period of 2,520 years. At the end of these 2,520 years, the vernal equinox would have reached 19 April, only to undergo after that another period of access and shift towards the beginning of the year.[100] For Cunestabulus, as for Peter, the message with regard to the Julian calendar and the whole system of Easter *computus* was a conciliatory one. Rather than being gravely mistaken about the length of the solar year, the founders of the ecclesiastical calendar had simply picked the most convenient parameter known to them, which happened to be the $365\frac{1}{4}$ days first proposed by the ancient astronomer Callippus. This number had the advantage of containing no fractions smaller than a quarter-day, making it easy to perform calculations with, but it nevertheless came reasonably close to the (sidereal) value that occupied the middle between the maximum and minimum year lengths produced by the motion of access and recess.[101]

[97] Cunestabulus, *Compotus* 36 (CCCM 272: 109, ll. 123–8).
[98] Ibid. (CCCM 272: 109–10, ll. 128–38).
[99] Ibid. (CCCM 272: 110, ll. 139–48). [100] Ibid. (CCCM 272: 110, ll. 148–53).
[101] Ibid. (CCCM 272: 111, ll. 174–7). See also ibid. 1 (CCCM 272: 62, ll. 9–16), for the reference to Callippus.

Far from undermining the traditions of the Church, the advanced astronomy from Greek and Arabic sources thus seemed to reinforce their authority, if only by diminishing the prospects for a swift and successful emendation of the calendar. While it was undeniable that the Julian calendar currently failed accurately to represent the actual length of the tropical year, any attempt to improve its intercalation pattern could appear overly cumbersome or simply ill-advised as long as the access and recess of the eighth sphere remained a possible scenario. If the model was correct in its predictions, it was only a matter of time until the errors caused by the altered date of the vernal equinox were going to vanish as a result of the natural change of the length of the tropical year. Perhaps this meant that the Church could simply wait for things to adjust themselves, although the model also implied that an error that grew in the opposite direction, with Easter Sundays being celebrated a month too early, was in the cards for the long run. All astronomers had ever been able to observe, however, was a shift of the equinoxes and solstices from late towards early dates, and some doubt still remained as to whether the reverse scenario would ever become actualized. From the vantage point of Peter, the computist who pioneered the argument in question, it was best to sit back and bide one's time until the astronomers had learned more about the long-term behaviour of the Sun's tropical course: 'But even if at some future time the Church should sigh for a correction of its calculation, it will now, if it acts advisedly, not hasten things, but wait until it has been ascertained whether the doctrine of Thābit is true or not.'[102] As the following chapters will show in more detail, the uncertainty that Peter expressed in 1171 was never fully assuaged during the four centuries of calendrical discussion that were to follow his cautious intervention. Its persistence among astronomers, and their resulting reluctance to advance a definitive solution to the 'equinox problem', is one of the key factors that help explain why no reform of the Julian calendar ever emerged during the Middle Ages.

3.4. ROGER OF HEREFORD'S SYNTHESIS

With their carefully considered thoughts on the fluctuating length of the solar year, the works of Peter and Cunestabulus, both written during the 1170s, are indicative of a new generation of Latin computists who had found their discipline enriched by a whole cluster of new astronomical ideas and concepts drawn from Arabic sources. At the same time, their use of these sources remained a conservative one, aimed towards defending established ecclesiastical custom against its potential critics, detractors, and innovators. This apologetic strand comes to the fore with particular force in the *Compotus* penned in 1175 by Master Cunestabulus, who used his preface to inveigh against:

[102] MS Paris, Bibliothèque Mazarine, 3642, fols. 19vb–20ra: 'Quodsi ad correctionem sui compoti aliquando post hoc tempus ecclesia suspiraverit, modo cum consilio non properabit, sed exspectet donec certum fuerit utrum sententia Thebit sit vera vel non.'

novelty-hunters and immoderate scorners of antiquity, who even in matters of Christian doctrine arrogantly reject the position sanctioned by authority, as if it were not sophisticated enough, and who, in relying on their own wit, wish to think otherwise than the entire Church, as if they were the only ones in the know. But, what is even worse, I have also seen, and was pained to see, it put into writing that some things are one way according to the Church and another way according to the truth.[103]

At least some of this ire was directed against Marianus Scottus and Gerland, who were both notorious for their campaigns to alter the accepted years of Christ's birth and crucifixion. Cunestabulus undermined their computistical arguments in the final chapter of his *Compotus*, where he drew on his knowledge of the Jewish calendar and his ability to perform advanced astronomical calculations to show that Passover could have easily coincided with Good Friday in AD 34 (see Ch. 3.2). This skilful demonstration greatly impressed at least one of his readers, the monk Salomon of Canterbury, who in *c.*1185 concluded an excerpt of Cunestabulus's chapter by encouraging others to:

seek out the writings of Cunestabulus and read them. There he illuminates Bede's darkness [and] exposes the errors made by the Lotharingian [*sc.* Gerland]—and also yours, O Marianus. And by calling the motion of the stars back to Christ's times he proves that the bride [*sc.* the Church] upholds the truth about the groom [*sc.* Christ].[104]

And yet the motions of the stars were not always as supportive of the traditions of the Latin Church as Cunestabulus and his admirer Salomon may have hoped. Some unresolved tensions can be detected in the 38th and penultimate chapter of Cunestabulus's work, in which the author made an attempt to exonerate the inventors of the *computus* from any error they might have committed in accepting a lunar cycle of 19 Julian years. In a manner reminiscent of Bede, he began by discussing various factors that accounted for occasional deviations of two days between the calendrical lunar count and the observable appearance of the thin crescent of the new moon. These ranged from the occurrence of four 30-day lunations in a row in embolismic years that also contained a Julian bissextile day to the very unequal rising and setting times of the zodiacal signs at northerly latitudes. In Cunestabulus's English environment, such seasonal differences meant that, at the vernal equinox, an elongation between Sun and Moon of 8° could be enough for the thin crescent to appear at sunset, whereas at other times the Moon remained hidden even at a distance of 29°. From a computist's perspective, one could always argue that it was better to saunter behind the observable new moon from time to time than to do a forbidden thing and call the evening *luna* 1 when the Moon was not yet visible.[105] Cunestabulus also knew, however, that there were reasons to regard the problem as structural and permanent, rather than just temporary:

Finally, to confess the truth, someone might be of the opinion that the authors of our *computus* failed to investigate the time of the Moon's rekindling by the Sun with

[103] Cunestabulus, *Compotus*, praef. (CCCM 272: 59, ll. 8–15).
[104] MS London, BL, Egerton 3314, fol. 8r (CCCM 272: xxxviii). On Salomon's computistical anthology, see Willetts 1965.
[105] Cunestabulus, *Compotus* 38 (CCCM 272: 112–13).

> complete exactitude. For the Arabs, with whom al-Farghānī agrees, make one lunation 160 atoms smaller than we do, whereas the Jews [make it smaller by] 139 atoms and the ninth part of an atom. Azarquiel, in whose tables modern astronomers put the greatest trust, agrees with the latter. If, therefore, our lunation stretches out longer than appropriate, it will also be the case that the new moon is called out later than appropriate. And this is the reason why the Arabs and Jews call out the new moon both earlier and closer to the Moon's kindling than we do.[106]

These very concise remarks were based on an accurate analysis of the mutual discrepancies between the mean synodic month implicit in the 19-year cycle of the Alexandrian Church and the corresponding values of the Jewish and Arabic calendars, if the latter were converted into a computistical notation involving moments and atoms. For the $29;31,50,8,20\mathrm{d} = 29\mathrm{d}\ 12;44,3,20\mathrm{h}$ that separated two *moladot* in the Jewish calendar, such a transfer yielded 29 days, 12 hours, 29 moments, and $208\frac{8}{9}$ atoms, which was $139\frac{1}{9}$ atoms below the 29 days, 12 hours, 29 moments, and 348 atoms that Hermann of Reichenau had established for the *computus naturalis* in the eleventh century. The Arabic calendar subtracted from this another $3\frac{1}{3}$ seconds or $20\frac{8}{9}$ atoms, arriving at 29 days, 12 hours, 29 moments, and 188 atoms. If this latter estimate was correct, the predictions made by the ecclesiastical calendar were bound to deteriorate at a rate faster than $M = 285y$, as had already been demonstrated earlier in the twelfth century by an anonymous West English computist.[107] Was it therefore necessary to admit that the Church had gotten it all wrong? Cunestabulus hoped otherwise and tried his luck one more time with the Toledan theory of access and recess, which implied that the length of the solar year had changed between antiquity and the present. From the 19-year cycles attributed to Meton and Callippus, but also from the findings of Hipparchus and Ptolemy recorded in the *Almagest*, one could gather that the ancient Greek astronomers had shared a widespread belief that 19 solar years comprised exactly 235 lunations, even though their respective estimates for the length of this period varied. If their basic assumption was correct, if $19y = 235m$ was indeed a constant relation, the variations in the length of the tropical year brought about by the access and recess of the eighth sphere should also affect the length of the Moon's synodic period, which could hence be expected to oscillate between two extremes just like the solar year. 'If this is true', Cunestabulus concluded, 'the truth of our whole *computus* will return once the equinoxes and solstices return to the later dates.' In other words: if the length of the tropical year was going to exceed $365\frac{1}{4}$ days at some point in the future, the length of the synodic month might do the same for the 29 days, 12 hours, 29 moments, and 348 atoms implied by the Alexandrian lunar cycle, in which case the currently observable 2-day error in the count of the lunar age would disappear over time. By contrast, the Arabs and Jews, 'although they are

[106] Ibid. 38 (CCCM 272: 113–14, ll. 51–60). For the reference to al-Farghānī, see ch. 1 of his *Elements of Astronomy* in the Latin translations of John of Seville (ed. Carmody 1943: 3) and Gerard of Cremona (ed. Campani 1910: 56–7).

[107] See n. 21 in this chapter.

now seen to make better pronouncements', would be 'charged with an even more manifest error' in the distant future.[108]

A greater willingness to abandon such wishful thinking and face the flaws of the 19-year cycle with an open visor is on display in a *Compotus* written only one year later, in 1176, by the astrologer Roger of Hereford, whose work represents the culmination point of some of the developments Walcher of Malvern had inaugurated more than a half a century earlier.[109] The preface to Roger's work, which comprised five books and a total of 28 chapters, contains some revealing remarks about the controversies that had erupted in the West of England with regards to the study of celestial motions. Not only had the newly emerging class of astronomer-astrologers begun to find fault with the *computus* for making false predictions, with some going so far as to argue that it should be excluded from the discipline of philosophy, but the computists themselves were divided between two factions, 'natural' and 'vulgar':

> the natural ones reject the vulgar *computus* for not conforming to their standard of subtlety and for following the opinion of the senses rather than the truth of reason; by contrast, the vulgar [computists] call the natural [*computus*] a vain and empty science that is remote from the senses and only accessible to reason, one which eyes cannot see and ears cannot hear. In addition, some who deal with this science insert many superfluous points without distinguishing between natural and artificial *computus*, whereas others, who want to address only the vulgar usage of the Church, have been found discarding many necessary things.[110]

Roger's own work was a conscious attempt to bridge this growing gap and to unite vulgar and natural *computus* under one roof together with the 'astronomical subtleties concerning the motion of Sun and Moon'. His stated goal was to examine 'the convergence and difference between these three [branches]' and to see 'why the vulgar new moon [*primatio*] has now receded so much from the truth', taking into consideration the opinions of the 'Chaldaeans and Hebrews' next to that of the Latins.[111] The first three books of Roger's *Compotus* were mostly devoted to the discipline's conventional strands, starting off with a meticulous presentation of the ecclesiastical computation of mobile and immobile feast days. In the second book, he concentrated on the 'vulgar' perspective on the motion of the Sun, while the third book was dedicated to lunar computistics. Book IV involved a switch to the *computus naturalis*, while Book V left the familiar tracks of the *computus* entirely and confronted readers with the insights provided by the art of astronomy. In

[108] Cunestabulus, *Compotus* 38 (CCCM 272: 114, ll. 68–71). Cunestabulus here ignores the fact that the motion of the sphere of the fixed stars would have affected the sphere of the Sun and Moon in equal measures, keeping the length of the synodic month constant. See n. 123 in this chapter for Roger of Hereford's critique of a similar argument.

[109] See now the edition by Alfred Lohr in CCCM 272: 125–232, with introduction and references to further literature on pp. xix–xxvi, xl–xlii. For a fuller discussion of Roger and his *Compotus*, see Nothaft (forthcoming-3).

[110] Roger of Hereford, *Compotus*, praef. (CCCM 272: 127–8, ll. 23–37). The beginning of the preface is also printed in Wright 1842–6: ii. 90–1, and in Russell 1932: 20–1.

[111] Roger of Hereford, *Compotus*, praef. (CCCM 272: 129, ll. 63–70).

Roger's estimation, only the latter discipline probed into the deepest truths about the heavenly motions by taking into account their true inequality. In this regard, astronomy differed conspicuously from both the vulgar *computus*, where periods were unequal merely as a matter of calendrical convention, and the natural *computus*, whose practitioners used nothing but equal intervals in tracking celestial phenomena.[112]

That the computists' urge towards uniformity was based on an oversimplification became clear from a look at the course of the Sun. As Roger showed in the first chapter of Book V, the brighter of the two luminaries surrounded the Earth on an eccentric orbit, such that its apparent motion was much faster when it was in Sagittarius than when it was found in Gemini. Another example of such inequality was the length of the 24-hour day, which underwent slight fluctuations over the course of a year owing not only to the Sun's eccentricity, but also to the obliquity of the zodiac relative to the equator and the concomitant differences in the rising times of individual signs.[113] Anomalies of this sort made it necessary to distinguish between the mean longitude of a celestial body and its true position, which could be obtained by adding the appropriate equation. In the case of the Moon, the difference between mean and true longitude could reach ±8;39° or the equivalent of approximately 15 hours in either direction.[114] Reflecting on these fundamental aspects of solar and lunar astronomy, Roger discerned another key difference that separated computists from astronomers. The latter were principally concerned with the prediction of longitudes of celestial bodies as a function of time, whereas computists tried to measure time spans as defined by cycles of celestial motions. Rather than worrying about the inequality of these motions, however, they always divided the periods thereby obtained into equal parts. This aspect of their practice could be compared to an astronomer who worked with nothing but mean motions, knowing that the difference between mean and true motion varied periodically and would reduce to zero in the longer run.[115] In an ideal situation, the mean values that natural computists managed to extract from their 19-year cycle were perfectly equivalent to the mean motion parameter astronomers used to chase after solar and lunar longitudes. And yet, calculations of the time of the new moon carried out by natural and vulgar computists regularly manifested discrepancies of two or even three days as soon as natural computists adjusted their calculations to observable reality. According to Roger's testimony, this evident state of affairs had caused much consternation among his contemporaries, to the extent that some had come up with rather fanciful attempts to explain the discrepancy away. According to one explanation, the delay in the vulgar lunation reflected the fact that the Moon had first been seen on the third day after its creation, since man was only formed on the sixth day (Genesis 1:26), whereas the luminaries had already been placed in the heavens on the fourth day (Genesis 1:14). Hence, 'since man

[112] Ibid. 5.23 (CCCM 272: 213, ll. 8–18).
[113] Ibid. 5.23 (CCCM 272: 213–17). [114] Ibid. 5.24 (CCCM 272: 217–21).
[115] Ibid. 5.25 (CCCM 272: 223, ll. 43–51). See also ibid. 1.1 and 5.24 (CCCM 272: 133–4, ll. 10–37; 221–2, ll. 98–131).

first saw the Moon on the third day after its creation...he calls it "first" on the third day after conjunction in accordance with this first sighting'.[116] A version of this scripturally inspired argument, which was to have a long afterlife, already occurs in Abraham Ibn Ezra's *Liber de rationibus tabularum*, which hinted at a 2-day difference between the lunar count of the Jewish calendar and the 19-year cycle of the Church. In years that had the paschal new moon fall on a Saturday or Sunday, this discrepancy caused Easter Sunday to be postponed by a full week compared to Passover:

> The reason for this is that the lunar month according to the Jews begins either with the conjunction of Sun and Moon or when [the Moon] becomes visible. The Christians, on the other hand, reason that one cannot speak of the first [day of the Moon] except in relation to how it was first sighted by Adam, who was created on the third day after the creation of the Moon. And this is why the Christians say that [the Moon] is in its first [day] about two days after its conjunction with the Sun, in accordance with the [difference between] the age of the Moon at its creation and at its first sighting by man.[117]

Not everyone was satisfied with this explanation, however, and some regarded it as downright frivolous, especially since the calendrical new moon was not always found on the third day after conjunction, as presupposed by the argument, but was sometimes delayed until the fourth day. These critics instead appealed to the known variations in the courses of the Sun and the Moon and the different times it took for the signs that hosted them to rise and set, which had the effect that the Moon could on occasions be hidden from sight for up to three days, as was the case in autumn when the Sun occupied the beginning of Libra. Since the ecclesiastical calendar identified the new moon with the first visibility of the Moon's crescent, which was done for the sake of the common people, 'who put their faith more in the senses than in reason', but at the same time tried to distribute its new moons as evenly as possible (in so far as the use of complete days permitted this), temporary discrepancies were impossible to avoid.[118] A third faction disagreed with this conciliatory attitude, insisting that in ancient times the calendrical new moon had sometimes been proclaimed ahead of the natural conjunction, sometimes after it, whereas in present times it always lagged two or three days behind. If that was the case, then the only sensible conclusion was to assume a systematic error inherent

[116] Ibid. 5.25 (CCCM 272: 224, ll. 66–75).

[117] Abraham Ibn Ezra, *Liber de rationibus tabularum* (ed. Millás Vallicrosa 1947: 100). See Nothaft 2012: 125, for a fuller version of this quotation. Later versions of this argument crop up in a commentary on Alexander of Villedieu's *Massa compoti* written in 1396 by a certain Master Conrad (MS Vienna, ÖNB, 3816, fol. 99va; Kaltenbrunner 1876: 295–6) and in a fourteenth-century Venetian merchant's notebook known as the *Zibaldone da Canal* (Stussi 1967: 78–9). It was still combated in 1437 by the calendar reformer Hermann Zoest in the second version of his *Phaselexis*, ch. 8 (ed. Solan 2016: 472–4). Roger of Hereford's contemporary Cunestabulus used the two-day difference between the Moon's and Adam's creation to justify his thesis that the Jews celebrated Passover two days after the astronomical full moon in the year of Christ's Passion. See Cunestabulus, *Compotus* 39 (CCCM 272: 121, ll. 214–26).

[118] Roger of Hereford, *Compotus* 5.25 (CCCM 272: 224, ll. 75–90). The point that the interval between conjunction and first visibility is greater in autumn than in spring was already made by Cunestabulus, *Compotus* 38 (CCCM 272: 113, ll. 32–43).

in both the vulgar and natural *computus*, which was caused by an overestimation of the length of the month.[119] In order to see more clearly on these matters Roger proposed to take into account the estimates of other nations, in particular those of the Arabs ('Chaldaeans') and Jews ('Hebrews'), whose systems of time reckoning he had already described in earlier parts of his treatise.[120] His ability to perform calculations in both of these exotic lunar calendars allowed him to use their projected new moons of September 1176 as a means of gauging the accuracy of the Christian calendar. It turned out that the Arabs placed their new moon on 5 September, at the 17th hour after sunset, which was roughly 7 hours ahead of the Hebrew *molad*. The vulgar *computus*, by contrast, assigned the start of the same lunation to 9 September.[121] Equally disturbing was the comparison between the natural *computus*, as represented by Gerland's cyclical lunar tables, and the mean syzygy tables contained in Arabic *zījes*, which Roger referred to as the 'natural *computus* of the Chaldaeans' (*Chaldaeorum naturalis computus*). Roger knew that Gerland had based his improved lunar calendar on the observed time of the solar eclipse of 23 September 1093, which had occurred during the 15th hour of the day. He claimed that, according to the Chaldaeans, the corresponding mean conjunction fell just 2 hours and 30 moments (= $2\frac{3}{4}$ hours) later, whereas by 1176 the discrepancy between both systems had widened to nearly 8 hours.[122]

This striking increase over just 83 years could be explained by examining the different estimates for the mean lunation that were implicit in the Latin, Jewish, and Arabic calendars, and for which Roger obtained the same mutual discrepancies as Cunestabulus in the year before him (see Figure 3.3). Unlike his mysterious predecessor, however, the astrologer from Hereford saw no merit in attempting to defend the 19-year cycle by appealing to theories of trepidation. In his response to the hypothetical objection that the Golden Number had been placed in the calendar at a time when the Sun's motion had been swifter than it was now, explaining why the Moon now needed less time to catch up with it than it used to, Roger dryly noted that any periodic speeding up and slowing down of the eighth sphere was bound to affect the spheres of the Sun and the Moon in equal measures, leaving their synodic period unaffected.[123] The reason for the observed discrepancies

[119] Roger of Hereford, *Compotus* 5.25 (CCCM 272: 224–5, ll. 90–108).
[120] Ibid. 2.5, 3.11, 3.15 (CCCM 272: 153, ll. 21–43; 164–5, ll. 9–41; 175–8, ll. 129–214).
[121] Ibid. 5.26 (CCCM 272: 225).
[122] Ibid. 5.27 (CCCM 272: 230). Roger's calculations are discussed in more detail in Nothaft (forthcoming-3). Another noteworthy reference to 'Chaldaean' calendrical wisdom appears in a fourteenth-century copy of Raymond of Marseille's treatise on the astrolabe, to which an unidentified scribe added a few passages on lunar computation, claiming to have found 'in the writings of a certain Chaldaean' (*in cuiusdam Caldei scriptis*) a doctrine to the effect that $M = 532$ years. He notes that more than $2 \times 532 = 1{,}164$ years have gone by since Christ's birth, but ultimately refuses to follow the 'Chaldaean' doctrine. See MS Vatican City, BAV, lat. 4083, fols. 68vb–69ra.
[123] Roger of Hereford, *Compotus* 5.26 (CCCM 272: 228, ll. 82–95). Roger here describes a non-Toledan access-and-recess model with an amplitude of 8° and a period of $2 \times 900 = 1{,}800$. The details are strongly reminiscent of two passages in the obscure twelfth-century *Liber de motibus planetarum* (Inc.: 'Investigantibus astronomie rationes'). See e.g. MS Oxford, BodLib, Selden supra 76, fols. 20r–29v, at fols. 20v and 21v. The second of these passages depends on Walcher of Malvern, *De Dracone* 3.3 (ed. and trans. Nothaft 2017e: 204–5).

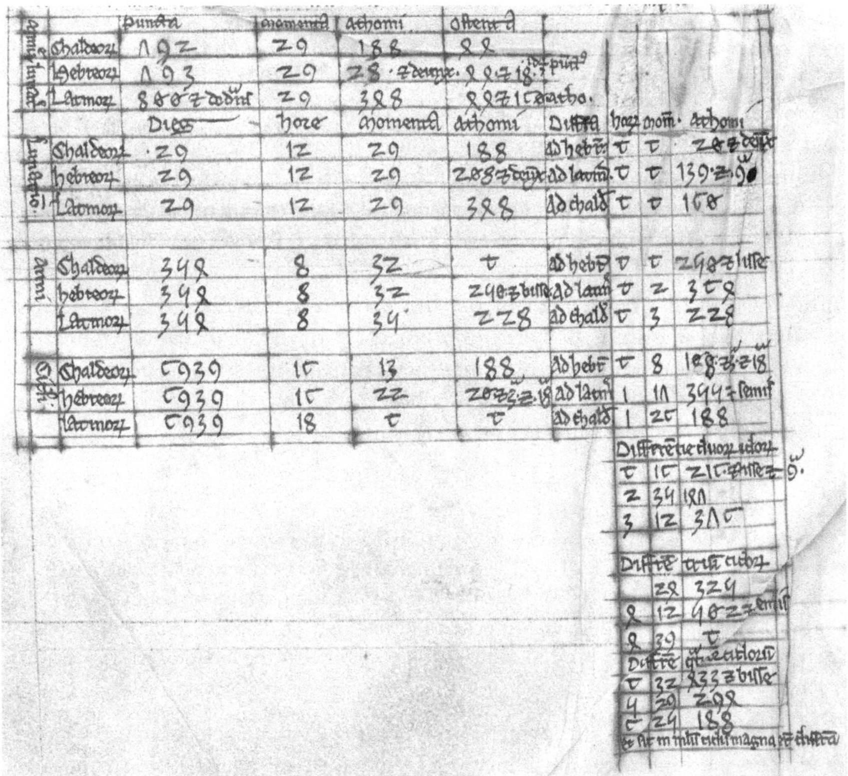

Figure 3.3. Roger of Hereford, *Compotus* (1176), comparative table for the length of the mean lunation according to the 'Chaldaeans', 'Hebrews', and 'Latins'. Cambridge, University Library, Kk.1.1, fol. 238r.

was hence not the access and recess of the eighth sphere, but the simple fact that 19 lunar years and 19 solar years were not really equivalent, as the mean motion parameters contained in the Toledan Tables made perfectly clear.[124] A slightly different perspective was provided by the fluctuating true motions of Sun and Moon, which in theory allowed for rare cases where the courses of the two luminaries converged after 19 years or a period very close to it. Roger even speculated that such a coincidence might have happened once or twice during antiquity, leading the ancients to assume that 19 years = 235 months described a constant relation. This was an easy mistake to make, since the difference from one 19-year cycle to the next was not very noticeable, 'but from a large number cycles, as is now

[124] Roger of Hereford, *Compotus* 5.26 (CCCM 272: 228–9, ll. 95–134). The numerical information provided here was excerpted into MS Oxford, BodLib, Ashmole 1796, fol. 172v (cf. North 1976: ii. 380).

obvious, a great divergence arises'.[125] The truth of this sentence could be confirmed by turning one's gaze from Gerland's natural *computus* back to the much cruder predictions made by the ecclesiastical calendar, which claimed *luna* 1 for 9 September 1176. If the new moon was assigned to the end rather than the beginning of this day, the Latin vulgar computation turned out to be 4 days and 2 hours behind the mean conjunction indicated by the astronomical tables of the 'Chaldaeans', thereby suggesting that the lunar calendar had been steadily deteriorating without any intervention for the past 1,124 years.[126] Roger's initial reaction to this revelation was not without a hint of righteous indignation: 'How unfitting it is that Latin men of the highest distinction should have erred so manifestly about a thing of such importance!'[127] In the final chapter of his treatise, however, he did his best to temper the negative impression by portraying the 19-year cycle of the computists as the product of a carefully deliberated compromise, which had been enacted in the knowledge that:

> the concordance of these motions [of Sun and Moon] can only be found through extremely difficult labours and over a great period of time. And since the Latin wise men tried to cater to the masses in transmitting to them some knowledge of these things, they put together a 19-year cycle that was as short as they could make it without digressing too much from the truth, lest the [the simple folk] be confused by the great numbers involved. And compelled by this kind of diligence they fell into an error, if it is even appropriate to call it that, by saying that the lunation has this exact length.[128]

Given the great success with which the 19-year cycle managed to maintain a balance between simplicity and convenience, on the one hand, and astronomical accuracy, on the other, Roger saw no reason to discard it in its entirety. Rather than chiding its makers for its flaws, it was more fitting to praise the way the Latins had managed to construct the cycle 'in such a brief manner, so artfully, and so close to the truth as a work of art worth marvelling at and one of incomparable economy (*diligentia*). For the discrepancy over such a long span of time is small and once [the cycle] strays from the truth it will be extremely easy to recover it.'[129] All anyone had to do was to reset the Golden Numbers that were regularly inscribed into contemporary *kalendaria* and thereby rectify an error that had been growing ever since the 19-year cycle's institution. The trick was to move all calendrical new moons three days upwards—for instance, shifting Golden Number XIX from 5 January to 2 January—while maintaining their customary intervals. Nothing prevented those born in later times to repeat such an adjustment whenever the gap between calculated *primatio* and natural conjunction had again increased beyond acceptable bounds. 'And this way the calendar will be able to

[125] Roger of Hereford, *Compotus* 5.26 (CCCM 272: 228, ll. 115–16).
[126] Ibid. 5.27 (CCCM 272: 230–1, ll. 17–52).
[127] Ibid. 5.26 (CCCM 272: 226, ll. 52–3).
[128] Ibid. 5.28 (CCCM 272: 231–2, ll. 14–23).
[129] Ibid. 5.28 (CCCM 272: 232, ll. 37–41).

last for many centuries.'[130] For Roger, who closed his treatise by outing himself as an admirer of the Christian calendar, there was hence no reason to despair. The lunar calendar was in error, but this error had been quantified and could be easily counteracted in the future, provided other computists—and, more importantly, the Church—found themselves willing to make some concessions to astronomical reality. Little did he know of the complications that were yet to arise on the way to such a consensus.

[130] See ibid. 5.28 (CCCM 272: 232, ll. 24–36), and the excerpt in MS Oxford, BodLib, Ashmole 1796, fol. 172v. Roger here mentions an error of roughly four days (CCCM 272: 232, ll. 28–9: 'cum iam fere per quattuor dies primatio a veritate discesserit'), but only a describes a change from 5 January (the Nones) to 2 January (the fourth day before the Nones), which clashes with his previous statement in ch. 5.27 that the 'vulgar' new moon lagged 4 days and 2 hours behind the mean conjunction.

4

The Consolidation of a Calendar-Reform Debate in the Thirteenth Century

4.1. 'VULGAR *COMPUTUS*' AND THE ERROR OF THE JULIAN CALENDAR

Roger of Hereford's account of the controversy that raged among the computists of his day, but also his reference to 'the governance of the schools, in which I have already sweated for many years',[1] may be taken as subtle hints at the existence of a larger institutional context in which problems relating to the ecclesiastical calendar could be taught, discussed, and debated. In Roger's case, the venue in question appears to have been the cathedral town of Hereford, which the local canon Simon du Fresne, in a poem written during the final years of the twelfth century, described as a site for the study of exotic disciplines such as astrology and geomancy next to law and the seven liberal arts.[2] Over the coming decades, the study of astronomy in medieval Europe received an even steadier footing thanks to the expansion of the universities, founded as autonomous corporate bodies of masters and students whose designation as *studium generale* gave expression to their supra-regional outlook and importance. The most significant example for England is the gradual emergence of the University of Oxford, first chartered in 1214, which became a place notable for the attention its teachers and curricula gave to the natural and mathematical sciences.[3] An important intellectual figure from Oxford's early period was Alexander Neckam (1157–1217), one of the first English scholars to engage fully with the newly translated works of Aristotle. Neckam's interest in the natural world is reflected most strongly by the first two books of his encyclopedic *De naturis rerum*, which were part of a five-book commentary on Ecclesiastes, and by his *Suppletio defectuum*, a poem of nearly 3,300 lines written in 1216, by which time he had become abbot of the Augustinian monastery of Cirencester. The *Suppletio* includes a section of 852 verses in elegiac couplets dealing with various aspects of astronomy. These reflect their author's familiarity with Roger of Hereford and the attempts made in the previous century

[1] Roger of Hereford, *Compotus*, praef. (CCCM 272: 128, ll. 42–3).

[2] The relevant part of Simon du Fresne's poem is edited in Hunt 1936: 36–7. See also Russell 1932: 19–25; French 1996: 460–2.

[3] Grant 1984; Southern 1984; Weisheipl 1984; Cobban 1988, North 1992a; 1992b; McCluskey 1998: 190–2; Shank 2013.

to create a synthesis between traditional computistical lore and new information drawn from Arabic sources.[4]

The palpable signs of Roger's influence on Neckam's *Suppletio* include a reference to the 'Chaldaeans, whose first day of the Moon | now precedes the conjunction, now follows upon it'.[5] Likewise, when Neckam dealt with the variations in the Sun's apparent speed caused by its eccentric path around the Earth, he did so in terms reminiscent of Book V of Roger's *Compotus*.[6] Other computists, such as Cunestabulus, would have been able to supply him with his references to Ptolemy, whose length of the solar year differed from the Julian one by the equivalent of 1 day in 300 years.[7] In response to those who feared that this difference would lead to Easter being celebrated in the wrong season, Neckam appealed to Thābit ibn Qurra and his discovery that the current motion of the eighth sphere in the direction of the heavens would eventually give way to a compensating period of *recessus*, which like the period of *accessus* took 900 years to complete. The number was supplied by Roger of Hereford, but Neckam's application of the theory of 'access and recess' to the problem of the receding equinoxes had much in common with the defensive arguments the computists Peter and Cunestabulus had deployed in 1171 and 1175 (see Ch. 3.3).[8] He was more openly critical of the lunar cycle incorporated into the ecclesiastical calendar, and followed Roger's lead in rebuking those who tried to excuse it with an appeal to the perspective of the first-created man, who had seen the Moon on the third day after its creation. Such a response was certainly 'charming to the ears' (*Aures demulcet responsio*), but nonetheless a 'vain fiction' (*fictio vana nimis*), as evident from the fact that the difference could nowadays reach up to four days.[9] Seeing that the vulgar *computus* erred so manifestly, the question arose why no corrective steps had yet been taken (*En, iam vulgaris manifeste compotus errat; | Hunc cur errorem non revocare student?*).[10] Defenders of the old regime might refer to the necessity of keeping the *termini paschales* in their accustomed places, lest the order of the mobile feasts be disturbed,[11] but in Neckam's view an intervention would have to be made sooner or later: 'Future ages will lament because of this! | An urgent crisis will call this error back [to mind]. | O, would that the love of

[4] Alexander Neckam, *Suppletio defectuum* 2.535–1386 (CCCM 221: 130–71). On Neckam and his encyclopedism, see Hunt 1984; Zahora 2014.

[5] Alexander Neckam, *Suppletio defectuum* 2.733–4 (CCCM 221: 139); Roger of Hereford, *Compotus* 5.26 (CCCM 272: 226, ll. 56–9).

[6] Alexander Neckam, *Suppletio defectuum* 2.877–932 (CCCM 221: 146–9); Roger of Hereford, *Compotus* 5.23 (CCCM 272: 213–15, ll. 19–62). Neckam's verses even include a description of the zodiacal diagram included in Roger's chapter. See *Suppletio defectuum* 2.891–6 (CCCM 221: ciii, 147).

[7] Alexander Neckam, *Suppletio defectuum* 2.808, 1069–72 (CCCM 221: 142, 156); Cunestabulus, *Compotus* 36 (CCCM 272: 109, ll. 133–6).

[8] Alexander Neckam, *Suppletio defectuum* 2.805–14 (CCCM 221: 142); Roger of Hereford, *Compotus* 5.26 (CCCM 272: 228, ll. 82–95).

[9] Alexander Neckam, *Suppletio defectuum* 2.835–8 (CCCM 221: 144).

[10] Ibid. 2.843–4 (CCCM 221: 144).

[11] Ibid. 2.845–6 (CCCM 221: 144). The issue of the *termini paschales* is also brought up in Roger of Hereford, *Compotus* 5.26 (CCCM 272: 226, ll. 45–7).

truth brought this about!' (*ob hoc secla futura gement.* | *Urgens articulus hunc errorem revocabit;* | *O utinam veri cogeret istud amor*).[12]

Such calls for reform, sincere as they may have been, could not inhibit the production of texts that stuck to expounding all the traditional ways of navigating the ecclesiastical calendar, the number of which only increased as the universities rose to prominence.[13] Although university statutes that would bear witness to the inclusion of *computus* in the curricula are lacking from the thirteenth century, the extant manuscripts attest to the increasing proliferation after 1200 of certain computistical 'school texts', which strongly suggests that instruction in calendars and time reckoning was already in full swing during this period. One case in point is the inclusion of *computus* in manuscripts of the so-called *Corpus astronomicum*, a type of astronomical-mathematical anthology intended for university use that became very common from the middle of the thirteenth century onwards.[14] A virtually indispensable element in any collection of this type was the *Liber de sphera* by John of Sacrobosco (*c.*1220/30), the immense influence of which is manifest in the way it was still used in higher education 400 years after its creation, leading to hundreds of manuscript copies and at least 222 printed editions between 1472 and 1650.[15] Sacrobosco managed to contribute further to his ubiquitous presence in the late medieval quadrivium with a treatise on arithmetic using Hindu-Arabic numerals (*Algorismus vulgaris* or *De arte numerandi*), which was easily the most popular text of its kind.[16] More voluminous than either the *Sphere* or the *Algorism* was his *Computus ecclesiasticus*, also known as *De anni ratione*, which became one of the most widely read medieval texts of the entire discipline.[17]

One is tempted to assume that the preserved Oxford statutes of the first half of the fourteenth century had all three of Sacrobosco's textbooks in mind when they listed a '*Compotus* together with the *Algorismus*' and 'the treatise *De spera*' among the books any bachelor was required to have heard lectures on before he could incept as a lecturer.[18] It is important to note, however, that Sacrobosco's *Computus* was similar to Roger of Hereford's work in that it borrowed elements from Graeco-Arabic astronomy, including the occasional use of sexagesimal notation (see Ch. 4.2), which is a trait that set it apart from most entry-level texts of the same genre.

[12] Alexander Neckam, *Suppletio defectuum* 2.846–8 (CCCM 221: 144). The passage is already quoted in Hunt 1984: 76–7.

[13] Although a comprehensive survey of later medieval *computus* literature is still lacking, some useful information may be gleaned from Thorndike 1954; Means 1992b.

[14] Pedersen (Olaf) 1975: 73–82; 1978: 314–20; 1981; McCluskey 1998: 192–3.

[15] See Thorndike 1949, for an edition of the text and several medieval commentaries. The printed editions are counted in Hamel 2014. For more on the reception of *De sphera*, see Simek 1987; Ludwig 2010; Crowther et al. 2015. On the diagrams, see Müller 2008: 203–51. The available evidence for John of Sacrobosco's identity and literary output is collected and analysed in Pedersen 1985; Knorr 1997: 196–200.

[16] Sacrobosco's *Algorismus* was last edited in Pedersen 1983–4: 174–201.

[17] The text was first printed as an appendix to John of Sacrobosco, *Libellus de Sphaera*, ed. Philipp Melanchthon (Wittenberg: Clug, 1538), sigs. Br–H3r. In what follows, I shall quote this edition as 'John of Sacrobosco, *Computus*'. A new critical edition by Alfred Lohr is currently in preparation. See also Pedersen 1985: 206–14; Moreton 1994.

[18] Gibson 1931: 33 (see also 200, 378, 390); Wordsworth 1904: 32–4, 108; Weisheipl 1964: 156–62, 170, 172–3.

While this degree of sophistication did not hinder Sacrobosco's success as a textbook writer,[19] the lion's share of *computi* produced during the thirteenth and fourteenth centuries offered far more elementary treatments, often using metre as a means to aid the memorization of numerical information and reckoning instructions. Of the various 'metrical textbooks' on calendar reckoning that were penned during this period, none was more popular than the *Massa compoti* attributed to Alexander of Villedieu, which appears to have originated soon after 1200.[20] The poem employed a variety of pedagogical gambits, among them the replacement of numbers with letters in the alphabet, which offered a recipe for the creation of mnemonic words. A simple example are the ferial letters for the first day of each Julian month, which could be represented by the following two lines of six words each: *Altitonans dominus divina gerens bonus extat* | *Gratuito celi fert aurea dona fideli*.[21] Another potential aid to calendrical computation was the principle of assigning dates and numerical values to the joints and fingers of one's hand, which gave rise to a popular sub-genre of computistical literature known as *Computus (Compotus) manualis* or *chirometralis*. Significant examples include a type of 'hand calendar' called the *Palma*, which a Franciscan monk named Blasius devised around 1224, and the metrical *Computus manualis* by a certain Master Anianus, which was to go through dozens of early printed editions.[22] Other popular works of this type were composed in 1289 by John of Brunswick (*Iohannes de Pulchro Rivo*) and in *c*.1281 by Balduin of Mardochio (or Marrochio), whose work typically came with elaborately drawn hand diagrams (see Figure 4.1).[23] For the later Middle Ages, one may cite the *Computus manualis*

[19] The number of manuscripts of Sacrobosco's *Computus* seems to approach 200 (private communication by Alfred Lohr) and the early modern printed editions are reported to number at least 35. See Rosen 1974; Pedersen 1985: 184–5, 187–90; Reich and Knobloch 2004: 31–2; Skemer 2010: 94–5. The text was also used as the basis for the *Computus* of Master Gordianus (see n. 55) and for the computistical sections of the *Mainauer Naturlehre*, datable to the first half of the fourteenth century. See Brévart 1987: 161–78; Mosimann 1994: 317–90; Kleine 1995: 103–6.

[20] For editions, see Steele 1926: 268–89; Wijk 1936: 52–69 (Latin text), 72–84 (French translation), 85–134 (commentary). See also Cordoliani 1952: 342; Borst 1998: 74, 338–9; Hübner 2000.

[21] Alexander of Villedieu, *Massa Compoti*, ll. 108–9 (ed. Wijk 1936: 55). The role played by mnemonic verses in medieval education is documented by Thorndike 1955. On their use in *computus* texts, see Bischoff 1940 (repr. in Bischoff 1967: 192–227); Cordoliani 1960–1, pt. 1: 108–17; pt. 2: 181–208; Wiesenbach 1986: 60–1; Wallis 1990a: 58; Borst 1998: 495–500; Silagiová 2013: 309–12; Skemer 2010: 89–103; Pasquale 2003, and the commentary on the *Massa compoti* in Wijk 1936: 74–125. Specifically on musical memory aids, see Irtenkauf 1957.

[22] Blasius the Franciscan's computistical works *Palma* and *Circius* (a system of mnemonic key words) appear with commentaries in MS Paris, BnF, lat. 14070, fols. 25va–27r, 97ra–103va. Additional commentaries are preserved in MS Cambridge, Gonville & Caius College, 137, pp. 51–70. See Thorndike 1947; 1954: 229–32. For the *Computus manualis* of Anianus, see Smith 1928, who counts no fewer than 62 editions printed between 1483 and 1540 (pp. 80–101). The edition Oxford 1519 was reprinted with explanatory notes in Wordsworth 1904: 149–94. See North 1992a: 129–30. For the commentary attributed to Peter Nightingale, see Pedersen 1983–4: 513–52.

[23] For the treatise by Balduin of Mardochio, see MSS Berlin, Staatsbibliothek, lat. fol. 610, fols. 9r–14v; Cambridge, University Library, Ii.3.3, fols. 11v–19r; El Escorial, Real Biblioteca de San Lorenzo, O.II.10, fols. 12rb–17v; London, BL, Harley 3647, fols. 4ra–9r; Munich, BSB, Clm 5538, fols. 44v–52r; Oxford, BodLib, Tanner 192, fols. 9r–18ra; Paris, BnF, lat. 7420A, fols. 30v–33v; Paris, BnF, lat. 7298, fols. 9bisr–15va, and the manuscripts listed in Thorndike and Kibre 1963: 42; Pedersen 1983–4: 221, 227, 239, 264, 288, 305; Thorndike 1954: 233–4; Cordoliani 1954: 136–7. For John of Brunswick and an earlier *Compotus manualis* written *c*.1275, see Nothaft 2014c.

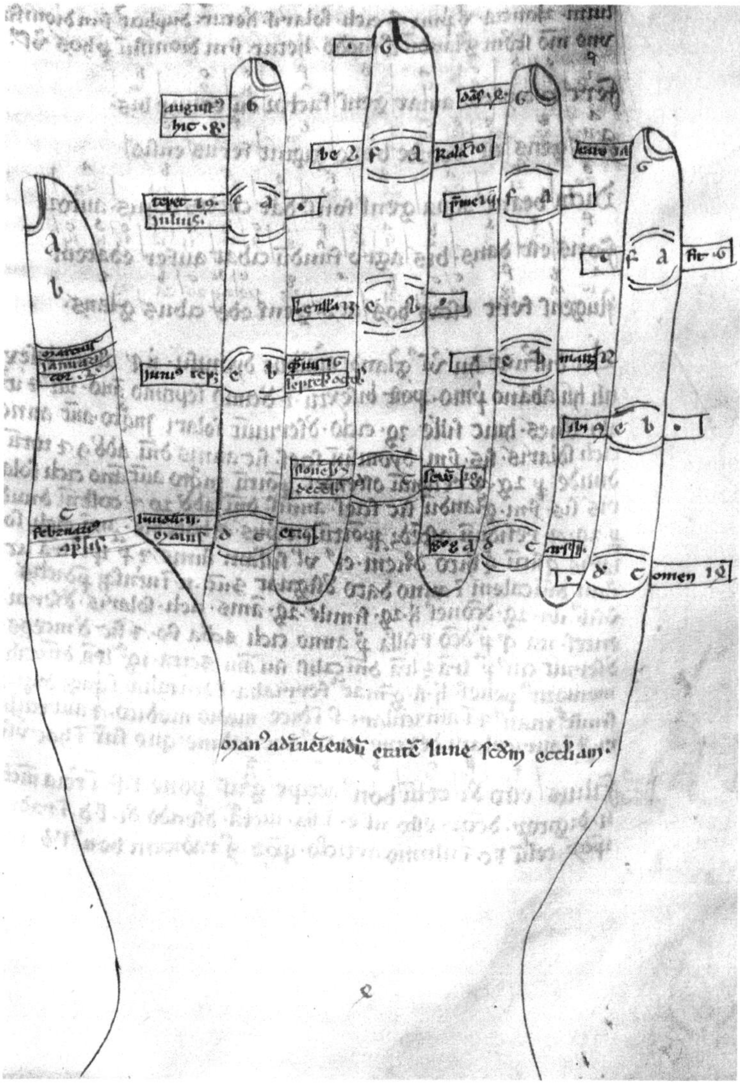

Figure 4.1. Balduin of Mardochio, *Computus manualis* (*c.*1281), hand diagram for finding the Golden Number. Munich, Bayerische Staatsbibliothek, Clm 5538, fol. 47r.
Source: Bayerische Staatsbibliothek

metricus by John of Poland, the *Computus chirometralis* by John Algeri of Erfurt (1330), and another work of this name by the Czech astronomer Christian of Prachatice (1400/30).[24]

The tedium many students must have felt while working through material of this kind was humorously captured by the astrologer, translator, and polymath

[24] Mütz 2003; Silagiová 2007.

Michael Scot, who derived the common thirteenth-century spelling of the word *com-potus* from the drink allegedly shared by the founders of the calendar or from the drinking parties (*compotationes*) necessary to entertain those who were being educated in the art.[25] In his *Liber introductorius*, written between 1220 and 1235 for Emperor Frederick II, Scot offered a farrago of astronomical and computistical doctrines both old and new, which in the case of the famous Munich codex of the work (Padua, *c.*1320) appears to have been inflated by multiple layers of scribal intervention and augmentation.[26] While some passages of the codex endorse the antiquated idea that the vernal equinox occurs on the eighth day before the Kalends of April, as supposedly confirmed by 'horological proof', others indicate that Scot knew of the annual disparity between the Julian calendar and the true solar year, which is here given as 8 moments (= 12 minutes) of an hour, leading to an equinoctial drift of $J = 120$ years.[27] This fairly accurate estimate, which implies a mean solar year of $365\frac{1}{4} - \frac{1}{120}\,\text{days} = 365\text{d }5;48\text{h}$ or 265.241666d, also crops up in a variety of other thirteenth-century sources, including Alexander of Villedieu's popular *Massa compoti*. In a section towards the end of this poem, Alexander versifies the notion that the solar year is 8 moments short of $365\frac{1}{4}$ days, creating a lag of 1 hour every 5 years and 1 day every $5 \times 24 = 120$ years. In line with this shift, the text asserts that by the year 1200 the solstices and equinoxes had all been found 10 days earlier than at the time of Christ's nativity, when the winter solstice had still coincided with 25 December.[28] The passage offers a seductive hypothesis as to how the estimate of $J = 120$ years might have been determined without the involvement of advanced astronomical considerations. If the winter solstice in AD 1200 had indeed been found to occur on 15 December and hence ten days before Christmas, which was the assumed solstitial date at the time of Christ's birth, medieval computists would have been able to infer the size of the error that had caused this discrepancy from the quotient between the number of years since the nativity of the Lord and the observed 10-day discrepancy: $1{,}200 \div 10 = 120$.[29]

[25] See ch. 20 in Michael Scot, *Liber introductorius*, lib. II = *Liber particularis*, as preserved in MS Oxford, BodLib, Canon. Misc. 555, fol. 10vb: 'Et dicitur compotus ex "cum" prepositione et "poto, -as", quod est "bibo, -bis", quia omnes congregati causa inveniendi hanc rationem de impedimento aliquid portionaliter potaverunt et kalendarium constituerunt, ut hodie observatur. Et est sciendum quod compotorum alius phylosophycus et alius vulgaris; phylosophycus est scientia certificandi tempus secundum astronomiam, vulgaris est scientia certificandi tempus secundum usum ecclesie. Vel dicitur compotus a "compoto, -as", vel a "compotando", non quia in compoto agatur de computationibus, sed quia compotationes neccessarie sunt ad doctrinam eorum que in compoto edocentur.' The critical edition in Voskoboynikov 2014: 283 elides the important spelling difference between *computationibus* and *compotationes*.

[26] See Michael Scot, *Liber introductorius*, lib. I = *Liber quatuor distinctionum* (dist. 1–3), MS Munich, BSB, Clm 10268, fols. 1ra–146vb, which was the exemplar for MS Oxford, BodLib, Bodley 266, fols. 1ra–218va (s. XV). For more on Scot and the manuscripts of his *Liber introductorius*, see Thorndike 1961; 1965; Bauer 1983; Edwards 1985; Ackermann 2001; 2009; Voskoboynikov 2014: 249–60.

[27] MS Munich, BSB, Clm 10268, fols. 53ra, 59rb–va, 60r, 62va. See also ibid., fol. 53rb, for a description of the lunar year as comprising $354 + \frac{1}{5} + \frac{1}{6}$ or $354\,\text{days} + 8\,\frac{4}{5}\,\text{hours} (= 354;22\text{d})$, and fols. 64vb–65ra, for a passage on lunar computations based on the Jewish *molad*-system.

[28] Alexander of Villedieu, *Massa Compoti*, ll. 275–89 (ed. Wijk 1936: 58–9).

[29] This is the rationale Kaltenbrunner 1876: 293–4 imputes on 'Master Conrad' (see n. 44 in this chapter). The actual date of the solstice in *c.* AD 1200 was 14 December.

Plausible as this may sound, the issue is complicated to no small degree by the existence of an obscure *Summa de computo*, which was written by a certain Master William in 1163.[30] That this otherwise unknown author was capable of furnishing his subject with a certain degree of originality is conspicuous from his treatment of the history of the Roman calendar, which appears to have no precedents in earlier *computus* literature. Against the standard account found in Macrobius's *Saturnalia*, William asserts that the Julian calendar originally paired 31-day and 30-day months in an orderly sequence, the exceptions being Quintilis and Sextilis, which both only had 30 days. When Julius Caesar decided to rename Quintilis after himself in order to perpetuate his fame, he saw fit to increase the month's dignity by taking away one day from the originally 30-day February and adding it to what was now July. His successor Caesar Augustus, unwilling to be overshadowed by his uncle in this respect, repeated the game by renaming Sextilis into August and snatching yet another day away from February, leaving the latter with only 28 days.[31] This creative just-so story, designed to explain the uneven lengths of the months in the Julian calendar, was subsequently repeated, in a somewhat modified form, in the *Compotus Petri* of 1171 as well as in John of Sacrobosco's work on the topic, which appears to be at the root of similar claims still encountered in the modern literature.[32] When it came to assessing the length of the year presupposed by this calendar, Master William deferred to the wisdom of the Arabic sages, '*Albugosor* and *Arquel*', whom he invoked as witnesses to the alleged fact that the true solar year exceeds $365\frac{1}{4}$ days by a certain amount. According to William, the superabundance grew to a whole day in the course of 120 years and made it so that the winter solstice, which had once occurred at the beginning of January, was now found on St Lucy's Day or 13 December. An equivalent drift had affected the other three cardinal points, which were now all found on the Ides of the respective month—15 March, 13 June, and 13 September. William added that if this change was allowed to continue it was eventually going to cause the winter solstice to occur on the feast of St John (24 June), whereas Christmas was going to be celebrated during the summer. 'But maybe the world will end before that' (*Sed forsitan prius mundus finietur*).[33]

There is an obvious contradiction between William's assumption of an abundant solar year, which should have caused the equinoxes and solstices to fall increasingly late in the Julian year, and his simultaneous assertion that the winter solstice had moved upward from 25 to 13 December. Given this confusion, it is difficult to

[30] The *Summa magistri Willelmi de compoto* is preserved in MSS Oxford, BodLib, Digby 56, fols. 212ra–217ra (the year 1163 is mentioned in a dating clause on fol. 216vb); Paris, BnF, lat. 10358, fols. 273v–283v (the year 1163 here appears on fol. 282v). Its existence was first noted by Haskins 1924: 86.

[31] *Summa magistri Willelmi de compoto*, MSS Oxford, BodLib, Digby 56, fol. 212vb; Paris, BnF, lat. 10358, fol. 275r.

[32] *Compotus Petri*, MS Paris, Bibliothèque Mazarine, 3642, fol. 14v (ch. 7); Sacrobosco, *Computus*, sigs. B8v–Cr; Bono of Lucca, *Computus lunaris* (ed. Arrighi 1991: 9–10). On the influence of Sacrobosco's passage, see Lamont 1919: 585–7. Modern authors who put stock in this legend include Philip 1921: 13–15; Richards 1998: 212; Steinmetz 2011: 27.

[33] *Summa magistri Willelmi de compoto*, MSS Oxford, BodLib, Digby 56, fols. 216vb–217ra; Paris, BnF, lat. 10358, fols. 283r.

tell if William's text had originally spoken of a year length of $365\frac{1}{4} - \frac{1}{120}$ days, as was the case in the *Compotus Petri* written just a few years later (1171),[34] or if the implied year length of $365\frac{1}{4} + \frac{1}{120}$ days had been intended all along. A sidereal year of $365\frac{1}{4} + \frac{1}{120}$ days had indeed been mentioned by Abraham Ibn Ezra, whose *Liber de rationibus tabularum* of 1154 identified it as the year length accepted among the Indians, adding that their tables were later followed by Arabic astronomers such as al-Khwārizmī.[35] It is not clear, however, if and how William's mention of *Albugosor* and *Arquel* relates to this passage. Both names are so heavily garbled that one can only guess at their original referents, with Abū Maʿshar (*Albumasar*) and Azarquiel (*Arzachel*) being two possible candidates. The latter name was generally associated with the Toledan Tables, where the length of the sidereal year was close to $365\frac{1}{4} + \frac{1}{153}$ days. At the same time, it is striking that William should have associated the current dates of the equinoxes and solstices with the Ides of March, June, September, and December. In the case of the winter solstice, this was not 10 but 12 days before Christ's birth on 25 December and still 1 or 2 days earlier than the true solstice, which in the 1160s would have fallen on 14/15 December. For an error of 12 days to accrue over a period of just 1,160 years would have required the mean tropical year to be *c.*$365\frac{1}{4} - \frac{1}{97}$ days in length, which does not agree at all with the statements in William's text.

A more coherent account of the shifting cardinal points, one which offers the same numbers as provided in Alexander of Villedieu's *Massa compoti*, can be found in an anonymous English text simply known as *Compotus ecclesiasticus*, which is essentially a simplified version of, and possible precursor to, Sacrobosco's *Computus*.[36] Its author did not forget to mention that the winter and summer solstices no longer fell on the birthdays of Jesus Christ and John the Baptist, as they had done in antiquity, due to a superabundance of $\frac{1}{120}$ of a day in the Julian calendar.[37] He considered this to be in line with present-day observation, and added two verses to commemorate the 10 days that now separated the equinoxes and solstices from the dates on which Christ and John the Baptist were born (*Solstitium decimo Christum preit atque Iohannem | Nuncia sic matris nox equa diemque Iohannis*).[38] An alternative version of the second verse expressly located the present autumn equinox on the Feast of the Exaltation of the Holy Cross, 14 September, and the vernal equinox on 15 March (*Equa crucis festum dant tempora Martis et idus*). It already appears in a *Computus lunaris* written in 1254 by Bono of Lucca and enjoyed a wider dissemination thanks to its inclusion in William Durand's *Rationale divinorum officiorum*, whose computistical passages, dated 1286, often replicate Bono's text verbatim.[39]

[34] *Compotus Petri*, MS Paris, Bibliothèque Mazarine, 3642, fols. 19ra–b, 20ra (chs. 17–18). See nn. 75 and 77 in Ch. 3.3.

[35] Abraham Ibn Ezra, *Liber de rationibus tabularum* (ed. Millás Vallicrosa 1947: 75).

[36] See Moreton 1994: 236–8; 1995b; 2003: 170–5.

[37] See ch. 13 ('De solstitiis et equinoctiis'), as found in MS London, BL, Add. 27589, fols. 19vb–20rb, and the textual comparison with Sacrobosco's *De anni ratione* in Moreton 1994: 242–4.

[38] MS London, BL, Add. 27589, fol. 20rb; Sacrobosco, *Computus*, sig. E2v. A slightly different version appears in Alexander of Villedieu, *Massa Compoti*, ll. 288–9 (ed. Wijk 1936: 59).

[39] Bono of Lucca, *Computus lunaris* (ed. Arrighi 1991: 8); William Durand, *Ratione divinorum officium* 8.3.25 (CCCM 140B: 140). Durand's year of writing is mentioned in ibid. 8.9.2 and 8.12.8

A variation of the first of the two verses, here changed to mark a gulf of merely 9 days between the summer solstice and St John's birthday (*Solsticium nono Christum preit atque Iohannem*), crops up in a *computus* written by a certain Conrad of Strasbourg,[40] which is noteworthy for quoting the same estimates of the length of the lunar month according to the Latins, Hebrews, and Chaldaeans as Roger of Hereford had done in 1176.[41] For the error of the solar year, Conrad's discussion is almost entirely parallel to that in the *Compotus ecclesiasticus*, including an acceptance of $J = 120$ years. Since the text was probably written no earlier than 1200, a difference of 10 days would have been more congruent with the general theory.

A precise year can be assigned to the *computus* text that served Vincent of Beauvais in drawing up the fifteenth book of his *Speculum naturale*, first completed in 1244. His source discussed the overestimation of the solar year in the Julian calendar in terms very similar to the *Compotus ecclesiasticus*, but was more fastidious in stating that the total error since Christ's nativity had now reached 10 days, 4 hours, and $\frac{3}{5}$ of an hour, which points to the year 1223.[42] A particularly detailed calculation appears in a *Compotus ecclesiasticus secundum Augustinum* of the year 1234, whose author accused Julius Caesar of having added 1 minute or $\frac{2}{3}$ of a moment to the time the Sun actually remained in each of the twelve zodiacal signs, concluding that Caesar's calendar had intercalated a surplus of 10 days, 6 hours, and 32 moments since the start of the Christian era.[43] By 1396 this surplus had begun to exceed $11\frac{1}{2}$ days, as was duly noted in a massive commentary on the *Massa compoti* written in that same year by a certain Master Conrad, who appears to have rounded down from 11 days, 15 hours, and 8 moments.[44] Another, somewhat less precise, comment appears in a Gaelic commentary written in 1406 on John of Brunswick's *Computus manualis* (1289), from which Irish readers could draw the

(CCCM 140B: 161, 171). For a medieval Italian translation of Bono's text, see Pezzella 1982: 99–153 (translated into modern Italian on pp. 65–95). In the anonymous fourteenth-century *computus* of MS Oxford, BodLib, Canon. Misc. 561, fol. 96v, the first verse is changed to 'Solstitium undecimo Christum preit atque Iohannem'. Solstices on 14 June and 15 December are still recorded in the calendar of Abbotsbury Abbey of *c.*1300. See Wormald 1939–46: i. 7, 13.

[40] MS Bruges, Openbare Bibliotheek, 528, fols. 1r–6v, at fol. 1v. Another copy of this text is in MS Trier, Stadtbibliothek Weberbach, 8° 1077/1263, fols. 1r–5v, 9r–10r (s. XIV; the *computus* text on fols. 6r–8v is unrelated).

[41] MSS Bruges, Openbare Bibliotheek, 528, fol. 6r; Trier, Stadtbibliothek Weberbach, 8° 1077/1263, fol. 9v.

[42] Vincent of Beauvais, *Speculum naturale* (15.72), ed. in Vincent of Beauvais, *Speculum quadruplex*, vol. 1 (Douai: Beller, 1624), col. 1137. See Kaltenbrunner 1876: 296–7; Thorndike 1923–58: ii. 457–76. A difference of '10 dies et plus' between the solstices of Christ's time and today is mentioned in a treatise *De spera* probably written by the Genoese astronomer Andalò di Negro (d. *c.*1340). See MS Paris, BnF, lat. 7272, fol. 66vb, ed. Thorndike 1949: 473.

[43] MS Glasgow, University Library, Hunter 467, fols. 21r–44r, at fols. 31v–32r, 33r. The text on fol. 32r has 'X dies et una hora et XXXII momenta', which must be emended to 'X dies et VI hore et XXXII momenta'. For a description of the whole manuscript, see Cordoliani 1949b.

[44] MS Vienna, ÖNB, 3816, fols. 51v–109r, at fol. 98r. In a later section (fols. 99va–100va), Conrad dealt with what is here described as a two-day discrepancy in the lunar calendar. Kaltenbrunner's misreading of this commentary gave rise to the legend that the estimate of $J = 120y$ was first introduced by Master Conrad around AD 1200. See Kaltenbrunner 1876: 293–6; 1878: 388–90, with many repetitions in the literature, e.g. Ginzel 1906–14: iii. 252; Duhem 1913–59: iv. 44; Steinmetz 2011: 46; Ribordy 2017: 32.

information that the solstice had gained 2 more days since the 'old computation' of 1200, meaning that it currently fell 12 days before Christmas, on 13 December.[45] This was St Lucy's Day, which computists had already begun to associate with the 'modern-day' solstice in the twelfth century. It was a tradition so persistent that it continued for centuries in the guise of various vernacular rules, from 'Lucy light, the shortest day and the longest night' to *Luzen macht den Tag stutzen*.[46] By 1406, of course, an accurate measurement would have urged putting the solstice 1 or even 2 days earlier.

4.2. 'PHILOSOPHICAL *COMPUTUS*' AND THE SCIENCE OF THE ARABS

In spite of their occasional criticism of the Julian calendar, most of the *computus* texts mentioned in the preceding section were pedagogical in their orientation. They focused on the conventional elements of the ecclesiastical calendar, and especially on the techniques needed to calculate the mobile feast days, without attempting to add much new astronomical information to the mix. According to a distinction made in the preface to Alexander of Villedieu's *Massa compoti*, texts of this kind could be grouped under the label *computus vulgaris sive ecclesiasticus*, as their authors discussed and divided time in accordance with the established use of the Church. They had to be distinguished from the more sophisticated *computus philosophicus*, which can be regarded as the thirteenth-century successor to the *computus naturalis*. Writings from the latter category divided, or purported to divide, time in an 'infallible' manner (*Compotus philosophicus est sciencia temporis discretiva infallibilis*) and, in doing so, confronted some of the same astronomical subtleties that authors such as Master Cunestabulus and Roger of Hereford had already pondered in the previous century.[47]

One thirteenth-century writer who managed to strike a successful balance between the two branches of the discipline was the aforementioned John of Sacrobosco, whose highly popular text on time reckoning appears as *Computus ecclesiasticus* in a great number of manuscripts, while others label it *Computus philosophicus*.[48] The influence of Roger of Hereford, who had attempted a similar kind of synthesis in the previous century, is revealed in the introductory portion of the work, which draws the line between astronomy and *computus* in the following way: 'Whatever the astronomer determines regarding the fractions of time, he does for the sake of celestial motions. For his is to consider motion itself, whereas the computists

[45] MS London, BL, Cotton App. LI, fol. 7v. See O'Grady 1926: 325. [46] Albertini 1967.

[47] Alexander of Villedieu, *Massa Compoti* (ed. Wijk 1936: 52). See also Michael Scot, *Liber particularis*, MS Oxford, BodLib, Canon. Misc. 555, fol. 10vb; John of Brunswick, *Compotus novus*, MS Leiden, Universiteitsbibliotheek, Scaliger 66, fol. 9rb–va. Silagiová 2013: 306–7, cites further examples, but erroneously subsumes the works of Bede and Rabanus Maurus under the category of 'philosophical' *computus*. The distinction does not apply to early medieval works.

[48] See n. 17 in this chapter.

[deal with] the divisions of time.'[49] One hallmark that justifies the designation of Sacrobosco as a 'philosophical' computist was his willingness to deal with the error of the lunar calendar, a topic studiously avoided by most authors on the vulgar or ecclesiastical side of *computus*, who were mindful of Dionysius Exiguus's claim that the Council of Nicaea had declared the 19 Easter full moons 'firm and immovable'.[50] In one of the later chapters of his treatise, Sacrobosco used data culled and rounded from Ptolemy's *Almagest* to argue that 235 lunations were approximately $1\frac{1}{3}$ hour short of 19 Julian years, with an implied error of $M = 342$ years. Over the more than 1,230 years that separated the birth of Christ from Sacrobosco's year of writing this had led to an accumulated difference of 3 days and 14 hours, with the result that the day marked as *luna* 1 in the calendar was frequently the fourth day of the Moon in reality.[51]

Another 'philosophical' trait of John of Sacrobosco's *Computus* was its preference for the sexagesimal division of time that characterized astronomical works in the Graeco-Arabic tradition. During the twelfth century, even progressive computists such as Cunestabulus and Roger of Hereford had shunned this new method, preferring instead to express the length of the mean lunations in the Latin, Jewish, and Arabic calendars with the moments and atoms already familiar to Hermann of Reichenau. Sacrobosco switched to sexagesimal fractions, but rendered them more congenial to computistical use by applying them not to the day but to the hour, thereby inaugurating the use of hours, minutes, and seconds that still prevails in the twenty-first century.[52] In Sacrobosco's case, such minutes and seconds of the hour came in handy especially when writing about the length of the tropical year, a period he defined as the Sun's return to one of the four equinoctial or solstitial points. For the relevant numerical values he again had recourse to Ptolemy, more specifically to his tropical year of $365\text{d } 5;55,12\text{h} = 365\frac{1}{4} - \frac{1}{300}$ days.[53] Dividing the year by 12, Sacrobosco obtained a mean solar month of 30d 10;29,36h, which was 24 seconds below the average month of the Julian year (30d 10;30h). The discrepancy between the Julian and the tropical year hence amounted to $0;0,24\text{h} \times 12 = 0;4,48\text{h}$ or 288 seconds. In a less than meticulous move, Sacrobosco equated these 288 seconds with $\frac{1}{12}$ rather than $\frac{2}{25}$ of a complete hour, asserting

[49] See Sacrobosco, *Computus*, sig. Br–v, based on Roger of Hereford, *Compotus* 5.25 (CCCM 272: 223, ll. 43–6).

[50] Dionysius Exiguus, *Epistola ad Petronium* (ed. Krusch 1938: 63).

[51] Sacrobosco, *Computus*, sigs. F2r–F3r. The printed edition gives the number of years elapsed as 1,232, whereas many manuscripts have 1,235, which is exactly 65×19 years. See e.g. MSS London, BL, Harley 3647, fol. 48ra; London, BL, Royal 12.C.IX, fol. 11vb, and the quotation in Nicholas of Cusa, *De correctione calendarii* 3 (ed. Stegemann 1955: 22). Either number would be compatible with the calculation offered, though according to Alfred Lohr (private communication) the best manuscript witnesses agree with the printed edition in giving 1232. The contrary was argued by Schulz 1945: 271; Pedersen 1985: 189. John's dependency on Ptolemy is clear from his value for the daily mean motion of the Moon in longitude, which is 13;10,34,58,33,30,30° (exactly as in *Almagest* 4.3), leading to a sidereal month of 27d 7;43,7,15,36,55h. See Sacrobosco, *Computus*, sig. E5v.

[52] See Thorndike 1923–58: iii. 123–4; Moreton 1994: 238–9.

[53] Sacrobosco, *Computus*, sig. C8r. A brief text in MS Oxford, BodLib, Ashmole 341, fols. 111r–112r (s. XIII$^{3/4}$), bases itself on Sacrobosco, but offers a much more detailed analysis of the length of the solar and lunar years. It begins by showing how to derive the length of the year (here: 365d 5;55,12,6,3,7h) from the daily solar motion (0;59,8,17,13,10,4°).

that $J = 288$ years (since $288 = 12 \times 24$). He thus more than halved the rate of retrogression established by earlier computists ($J = 120$ years), but was still happy to accept their frequently uttered statement that the winter solstice currently preceded Christmas by a full 10 days. As a result of this combination of commitments, he saw himself forced to argue that Christ and John the Baptist had not been born on the day of the solstice, but six days after it, given that 1,200 years contained no more than four complete 288-year periods.[54] This bold departure from received tradition, which amounted to putting the solstice in the year of Christ's birth on 19 December, may explain why Sacrobosco's argument was often ignored by later computists familiar with his text. These included a certain Master Gordianus, who, in spite of his heavy reliance on Sacrobosco, preferred to espouse the standard argument, with an equinoctial drift of $J = 120$ years and a total gap of 10 days.[55]

Another work Gordianus tacitly recycled for the purposes of his own *Compotus*, which dates from 1269–78,[56] was the *Compotus* composed during the 1220s by Robert Grosseteste (1168/75–1253), who took the incorporation of Arabic astronomical influences into computistical theory even further than Roger of Hereford had done half a century before him.[57] In a manner similar to Roger and Cunestabulus as well as some of his own contemporaries, Grosseteste defined the discipline of *computus* as a 'science' of counting and dividing time (*computus est scientia numerationis et divisionis temporum*).[58] Rather than subdividing this science into a *computus naturalis* and a *computus vulgaris*, however, his work reduced

[54] Sacrobosco, *Computus*, sig. E2r–v. See also the comparison of the relevant passages in Sacrobosco and the English *Compotus ecclesiasticus* in Moreton 1994: 242–4.

[55] MS Lüneburg, Ratsbücherei, Ms Miscell. D 4° 46, fols. 25r–42r, at fol. 33rb–va. Kaltenbrunner 1876: 308–9 used the copy in MS Vienna, ÖNB, 5239, fols. 35r–52v, which derives from the Lüneburg codex. A magnificently produced copy of the same work appears in MS Munich, BSB, Clm 17703, fols. 27v–83v (cf. ibid., fols. 64r–65v). It breaks off several pages before the proper ending and lacks an ascription. The manuscript also includes a list of the bishops of Prague ending with Johann of Dražic (fol. 89v), whose year of death, 1278, may be taken as *terminus ad quem* for the *Compotus* contained in it.

[56] In MS Vienna, ÖNB, 5239, fol. 43v, Gordianus's discussion of the Julian error states that more than 1,300 years had gone by since Christ. Kaltenbrunner (1876: 309) inferred from this passage that Gordianus wrote after 1300 but prior to 1320, since the total error would have reached 11 rather than 10 days after that date $(1320 \div 120 = 11)$. The exemplar in MS Lüneburg, Ratsbücherei, Ms Miscell. D 4° 46, fol. 33rb, instead shows that the text originally had 1,200 years, which was only later corrected to 1,300 and furnished with a marginal note: 'Istud quod hic dicitur verum fuit eo tempore quo liber iste compositus fuit. Iam autem sunt anni 1301 a nativitate domini transacti.' Gordianus used a copy of Robert Grosseteste's *Compotus* that incorporated tables from the *Compotus* of Campanus of Novara written in 1269 (see nn. 79–80 in this chapter). This offers a *terminus post quem* in addition to the *terminus ad quem* posed by the earliest known manuscript (see n. 55).

[57] The Grossetestian *Compotus* survives in 38 manuscripts and was edited (deficiently) by Steele 1926: 212–67, on the basis of MSS London, BL, Add. 27589, fol. 77ra–100vb; London, BL, Harley 3735, fols. 82va–105rb. A new critical edition and English translation by Alfred Lohr and myself is in preparation. On the text and its manuscripts, see Baur 1912: 65*–68*; 1917: 46–61; Welborn 1932: 24–33; Thomson 1940: 96; McEvoy 1983: 619–20, 655; Borst 1984: 468–9; 1998: 75, 339–40, 701, 720; Southern 1986: 131, 139; Dales 1989: 77–8; McCluskey 1998: 198–9; Panti 2001: 47–9; Dowd 2003: 219–80, 292, 306, 308. On the misascription of other computistical texts to Grosseteste, see Moreton 1995b; 2003.

[58] Robert Grosseteste, *Compotus* 1 (ed. Steele 1926: 213, l. 6). See also Cunestabulus, *Compotus* 1 (CCCM 272: 61, ll. 2–3); Roger of Hereford, *Compotus* 1.1 (CCCM 272: 133, ll. 7–10); Alexander

the familiar distinction to a more general one between ways of measuring and dividing time that were oriented towards the natural motions of celestial bodies and ways that were grounded in the customs and conventions prevalent in a particular society or religion.[59] Grosseteste's discussion of the solar year, as found in the first of twelve chapters, prioritized the former approach and hence began by outlining the difference between the tropical and the sidereal year, which involved distinguishing between the fixed zodiac of the ninth and the mobile zodiac of the eighth sphere. In trying to establish a measure for the former type of year, which corresponded to the Sun's return to a point on the fixed zodiac (such as an equinox or solstice), the 'founders of our calendar' had strayed from the truth by following the opinion of Hipparchus, who believed it to last $365\frac{1}{4}$ days. This compared unfavourably with Ptolemy, who subtracted $\frac{1}{300}$ of a day, and even more so with al-Battānī, who recognized the year to be only $365\frac{1}{4}-\frac{1}{100}$ days.[60] In Grosseteste's estimation, the latter value agreed 'more strongly with what we find from contemporary experience concerning the amount by which the solstice goes ahead', as the solstice now allegedly preceded 25 December by as many days as centuries had gone by since Christ's nativity.[61]

Far from remaining content with this straightforward result, however, Grosseteste proceeded to challenge his previous reliance on the tropical definition of the year by offering a brief description of the theory of access and recess commonly attributed to Thābit ibn Qurra, which implied that only the sidereal year could be treated as a fixed measure, whereas the tropical year underwent permanent variation. To Thābit, Grosseteste attributed a sidereal year of $365\frac{1}{4}$ days + 23 seconds of a day, or 365;15,23d, which agreed with the value implicit in the Toledan Tables (*c.*365;15,23,29d).[62] He also pointed out that a calendar based on this year length would need to intercalate an additional day every 156 years, but then subtract one day at the end of 46,800 years in order for the head of the year to stay aligned with the appropriate reference star. What a sidereal calendar of this type could not offer was a fixed set of dates for the equinoxes and solstices. Depending on which 90° segment of its circular path the head of Aries was currently travelling, a given solstice would increasingly either anticipate or fall behind the date initially

of Villedieu, *Massa compoti* (ed. Wijk 1936: 52); Michael Scot, *Liber introductorius*, MS Munich, BSB, Clm 10268, fol. 53rb; Sacrobosco, *Computus*, sig. Br.

[59] Robert Grosseteste, *Compotus* 1 (ed. Steele 1926: 213, ll. 7–22). In line 10, read *religionum* for *regionum*.

[60] Ibid. (ed. Steele 1926: 213–15). Grosseteste's likely source for these historical estimates was the *Almagestum parvum*, where al-Battānī's year is given as $365;14,24d = 365\frac{1}{4}-\frac{1}{100}$ days instead of $365;14,26d = 365\frac{1}{4}-\frac{1}{106}$ days. See *Almagestum parvum*, bk. III.1, in MS Paris, BnF, lat. 16657, fol. 96v (https://ptolemaeus.badw.de/ms/2/96v), and the study of the text by Zepeda 2015. The claim that Hipparchus's year was $365\frac{1}{4}$ days long already appears in al-Battānī's *De motu stellarum*, trans. Plato of Tivoli (Nuremberg: Petreius, 1537), fols. 26v, 80v (chs. 27, 52).

[61] Robert Grosseteste, *Compotus* 1 (ed. Steele 1926: 215, ll. 29–33). The actual winter solstice at Grosseteste's time of writing (*c.*1225) fell on 14 December (14/15 December according to the Toledan Tables).

[62] As with his values for the tropical year (n. 60), Grosseteste probably took this figure from the *Almagestum parvum*, bk. III.1 (MS Paris, BnF, lat. 16657, fol. 97r), where it is linked both to Thābit and the Toledan Tables ('...et super hoc Arzacel tabulas motuum Toleti novissime composuit').

assigned to it, although the cyclical nature of this process guaranteed that everything would return to start after a full period of access and recess.[63] For all the mathematical complications it may have caused, Thābit's model hence reassured its Christian users that the current overestimation of the tropical year length by the Julian calendar was only temporary in nature. In the long run, the two years were bound to regain their former congruity, if only for a short time, and even the vernal equinox could be expected to return to its ancient and canonical dates (25 or 21 March) with the passage of time. If this was one reason to look favourably at the model in question, another was suggested by Grosseteste's widely copied treatise *De sphera* (*c.*1215–20), in which he commented on the consequences the eccentricity of the Sun's circular path implied for life in the southern hemisphere, which was rendered uninhabitable due to the location of the Sun's perigee in Sagittarius:

> For when the Sun is at the perigee it is five degrees closer to the Earth than when it is at the apogee. For when the Sun is found in the southern signs it draws very near to the Earth and thus, when it is directly above the southern regions, the cause for the heat during their summer is doubled. But when it is in the northern signs, it draws away from the zenith of the southern regions and grows distant from the Earth, which doubles the cold in their dwelling place, whereas when the Sun approaches the zenith above our heads, it grows distant from the Earth, and when it draws away from our zenith, it comes close to the Earth. And this is the reason why the northern region is temperate.[64]

For Master Cunestabulus, who had already dealt with this issue in his *Compotus* of 1175, the insight that the Sun's position at perigee caused the sea beyond the equinoctial circle to 'seethe like a kitchen pot' was welcome news, since it closed the door behind the theologically troubling possibility that the southern hemisphere might be populated by Antipodeans.[65] Grosseteste, for his part, was interested in a different aspect of this particular application of astronomy to geography, which was related to the discovery, made in the ninth century by the Banū Mūsā and expounded in the treatise *De anno solis* (see Ch. 3.3), that the Sun's line of apsides moved along with the sphere of the fixed stars. If this was indeed the case and precession was a linear process, as Ptolemy had believed, it was only a matter of time until the perigee made its way from the southern to the northern hemisphere, rendering humanity's present habitat unsuitable for life.[66] In what appears to have been a conscious reaction to this long-term prospect, Grosseteste went on to outline in some detail the theory of pseudo-Thābit, who imagined that the whole sphere of

[63] Robert Grosseteste, *Compotus* 1 (ed. Steele 1926: 215–17).

[64] Robert Grosseteste, *De sphera* (ed. Panti 2001: 309, ll. 376–85). The origin, date, and content of Grosseteste's *De sphera* are discussed in Thomson 1940: 115–16; Thorndike 1949: 10–14; McEvoy 1983: 617–18, 655; Southern 1986: 142–6; Panti 2001: 67–132; 2003: 135–44; Dowd 2003: 148–99.

[65] Cunestabulus, *Compotus* 12 (CCCM 272: 75, ll. 27–51). The background to this passage is discussed in Nothaft 2017a: 33–8.

[66] Robert Grosseteste, *De sphera* (ed. Panti 2001: 311–12, ll. 420–8). Grosseteste here mistakenly ascribes the idea of apsidal precession to Ptolemy, who had instead assumed the line of apsides to be tropically fixed.

the fixed stars as well as the apogees of all the planets were moved in accordance with the circular motion of the heads of Aries and Libra. Although Grosseteste did not make this point explicit, one of the consequences of the described model was that it did not allow the perigee to stray from its mean position in the southern hemisphere by more than 10;45° in either direction. The eighth sphere's access and recess could therefore be regarded as insurance against the kind of global climate change a linear model of precession held in store.[67] Grosseteste's *Compotus* did not return to this cosmological point, but instead confronted the theory of access and recess with the opposing doctrine of Aristotle, according to whom:

> these aforementioned theories of celestial motions are possible only in the [geometrical] imagination, but impossible in nature, because according to him all nine spheres are concentric and each sphere has its own motion from east to west around its axis and poles.[68]

It would appear that Grosseteste felt motivated to include this statement in his work because he had recently been made aware of the ideas of Nur ad-Din al-Biṭrūjī, an Andalusian author of the last quarter of the twelfth century, whose *Kitāb fi-l-hay'a* ('Book on astronomy') had recently become available to Latin readers in a translation made by Michael Scot in Toledo in 1217. Eager to uphold the Aristotelian dogma of strictly homocentric celestial motions against Ptolemy's system of epicycles and eccentric deferents, al-Biṭrūjī invested a lot of ingenuity in explaining anomalous planetary motions in terms of the spirals that resulted if the diurnal rotation of the outermost sphere transmitted its force downwards to planetary spheres that moved around their own sets of poles.[69] Grosseteste gave no detailed account of al-Biṭrūjī's attempt, but simply remarked that the Arabic astronomer 'recently invented a way and explained how it is possible to save the forward motion and stations and retrogradations of the planets...without eccentrics or epicycles'.[70] Whether he found the resulting models convincing or not, the English philosopher was well aware that al-Biṭrūjī had simply taken his mean motion parameters from Ptolemy and that his work accordingly added nothing of relevance to the calendrical issues at hand.[71] This may be the reason why Grosseteste remained silent on al-Biṭrūjī's model for the irregular motion of the fixed stars, which sought to replace Azarquiel's access-and-recess theory, but failed to provide any of the quantitative data a computist might have been interested in.[72]

Compared with the complications wrought by the solar year, an astronomical analysis of the errors of the lunar calendar looked like a relatively straightforward affair. Like Reinher of Paderborn before him, Grosseteste deemed it unnecessary

[67] See Robert Grosseteste, *De sphera* (ed. Panti 2001: 312–14, ll. 429–71), and Nothaft 2014b: 49–51. Grosseteste, in his description of the access-and-recess model, confused the maximum elongation of 0° Aries, which is ±10;45°, with the radius of the circle on which it travelled, which is 4;18,30°.

[68] Robert Grosseteste, *Compotus* 1 (ed. Steele 1926: 217, ll. 9–13).

[69] See the edition of Michael Scot's Latin translation in Carmody 1952, and the study of its thirteenth-century reception by Avi-Yonah 1985.

[70] Robert Grosseteste, *Compotus* 1 (ed. Steele 1926: 217, ll. 21–4).

[71] Ibid. (ed. Steele 1926: 217, ll. 27–33). See Avi-Yonah 1985: 130–1; Laird 2012: 135–8.

[72] Al-Biṭrūjī's model for the fixed stars is elucidated by Mancha 2004.

and unhelpful to try and determine the true time of the conjunction or to calculate the new moon based on the fluctuating true length of the synodic month. Having assured his readers that the differences between mean and true syzygy would cancel out in the long run, with 251 mean lunations being exactly equal to 251 true ones, he insisted that 'both because of the non-uniform true motion of the Sun and because of the true motion of the Moon, which is even less uniform, it would be much too longwinded and laborious to define the precise points in time of successive true lunations'. Investigations of the precise time of the true conjunction were therefore best 'left to those who find the locations of all stars as a function of time using astronomical tables', which fell outside the remit of the *computus*.[73] As far as the mean interval between two conjunctions was concerned, astronomical tables had long supplied computists with two alternative values: the 29;31,50d underlying the Arabic lunar calendar and the 29;31,50,8,20d used in most tables for mean syzygy as well as the Jewish calendar. Grosseteste, who based himself on the Toledan Tables, treated the former value as Azarquiel's deliberate truncation of the more precise one. For the latter, he cited the pseudo-Ptolemaic variant 29;31,50,8,9,20d, as found in interpolated Arabic and subsequent Latin versions of the *Almagest*.[74] One of the principal points Grosseteste set out to prove in his *Compotus* was that neither of these values fit into 19 Julian years a whole number of times. Accepting the simplified Arabic value (29;31,50d), the discrepancy relative to the ecclesiastical 19-year cycle turned out to be:

per 19-year cycle: 0;4,10d (= 1;40h)
per 76-year cycle: 0;16,40d (= 6;40h)
per 4×76 years $= 304$ years: 1;6,40d (= 1d 2;40h).[75]

For the pseudo-Ptolemaic value, which added sexagesimal *tertiae*, *quartae*, and *quintae*, the results came up as:

per 76-year cycle: 0;14,32,13,46,40d (= 5;48,53;30,40h)
per 4×76 years $= 304$ years: 0;58,8,55,6,40d (= 23;15,34,2,40h).[76]

Here, too, the deviation from the 19-year cycle approached something close to $M = 304$ years, which number was to become a staple element in subsequent debates. The question remained how one was supposed to interpret the discrepancy in question. Ardent defenders of the ecclesiastical *computus*, who shared Cunestabulus's bent for apologetics, might have pointed to Ptolemy's own

[73] Robert Grosseteste, *Compotus* 6 (ed. Steele 1926: 240, ll. 14–24). See already Reinher of Paderborn, *Compotus emendatus* 1.23 (CCCM 272: 34–5).

[74] Robert Grosseteste, *Compotus* 4 (ed. Steele 1926: 232, ll. 9–24). The value derives from Ptolemy's equation of 126,007d 1h with 4267y (*Almagest* 4.2), which implies 29;31,50,8,9,20d rather than 29;31,50,8.20d. See Rawlins 2002; Depuydt 2002: 92–4; Mancha 2002–03; Goldstein 2003: 67–8.

[75] Robert Grosseteste, *Compotus* 4 (ed. Steele 1926: 233, l. 17–235, l. 5). Grosseteste adds that it will take $14 \times 304 = 4{,}256$ years for the calendrical new moon to end up on the day of the full moon.

[76] Ibid. (ed. Steele 1926: 235, ll. 6–19). Steele's text, which is based on MS London, BL, Add. 27589, fol.87ra, gives the second value as 0;58,8,4,46,40d. The correct value is preserved in a majority of witnesses, e.g. MS Florence, Biblioteca Riccardiana, 885, fol. 234r.

admission (*Almagest* 4.2) that the determination of the mean synodic month was highly sensitive to observational error. Grosseteste was aware of this sort of argument, but defused it by recalling the successful predictions of eclipses astronomers in his own day were able to make by using Ptolemy's and Azarquiel's tables. This ongoing success proved that it was the Christians who had chosen the wrong length of the lunar month, based on an inadequate equation of 940 months with 27,759 complete days.[77] Grosseteste solidified this point by showing that the Arabic and the Latin ecclesiastical calendar could be joined together in a cycle that lasted $10{,}631 \times 27{,}759 \div 354;22 = 832{,}770$ Arabic years or $10{,}631 \times 27{,}759 \div 365;15 = 807{,}956$ Julian years. This gigantic time span comprised $27{,}759 \times 360 = 9{,}993{,}240$ times the Arabic lunation of 29;31,50 days, but only $10{,}631 \times 940 = 9{,}993{,}140$ times the average Latin lunation of 6939; $45 \div 235 = 29;31,51,3,49,47,14,2$ days, which made for a discrepancy of 100 lunar months.[78] Alarming as this result certainly was, it also pointed the way to an attractive alternative:

> We can always find the day of the new moon according to the astronomical truth if we reckon the times according to the years and months of the Arabs, because the first day of any given month of the Arabic years is the day of the conjunction of Sun and Moon according to either [luminary's] mean course.... Knowledge of the Arabic years and the beginnings of the months of the Arabic years will hence give us true knowledge of the dates of the new moon.[79]

Readers willing to follow Robert Grosseteste on this point were offered a chapter's worth of tables and excerpts drawn from the Toledan Tables (or 'tables of Azarquiel') and their accompanying canons, which provided the information necessary to convert Arabic into Christian dates, and vice versa. Key elements included a table of collected years, the first line of which conveniently showed that the first 600 years of the Arabic calendar (as reckoned from the epoch of the era of the Hijrah, which astronomers put on 15 July AD 622) contained 1,203 completed years of the Christian era as well as 7 months of 30 days, another 29 days, and 2 quarters of a day (see Figure 4.2).[80] The total number of days was $239\frac{1}{2}$, pointing to 27 August

[77] Robert Grosseteste, *Compotus* 4 (ed. Steele 1926: 235, ll. 20–30).

[78] Ibid. (ed. Steele 1926: 236, ll. 5–35). For Grosseteste's detailed discussion of the average lunation in the Latin calendar, see ibid. 7 (ed. Steele 1926: 241–2).

[79] Ibid. 4 (ed. Steele 1926: 237, ll. 1–13). This passage, together with some of the surrounding text, was copied into the *Computus* of Master Gordianus (see n. 55), who used Grosseteste's advocacy of the Arabic calendar to replace John of Sacrobosco's correction of the 19-year cycle based on Ptolemy. See MSS Lüneburg, Ratsbücherei, Ms Miscell. D 4° 46, fol. 35rb–va; Munich, BSB, Clm 17703, fols. 70v–71r. The excerpt ends with a reference to the 'tabulas Arzachelis et magistri Campani infra scriptas', which were apparently both included in the Grosseteste manuscript the compiler had seen (see next footnote).

[80] Robert Grosseteste, *Compotus* 5 (ed. Steele 1926: 237–40). The instructions contained in this chapter are taken from an early set of canons to the Toledan Tables, for which see Pedersen 2002: 193, 198, 228, 230. These canons are also included in MS Oxford, BodLib, Savile 21, fols. 27ra–40va, which contains sections in Grosseteste's hand (fols. 142v–160v). Tables similar to those in Grosseteste's work circulated both as part of the Toledan Tables (ed. Pedersen 2002: 907–15) and the widely diffused eclipse tract 'Ut annos Arabum', which dates from the third quarter of the thirteenth century (ed. Pedersen 2002: 552–68, 916). Note that later copies of the Grossetestian

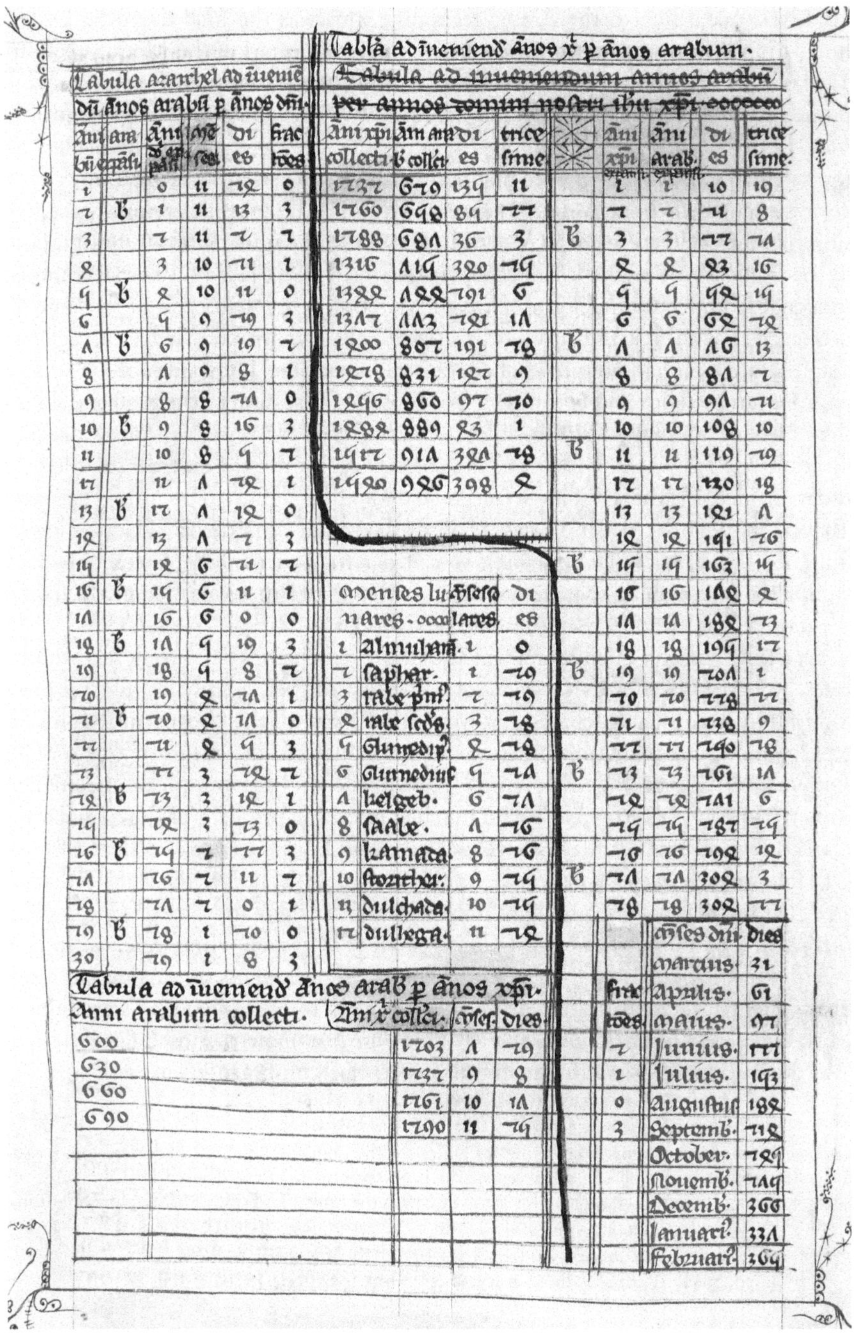

Figure 4.2. Robert Grosseteste, *Compotus* (1217/32), conversion tables for the Arabic lunar calendar. Oxford, Bodleian Library, Laud. Misc. 644, fol. 155r.

AD 1204. According to the Arabic reckoning, this was the final day of a 29-day lunar month, whereas in the ecclesiastical lunar computation the Moon was only 27 days old that day. In this way the table confirmed Grosseteste's more general point according to which the full moon in present times coincided with calendrical *luna* 12 or 13, but never with *luna* 14, as presumably envisaged by the makers of the calendar. 'And this becomes manifest through lunar eclipses, which nowadays always occur before the Moon is fourteen days old.'[81] This had serious repercussions for the calculation of the date of Easter, as did the undeniable overestimation of the tropical year by the Julian calendar. The latter had caused the vernal equinox to recede from its original location on 21 March, where it had been 'at the time of the teachers who first transmitted this set of rules, whom knowledge of the true length of the year had not yet reached, nor [did they have] knowledge of the anticipation of the solstice and equinox'.[82] For those in possession of the requisite skills, the Toledan Tables made it possible to confirm that the vernal equinox's present location was 14 March, which was the correct result once the sidereal solar longitudes yielded by these tables were converted into tropical ones using pseudo-Thābit's model of access and recess.[83] Since the vernal equinox provided the lower limit for the Easter full moon, the so-called *terminus paschalis*, Grosseteste was compelled to conclude that it was 'necessarily the case that we often err in fixing the day of Easter and of the other mobile feast days'.[84]

Of all the texts on calendar reform and improvement written during the Middle Ages, Robert Grosseteste's *Compotus* was the first to exert a truly lasting and profound influence on the debates that were to follow until the Gregorian Reform of 1582. Among the many prominent names who invoked the authority of *Robertus Lincolniensis* during the intervening centuries was the Dominican theologian Robert Holcot (*c*.1290–1349), who touched upon the problem of the calendar in his lectures on Peter Lombard's *Sentences*, given at Oxford between 1331 and 1333. Of Holcot's *quaestiones* on Book II, one was entitled: 'Whether the stars were created that by their motion and light they might be for signs and times' (*Utrum stelle sint create ut per lumen et motum sint in signa et tempora*), which referred to the creation of the Sun and Moon in Genesis 1:14. His thoughts on this matter were of sufficient interest to fourteenth-century readers to be copied as a stand-alone work known as *Tractatus de stellis*.[85] In this *Tractatus* Holcot quoted Grosseteste's assertions about the length of the solar year, but combined them with a more

Compotus sometimes add a second group of tables for the opposite direction of conversion, based on collected-year cycles of 28 Julian years, which correspond to those devised by Campanus of Novara (see n. 109 in this chapter). This is the case with the right-hand side of MS Oxford, BodLib, Laud. Misc. 644, fol. 155r, as shown in Fig. 4.2.

[81] Robert Grosseteste, *Compotus* 10 (ed. Steele 1926: 259, ll. 4–6).

[82] Ibid. (ed. Steele 1926: 259, ll. 22–4).

[83] Ibid. (ed. Steele 1926: 258, ll. 27–30).

[84] Ibid. (ed. Steele 1926: 259, ll. 7–10).

[85] See e.g. MS Oxford, BodLib, e Mus. 167, fols. 19v–41r. Holcot's *Tractatus de stellis* was summarized by Thorndike 1957a, who treated it as a separate work. Its original context as part of the *quaestiones* on Book II was established by Vennebusch 1978. See also Streveler and Tachau 1995: 39 (n. 93), 198. It appears as *quaestio* no. 5 on Book II in MS London, BL, Royal 10.C.VI, fols. 7ra–141va, at fols. 64va–76rb.

detailed exposition of pseudo-Thābit's access-and-recess model than his source had been willing to provide. This model, as Holcot was well aware, strongly suggested that the solar year was variable and that no stable set of numbers could be assigned to it.[86] Grosseteste's sympathies instead lay with the fixed tropical year espoused by al-Battānī, whose estimate of $365\frac{1}{4}-\frac{1}{100}$ days seemed to agree best with the calendrical regression of the winter solstice since Christ's nativity.[87] An earlier reader of Grosseteste's work who showed himself preoccupied with understanding the nature and length of the solar year was the Italian astronomer Campanus of Novara (d. 1296), whose *Compotus maior* dates from the year 1269.[88] Grosseteste's influence on Campanus's work is immediately evident from the opening of Chapter 1, which repeated the English philosopher's definition of *computus* as the 'science of counting and dividing time', but not without limiting its remit to the modes of reckoning in common or vulgar use (*vulgares observantias*), as opposed to the intellectually more demanding divisions of time employed by astronomers. Adding to this distinction, Campanus pointed out that 'different nations divide time in different ways', which meant that the Latin ecclesiastical *computus* was only one of several potential forms the discipline could take, quite apart from the fact that the Church owed most of its own version to the 'ancient nations who were the inventors of these divisions and the founders of the calendar we have in use'.[89]

Before he could proceed to discuss the calendars and years known to other nations, however, Campanus found it necessary first to devote one lengthy chapter to the solar year from an astronomical perspective, in which he considerably expanded upon the remarks Robert Grosseteste had made at the beginning of his *Compotus*.[90] The lion's share of his treatment was accorded to the theory of access and recess, the reliability of which could be tested by appealing to the widespread notion that Christ's nativity on 25 December had originally coincided with the winter solstice. Grosseteste had already used this historical assumption to argue for al-Battānī's estimate of the solar year ($365\frac{1}{4}-\frac{1}{100}$ days) in preference to Ptolemy's ($365\frac{1}{4}-\frac{1}{300}$ days), but the implication of his argument was that the solstice had regressed by twelve days since the beginning of the Christian era. Without specifically criticizing the bishop of Lincoln, Campanus offered a diverging analysis of this

86 See the seventh 'articulus' in MS Oxford, BodLib, e Mus. 167, fol. 31v–33v, which partly reproduces Robert Grosseteste, *Compotus* 1 (ed. Steele 1926: 215–16).

87 MS Oxford, BodLib, e Mus. 167, fol. 33v.

88 See the two early modern printed editions of the *Compotus maior* in *Sphera cum commentis* (Venice: Scotus, [19 Jan.] 1518), fols. 159r–177r, and *Sphera mundi noviter recognita* (Venice: Giunti, [30 June] 1518), fols. 159ra–177rb. I will use the latter for the citations that are to follow, which I have checked against MS Oxford, BodLib, Canon. Misc. 561, fols. 1r–38v. The same manuscript contains a copy of Campanus's *Compotus abbreviatus* (fols. 44r–53v), which rearranges the chapters on the vulgar *computus* while leaving out the astronomical content. For biographical information on Campanus of Novara, see Paravicini Bagliani 1973; 1991: 85–115; Busard 2005: 41–5.

89 Campanus of Novara, *Compotus maior*, fol. 159rb (ch. 1). Most of this passage was recycled for the introduction to the *Compotus abbreviatus*, MS Oxford, BodLib, Canon. Misc. 561, fol. 44r.

90 See Campanus of Novara, *Compotus maior*, fols. 161va–4vb (ch. 10). A brief discussion of the lengths of the solar year among various nations appears at the end of this chapter (fol. 164va), which also circulated independently of the *Compotus maior*, e.g. in MSS Florence, Biblioteca Nazionale Centrale, Conv. Soppr. J.X.40, fols. 28v–46r; Oxford, BodLib, Digby 178, fols. 9r–10v (extract); Utrecht, Universiteitsbibliotheek, 725 (shelfmark: Eccl. 464), fols. 193v–199v (extract).

shift, one which assumed that the solstice currently fell on 15 December. If this was the case, then the solstice in the year of Christ's nativity should have taken place on 27 December, if al-Battānī was right (1,200 ÷ 100 = 12), or on 19 December, if one instead accepted Ptolemy's value (1,200 ÷ 300 = 4).[91] A better result followed from using the Toledan Tables, which put the true vernal equinox and winter solstice in AD 1 on the dates that preceded Jesus's annunciation by the angel Gabriel (25 March) and his birth in Bethlehem (25 December).[92] No wonder, then, that the majority of astronomers in Campanus's day favoured the Toledan model of the access and recess of the eighth sphere, which made the sidereal heads of Aries and Libra revolve on small circles centred on the celestial equator, with the consequence that the vernal point oscillated between sidereal longitudes of 19;15° Pisces and 10;45° Aries. The completion of the whole cycle took 1,481,784 days, 18 minutes, and 44 minutes, which was the same as 4,056 Julian years and 330;46,50d.[93] It was a motion so slow and subtle that it had even escaped the notice of the great Ptolemy, not because of any specific weakness on the part of his science, but because the intervals between the observations at his disposal had not been large enough for a proper analysis of precession.[94] For his own time, Campanus found the head of Aries to have reached a *motus* of 58;57° on its small circle, from which one can gather that he wrote his treatise in early 1269.[95]

A perspective less firmly entangled with the Toledan access-and-recess model had been taken only a few years earlier by the Dominican philosopher Giles of Lessines, who tackled the flaws of the ecclesiastical calendar in the third and final book of his gargantuan *Summa de temporibus*, completed in Paris *c.*1264.[96] Giles's *Summa* was unique among the works on time reckoning produced during the thirteenth century, both for its size and for its author's effort to achieve a synthesis

[91] Campanus of Novara, *Compotus maior*, fol. 161vb (ch. 10).

[92] Ibid., fol. 164va. The Toledan Tables place the true vernal equinox of AD 1, on 24 March, 05:24h (from previous noon) and the true solstice on 24 December, 07:53h (from previous noon).

[93] Ibid., fols. 162ra–164va.

[94] Ibid., fol. 163ra.

[95] See ibid., fol. 162vb, where the corresponding equation of the eighth sphere is given as 9;12,24°, as is also the case in MS Oxford, BodLib, Digby 215, fol. 73va. It is 9;12,54° in MS Oxford, BodLib, Canon. Misc. 561, fol. 9v, but the correct reading appears to be 9;11,45°, as found in MSS Florence, Biblioteca Riccardiana, 885, fol. 263v; Oxford, BodLib, Rawlinson C.117, fol. 12v. Both the Toledan Tables (ed. Pedersen 2002: 1545) and the Novara Tables (MS Florence, BML, San Marco 194, fol. 13v) agree that a *motus* of 58;57° was reached in February or March 1269. The claim in Benjamin and Toomer 1971: 13–14 (n. 46) that the value points to 1268 was probably the result of a confusion between completed and current years.

[96] Giles of Lessines, *Summa de temporibus*, bk. III = (pseudo-)Roger Bacon, ed. in Steele 1926: 1–211. Giles's authorship of the '*Compotus* of Roger Bacon' is demonstrated in Delorme 1939. Steele's edition of Book III was based on the stand-alone copies in MSS London, BL, Royal 7.F.VIII, fols. 99r–163r; Oxford, BodLib, Ashmole 341, fol. 1r–60v. Other witnesses include MS Vatican City, BAV, Reg. lat. 1191, fols. 51r–171v, and the fragments in MSS Bergamo, Biblioteca Civica Angelo Mai, 571, pp. 25–6sexies; Erfurt, UFB, CA 2° 394, fols. 145r–147v. The only known manuscript to include all of Books I–III is in MS Bologna, Biblioteca Universitaria, 1845, fols. 1r–88v. A different recension of the text, which includes a heavily reworked version of Book II (attaching a chronicle), but omits all of Book III, is found in MSS Arras, Bibliothèque Municipale, 674 (722), fols. 3r–123r; Paris, BnF, lat. 15268, fols. 211r–229r. Two chapters from Book I are edited and discussed in Nothaft 2014b; 2014f. See also the biographical article in Grabmann 1936: 512–30.

not just between astronomy and computistics, but also between astronomy and the chronological puzzles raised by biblical and profane history, which took centre-stage in the first two books.[97] When it was finally time to turn towards the astronomical underpinnings of calendar reckoning in Book III, Giles chose as one of his main guides Abraham Ibn Ezra's *Liber de rationibus tabularum*, which put him in a position to sketch a potted history of Greek and later Arabic astronomy with a particular focus on lunar and solar parameters.[98] Although Giles was aware that there were three possible definitions of the solar year (tropical, anomalistic, and sidereal), he also knew that most astronomers only ever acknowledged two of these, since they considered the anomalistic year, that is, the Sun's return to a given point on its eccentric orbit, to be the same period as the sidereal year. This position had been common since the days of Thābit, the putative author of the ninth-century treatise *De anno solis* (see Ch. 3.3), according to whom the apogee of the Sun, together with all other planetary apogees, moved in unison with the sphere of the fixed stars. Azarquiel later equipped the solar apogee with an additional independent motion, motivating him to assign a separate length to the anomalistic year. Ptolemy, by contrast, had assimilated the Sun's anomalistic period to the tropical rather than the sidereal year, assigning a fixed quantity to both of these year lengths. Of the sages who came after him, however, many believed that the tropical year varied periodically due to the way the eighth sphere was subject to a periodic motion of access and recess.[99] Giles, unlike Campanus of Novara, displayed no full commitment to this theory, despite its ability to explain the astounding diversity of estimates for the length of the tropical solar year recorded in Greek, Hebrew, and Arabic sources, including the following:

Callippus (329 BC): $365\frac{1}{4}$ days
Hipparchus (145 BC): $365\frac{1}{4}-\frac{0.5}{1.45}$ days
Jewish calendar: $365\frac{1}{4}-\frac{1}{320}$ days
Ptolemy (AD 140): $365\frac{1}{4}-\frac{1}{300}$ days
al-Ṣūfī: $365\frac{1}{4}-\frac{1}{131}$ days
Thābit, al-Battānī, *Abencine*, and the *magistri probationum*: $365\frac{1}{4}-\frac{1}{106}$ days.[100]

While the first of these values ($365\frac{1}{4}$ days), which had become the basis of the Roman calendar, could be safely dismissed as the one option that was 'furthest from the truth', the diversity among the remaining estimates seemed to be beyond complete resolution. If anything, it confirmed a suspicion once uttered by al-Battānī, who wondered about the existence of some long-term motion in the

[97] See Nothaft (forthcoming-4), for further discussion.

[98] Compare Abraham Ibn Ezra, *Liber de rationibus tabularum* (ed. Millás Vallicrosa 1947: 74–83), with Giles of Lessines, *Summa de temporibus* III.1.2–3 (ed. Steele 1926: 9–17). Some of the closer textual parallels were already noted in Steele's edition (pp. xxvii, 292–4).

[99] Giles of Lessines, *Summa de temporibus* III.1.2 (ed. Steele 1926: 9, l. 22–p. 11, l. 15).

[100] Ibid. III.1.3 (ed. Steele 1926: 12, l. 16–p. 14, l. 27). Giles's main sources were Ptolemy's *Almagest* 3.1, and Abraham Ibn Ezra, *Liber de rationibus tabularum* (ed. Millás Vallicrosa 1947: 75–7). The *Abencine* mentioned in these passages must probably identified with Ibrāhīm ibn Sinān, the grandson of Thābit, as pointed out by Mancha 2012: 55 (n. 21).

heavens, as yet undiscovered.[101] Whether this motion corresponded to the access or recess of the eighth sphere postulated by the Toledan Tables remained unclear, however, especially since divergences between measurements could have been caused by 'the imperfections of the art of constructing instruments', a point that, according to Giles's source Ibn Ezra, had already been raised by Thābit. More than that, al-Battānī's worries about undiscovered motions could be seen to nourish a general scepticism about man's ability fully to comprehend the secrets of heavenly motions, as expressed by the author of the Book of Wisdom (9:16): 'we must struggle to learn about things that are close to us. Who, then, can ever hope to understand heavenly things?'[102] In spite of all this uncertainty, Giles was still willing to proclaim that his own tests based on 'instruments and eclipses' had shown the true length of the solar year to be closest to the value allegedly upheld by Thābit and al-Battānī, with $J = 106$ years.[103] If this estimate was trustworthy, it had the obvious consequence of rendering futile any attempt to assign permanent dates to the equinoxes and solstices in the Julian calendar, as confirmed by the fact that the winter solstice no longer fell on 25 December. That this had not even been the correct date for the year of Christ's nativity became clear, moreover, from a look at the observations carried out by Ptolemy in AD 140, which included a vernal equinox on 22 March and a winter solstice on 22 December. In this case, the solstice at the start of the Christian era could hardly have occurred later than 23 December.[104] Giles, despite his previous endorsement of $J = 106$ years, decided to shed some further light on this situation by using the parameters provided by al-Ṣūfī, which included a tropical year of $365\frac{1}{4} - \frac{1}{131}$ days. Following the lead of Abraham Ibn Ezra, who had introduced al-Ṣūfī's name and work to a Latin audience, he asserted that this value had received empirical confirmation from 'Azarquiel the Spaniard' (*Azarchel Hispanus*) and had subsequently become the basis for a number of astronomical tables used in the Latin West, including those for the meridians of Pisa, London, and Marseille.[105] Some such set of tables appears to have been at Giles's disposal when he established that in AD 1264 the Sun had crossed the vernal point on 13 March, 10 hours after noon. If $J = 131$ years, the equinoctial drift over the

[101] Giles of Lessines, *Summa de temporibus* III.1.3 (ed. Steele 1926: 14, l. 32–p. 15, l. 2). For the reference to al-Battānī, see *De motu stellarum*, trans. Plato of Tivoli (Nuremberg: Petreius, 1537), fol. 81r (ch. 52); Ragep 1996: 295. The same passage is already alluded to in Cunestabulus, *Compotus* 36 (CCCM 272: 110, ll. 138–9).

[102] Giles of Lessines, *Summa de temporibus* III.1.3 (ed. Steele 1926: 15, ll. 6–18). Cf. Abraham Ibn Ezra, *Liber de rationibus tabularum* (ed. Millás Vallicrosa 1947: 81).

[103] Giles of Lessines, *Summa de temporibus* III.1.3 (ed. Steele 1926: 15, ll. 2–6). See also ibid. (ed. Steele 1926: 15, l. 23–p. 18, l. 6), on different estimates for the sidereal year, and ibid. III.1.20 (ed. Steele 1926: 81–3), where Giles compares different rates of precession without making a decision between linear motion and the theory of access and recess. Either model could be used to make estimates as to the age of the world, as Giles himself showed in Book I.3.7. See Nothaft 2014b, for further details.

[104] Giles of Lessines, *Summa de temporibus* III.1.7 (ed. Steele 1926: 31, l. 5–32, l. 13).

[105] Ibid. III.1.3 (ed. Steele 1926: 14, ll. 23–6). Cf. Abraham Ibn Ezra, *Liber de rationibus tabularum* (ed. Millás Vallicrosa 1947: 76, 78, 83, 87). Azarquiel's confirmation of al-Ṣūfī's solar year is also discussed at some length in the mid-twelfth-century treatise in MS Cambridge, Fitzwilliam Museum, McClean 165, fols. 67r–76v, at fol. 68r–v. On the tables for Pisa and London, see Ch. 3.1 and 3. The only tables for Marseille to be extant from the twelfth and thirteenth centuries are based on the Toledan Tables. See the index entries in Pedersen 2002: 1654–5.

past 1,264 years amounted to 9 days and *c.*15 hours, which implied that the vernal equinox in the year before Christ's birth had occurred on 22 March, around 1 p.m. The notion according to which the annunciation of this birth, on 25 March, had taken place at the astronomical beginning of spring was thereby conclusively refuted.[106] Giles reiterated his support for al-Ṣūfī's implied drift rate of $J = 131$ years in the preface to a monumental table included in the third and final part of Book III, which displayed the exact date, hour, and even minute of all four equinoxes and solstices between the beginning of the Christian era and AD 1581, using steps of 20 years and the meridian of Paris as a reference point for the calculation (Figure 4.3).[107]

An area where Giles of Lessines was clearly on the same page with his contemporary Campanus of Novara was the calculation of the lunar month. Campanus's approach was broadly Grossetestian, in that he treated the 30-year lunar calendar of the Arabs as a superior alternative to the increasingly inaccurate 19-year cycle currently in Church use. As he explained to his readers, the former counted their years and months not from the beginning of the Christian era, but 'from the time of the abominable transgression of Muhammed', which began 621 solar years and 112 days after Christ's conception, on a Thursday.[108] At the heart of Campanus's engagement with this lunar calendar were tables for calendrical conversion based on a 28-year period of Julian 'collected years', which also characterized Campanus's astronomical tables for Novara.[109] Having introduced the conversion tables, Campanus invested his usual diligence into demonstrating that the lowest number to contain both a whole number of Arabic and a whole number of Latin-Christian lunations was the product of 27,759 days (a 76-year cycle) and 10,631 days (a 30-year cycle), which was equal to 295,105,929 days or 807,956 Julian years. As Grosseteste had already discovered some decades earlier, the number of Arabic lunations contained in this cycle exceeded the number of Christian lunations by exactly 100.[110]

106 Giles of Lessines, *Summa de temporibus* III.1.7 (ed. Steele 1926: 32–3). Both Steele's text (32, l. 29) and MS Bologna, Biblioteca Universitaria, 1845, fol. 45vb, give the recession over 1,264 years as 'novem dies et dimidium et 5 horas fere', or 9d 17h, but 9d 15h are clearly implied by the other numbers involved.

107 Giles of Lessines, *Summa de temporibus* III.3.4 (ed. Steele 1926: 159–64); MS Bologna, Biblioteca Universitaria, 1845, fols. 81v–82v. There are numerous incongruities between this table and its auxiliary table for expanded years and the statements made elsewhere in the work. Similarly, the following chapter (III.3.5) offers tables for calculating the length of day and night, based on the winter solstice in the year 1264, which Giles here places on 12 December, $12\frac{1}{2}$ hours after noon (ed. Steele 1926: 165–6). This agrees neither with the preceding table nor with the statement in III.1.7 (ed. Steele 1926: 34, ll. 32–3), according to which the winter solstice that year happened on 12 December, 8 hours after noon.

108 Campanus of Novara, *Compotus maior*, fols. 167vb (ch. 15). See Marzi 1896: 27–8, who documents the use of this part of Campanus's discussion by Pellegrino Prisciano (1508).

109 Campanus of Novara, *Compotus maior*, fols. 168rb–169rb (ch. 15). A version of Campanus's calendar tables circulated separately together with a brief canon. See Thorndike and Kibre 1963: 1234; Benjamin and Toomer 1971: 3 (n. 1), 16 (n. 58), and the edition in Pedersen 2002: 505–6, 919–20. These tables are represented in the right half of Fig. 4.2 above. On Campanus's astronomical tables, see Benjamin and Toomer 1971: 4 (n. 4), 5 (n. 14), 15–16; Pedersen 2002: 1205–6, 1548. Specifically on his tables for mean syzygy, which circulated in stand-alone copies, see Pedersen 2002: 1347; Zinner 1925: nos. 10866–71; Thorndike and Kibre 1963: 241, 486. Campanus's name is also attached to an almanac of mean conjunctions for the years 1254–1329. See e.g. MSS Oxford, BodLib, Digby 215, fols. 93r–94r; Vatican City, BAV, lat. 3118, fols. 6v–8r.

110 Campanus of Novara, *Compotus maior*, fol. 169va–b (ch. 16).

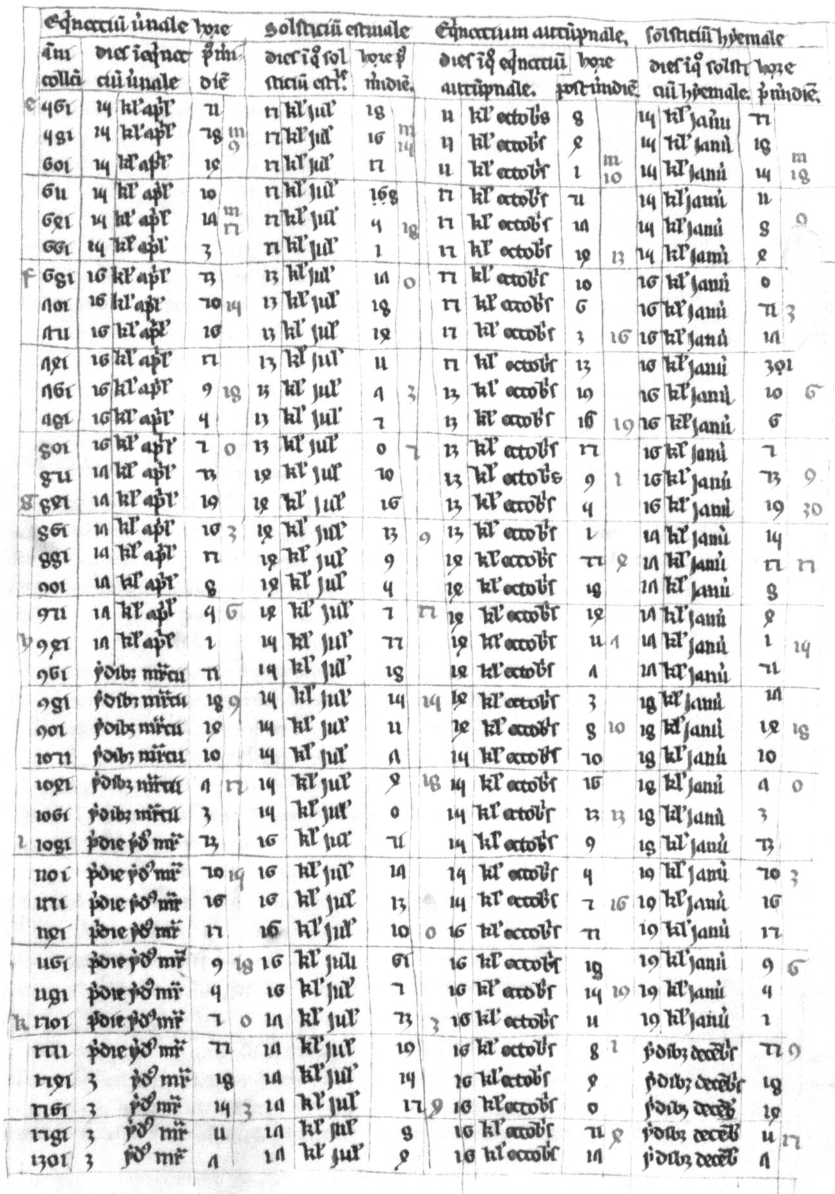

Figure 4.3. Giles of Lessines, *Summa de temporibus*, bk. III (*c.*1264), table for equinoxes and solstices, AD 561–1301. Bologna, Biblioteca Universitaria, 1845, fol. 82r.

Reproduced by kind permission of the University Library of Bologna (© 2017)

Even the Arabic calendar, however, was open to further refinement, as its average lunation omitted the 0;0,0,8,20d appended to it by Azarquiel.[111] The minor shortfall of 29;31,50d compared to 29;31,50,8,20d amounted to a day in $72 \times 30 = 2{,}160$ years and was easily compensated for if another day was added to the lunar count after every 72nd repetition of the Arabic 30-year lunar cycle.[112] For the ecclesiastical calendar, this meant that the error committed over the course of the aforementioned 807,956 Julian years shrank from 100 lunations to close to 87 lunations. At the same time, the minimum period to unite a full number of mean synodic months and a full number of Latin-Christian lunations now increased to 58,172,908 Julian years, during which the Church counted 6,260 lunations too few.[113] Another way of putting it was to say that, according to Azarquiel's mean lunation (29;31,50,8,20d), it was 'only' a matter of 9348 years until the Christian calendar had fallen more than a full month behind, the precise error being 29;42,21,40d.[114] Campanus of Novara, who made no secret of the fact that he was appalled by this 'disgraceful error' (*fedus error*), insisted that to deviate from Azarquiel and the Toledan Tables was to err against the truth. 'For the tables for solar and lunar eclipses that are based on the aforementioned parameters do not deceive us to any extent that could be grasped by the senses.' More than that, the predictions these tables made about the time, magnitude, and location of these eclipses were so accurate 'that it is even possible for the astronomer to make a drawing of a future eclipse of the Sun or Moon that is similar in all respects to this [eclipse] and to its circumstances'.[115]

An even greater amount of astronomical detail was on display in the work of Giles of Lessines, who analysed the error of the 19-year cycle by comparing it with three different systems of lunar reckoning: that of the ancient Chaldaeans, that of Hipparchus and Ptolemy, and that of the Arabic calendar. In contrast to previous authors such as Roger of Hereford, who used 'Chaldaean' as a synonym for 'Arabic', Giles's Chaldaeans actually referred to the astronomers of ancient Babylon, whose observations of eclipses had taught them that 223 lunations were equal to $6{,}585\frac{1}{3}$ days (the so-called Saros cycle). This span of 18 years, 10 days, and 20 hours implied a mean lunation of 29;31,50,18d and a lunar year of 354;22,3,36d. If the mean lunation was multiplied by 235, the corresponding 19-year period turned out to comprise 6939;42,27,10d, at least according to Giles, who concluded that 235 Chaldaean lunations were 0;2,32,50d below the length of 19 Julian years (6939;45d). After 24 such 19-year cycles, that is, after 456 Julian years, this difference was bound to exceed a full day.[116] Analogous calculations could be presented for the pseudo-Ptolemaic 29;31,50,8,9d and the Arabic 29;31,50d, which made for discrepancies of a day in a little more than 300 years and a little less than 275 years

[111] Ibid., fol. 167vb (ch. 15).
[112] Ibid., fols. 167vb–168ra.
[113] Ibid., fol. 169vb (ch. 16).
[114] Ibid., fol. 169va.
[115] Ibid., fol. 169vb. See also ibid., fols. 174va–175ra (ch. 28), where Campanus offers an analysis of the consequences of the error for the calculation of Easter, which adds only little to Grosseteste's remarks.
[116] Giles of Lessines, *Summa de temporibus* III.1.12 (ed. Steele 1926: 52, l. 8–53, l. 14). The passage is based on numerical information found in Abraham Ibn Ezra, *Liber de rationibus tabularum* (ed. Millás Vallicrosa 1947: 98). Better results would have been: $29;31,50,18\text{d} \times 235 = 6939;42,0,30\text{d}$ and $6939;45\text{d} - 6939;42,0,30\text{d} = 0;2,59,30\text{d}$, with $M \approx 381$ years.

respectively.[117] He refrained from spelling out the practical consequences of these and other calculations until the second part of Book III, in which he left the world of mathematical astronomy and fully delved into the rules and intricacies of the ecclesiastical *computus*, where moments, ostents, and atoms stood ready to replace the orderly sexagesimal fractions used in astronomical tables. Among the names most frequently mentioned in these chapters was Gerland, whom Giles had taken to task in the second book of his *Summa de temporibus* for shifting the death of Jesus to the year AD 42 on computistical grounds. Like Reinher of Paderborn and Cunestabulus before him, the Dominican from Lessines underlined the futility of Gerland's approach, which substituted the Jewish calendar of the first century for the erroneous 19-year cycle of the present-day Church.[118] The evident delay of the new moon predicted by this cycle relative to the evening of first visibility, which now reached up to three days, explained why it had not been possible to confirm that 'in the year of the Lord's Passion our Lord Jesus Christ suffered on a Friday and on the 15th day of the lunar month'. Attempts to apply the 19-year cycle to biblical history instead continued to produce divergent dates, 'which either contradict the Gospels, which to impugn is not allowed to any Catholic, or go against some other famous assumption as to the age until which Christ remained in the flesh or against the whole original opinion of the Saints on the month and day on which He is said to have suffered'.[119]

For Giles, this chronological failure posed a major—perhaps the most important—reason to reject the ecclesiastical calendar in its present form. He considered it worse, in fact, than the mistaken assignment of the vernal equinox, which currently fell 9 days earlier than assumed by the *computus*, causing the Church to celebrate Easter in the second rather than the first spring month in years 3, 8, 11, 14, and 19 of each 19-year cycle. This latter problem he classified as 'tolerable, because it does not incline towards any detriment to the faith', even though 'it is a sign of ignorance and a cause for derision by those who do not walk in the simplicity of faith'.[120] A much more serious issue were the imprecisions inherent in the calendar's distribution of Golden Numbers, which showed that its inventors had failed to test their doctrines against experience in a sufficient manner.[121] According to Giles's historical conjecture, the Church Fathers had inherited the idea of a 19-year cycle from the ancient Hebrews, from whom they learned that 235 synodic months were exactly equal to 19 solar years. The problem with their supposition

[117] Giles of Lessines, *Summa de temporibus* III.1.12 (ed. Steele 1926: 53, l. 24–56, l. 26).

[118] See the lengthy discussion in Giles of Lessines, *Summa de temporibus* II.2.1–3.5, MS Bologna, Biblioteca Universitaria, 1845, fols. 24va–31va, and the alternative version of Book II found in Giles of Lessines, *Summa de temporibus*, MS Arras, Bibliothèque Municipale, 674 (722), fols. 43r–70v. See, in addition, the references to Gerland in Book III.2.4, 2.9, 2.12 (ed. Steele 1926: 96–7, 111–12, 121–2).

[119] Giles of Lessines, *Summa de temporibus* III.2.19 (ed. Steele 1926: 150, ll. 3–11). See also III.1.13 (ed. Steele 1926: 59, ll. 4–15), for Giles's affirmation that the ecclesiastical calendar is supposed to count the age of the Moon from first visibility, and III.1.15 (ed. Steele 1926: 64–6), for his detailed discussion of the fluctuation in the interval between conjunction and first visibility.

[120] Ibid. III.2.19 (ed. Steele 1926: 149, ll. 21–6).

[121] Ibid. III.2.18 (ed. Steele 1926: 146–8).

was not simply that this equivalence failed to reflect nature. Worse, the original Jewish estimate of the length of the solar year had been $365\frac{1}{4}-\frac{1}{320}$ days, which meant that the Church, by adopting the Julian calendar, had unwittingly increased the average length of the lunation.[122]

Eager to equip his readers with the means to arrive at a more accurate prediction of the new moon, Giles used the third and final part of Book III to display and describe a cornucopia of different calendrical and astronomical tables, most of which revolved around the relation and conversion between the Julian calendar and the fixed lunar calendar of the Arabs. Perhaps the most innovative among these was a double-page table for the first day of each Arabic 30-year cycle, which went back proleptically to the year AD 11, in which the first day of the first month (Muḥarram) would have fallen on 23 April. For this and all subsequent first years of the cycle up to AD 1553, the facing part of the table showed the corresponding Arabic lunar age on the first day of each Julian month. A second table, for expanded years, made it possible to extrapolate from the previous table the lunar ages of all other 29 years of the cycle. Both tables were in turn mirrored by another full set, this time constructed for solar cycles of 28 years, which created a technically sufficient framework to transfer the entire business of calculating lunar ages from the traditional tools sanctioned by the Church of Alexandria to the Arabic *computus* commonly associated with astronomical tables.[123] In Giles of Lessines's imposing *Summa de temporibus*, one of the core ideas first advanced by English and Bavarian computists in the middle of the twelfth century—that of co-opting the Arabic calendar as a natural *computus*—had thus reached a stage of refinement at which it could be put into useful practice, provided his fellow computists, and most importantly his Church, could muster the will to adopt completely new tables and rules.

4.3. FRANCISCANS, THE POPE, AND THE HEBREW CALENDAR

A look back at the literature produced during the first three quarters of the thirteenth century leaves us in no doubt that the insight of something being wrong with the ecclesiastical calendar had been widely disseminated in both computistical and astronomical literature—widely enough, in fact, to make it difficult for the educated person to remain completely ignorant of the problem. And yet, even though proponents of the *computus philosophicus* such as Robert Grosseteste, Giles of Lessines, and Campanus of Novara were very outspoken about the troubles the errors of the Julian calendar and Alexandrian 19-year cycle caused for the celebration of Easter, their works offered few truly concrete measures for their removal beyond suggesting the Arabic calendar as an alternative to the existing lunar computation. The only significant exception can be spotted in the computistical work of John of Sacrobosco,

[122] Ibid. III.2.12–13, 18 (ed. Steele 1926: 119–20, 123–4, 147, ll. 2–31).

[123] Ibid. III.3.6 (ed. Steele 1926: 167–79). More conventional conversion tables (Arabic/Christian and Christian/Arabic) are presented in III.3.1–3 (ed. Steele 1926: 151–9).

who went further and pondered the possibility of dropping a day from the calendar every 288 years, removing either the last day of February or, better still, the bissextile day. If $J = 288$ years, as Sacrobosco's own calculation suggested (see Ch. 4.2), this was indeed enough to halt the retrogression of the equinoxes and solstices. Such an intervention had the welcome side-effect of also neutralizing the lunar 'slippage' inherent in the 19-year cycle ($M = 342$ years according to Sacrobosco), albeit at the price of a slight overcompensation, which would have shifted the actual new and full moons towards later calendrical dates over longer periods time. Still, it was true that 'this same subtraction would stabilize the first day of the Moon for several centuries', leading Sacrobosco to suggest that years affected by it should henceforth be designated as 'Years of Jubilation'.[124] As far as the present-day discrepancy between calendrical and astronomical new moons was concerned, Sacrobosco discerned the necessity of resetting the Golden Numbers by at least 3 days, such that the Golden Number I in the month of March was going to appear next to 20 March instead of 23 March. He also admitted, however, that such a correction had to remain a hypothetical scenario as long as the Church felt bound to the decrees of the General Council of Nicaea, which had prohibited any change to the calendar. It followed that Sacrobosco's contemporaries, the *moderni*, had to tolerate the error for the time being.[125] The same worry is implicit in the work of Robert Grosseteste, who dedicated a whole chapter to explicating tables for the mobile feast days that had been constructed according to traditional rules,[126] but was remarkably terse when it came to the steps that had to be implemented to keep these feasts from occurring on the wrong dates. His insights on this point filled no more than a paragraph:

> The way to correct this error is to determine the length of the year and to correct the calendar accordingly. [But] even without determining the length of the year, one can always find the day of the vernal equinox with the help of observational instruments or accurate astronomical tables. And one may find the day of the first full moon after this equinox in accordance with what we have taught above with regard to placing the new moon dates in line with the truth [i.e. according to the Arabic calendar], and accept this [full moon] as the paschal boundary. Yet since the Holy Church has not yet changed the old set of rules of how to find the mobile feasts, we shall explain these rules [in what follows].[127]

One of Grosseteste's readers, the aforementioned Master Gordianus, elaborated on this point later in the same century by adding that the faithful were not supposed to bemoan any error or blunder, but were required 'obediently and steadfastly to maintain [the rules] according to the our holy mother, the Roman Church, until

[124] Sacrobosco, *Computus*, sig. E2v.
[125] Ibid., sig. F3r. An extended excerpt of this passage, updated to match the situation in 1368, by which time the hypothetical error had reached exactly 4 days ($4 \times 342 = 1{,}368$), appears in MS Wilhering, Stiftsbibliothek, IX.81, fol. 24r–v. See already Roger of Hereford, *Compotus* 5.26 (CCCM 272: 226, ll. 45–7), who was among the sources used by Sacrobosco.
[126] Robert Grosseteste, *Compotus* 11 (ed. Steele 1926: 262–6).
[127] Ibid. 10 (ed. Steele 1926: 259, ll. 11–19).

it pleases her to approve of the contrary'.[128] Such reluctance to accuse the Church of any misjudgement in the area of *computus* was no doubt increased in the minds of all those familiar with the canon law and the available collections of papal letters and conciliar canons, in which the validity of the 'Nicene' date of Easter was underlined with some frequency.[129] Dionysius Exiguus, who had been the translator of several such documents from Greek into Latin, contributed to this stance by referring back to the Council of Antioch (346), whose fathers had threatened anyone with excommunication who dared to deviate from the decisions the Council of Nicaea had made with regard to the celebration of Easter.[130] That legal barriers of this sort had been hampering the prospects of calendar improvement was a point explicitly made in 1267 by the Franciscan friar Roger Bacon, who is perhaps best known for the series of learned writings he prepared between 1266 and 1269 for the eyes of Pope Clement IV (1265–8). Two of these, the *Opus maius* and *Opus tertium*, contain almost identical versions of a lengthy disquisition on the problems of the calendar, in which Bacon summarized many of the complaints of his predecessors but gave them a new polemical spin in his spirited attempt to stir the pope into action.[131] The general tone may be gleaned from the opening words added to this section in the *Opus tertium*, which branded the present state of the ecclesiastical calendar as:

> unbearable for any wise person, horrible to any astronomer, and worthy of ridicule from any computist, which is why all those educated in astronomy and in the *computus* and in such things are amazed that such abominable falsity is sustained.... Any intelligent Christian who writes on this [topic] demonstrates the elements of this corruption and teaches the remedies. Still, no one dares to deliver a corrected calendar because the general council prohibits anyone from changing the calendar without special permission from the apostolic see. And this is just. Yet this most blessed see should relieve the Church from this monstrosity.[132]

The unprecedented boldness with which Bacon felt permitted to address the problem must be understood against the background of his personal contacts with Cardinal Guy le Gros de Foulques, a noted lawyer and adviser to the King of France (Louis IX), who took the name Clement IV upon his election to the papacy in February of 1265. Even before this election, le Gros de Foulques had received news of Bacon's unusual philosophical ideas and expressed his desire to see them in written form. He renewed this request in June 1266, now in the shape of a papal mandate, which ordered the Franciscan to 'explain to us the remedies that you think we should

[128] Gordianus, *Compotus*, MS Lüneburg, Ratsbücherei, Ms Miscell. D 4° 46, fol. 39rb–va: 'Verumtamen, quia sancta ecclesia nondum mutavit antiquam doctrinam de inveniendis festis mobilibus, illam doctrinam explanemus nec errorem vel peccatum aliquid putare debemus, sed tenere volumus et debemus obedienter et firmiter secundum sanctam matrem nostram Romanam ecclesiam quousque ei placuerit contrarium approbare.'

[129] See e.g. *Decretum gratiani* III, dist. 322–3, 26 (ed. Friedberg 1879–81: i. 1358–60).

[130] Dionysius Exiguus, *Epistola ad Petronium* (ed. Krusch 1938: 66–7).

[131] See Roger Bacon, *Opus maius*, pars IV (ed. Bridges 1897: 269–85). This section is largely identical with Roger Bacon, *Opus tertium* 67–71 (ed. Brewer 1859: 272–95). On the genesis and content of these works, see Easton 1952: 144–66; Massa 1955: 7–80; Power 2013a: 84–125.

[132] Roger Bacon, *Opus tertium* 67 (ed. Brewer 1859: 272–3).

adopt to address those issues that you have recently described as the occasion of such great danger, and do this quickly and as secretly as possible'.[133] From the *Opus maius* and other writings Bacon completed in response to this request, it emerges that most of the proposed 'remedies' were tied to an expectation that the end times were imminent and that the one true Church was soon going to become embroiled in an intense struggle against rival religious sects. In Bacon's richly textured view of the present and future, Roman Christianity had become a marginal and endangered force in a world overflowing with Saracens, Tartars, pagans, and schismatics, to which the arrival of the Antichrist was soon going to add an even greater threat. The only way for the Latin Church to prevail in these clashes was to restore the ancient sciences to their former glory, in particular those concerned with mathematics and the prediction of natural events. Alchemy, magic, and optics were needed to construct powerful new weaponry, based on the use of fireworks, poison, or mirrors; knowledge in geography and foreign languages was supposed to improve missionary activity in far-flung regions, while astrology allowed strategic planning by predicting the future. Many of these sciences, Bacon believed, had survived in the books of the Greeks, Arabs, and Hebrews, who appeared to possess knowledge superior to that of the Latins in a great number of different field, the art of astronomy being one of the chief examples.[134]

In his drawn-out section on calendar reform, Bacon was exceptionally frank in denouncing the prelates of his Church for the neglect with which they continued to treat the science of the stars, a neglect he traced back to the ancient days of the 'primitive Church', when mathematics and astronomy were still viewed with suspicion and deemed to be of no particular use to the divine service. Ignorance of this kind had burdened the modern-day Church with a calendar defaced by inherent flaws, a *multiplex error* that was known not just to a few expert astronomers, but also to the mass of computists, who continued to write about the problem in terms not substantially different from the former.[135] One previous author on the calendar problem whose influence on Bacon's thoughts is difficult to overlook is Robert Grosseteste, to whom he owed the insight that the difference between the calendar of the Church and that of the Arabs amounted to 0;16,40d per 76-year cycle and 1;6,40d per 304 years.[136] As a result of this shortcoming, the new moons indicated by the standard Golden Number no longer conformed to what presented itself even to the untrained eye. 'And that this is all true is not in doubt to skilled astronomers; and any computist knows that the new moon nowadays is incorrect by three days

[133] *Epistola Clementis Papae IV. ad Rogerium Baconem* (ed. Brewer 1859: 1). I follow the translation in Power 2013a: 69. On the biographical background, see Thorndike 1923–58: ii. 616–30; Lindberg 1983: xv–xxvi; Bartlett 2008: 111–48; Power 2013a: 29–83; Hackett 2016.

[134] See the accounts of Bacon's thought in North 1999; Bartlett 2008: 129–60; Power 2008; 2011; 2013a: 164–264; 2013b; Matus 2012.

[135] Roger Bacon, *Opus tertium* 67–8, 71 (ed. Brewer 1859: 273–4, 292–5).

[136] See ibid. 70 (ed. Brewer 1859: 281–8), which is mostly a paraphrase of Robert Grosseteste, *Compotus* 4 (ed. Steele 1926: 232–7) and 7 (ed. Steele 1926: 243). Grosseteste's influence on Bacon was already recognized by Baur 1917: 52–63. See also the discussions in North 1983: 82–3; Borst 1998: 74–5; Aufgebauer 2006: 73–4, 77–8.

or four; and any peasant can notice this error in the heavens.'[137] A little more independence of thought, but also a good deal less clarity, is exhibited by Bacon's discussion of the Julian calendar and its evident failure to keep the equinoxes and solstices fixed to their accustomed dates. Like many authors before him, Bacon discerned a constant reminder of this shift in the ancient tradition of placing all four cardinal points on the eighth day before the respective Kalends, assuming that Christ's nativity (25 December) and the birth of John of Baptist (24 June) had both coincided with the day of the solstice. A completely different picture presented itself in the present year 1267, which according to Bacon showed that the equivalent points of the solar year already fell on 13 March, 15 June, 16 September, and 13 December. In line with the observational method described in the eleventh century by William of Hirsau and occasionally referenced by later authors, Bacon was quick to add that this sort of result could be supported not just with astronomical tables, 'but any layman can perceive it with his own eyes through the way a sunbeam falls on a wall or other object, sometimes higher, sometimes lower'.[138] When it came to estimating the size of the error that was at the root of the equinoctial and solstitial drift, Bacon showed no real consistency, as he switched from $J = 130$ years to $J = 125$ years and then to $J = 120$ years in the course of his discussion. The last of these estimates was a widespread axiom among thirteenth-century computists (see Ch. 4.1), whereas the solar year of $365\frac{1}{4} - \frac{1}{130}$ days appears to have been a rounded version of the $365\frac{1}{4} - \frac{1}{131}$ days associated with al-Ṣūfī and known from Latin authors such as Giles of Lessines, whose *Summa de temporibus* Bacon had read.[139] By contrast, the alternative estimate of $365\frac{1}{4} - \frac{1}{125}$ days had come to the English friar as a result of his own investigation, which involved looking at the 9-day shift in the dates of the vernal equinox and winter solstice that had occurred between the time of Ptolemy (22 March and 22 December) and his own year of writing (13 March and 13 December). Since Ptolemy made his observations in AD 140, it was clear that these 9 days had accumulated over a period of $1{,}267 - 140 = 1{,}127$ years, making for an approximate shift of 1 day every $1{,}127 \div 9 \approx 125$ years.[140]

This neat calculation remained difficult to reconcile with Ptolemy's own estimate for the tropical year, which at $365\frac{1}{4} - \frac{1}{300}$ days was considerably longer than the three values invoked by Bacon. The English friar refrained from making a definitive pronouncement, indicating to his readers that 'a final demonstration of the length of the year and the change of the solstices' was as yet outstanding.[141]

[137] Roger Bacon, *Opus tertium* 70 (ed. Brewer 1859: 282).

[138] Ibid. 69 (ed. Brewer 1859: 277). Compare this statement with the one made in the *Compotus Petri* of 1171, cited in n. 73 of Ch. 3.3. Bacon's equinox and solstice dates are the same as those indicated for 1261 in the table included in Giles of Lessines, *Summa de temporibus* III.3.4 (ed. Steele 1926: 159–64).

[139] Giles of Lessines's influence on Bacon is documented in Delorme 1939 and Nothaft 2014b. Some conspicuous parallels exist between the accounts of the dates of the equinoxes and solstices in Giles of Lessines, *Summa de temporibus* III.1.7 (ed. Steele 1926: 29–33), and Roger Bacon, *Opus tertium* 69 (ed. Brewer 1859: 275–9).

[140] Roger Bacon, *Opus tertium* 69 (ed. Brewer 1859: 277–8).

[141] Ibid. (ed. Brewer 1859: 278). See also 71 (ed. Brewer 1859: 292–3).

Another sign of Bacon's lack of certainty with regard to this crucial parameter may be spotted in his failure to address the question whether the variety of existing estimates for the tropical year might be explained by its inherent mutability, which was one of the predictions made by the theory of access and recess classically ascribed to Thābit ibn Qurra and still *en vogue* among the astronomers of his day. Bacon's own familiarity with the theory is evidenced very clearly by his criticisms of Giles of Lessines included in the *Opus maius*, where he dealt with the Dominican author's attempt to calculate the age of the world based on the precessional motion of the apogee and perigee of the Sun. In line with an argument pioneered by Robert Grosseteste, Bacon responded by highlighting the unacceptable consequences that a full linear precession of the line of apsides spelled for the long-term habitability of the lands of Christendom in the northern hemisphere. In the scenario described by Friar Giles, 'the habitable regions above which is the apogee would become uninhabitable in the succession of time, namely when the perigee arrives over them; and, contrariwise, the uninhabitable regions would become habitable, which is absurd'. The unsavoury prospect of a northern hemisphere scorched by the heat of the Sun at perigee could be seen to speak in favour of the classical 'Toledan' access-and-recess theory, which had the solar apogee move back and forth by 10;45° in either direction without letting it 'depart from the sign of Gemini, which is why it does not travel in a circuit around the earth, such that the perigee should at some time affect the habitable lands'.[142]

The seemingly imponderable motion of the fixed stars could fortunately be left aside when it came to assessing the fundamental error that arose whenever the 14th day of the lunar month fell between 13 March, the date of the vernal equinox observable in the here and now, and 21 March, which remained in use as the earliest permissible *terminus paschalis*. A problematic case of this kind could be encountered in the present year 1267, which was the 14th year of the 19-year cycle. In this year, the Golden Number XIV showed a new moon on 28 February, implying that 13 March was the seat of *luna* 14. If the Alexandrian rule of the equinox was to prevail, this should have been the official *terminus paschalis* followed by Easter Sunday on 20 March. This date was outside the valid Easter limits, with the result that 'the glorious Pasch will not be celebrated at its time', whether in 1267 or in any of the other years affected by the same calendrical error.[143] In the long run, the situation threatened to become even worse, as the receding equinox was bound to move the legitimate date of Easter Sunday closer and closer towards the beginning of the year. As Bacon was careful to remind the pope, this drift had an inevitable knock-on effect on all the other mobile feast days, including Quadragesima and Pentecost, leading to the absurd and worrisome consequence that Christians gave in to carnal pleasures at a time of the year when the astronomical situation would have prescribed the observance of Lent. 'And while these things are horrible in themselves, they are all the more stupid and worthy of derision because it was due

[142] Roger Bacon, *Opus maius*, pars IV (ed. Nothaft 2014b: 59).

[143] Roger Bacon, *Opus tertium* 69 (ed. Brewer 1859: 279). See also 70 (ed. Brewer 1859: 289–90), concerning the error inherent in the lunar cycle.

to ignorance and negligence that the Devil was able to make this befall the Church of God.'[144] Deplorable as the situation certainly was, a solution appeared to be available, provided that the Church was ready to leave the matter in the hands of astronomers. Their most pressing task was to verify the length of the tropical year and to draw up tables and explanatory canons that showed how to find the date of the vernal equinox for many centuries in the future. Such tables and lists, Bacon suggested to Clement IV, could be disseminated throughout Latin Christendom and placed alongside all conventional calendars.[145] In a similar vein, the traditional Golden Number was best replaced by astronomical calculations of the new moon to help ensure that Easter was truly the first Sunday after the first 14-day old moon that fell on or after the equinox of spring.[146]

Having come this far in his argument, Bacon still saw the need to remove one potential obstacle from the path to reform, which was posed by the question of authority in calendrical matters. The idea that the Council of Nicaea was the locus of this authority and that its endorsement of the Alexandrian 19-year cycle remained binding even today was hardly a new one in his time. It already crops up in Bede's *De temporum ratione*, chapter 43, in which the venerable monk supported the validity of the Alexandrian Easter date by citing Paschasinus of Lilybaeum, who had used the story of a miraculously filling baptismal font to sway Pope Leo. At the same time, however, it could not be overlooked that Bede himself had traced the history of the 19-year cycle as far back as Eusebius of Caesarea, which suggested that its actual existence preceded the Nicene Council. In Bacon's view, the conclusion to take home from the sources at his disposal was that the bishops who convened at Nicaea during the reign of the emperor Constantine endorsed Eusebius's cycle provisionally in order to guarantee uniformity of worship, intending to keep it for as long as it yielded adequate predictions. What they did not insist upon was the cycle's eternal validity. Indeed, it was only the weight of long-standing custom as well as the absence of astronomical expertise within the Latin Church that had prevented its prelates from carrying out a reform centuries ago.[147] Given these circumstances, it was up to the present-day Church or, rather, the current pope, whether or not it was time to introduce a new method of lunar reckoning and thereby to provide a remedy for the errors that plagued the ecclesiastical *computus*. Not only was this a step long overdue, but it had the potential of rescuing the Church from the embarrassment it currently suffered at the hands of 'infidel philosophers', whether they be Arabs, Hebrews, or Greeks,

> who live among Christians, as is the case in Spain and Egypt and in parts of the Orient and in many other regions of the world, [and who] shudder at the stupidity which they discern in the ordering of time used by the Christians in their feasts. And now the Christians have the knowledge of astronomy by means of which a

[144] Ibid. 69 (ed. Brewer 1859: 280–1). I here follow the English translation in Power 2013a: 107–8.

[145] Roger Bacon, *Opus tertium* 69 (ed. Brewer 1859: 281).

[146] Ibid. 70 (ed. Brewer 1859: 290–1).

[147] Ibid. 71 (ed. Brewer 1859: 291–4). Bacon here refers to Bede, *De temporum ratione* 43–4 (CCSL 123B: 412–19).

> proper rectification can be made. As a result, your honour can command it and you will find men who apply excellent remedies in this regard and not only in the points mentioned, but in all the defects of the calendar.... Hence, if this glorious work should be carried out during the reign of Your Holiness, it would mean that one of the greatest and best and most beautiful things ever attempted in the Church of God has been accomplished.[148]

Roger Bacon's impression that the calendar's ailments were best overcome by having the pope command a cure was shared by the author of a previously unstudied text written in 1276 (*Assignacio errorum kalendarii et eorundem correctio*), which only survives in a single manuscript of northern Italian origin, notable for containing several works by Roger Bacon, Robert Grosseteste (including the *Compotus*), and Campanus of Novara (including the *Compotus maior*).[149] The anonymous author of this crisp 2,000-word treatise began by depicting the calendar as a potential source of public scandal and unrest, not because of the growing discrepancy between calendrical and astronomical Easter dates, but because of the simple fact that the calendar was no longer in alignment with the annual course of the Sun. Like many computists of his day, he traced the invention of the Roman calendar back to King Romulus, whose 10-month-long year Numa Pompilius later expanded by adding the months of January and February. Contrary to ancient and commonly received wisdom, however, the anonymous writer of 1276 asserted that Numa's calendar had been created to track the course of the Sun rather than the Moon. Its purpose was to 'instruct the people about the dates and times when they would have to cultivate the soil, plant trees, sow seeds, frequent markets, or set aside time for games or leisure', thus making sure that everything was done at the optimal time and to the best possible outcome.[150] Praiseworthy as Numa's initiative certainly was, he failed at finding the correct estimate for the length of the solar year. At 365 days per year, his calendar suffered a small, but noticeable error, which caused the seasons gradually to slip away from the months customarily assigned to them. According to the narrative's continuation, the only improvement Julius Caesar contributed to this Roman calendar was the bissextile day, which eliminated the original error, but also created a minor deviation in the opposite direction. Ever since Caesar's reform, the Julian calendar had been burdened by a surplus of 12 minutes per year, causing the solstices and equinoxes to drift at a rate of $J = 120$ years.[151]

On the level of astronomical analysis, this assessment contained nothing that had not already been spelled out by vulgar computists during the preceding decades.

[148] Roger Bacon, *Opus tertium* 71 (ed. Brewer 1859: 295).

[149] MS Florence, Biblioteca Riccardiana, 885, fols. 215r–223v (s. XIII^{ex}/s. XIV$^{1/2}$; 'Incipit prologus in assignacionem errorum kalendarii et eorundem correctionem. Quoniam ex diversitate cursuum solis et kalendarii...'). The existence of this text was first noted by Marzi 1906: 640. See also Thorndike 1959: 38–40, for a description of the manuscript.

[150] MS Florence, Biblioteca Riccardiana, 885, fol. 215r: 'Intendebat enim, ut dicitur, idem Numa per coequalem cursum solis et kalendarii populum instruere quibus terminis et temporibus terras deberent excolere, arbores plantare, seminare semina, nundinas frequentare, ludis etiam vacare et ociis, ut sic agenda omnia suis temporibus effectus congruos sortirentur.'

[151] Ibid., fol. 215v.

What was new, and indeed ground-breaking, were the specific consequences our author felt justified to draw from it. Julius Caesar, he mistakenly asserted, had originally taken care to start his calendar year on the day following the winter solstice, making the Kalends of January the legitimate *ianua anni* or 'gate of the year'. The original reform had thus taken place 720 years before the birth of Christ, which was the time it would have taken the solstice to move from its original position on 31 December to 25 December. While he failed to cite any source for this idiosyncratic view of the history of the Roman calendar, it is not unlikely that our author was mindful of a passage in Ovid's *Fasti* (1.164), which seemed to associate the winter solstice and the inception of the Sun's annual course with the Kalends of January. The interpretation of Ovid's words along these very lines can be traced as far back as the 1160s, when the computist known as Master William used them to conjecture that the winter solstice once fell on 1 January.[152] For the anonymous critic of 1276, January's original role as the *ianua anni* called for an intervention more radical than anything contemplated in writing before his time. In order to achieve a full restoration of Caesar's calendar, it was necessary to erase 16 days from a given year, such that the winter solstice, now located on 15 December, was once again immediately followed by the Kalends of January. If this remedy was withheld until AD 1320, the correction would have to be increased even further to 17 days.[153] It was clear, moreover, that if the calendar was supposed to remain aligned to the Sun, the bissextile day had to be suppressed once every 120 years, starting in AD 1320, followed by equivalent steps in 1440, 1560, and so on.[154] Yet, since the night of Jesus's birth had coincided with the winter solstice, a perfect calendar was naturally also required to maintain this historical correlation, calling for Christmas to be moved by seven days, from 25 December to 1 January. The new calendar hence dissolved the accustomed connection between Christ's nativity and the eighth day before the Kalends of December and instead took the course of the Sun from solstice to solstice as the ultimate reference frame for locating saint days and other liturgical or commemorative dates:

> And thus it will be possible to restore the calendar to its pristine state and preserve it in perpetuity without any error; and in the year when the bissextile [day] is omitted the winter solstice and the hour in which the Son of God was born will always come together at midnight at the end of the year and the aforementioned solstice and the calendar will complete their annual course and jointly embark on a new course.[155]

[152] *Summa magistri Willelmi de compoto*, MS Oxford, BodLib, Digby 56, fol. 217ra. See also MS Paris, BnF, lat. 10358, fols. 283r. A similarly worded passage, with the same reference to Ovid, appears in the undated thirteenth-century *Computus* of Peter of Berne, found in MS Oxford, BodLib, Canon. Misc. 71, fols. 5r–14v, at fols. 5v–6r. See also Bono of Lucca, *Computus lunaris* (ed. Arrighi 1991: 8), and the note in MS Oxford, BodLib, Ashmole 1796, fol. 172v, cited in North 1976: ii. 380.

[153] MS Florence, Biblioteca Riccardiana, 885, fols. 215v–216r.

[154] See ibid., fol. 216r, for the proposal, and fol. 217r, for a corresponding table of 120-year intervals from AD 120 to 12000.

[155] Ibid., fol. 216r–v: 'Et sic kalendarium in statum pristinum poterit restitui et sine errore aliquo in perpetuum custodiri semperque in anno cessationis bisexti hyemale solsticium et hora qua filius Dei natus fuit in fine anni in media nocte simul convenient dictumque solstitium et kalendarium cursum suum annualem simul perficient et simul reincipient novum cursum.' The phrasing here suggests that

The seriousness with which the author viewed this suggestion is underscored by his inclusion of a complete *kalendarium*, which, in full agreement with the corrections described in the text, marks the first day of the year as *nativitas Domini*, while the autumn equinox appears on the final day of September. More than that, all fixed celebrations of the ecclesiastical year appear uniformly shifted by 7 days, with Epiphany appearing on 13 January, rather than 6 January, and the feast of St Benedict on 28 March, rather than 21 March (see Figure 4.4).[156] The question that remained to be addressed was how such a far-reaching change in calendrical practice could become binding and effective for all of Christendom. Faced with this question, our author's only hope was that wiser men than himself were going to read his little work, correct it where necessary, and pass it on to the supreme pontiff, so that he might:

> deign to administer the remedy for the mentioned errors of the calendar, lest the nations of unbelievers perhaps take this as an opportunity to suspect that some deformity appears on the face of Christ's bride and dare rashly to abuse the customs of Christians and of the holy Church of God with their perverted mockery.[157]

His brief but frank warning regarding the laughter of infidels at the Church's expense is reminiscent of the opening lines of Reinher of Paderborn's *Compotus emendatus*, written in the previous century, and even more so of some of the words Roger Bacon had addressed to Pope Clement IV only a decade earlier. In spite of this good company, however, our author's proposal to subtract 16 days from the calendar and to move Christmas from 25 December to 1 January was almost certainly too extreme in its implications to earn it serious consideration from the pope, or indeed anyone else. It is true that at least one of his specific demands, that of having the start of the year coincide with the winter solstice, was to crop up sporadically in early modern writings on the calendar problem. Known advocates of this idea during the sixteenth and seventeenth centuries include the Spanish humanist Juan Ginés de Sepúlveda, the Italian mathematician Giambattista Benedetti, and the German astronomer Johann-Heinrich Voigt, yet none of these authors was ever close enough to the mainstream of the contemporary debate to influence actual calendrical reform policy.[158] Indeed, the desire of these authors to see the Roman calendar restored to a putative original setting was at odds with the more widespread goal of celebrating Easter in line with the decrees of the late antique Church.

the solstice fell at midnight between 24 and 25 December in the year of Christ's birth, but between 31 December and 1 January at the time when the calendar was founded. This conflicts with the author's claim that the solstice had moved by only six days during the period that separated the beginning of the calendar from Christ's birth.

156 Ibid., fols. 218r–223v.

157 Ibid., fol. 217r: 'Cum vero sapientes dictum opusculum examinaverint pleneque . . . correxerint, summo pontifici totius orbis principi sugger<re>e, si eis placuerit, poterunt quod dictis erroribus kalendarii remedium dignetur apponere, ne forte nationes incredule occasionem concipiant suspicandi quod aliqua deformitas appareat in facie sponse Christi moresque Christianorum et ecclesie sancte Dei suis perversis derisionibus presummant temere depravare.'

158 See Kaltenbrunner 1876: 408 (Sepúlveda); Ziggelaar 1983: 211, 214 (Benedetti); Koller 2014: 319 (Voigt).

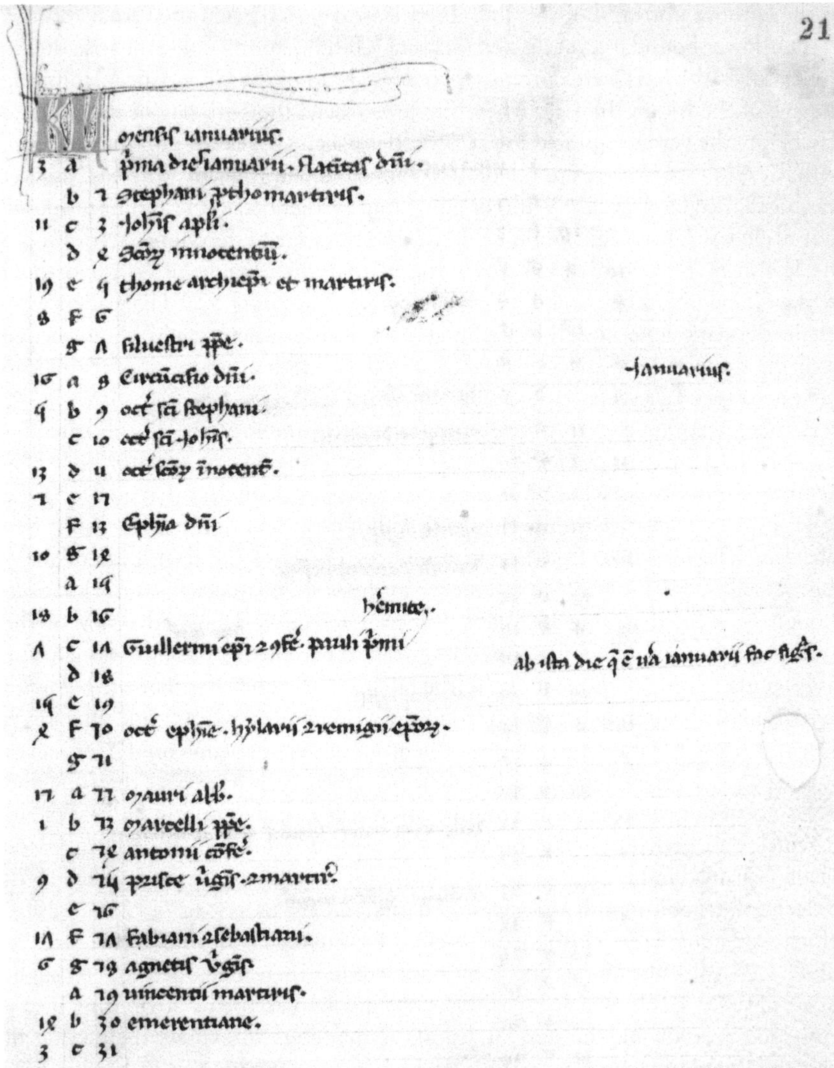

Figure 4.4. *Assignacio errorum kalendarii et eorundem correctio* (1276), reformed calendar with Christmas on 1 January. Florence, Biblioteca Riccardiana, 885, fol. 218r.

Those who were chiefly concerned about the latter issue tended to frame the error of the Julian calendar not as some misalignment between the winter solstice and 1 January, but as the regression of the vernal equinox from its canonical date on 21 March, where the earliest *terminus paschalis* resided. The author responsible for the proposal of 1276 devoted only a single brief chapter to this question, in which he mostly emphasized the tight historical link that existed between the date of Passover in the Jewish calendar and the calendrical *terminus* for Easter. According to the author's portrayal of this link, both feast days respected the vernal equinox as their lower boundary, yet discrepancies of a full lunation could still arise due to the fact that the Jews were currently two days ahead of the Christians in counting the age of the Moon. In cases where the Jews placed the 14th day of the Moon a day before the vernal equinox, the ecclesiastical calendar placed it on the day after, with the result that Easter was celebrated on the following Sunday, whereas Passover was postponed by a month. Without going into further detail, the author proposed that his fellow Christians should begin to count the age of the Moon 'according to the Jewish manner', which effectively meant that any calendrical *luna* 1 marked by the Golden Number would have to be converted to *luna* 3.[159] His readers were left in the dark as to how this seemingly simple adjustment could be reconciled with the projected omission of 16 days or the additional suppression of the leap day once every 120 years.

A more accurate and technically proficient treatment of the Jewish calendar vis-à-vis the Christian *computus* had already been offered one century earlier by Reinher of Paderborn, whose *Compotus emendatus* made it clear that adopting the 'Jewish manner' involved more than just adding two days to the lunar count (see Ch. 3.2). The idea that the Church should turn towards the Jewish calendar for counsel and inspiration was to re-emerge in the work of Roger Bacon, who, even though he seemed willing to follow Robert Grosseteste in his advocacy of the Arabic calendar, also drew the pope's attention to two further options, namely 'the tables of the Greeks and, most of all, [the tables] of the Hebrews. For the Hebrews were from the beginning highly skilled in the wisdom of astronomy and all nations have this science from them.'[160] It is difficult to see what Bacon could have meant by the 'tables of the Hebrews' other than the tables used to extract the time and week-day of the *molad* or mean conjunction for a particular month, Latin versions of which had been in circulation since the days of Reinher of Paderborn.[161] As Bacon explained in his *Opus maius*, the Jewish lunar reckoning measured the length of the mean synodic month as 29 days, 12 hours, and 793 parts (or 29d 12;44,3,20h), which was remarkably close to the pseudo-Ptolemaic 29;31,50,8,9,20d (or 29d 12;44,3,15,44h) already accepted by Grosseteste. There remained a very slight excess of 0;0,0,4,16h, which Friar Roger was quick to excuse by attributing it to a conscious decision on the part of Jewish astronomers to express their value in whole numbers of Hebrew parts, one of which was equivalent to $\frac{1}{1080}$ of an hour.[162] His admiration for the present-day Jewish calendar, which had been entirely ignored by Grosseteste and received only marginal attention in Giles of Lessines's *Summa*

[159] MS Florence, Biblioteca Riccardiana, 885, fol. 217r: 'Unde, cum liquide constet nos aut ipsos Iudeos a vero pascali termino deviare, mihi visum est, salvo iudicio meliori, nisi ipsi in observatione etatis lune decepti sint, quod nos etatem lune deberemus secundum morem iudaicum numerare, ut videlicet ubicumque eam primam esse dicimus, tertiam diceremus.'

[160] Roger Bacon, *Opus tertium* 70 (ed. Brewer 1859: 288).

[161] See n. 55 in Ch. 3.2. Reinher's tables for *molad* reckoning, together with some of the accompanying text, also circulated separately in some thirteenth-century manuscripts. See MSS Paris, Bibliothèque de l'Arsenal, 877, fol. 3r–v; Paris, BnF, lat. 7434, fols. 104r–105r; Florence, BML, San Marco 185, fols. 84v–85r (tables only); Vatican City, BAV, lat. 3118, fol. 3r–v (tables only).

[162] Roger Bacon, *Opus maius*, pars IV (ed. Bridges 1897: 196–7).

de temporibus, is further revealed by the fact that the lost 'master copy' of the *Opus maius*, which he sent to the papal curia in 1267 or 1268, was accompanied by a Hebrew manuscript containing calendrical tables. To the intended recipient, Pope Clement IV, Bacon recommended these tables as an 'extraordinary work of astronomical art', adding that their knowledge could be useful in the Church's efforts to 'convince' the Jews to let go of the faith of their ancestors and convert to Christianity.[163]

That Roger Bacon might have been something of a catalyst for a renewed interest in the Jewish calendar among late thirteenth-century computists, at least among those who belonged to his own monastic order, is indicated by the writings of three further Franciscan friars, each of whom allowed the Jewish calendar to play a central role in their approach to time reckoning. The earliest example worth mentioning is a *Compotus philosophicus* written in the year 1273, or very shortly thereafter, by an otherwise unknown friar named John. In the earliest copy of this text, once owned by the Franciscans in Lüneburg, his name is given as *Iohannes de Gu*..., with the rest of his name inexplicably deleted.[164] It seems possible that this Friar John was himself once a member of the Lüneburg Franciscans, whose friary is said to have been founded in 1235, but the only true hint as to his location of writing are two references to the city of Magdeburg in Saxony, which his treatise mentions as a location of reference for both geographical longitude (2;40h or 40° west of Jerusalem) and latitude (52° N).[165] Even more so than the 'philosophical' *computi* written by his predecessors Robert Grosseteste and Campanus of Novara, Friar John's treatise was an almost exclusively astronomical affair, one that bore only little resemblance to the conventional calendrical treatises that had populated manuscripts up to this point in time. John highlighted the distinctive nature of his work by declaring from the outset that its purpose was 'to demonstrate the falsity of our calendar and to offer a way by which it can be brought in line with the truth', as well as 'to fix with certainty the times of the new moons and the conjunctions of Sun and Moon according to the truth afforded by philosophy'.[166] In fact, only the first part of the *Compotus philosophicus* was at all dedicated to matters calendrical, whereas the second part assembled all the knowledge necessary to calculate true

[163] Roger Bacon, *Opus tertium* 56 (ed. Brewer 1859: 220). See also 54 (ed. Brewer 1859: 214–15), and the more detailed discussions in Nothaft 2012: 178–96; 2014d: 132–40.

[164] See MS Lüneburg, Ratsbücherei, Ms Miscell. D 4° 46, fols. 1ra–19r (s. XIV^{in}), where the text ends 'Explicit conpotus novus philosophicus compositus per fratrem Johannem de Gu...' (fol. 18vb). Zinner 1939: 70–1; Zinner 1988: 77–8, gives the name as 'Johannes von Gustedt', but without explaining his reasoning. All references to the *Compotus philosophicus* in what follows will be to the Lüneburg codex, which was the exemplar for MS Vienna, ÖNB, 5239, fols. 10r–28v, used in Kaltenbrunner 1876: 307–8. Two more copies, both of them anonymous, are in MSS Hannover, Niedersächsische Landesbibliothek, IV 389, fols. 1r–15v (dated 1342), and Vatican City, BAV, lat. 3112, fol. 1r–28v. For further discussion, see Nothaft 2014c: 229–31; 2014d: 573–611.

[165] *Compotus philosophicus*, MS Lüneburg, Ratsbücherei, Ms Miscell. D 4° 46, fols. 10vb, 13v. Magdeburg had been the site of a Franciscan friary since 1225. On the beginnings of the Lüneburg friary, see Logemann 1996: 12–18.

[166] *Compotus philosophicus*, fol. 1ra (pt. 1, ch. 1): 'Cum sit intencio ostendere falsitatem kalendarii nostri et dare viam per quam possit verificari, determinare certa tempora primacionum et coniunctionum solis et lune secundum philosophicam veritatem, et cetera huiusmodi, necessarium fuit pauca premittere de cursu solis et lune, quem faciunt in suis circulis.'

longitudes and syzygies of the Sun and Moon as well as the circumstances and magnitudes of their eclipses, to which end John interspersed his text with a comprehensive set of tables as well as geometrical diagrams. Although most of this material was derived from the Toledan Tables, the Franciscan astronomer chose to arrange his mean syzygy tables as well as the tables for the access and recess of the eighth sphere in a distinct manner, basing them on 'collected year' cycles of 19 years in imitation of the Easter *computus* and the calendar of the Jews. The starting epoch for these cycles was keyed to the mean conjunction, or *molad*, of Nisan in 1273, which the Jewish calendar placed on Monday, 20 March, at 12 hours and 456 parts after sunset. John converted the Hebrew parts (or *ḥalaqim*) into sexagesimal minutes and changed the day-epoch from sunset to midnight, putting the *radix* for mean conjunctions on Monday, 06:25h, but otherwise his syzygy tables fully adhered to the Jewish *molad*-system, which was supposedly based on the longitude of the holy city of Jerusalem.[167]

The same organizing principle can be discerned in the various tables John saw fit to include in his work's first part, which put users in a position to find the mean conjunction of Sun and Moon according to four different sets of parameters, associated with Ptolemy and Azarquiel as well as with the distinctive calendar systems of the Jews and Arabs. One particularly innovative example was John's conjunction tables 'founded on the [month] length of Ptolemy' (*fundata super quantitatem Ptolomei*), which employed a unique subdivision of the hour into 64 parts and a lunation containing 29 days, 12 hours, and 47 such parts (see Figure 4.5). Whether it was John's own invention or not, it seems clear that this odd definition of the 'part' was rooted in an attempt to adhere closely to a pseudo-Ptolemaic mean synodic month of 29;31,50,8,9,20d, as previously invoked by Robert Grosseteste and Roger Bacon, but without resorting to sexagesimal fractions below the level of a minute. To achieve this goal, the table subtracted two parts over the course of 19 years, thus raising the implied mean value of the entire table to 29;31,50,8,10,41 . . . d.[168] In the accompanying text, Friar John revealed that Ptolemy's 29;31,50,8,9,20d, if multiplied by 3,760, yielded 111,035 full days + 0;1,51,4,53,20d, whereas 304 Julian years contained 111,036 complete days.[169] Apart from the evident necessity of remedying this defect by suppressing the bissextile intercalation once every 304 years, it also seemed advisable to omit 2 days from a year in the present. These 2 days had been superfluously inserted since the beginning of the present calculation, which Friar John located 600 to 700 years before his time, 'when the Church first began to have peace' (*quando ecclesia primo incepit habere pacem*).[170] An even more reliable way of improving the accuracy of one's new-moon computations was opened up by the aforementioned tables of mean syzygy drawn up according to Ptolemy, Azarquiel, the Jews, and the Arabs. These tables closely resembled each

[167] See Nothaft 2014d: 608, and the tables in MS Lüneburg, Ratsbücherei, Ms Miscell. D 4° 46, fols. 11r–v, 13v, 19r.

[168] See *Compotus philosophicus*, fols. 3rb–4rb (pt. 1, ch. 6), with the described table on fol. 3v.

[169] Ibid., fol. 2rb–vb (pt. 1, ch. 4).

[170] Ibid., fol. 2vb. Friar John here argues that the present 19-year cycle could not have been in use since Caesar's time, since otherwise the presently observed error would be 4 days or more.

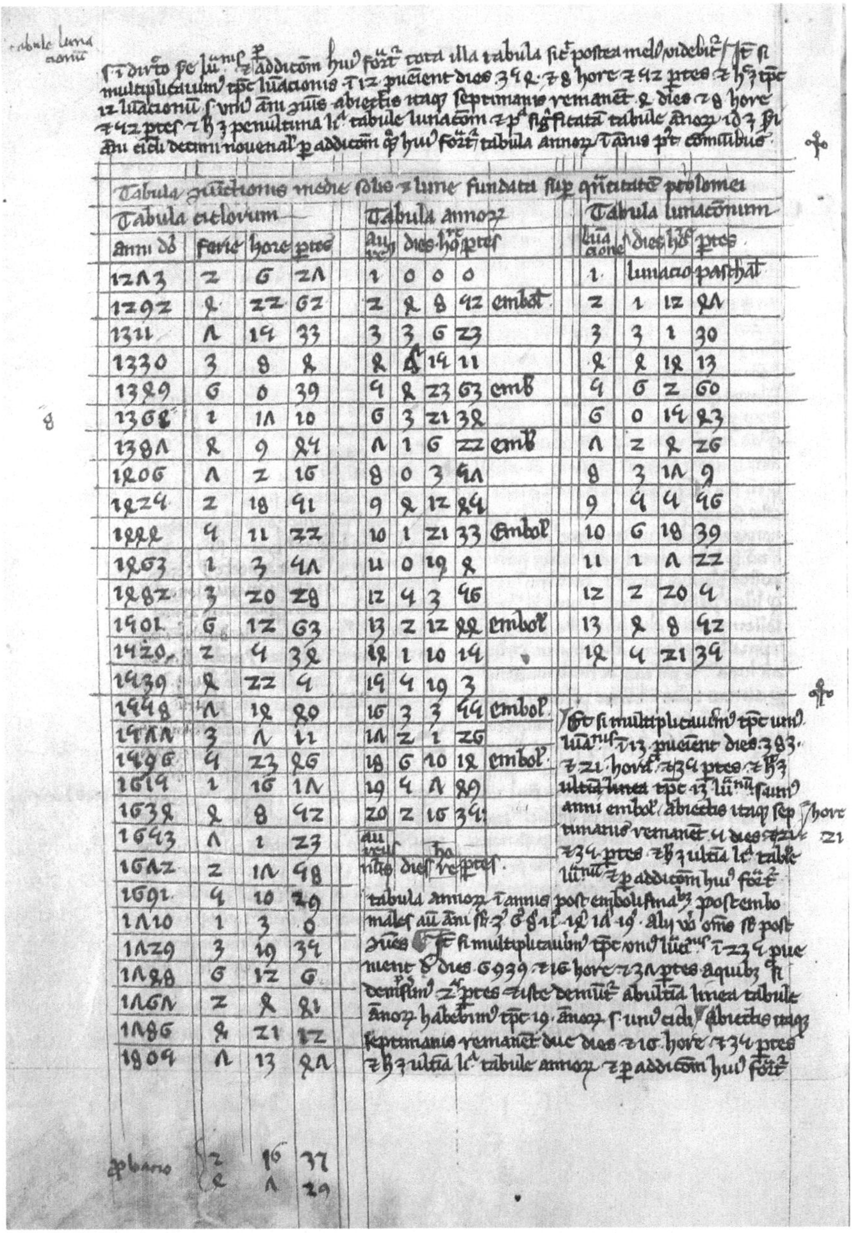

Figure 4.5. Friar John, *Compotus philosophicus* (1273), table of conjunctions 'according to Ptolemy'. Lüneburg, Ratsbücherei, Ms. Miscell. D 4° 46, fol. 3v.

Source: Ratsbücherei Lüneburg

other in displaying no calendrical dates but only the day of the week, hour, and hour-fraction of the conjunction in question, which was sufficient information to correct the new-moon dates indicated by the Golden Number in the ecclesiastical calendar as long as one followed the methods John outlined at great length in his treatise.[171]

Like Roger Bacon before him, Friar John showed clear signs of familiarity with Grosseteste's *Compotus*, in particular with its discussion of the length of the solar year, in which Ptolemy's solar year of $365\frac{1}{4}-\frac{1}{300}$ days was contrasted with al-Battānī's $365\frac{1}{4}-\frac{1}{100}$ days. The former value matched the approximate rate at which the equinoxes and solstices could be seen to regress from their calendrical dates in present times, whereas al-Battānī's much swifter tropical year was in better agreement with the total regression since the time of Christ's birth. A third opinion had been thrown into the ring by certain computists, who claimed that the equinoxes and solstices had moved by 10 days and 14 hours over the past 1,270 years, supporting an average solar year of $365\frac{1}{4}-\frac{1}{120}$ days.[172] John claimed that this position received support from the 'instruments of investigation' (*instrumenta consideracionis*), by which he appears to have meant nothing more than the Toledan Tables, which showed that the winter solstice in the previous year 1272 had fallen on 14 December, at noon.[173] In a move very reminiscent of the anonymous author of 1276, John even suggested that it was possible to create an ideal solar calendar, where the first day coincided with the day of the winter solstice and the feast of Christ's nativity and where the lengths of the individual months were made to correspond more closely with the 12 tropical signs of the zodiac.[174] What he failed to do in his extended discussion was to affirm or deny any particular opinion on the actual length of the tropical year, a reticence that is understandable in light of his simultaneous endorsement of the Toledan theory of access and recess. A roughly contemporary effort to spell out this theory's implications for future calendrical shifts of the equinoxes and solstices has survived in the form of a *Tabula solstitiorum et equinoctiorum*, which accompanies an eclipse tract in a fifteenth-century astronomical anthology, the compiler of which incorporated several of Friar John's tables. According to the accompanying explanation, the *Tabula solstitiorum et equinoctiorum* was made for the Julian leap-year cycle 1281–4, starting with the vernal equinox on 13 March 1281, at 07:05h. For the following three years, the equinox was indicated to shift to 13:03h, then to 19:01h, and finally—after the intercalation of a bissextile day in year III—to 00:59h, all of which suggested a year length

[171] Ibid., fols. 3ra–b, 4rb–5ra (pt. 1, ch. 5, 7).

[172] Ibid., fol. 2ra–b (pt. 1, ch. 4). Cf. Robert Grosseteste, *Compotus* 1 (ed. Steele 1926: 215, ll. 3–33).

[173] *Compotus philosophicus*, fol. 2rb (pt. 1, ch. 4). Modern calculation places the winter solstice of AD 1272 on 13 December, at *c.*7:30 p.m., whereas the Toledan Tables give 14 December, 8:45 a.m. An application of the Jerusalem meridian, which John believed to be 3;27h or 51;45° east of Toledo, would have shifted this result almost exactly to noon. In a later passage, John claims that the Sun at noon on 14 March 1273 had traversed 0;40° in the sign of Aries, again according to the *instrumenta consideracionis*. This is in fairly good agreement with the Toledan true solar longitude (0;42°) for the standard meridian. See *Compotus philosophicus*, fol. 5rb (pt. 1, ch. 8). Zinner 1988: 77 is hence mistaken in treating John's data as based on observation.

[174] *Compotus philosophicus*, fol. 4va–b (pt. 1, ch. 7).

of 365d 5;58h or $365\frac{1}{4}-\frac{1}{720}$ days. The anonymous author confirmed this by predicting that the Julian calendar would continue to lose 2 minutes per year until 1337, followed by a 56-year period (to 1393) during which the equinoxes and solstices stood still on their current dates. Afterwards, they were going to move in the opposite direction, as the tropical year was set to exceed the Julian year by 2 minutes until 1450.[175]

Even with this long-term perspective in mind, critics such as Friar John could still insist that the bishops at the Council of Nicaea had erred when fixing the rules for the celebration of Easter. According to John's analysis, the chasm that gaped between the current date of the spring equinox, on 13/14 March, and the Nicene *terminus paschalis* caused the Church to pick the wrong Easter lunation in years 8, 11, and 19 of every lunar cycle. In these years Christians ended up celebrating their most important feast in the second rather than the first lunation of the year, in breach of the biblical commandment to celebrate Passover in the first spring month. The resulting error manifested itself most clearly in years 8 and 19, which were the two years in the cycle in which the Jews always convened for their Passover meals one month ahead of Easter.[176] How the Jews actually ordered their ritual life in the here and now was a subject to which John devoted a surprising amount of space, explaining not just the familiar *molad*-system but also other, more exotic elements of the Jewish calendar, such as its postponement rules, the four *tequfot* (equinoxes and solstices), and the dates and origins of various feast days, both biblical and non-biblical. One symptom of John's unusual depth of knowledge in this area was his careful presentation of a bipartite table that made it possible to determine which of the 14 possible *keviyyot* or types (*genera*) a given year in the Jewish calendar was going to represent and, by extension, to find the day of the week at the start of each of its individual months.[177] A sizeable portion of this material was reused two decades later by John of Brunswick, whose *Compotus novus* of 1297 devoted ample room to the findings of 'reformist' computists, from John of Sacrobosco and Robert Grosseteste to the enigmatic Friar John.[178] That the man from Brunswick in Lower Saxony was more than a mindless compiler can be seen from the commentary he wrote in 1298 to elucidate his own *Compotus novus*, which contains observations on the Jewish calendar and its terminology not found in any of his sources, among them some intriguing quasi-ethnographic remarks on the Jewish custom of not drinking water at the turn of the *tequfot*.[179]

[175] MS Bernkastel-Kues, Bibliothek des St. Nikolaus-Hospitals, 212, fol. 258r. The table and explanation are placed underneath Friar John's *Tabula ciclorum ad mediam coniunctionem*, here ranging from 1273 to 1634.

[176] *Compotus philosophicus*, fol. 5ra–va (pt. 1, ch. 8).

[177] Ibid., fols. 6r–8vb (pt. 1, ch. 10–11). The corresponding tables are transcribed in Nothaft 2014d: 586, 588.

[178] See esp. John of Brunswick, *Compotus novus*, MS Leiden, Universiteitsbibliotheek, Scaliger 66, fols. 22ra–35ra (chs. 13–15), and the detailed comparison with Friar John's *Compotus philosophicus* in Nothaft 2014d: 573–94.

[179] Full copies of this commentary are preserved as *Scriptum super novum compotum*, in MS Florence, BML, Plut. 30.24, fols. 87r–99r, and as *Sententia compoti nove compilationis*, in MS Vatican City, BAV, 3112, fols. 27r–67v. The Hebraistic remarks are discussed in Nothaft 2014d: 594–607.

That alternative calendar systems were attentively studied in Franciscan circles during the 1270s is further suggested by the existence of a rudimentary *Summa astrologiae* written in 1275 or 1276 by another Franciscan author known to us only as Friar John (*frater Iohannes de ordine minorum*).[180] Before he came to speak about the motions and influences of the heavens, which was the main topic of his treatise, John dedicated the first of three parts to the subject of time reckoning. After a brief introduction to the Julian calendar and its shortcomings and a discussion of the errors currently committed in calculating Easter, he spent no fewer than three chapters on the calendar and world era of the Jews, followed by two chapters on the Arabic lunar calendar and its conversion into other systems.[181] Commenting on the relationship between the Jewish and Christian lunar calendars, John pointed out that the ecclesiastical version of the 19-year cycle surpassed the Jewish one by 1 hour and 485 parts, accounting for a discrepancy of roughly $M = 304$ years.[182] Similar to his northern German namesake, he offered readers all the tables they needed to determine the time of the current *molad*, with the difference that his were based on the actual epoch of the Jewish calendar (Monday, 7 October 3761 BC, 5h 204p), with a beginning of the year in autumn (Tishri).[183] What both authors could agree on was that the Jewish system promised a more reliable calculation of the beginning of the first lunar month in spring, which the Book of Exodus and other parts of the Pentateuch had designated as the appropriate month for Passover. Much like the other Friar John, the author of the *Summa astrologiae* voiced the complaint that 'those who composed our calendar make us err to a great degree' by failing to anticipate the regression of the equinoxes and solstices this calendar entailed.[184] In gauging the rate of this regression, he relied on the value worked out by Ptolemy, assuming that the cardinal points were continuously moving through the calendar at $J = 300$ years. Ptolemy's value urged the conclusion that the vernal equinox and winter solstice at the start of the Christian era had still resided on 20 March and 20 December, meaning that they preceded the conception and birth of Jesus by six inclusive days. In accordance with Jewish tradition, he put the date of the world's creation 3,760 years before these events, from which it followed that by the year of the world 4800, which overlapped with AD 1040, the equinoxes and solstices had reached a date 16 days earlier than where they had been found at the beginning $(4{,}800 \div 300 = 16)$. Had the Julian calendar already been in place at the moment of creation, the Sun's entry into the degrees representing the cardinal points (0° Aries, 0° Cancer,

180 The text survives in MSS Paris, BnF, lat. 7293A, fols. 48r–69r (s. $\text{XIII}^{4/4}$ or s. $\text{XIV}^{1/4}$); Vatican City, BAV, Urb. lat. 1399, fols. 29rb–31rb (s. $\text{XIV}^{2/4}$; pt. 1, chs. 1–4 only); Vienna, ÖNB, 5309, fols. 127ra–55va (s. $\text{XV}^{2/4}$; entitled *Summa Alberti*). See Thorndike 1957c: 126–7; Nothaft 2014d: 150–1, 617–21.

181 Friar John, *Summa astrologiae*, MS Paris, BnF, lat. 7293A, fols. 48r–52r.

182 Ibid., fol. 49v (pt. 1, ch. 4).

183 Ibid., fol. 50r–v (pt. 1, ch. 5).

184 Ibid., fol. 48v (pt. 1, ch. 2): 'Sed illi qui composuerunt kalendarium nostrum faciunt nos multum errare, quia enim tempore ipsorum erat plenilunium post equinoctium in festo sancti Benedicti [= 21 March] instituerunt quod semper facias pascha in primo plenilunio post festum dictum, acsi semper esset equinoctium in illo in festo sicut tunc erat. Et cum equinoctium antecedat plus et plus faciunt nos a veritate deficere.'

0° Libra, 0° Capricorn) would have been found on the first days of January, April, July, and October, at least according to John, who claimed that these entries now all occurred on the 17th day before the Kalends, that is, on 16 March, 15 June, 15 September, and 16 December.[185] None of this compared favourably with the much more accurate dates and values cited by contemporary authors such as Giles of Lessines, Roger Bacon, Campanus of Novara, and the Friar John responsible for the *Compotus philosophicus*.

An exposition of Jewish calendation and chronology even richer than any of those mentioned thus far became available to English readers in 1294, when a Franciscan named Robert of Leicester completed a whole 'Treatise on the *Computus* of the Hebrews adapted to [Our] Calendar' (*Tractatus de compoto Hebreorum aptato ad kalendarium*) in four parts, thereby fulfilling an assignment he had received from Richard Swinfield, bishop of Hereford from 1282 to 1317.[186] In its description of the 'Hebrew' manner of time reckoning, the resulting work was nothing if not systematic. The first part approached the topic from an astronomical perspective, explaining the basics of the *molad* calculation together with a method to determine the precise times of the equinoxes and solstices (*tequfot*) according to a system Jewish authors were wont to attribute to Rav Ada. In this system, the length of the tropical year was simply obtained by dividing the total length of 235 mean lunations by 19 solar years, one of which was hence found to last 365 days, 5 hours, 997 parts, and 48 'moments' (*regaim*), with one moment defined as $\frac{1}{76}$ of a part or *ḥeleq*.[187] In the following section of the *Tractatus*, Robert introduced his readers to the intricate rules that governed the Jewish calendar, where the lunar months and years of the *molad*-system were bound into periods comprising 353–5 complete days for common and 383–5 days for embolismic years. There were also rabbinic restrictions for the days of the week on which the year was allowed to begin. Combining these with the other rules produced 14 different year types, which Robert displayed in two tables very similar to those that had previously adorned Friar John's *Compotus philosophicus*. Borrowing a distinction established by twelfth-century computists, he characterized this system of rules as the *compotus usualis* or *vulgaris* of the Jewish calendar, associating it with the 'common folk of the Hebrews' (*vulgus Hebreorum*). Its counterpart was the Hebrew *compotus naturalis*, in which 19 lunar years comprised not a complete number of days, but 6,939d 16h 595p. This made it impossible for the same sequence of year types to recur exactly in the neat cycle of 247 years that some of the 'vulgar' tables employed.[188] Part III of Robert's treatise compared the lunar cycle of the Jews with that used by the Roman Church in an effort to obtain a conversion method between the two systems. One key ingredient in his method was a detailed tabular listing of the parts, hours, Julian dates, and ferials of the *moladot* of Tishri and Nisan of the first 76 years of

[185] Ibid., fol. 48r (pt. 1, ch. 1).

[186] See Nothaft 2013c, and the edition and translation of Robert's treatise in Nothaft 2014d: 206–335, with extensive introduction and commentary on pp. 128–205.

[187] Robert of Leicester, *Tractatus de compoto Hebreorum aptato ad kalendarium* 1.2–8 (ed. Nothaft 2014d: 212–22).

[188] Ibid. 2.1–5 (ed. Nothaft 2014d: 224–33).

the Jewish world era.[189] Another element could be reduced to the insight that the Church's 19-year cycle made each lunation last for an average of 29 days, 12 hours, 799 parts, $50\frac{30}{235}$ moments. This exceeded the Hebrew measure by 6 parts and $50\frac{30}{235}$ moments, such that two *moladot* spaced by 76 years were bound to lie exactly 5 hours and 860 parts apart. The growth of this excess over time could be tracked with the aid of another table, which in Robert's version covered 7,600 years. By adding or subtracting an adequate multiple of 5h 860p from the entries in the great table of 76 years, it was possible to obtain the time and date of any *molad* past, present, or future.[190]

In striking contrast to Reinher of Paderborn, who had drawn attention to this value 120 years earlier, Robert of Leicester refrained from expressly linking the calendrical wisdom of the *Hebraei* to the problem of Easter reckoning. Far from turning his treatise into a contribution to the ongoing calendar-reform discussion, he marketed it as a study of biblical chronology, dedicating the fourth and final part to what he referred to as the 'flow of history' (*decursus saeculi*). Even so, no attentive reader could overlook the fact that the ancient Hebrews, whose calendar Robert identified with the calendar of the modern-day Jews, tracked the mean motions of Sun and Moon with much greater success than the Christian Church could claim. Where his predecessor and likely source Roger Bacon had blamed this situation on the general ignorance of astronomy within the early Church, Friar Robert preferred to 'think piously about the saintly teachers' and assume that they 'did not want to plunge into an abyss of difficulties and thus presupposed the computistical use of the Church to be fixed and stable', in so far as the dates of the equinoxes and solstices were concerned. In a similar vein, they had assumed that 'the Golden Number, as inscribed into the calendar, has uniformly kept its place since the world's beginning and will always continue to do so', even though Robert's own treatise proved the contrary.[191] In his preface addressed to Bishop Swinfield, he expressed the worry that his little work on Hebrew time reckoning might appear to some as the

> subversion of the ecclesiastical reckoning hitherto used by very saintly and experienced men and therefore the invention of a profane novelty. For how could it be deemed credible that the observations of the Hebrews regarding the lunar cycle or what the astronomers hand down about the length of the solar year or the agreement of the lunar orbit with that of the sun could have escaped the notice of the Venerable Bede and other calculators of the Church for all this time?[192]

No such anxieties appear to have troubled the otherwise unknown Franciscan theologian John of Presorio, who used the Jewish calendar to create a tabular record of conjunctions covering the 67th 76-year period since the start of the Jewish world

[189] See ibid., table 8 and Robert's commentary in Nothaft 2014d: 310–26.
[190] Ibid. 3.2–6 (ed. Nothaft 2014d: 234–40). See also Robert's commentary on table 7 (pp. 306–8).
[191] Ibid. 4.2 (ed. Nothaft 2014d: 247).
[192] See the translation of Robert's prologue in ibid., ed. Nothaft 2014d: 207.

era, which ran from September AD 1257 to August AD 1332.[193] His lunar tables were attached to a standard calendar using the four-column layout made popular by Gerland, but with conjunctions timed to the sexagesimal minute, second, and third (as opposed to *ḥalaqim*). The work came with an extensive introduction, which reproduced most of the technical information also found in Robert of Leicester's work and offered a worked example for the *molad* of January 1301, the probable year of writing.[194] Tables for the calculation of *moladot* were included too, as was a double-page table for the mobile feast days of the liturgical year and their intervals from Advent and Christmas. Despite this juxtaposition of two different calendar systems, Johannes de Presorio made no visible effort to draw a connection between them or to discuss the consequences that an ability to fix the time of the *molad* might have for the calculation of Easter. Thanks to the groundwork laid by other members of his order, most notably Roger Bacon and Friar John (the author of the *Compotus philosophicus*), the message could be left implicit without escaping too many readers.

[193] See the calendrical tables with extensive introduction in MS London, BL, Add. 17368, fols. 36r–51v. The colophon (fol. 51v) reads: 'Explicit canon kalendarii fratris Iohannis de Presorio magistri in theolagia [*sic*] de ordine fratrum minorum.'

[194] Ibid., fol. 36v (final two lines).

5

Astronomers and the Calendar, 1290–1500

5.1. 'ENHANCED' CALENDARS IN LATE MEDIEVAL EUROPE

The medieval debate surrounding the flaws of the ecclesiastical calendar went through a crucial formative phase during the 1170s, when it was pushed forward by a small avant-garde of *computus* experts that included Reinher of Paderborn and Roger of Hereford. It received consolidation and gained wider currency in the course of the first three quarters of the thirteenth century, aided in part by the immense popularity of certain 'philosophical' or 'academic' works on time reckoning, like those by John of Sacrobosco and Robert Grosseteste. One of Grosseteste's admirers and readers, the Franciscan philosopher Roger Bacon, introduced an important new dimension by moving the debate from a purely theoretical to a practical-political level. This holds true at least in the sense that Bacon addressed his demands for calendrical change directly to the reigning pope, Clement IV, in whose commitment to the general reform of Christian learning the English friar put a great deal of hope.[1] In addition, Bacon deserves credit for having reasserted or rediscovered the Jewish calendar as a potential model for reform, which had already underpinned the proposal made by Reinher of Paderborn in 1170–1. As seen at the end of the previous chapter, this idea was to undergo a remarkable resurgence during the final three decades of the thirteenth century, most notably among Franciscan writers, quite as if Bacon's ideas had found an appreciative audience within his own mendicant order. Be that as it may, the fact remains that Pope Clement IV died on 29 November 1268, just one year after Bacon had begun to pen the *Opus maius* and other works for his eyes, making it doubtful if the pope ever had a chance to assess any of the arguments in favour of a calendar reform that had been presented to him. Subsequent occupants of the Chair of St Peter failed to rise to the occasion. The year 1276, when a particularly audacious author envisioned the erasure of 16 days from the calendar (*Assignacio errorum kalendarii et eorundem correctio*; see Ch. 4.3), saw the death of three consecutive popes (Gregory X, Innocent V, Adrian V) and the election of a fourth one (John XXI), none of whom is likely to have spent much time on the issue. Those who came after them, from Nicholas III (1277–80) to Boniface VIII (1294–1303), would have had the opportunity to receive expert advice from Campanus of Novara, who first appears as apostolic chaplain in 1264 and seems to have retained this position without

[1] See on this point Power 2013a: 65–6.

interruption until his death in 1296. And yet, the extant sources fail to reveal any connection between Campanus's curial position and his work on the *computus*.[2]

Such inactivity on the part of the official Church did not condemn to idleness the many capable astronomers that could be found within its ranks. A significant, if little-studied, phenomenon in this regard is the surging production of astronomically enhanced calendars in Latin and the vernaculars, which married the traditional format of the *kalendarium* to the superior methods of prediction furnished by astronomical tables. One result of this new trend, which began to gain momentum during the 1290s, was a veritable calendar reform 'from below', which decisively weakened the grip of the traditional 19-year cycle on Western lunar computations, even if it did not effect any changes to the dating of Easter.[3] A relatively early example of this sort of activity is a calendar for the years 1292–1376 attributed in some manuscripts to Roger Bacon (Bacoun), although this may refer to an author different from the famous English friar (who is thought to have died in 1292 or not long thereafter).[4] As with numerous other astronomically enhanced calendars from the later Middle Ages, it displays the dates, hours, and minutes of 940 consecutive new moons in a four-column format familiar from Gerland and the *Kalendarium* of pseudo-Grosseteste (see Figures 5.1a and 5.1b). Each recto page facing the lunar tables offers the corresponding calendar page of the month complete with traditional elements such as liturgical feasts and saint days, a column for the ferial letters from A to G, and another for the conventional Golden Number, which differs by an average of four days from the conjunctions displayed on the verso page. More unusual is the inclusion of columns whose purpose is to show for each day of the year the Sun's tropical longitude (in degrees), the length of daylight (in hours and minutes), and the Sun's noon altitude (in degrees and minutes). Whereas calendars in the tradition of Gerland and pseudo-Grosseteste had been based on the traditional 19-year cycle and were by implication perpetual, the calendar of pseudo-Bacon was supposed to be strictly valid only for the 76 years from January 1292 to December 1367, as seen from the way its author advised prospective users to subtract 5;44h from the new-moon times on display in case they wanted to adjust them to the following 76-year period.[5] As the accompanying text makes clear, moreover, the new-moon times listed were mean conjunctions calculated according to the Toledan Tables, but without any adjustment for the local meridian due to the fact that Toledo's longitude 'is not removed by much from the meridian of the

[2] On Campanus's curial career, see Paravicini Bagliani 1973.

[3] In what follows, I shall merely mention some examples from a vast body of literature that has never been adequately surveyed. An important resource for future studies on 'enhanced calendars' is Zinner 1925: 172–83 (nos. 5320–709), who lists hundreds of late medieval calendar manuscripts from Central Europe. See, in addition, Kaltenbrunner 1876: 355–9; Wijk 1936: 38–45; Pfaff 1944.

[4] Copies survive in MSS London, BL, Cotton Vespasian A.II, fols. 2r–8v; Oxford, BodLib, Bodley 464, fols. 58r–71r; Oxford, Corpus Christi College, 221, fols. 2r–8v; Erfurt, UFB, CA 4° 351, fols. 73r–78v. See also the notes in Wordsworth 1904: 137; Little 1914: 417–18; Pedersen 1983–4: 258, 419; Pedersen 2002: 761.

[5] MS Oxford, BodLib, Bodley 464, fol. 58v: 'Sciendum est quod completis quatuor ciclis redit luna ad principium primi cicli, exceptis 5 horis et 44 minutis, que [*sic*] oportet subtrahere tunc a coniunctione cuiuslibet mensis et secundo redierit duplum.'

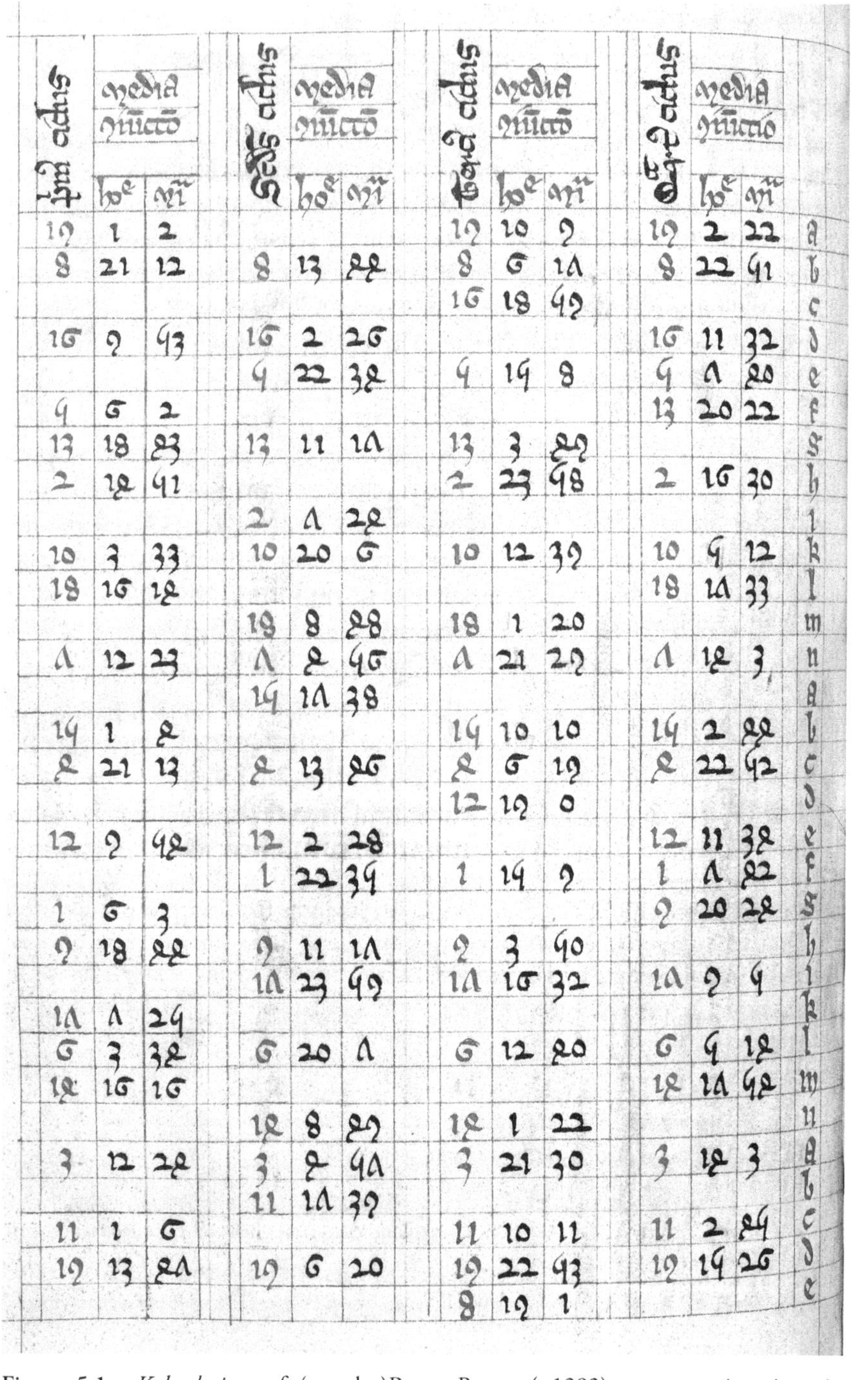

Figure 5.1a. *Kalendarium* of (pseudo-)Roger Bacon (*c.*1292), mean conjunctions in January for 1292–1367. Oxford, Bodleian Library, Bodley 464, fol. 59v.

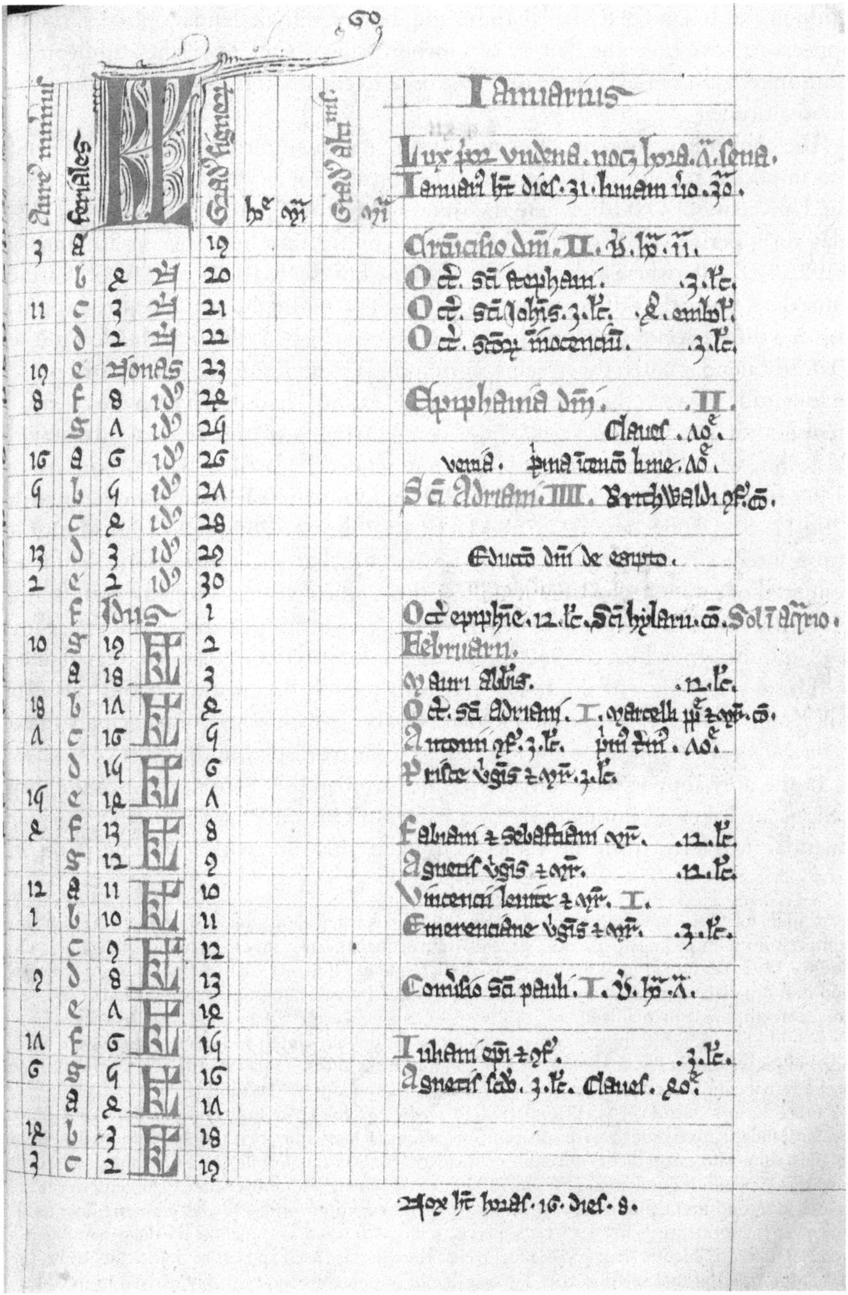

Figure 5.1b. *Kalendarium* of (pseudo-)Roger Bacon (*c.*1292), calendar page for January. Oxford, Bodleian Library, Bodley 464, fol. 60r.

midpoint of Ireland'.[6] Rather than Ireland, however, the calendar's place of origin appears to have been the diocese of Lincoln, which is the town the author mentions in relation to the latitude of 53° he used to construct his column for the Sun's noon altitude.[7]

The convenient shortcut of using Toledo's meridian for tables cast in England was implicitly renounced by the author responsible for a calendar made in 1292 at the University of Oxford, where the times of mean conjunction are again on display for a period of 76 years.[8] The starting conjunction here falls on 20 January 1292, at 18:18h, whereas the calendar attributed to Roger Bacon has 06:03h for the same date. Once the difference in epochs (noon vs. midnight) is taken into account, this is a difference of no more than 15 minutes, which, as the introduction to the Oxford calendar correctly explains, corresponds to an assumed difference of 3;45° in longitude between the cities of Toledo and Oxford.[9] In order to keep the calendar in use after the first 76 years, the accompanying text prescribes a subtraction of 5;48h, which exceeds the 5;44h recommended by pseudo-Bacon, but comes closer to what would follow from the Toledan value for the mean synodic month (29d 12;44,3,20h × 940 = 27,758d 18;12,13,20h = 27,759d − 5;47,46,40h). Of some interest is the author's follow-up remark that such corrections could be rendered obsolete if one were to follow the tropical year of Ptolemy and omit a bissextile day once every 300 years, which would leave a discrepancy of only *c.*48 minutes by which 4 × 940 lunations are longer than 304 solar years (since 5;48h × 4 = 23;12h = 1d − 0;48h).[10] As with pseudo-Bacon, the Oxford calendar still retains a column for the old Golden Number, which is here disparagingly titled *primacio falsa*, in contrast to the *verae primaciones* on display in the other columns.

In the only copy to have surfaced of this Oxonian production, the introductory canons are followed immediately by a paragraph introducing yet another 76-year calendar 'extracted from the Tables of Toledo', this time with a starting point in

[6] Ibid., fol. 58r: 'Canon super sequens kalendarium ad veram coniunctionem solis et lune accipendam compositum a fratre Rogero Bacoun. Kalendarium sequens extractum est a tabulis Tholetanis anno domini 1292, factum ad meridiem civitatis Tholeti, que in Hyspania scita [*sic*] est, cuius meridianus non multum distat a meridiano medii puncti Hybernie.' In reality, most places on Ireland's east coast are *c.*2° further west than Toledo.

[7] Ibid., fols. 58v–59r. In this copy, the columns for the day length and noon altitude were never filled in (as seen from Fig. 5.1b), but in MS London, BL, Cotton Vespasian A.II, fol. 5v, the altitude at the vernal equinox is indeed given as 37°, as is appropriate for 53° latitude.

[8] See the introduction in MS Oxford, BodLib, Digby 149, fol. 124r: 'Incipiendo annum incarnationis a capite Ianuarii precedente diem incarnacionis vel annunciacionis, quia ibi communiter annos incipimus usuales, supposito etiam dicto Abrachim mathematici, scilicet quod quilibet quartus annus solaris sit bisextilis. Sciendum quod anno Domini 1292 facta est in Anglia in universitate Oxoniense verificata primacio secundum medios motus solis et lune ponendo continue omnes medias coniunctiones unius magni cicli continenti ex 4^or^ 19^lis^ ciclis parvis, scilicet 76 annorum, incipiendo diem naturalem a noctis medio precedente, sicut ecclesia Dei facit.' The calendar itself appears on fols. 128r–133v. The (false) information that the bissextile day was due to Hipparchus was probably derived from Robert Grosseteste, *Compotus* 1 (ed. Steele 1926: 214, ll. 2–5).

[9] MS Oxford, BodLib, Digby 149, fol. 124r–v. The correct longitudinal difference between Oxford and Toledo is closer to 2;45°, but the text's stated northern latitude of Oxford (52°) is a very good approximation to the correct value (51;45°).

[10] Ibid., fol. 125r. The same page offers a computation to the effect that the equation of 940 lunations and 27,759 days implies a mean lunation of 29d 12;44,25,31,54,53,37,1h.

1289.[11] While no such calendar appears in the codex in question, it has survived elsewhere, including in a fourteenth-century manuscript at the British Library where the canon has been translated into Anglo-Norman. The attached calendar closely adheres to the format of the popular *Kalendarium* of pseudo-Grosseteste, but the hour of each conjunction has indeed been recomputed using the Toledan Tables with their standard meridian.[12] The same tables were still being used at the start of the next century by the Flemish astronomer and physician Alard of Diest, who calculated the nearest half-hour of all conjunctions from 1301 to 1376, and a decade later in southern France by the Franciscan friar Raymond Bancal, whose *Correctio calendarii* mapped them out for 1311–86.[13] A calendar for the latter range of dates is also extant from the hands of an anonymous Praemonstratensian monk, who calculated his new moons in 1313 for the nearest *triens* or third of an hour, placing his epoch for each day at equinoctial dawn (i.e. six hours before noon). Readers interested in keeping this calendar in long-term use were told to shift this reference point back to midnight once the first 76 years were over, such that after $4 \times 76 = 304$ years the mean conjunctions could all be found a whole day earlier.[14] Yet another calendar for the 76-year period from 1311 to 1386 was cast for the meridian of Milan by a friar named Fucus of Ferrara, who again assumed a somewhat crude correction of 6 hours per 76 years.[15] Campanus of Novara's tables for the meridian of Novara, themselves a variant of the Toledan Tables, were the acknowledged source behind a calendar for the years 1330 to 1405 devised by John of Hameln for the meridian of Riddagshausen, which is today a part of Brunswick.[16]

More influential than any of these efforts was a calendar with lunar columns valid for 1293–1368, which again stuck close to the design of pseudo-Grosseteste's *Kalendarium* in assigning to each conjunction a letter from A to T to denote the position in the 19-year cycle, as well as a number from 1 to 24 to denote the current hour from midnight. An innovative feature of this calendar was the way it added

[11] Ibid., fol. 125v: 'Kalendarium istud extractum est a tabulis Toleti anno domini 1289. In quo ponuntur 4 cicli 19les lune sive aurei numeri quibus etas lune docetur omni tempore.'

[12] See MS London, BL, Arundel 220, fols. 308r–313v. The introduction follows on fol. 314v: 'Ce kalender est estret de la table tholet le an nostre singur mil.cc.lxxxix en ke sunt mis quater cicles decenovenales du numbre orre par quel numbre les primes sunt cuntes.' Another version of the same calendar appears in MS Cambridge, Gonville & Caius College, 512, fols. 179vb–179Cv.

[13] Alard of Diest's calendar for 1301–76 is preserved in MS Erfurt, UFB, CA 4° 370, fols. 1r–7v, and in a folded book from a private collection described in Gumbert 2016: 93–6. On the author, see Vercauteren 1951: 87, 90, and n. 71 in this chapter. On Bancal, who in 1326 became the provincial minister to the Franciscans in Aragón, see Nothaft 2015e.

[14] MS Erfurt, UFB, CA 4° 362, fols. 63r–69v. Nothing is known about the author's place of writing, although the column for the solar noon altitude presupposes a geographic latitude of 48;40°. The calendar also contains information on solar longitude, solar declination, and the lengths of day and night.

[15] MSS Bologna, Biblioteca Universitaria, 132, fols. 142r–148r; Paris, BnF, lat. 13014, fols. 1v–8r; Vatican City, BAV, Ross. 732, fols. 131r–137v. See Thorndike 1959: 45.

[16] MS Prague, NKCR, XIII.C.17 (2292), fols. 77r–83v. The calendar features conjunction dates, the Moon's mean motion in anomaly, the Sun's true longitude, and the length of daylight. The canons (fol. 83r–v) are more detailed than usual, with information on the calculation of both mean and true syzygy. Further copies are listed by Zinner 1939: 71, 105, who identifies the author as Jan von Hameln of Riddagshausen Abbey (OCist).

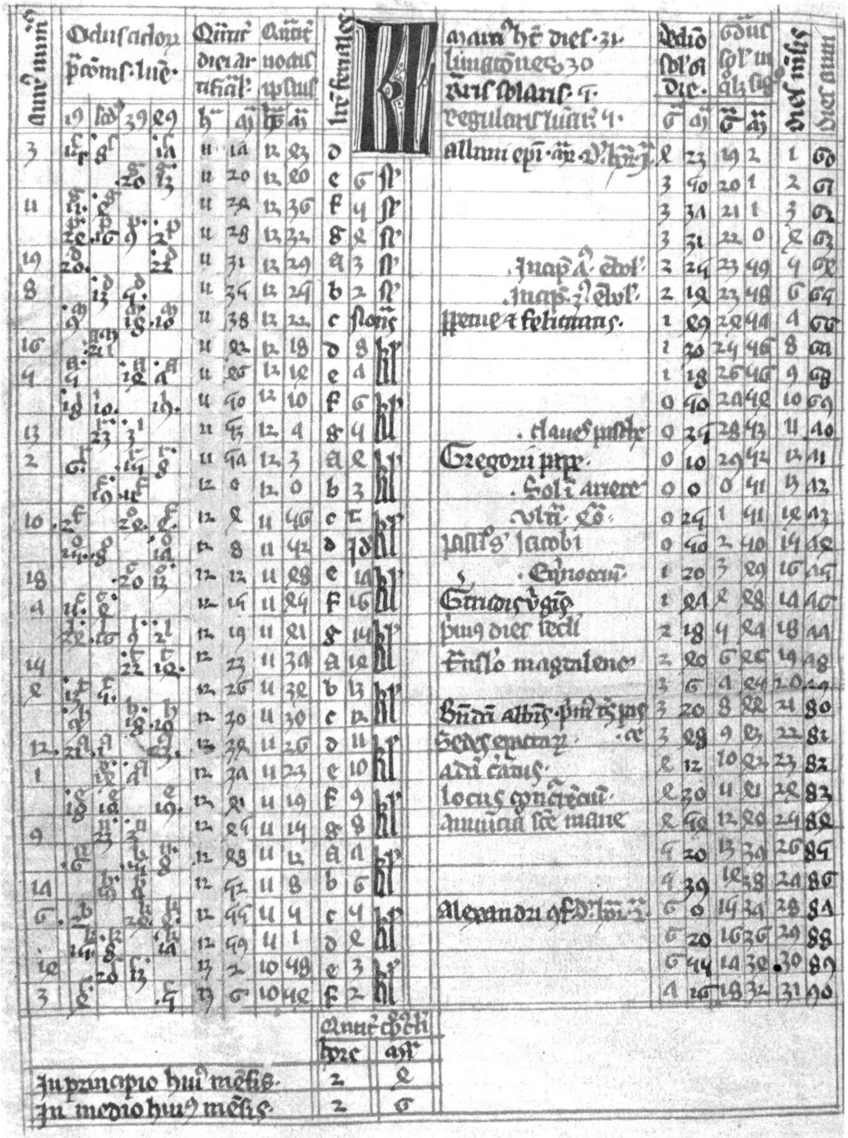

Figure 5.2. *Kalendarium* of Peter Nightingale (*c.*1293), page for March. Glasgow, University Library, Hunter 444, p. 4.

to each hour a small dot, the position of which in one of the four corners of each box was supposed to indicate the current quarter-hour (see Figure 5.2). The man responsible for this adjustment was the Danish astronomer and mathematician Peter Nightingale (Petrus Philomena de Dacia), who was active in Bologna and Paris during the final decade of the thirteenth century before returning as a cathedral canon to Roskilde, where his presence is last documented in 1303. His *Kalendarium* survives in more than 60 manuscripts, which reflect a remarkably wide circulation radiating outward from northern France to the rest of Latin Europe. An analysis of its new-moon times reveals that he worked with a standard lunation length of 29d 12;44,3,20h and used the epoch values provided by the Toledan Tables without any longitude adjustment, even though some of the manuscripts declare the meridian of reference to be that of Paris.[17]

Next to conjunctions, many extant copies of Peter's *Kalendarium* feature columns for solar data, in particular the Sun's daily longitude, noon altitude and/or declination, as well as the lengths of day and night. In the case of the noon altitude and day- and night-time lengths, the numbers employed are usually identical with those in the magnificent *Kalendarium regine*, which an astronomer named William of Saint-Cloud composed in 1296 in honour of Marie of Brabant, widow of the French king Philip III (1270–85). Besides the original Latin version, which is preserved in at least eight manuscripts, there is also a French translation made by William himself at the request of Joan of Navarre, the wife of the new king Philip IV (1285–1314), suggesting that the astronomer maintained good connections to the French royal court.[18] To those interested in the daily course of the Sun, the *Kalendarium regine* or *Kalendrier de la royne* offered columns specifying the lengths of day and night (in hours and minutes) as well as the Sun's altitude at noon (in degrees and minutes), while the lower margins of each page record the duration of the morning and evening twilight for the beginning and the middle of a given month (see Figure 5.3). Instead of adding a column for the daily solar longitude, as was done by several of the aforementioned calendars, William supplemented his work with a separate set of tables to aid the reader in finding the exact date and hour of the Sun's entry into each zodiacal sign for the bissextile cycle from 1296 to 1299 (see Figure 5.4). An auxiliary table showed how to adapt these entries to other years of the Lord, by adding or subtracting one hour for every five years in the past or future, suppressing this step only once every 60 years.[19]

This 60-year pattern reveals that William worked with a tropical year of 365d 5;49h or very nearly $365\frac{1}{4} - \frac{1}{131}$ days, which had entered Latin astronomy in

[17] See the edition with extensive introduction and presentation of supplementary material in Pedersen 1983–4: 203–456. The available biographical information is conveniently collected and summarized on pp. 26–41, 50–63. See also the older accounts in Duhem 1913–59: iv. 29–30; Zinner 1936; Pedersen (Olaf) 1967; 1976. A copy of Peter's *Kalendarium* not mentioned in the literature is MS Berlin, Staatsbibliothek, lat. oct. 416, fols. 78v–84r.

[18] The Latin and French texts of the *Kalendarium regine* are printed in Harper 1966: 73–239. Further MSS are noted in Pedersen 1983–4: 419; Shore 1989: 300, 305–6. On the annotated copy in MS Paris, BnF, lat. 7281, see Boudet 1994: 179–80. Additional remarks on William of Saint-Cloud appear in Duhem 1913–59: iv. 10–19; Zinner 1936: 322–4; Millás Vallicrosa 1943–50: 392–4; Pereira 1976: 196–9; McCluskey 1998: 200–2, 204–6; Chabás and Goldstein 2003b: 245.

[19] William of Saint-Cloud, *Kalendarium regine* 11–12 (ed. Harper 1966: 124–34). Stand-alone versions appear in MSS Bamberg, Staatsbibliothek, R. B. Msc. 176, fols. 14v, 16r; Erfurt, UFB, CA 4° 366, fol. 53r–v; Cambridge, University Library, Gg.6.3, fol. 129r (second table and canon only); Los Angeles, J. Paul Getty Museum, Ludwig XII.6, fols. 13v–14r. See also Pedersen 1983–4: 34, 322–3, 363, 416.

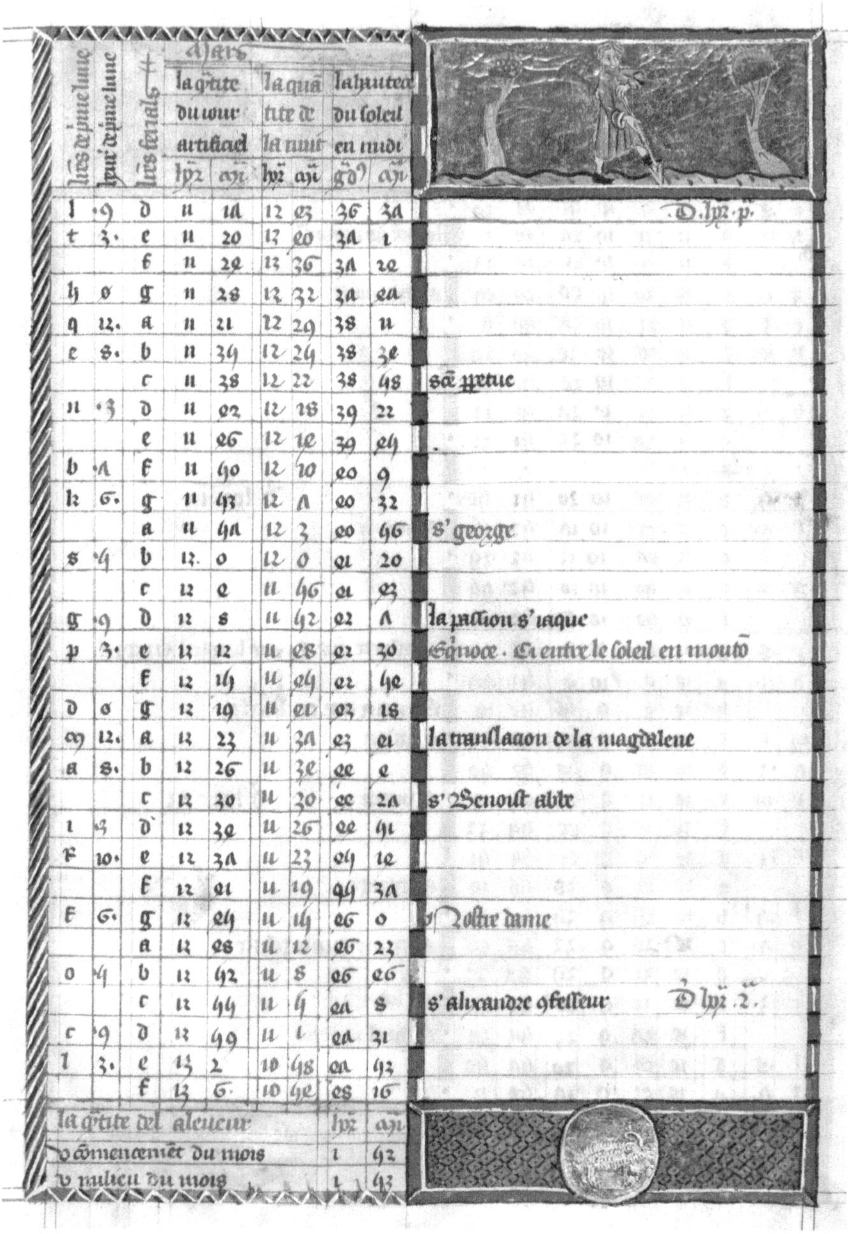

Figure 5.3. William of Saint-Cloud, *Kalendrier de la royne* (1296), page for March. Paris, Bibliothèque de l'Arsenal, 2872, fol. 2r.

Source: BnF

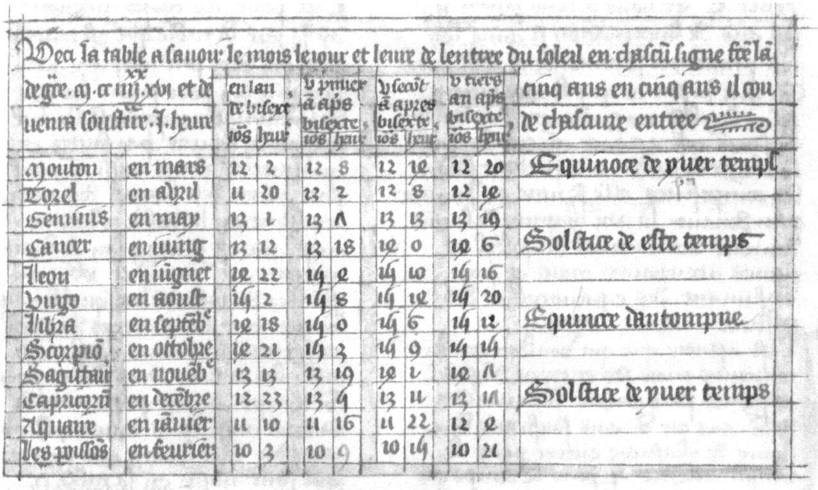

Figure 5.4. William of Saint-Cloud, *Kalendrier de la royne* (1296), table of equinoxes and solstices. Paris, Bibliothèque de l'Arsenal, 2872, fol. 17r.
Source: BnF

the twelfth century through Abraham Ibn Ezra's adaptation of the *zīj* of al-Ṣūfī. A preference for this parameter is also expressed in some contemporary astronomical notes from Paris, which predict the entry of the Sun into Aries, Cancer, Libra, and Capricorn in 1293 in a way that is congruent with William's own tables. The author expressly endorses a tropical year of $365\frac{1}{4} - \frac{1}{131}$ days or 365d 5;49h, stating that, in his estimation, 'the opinion of *Alzophi* [i.e. al-Ṣūfī] on the length of the solar year is closest to the truth'.[20] From the *Kalendarium regine*, the value made its way into a work on astro-meteorology composed in 1317 by Richard of Wallingford, who incorporated William's solar tables in unchanged form, albeit without indicating their true origin.[21] An important hint as to the basis for their construction comes from William's own record of the solar observations he carried out in Paris in 1290, which showed the vernal equinox to have occurred on 12 March, 16 hours after noon (= 13 March, 04:00h). When he drew up his calendrical tables a few years later, he decided to begin in 1296 and place the vernal equinox of AD 1298 on 12 March, 14 hours after noon (= 13 March, 02:00h). This two-hour decrease over eight years agrees reasonably well with William's own rule of subtracting 1h

[20] MS London, BL, Harley 3735, fol. 73ra: 'Et puto quod opinio Alzophi de quantitate anni solaris sit propinquior veritati.' See Nothaft 2017c: 219–23, for further details.

[21] Richard of Wallingford, *Exafrenon pronosticacionum temporis* (ed. North 1976: i. 186–91). Richard claims that the tables apply to the meridian of London. He also repeats William's statement that the winter solstice had receded by 11 days since the birth of Christ, assigning the latter to Saint Lucy's Day (13 December).

for every five-year interval, suggesting that he relied on his own observational data, which also became the basis for his *Almanach planetarum* for the years 1292–1311 (see below).[22]

Like this *Almanach*, William's *Kalendarium regine* came with a lengthy introduction, in which the astronomer explained to the queen the various uses to which his calendar might be put: reliable knowledge of the lengths of day and night as well as the duration of twilight were of particular value to the calibration of clocks and the timing of ecclesiastical services and daily prayers. Practitioners of medicine could use the current length of the daylight period to determine the times when a particular humour ruled the body, while travellers could estimate the distance they had journeyed north- or southward by comparing the solar noon altitude observed at their respective local latitude with that recorded in the calendar.[23] When it came to outlining the *Kalendarium*'s lunar features, he began by repeating the apocryphal story according to which the Golden Numbers had been introduced by Julius Caesar and the attribute 'golden' was due to the particular usefulness these numbers possessed in the eyes of their ancient Roman inventors.[24] To William's astonishment, the cyclical assignment of the new moons to particular days in the calendar had not undergone any correction since antiquity, with the unfortunate result that the lunar ages read out in churches on a daily basis and used for fixing mobile feasts presently fell at least two days behind the observable new-moon crescent. Echoing complaints made by earlier authors such as Roger Bacon, he concluded that the evident error of the lunar cycle 'and many other things contained in the commonly used calendar' threatened to turn the Church into a laughing-stock for heathens and Jews.[25] His response was to recast the new-moon dates in his calendar by calculating the hour of the mean conjunction for a period of 19 years or 235 lunations spanning the years 1292–1310. In doing so, he increased each conjunction time by a full day, thus defining the calendrical new moon as the point in time 24h after conjunction. This step helped push the results closer towards the evening of first visibility, which also had the side-effect of reducing the discrepancy between the ecclesiastical Moon and reality. As with the contemporary *Kalendarium* of Peter Nightingale, the new moons in William's calendar were represented by letters from A to T as well as by a Hindu-Arabic numeral for the completed number of hours. These had to be subtracted from noon of the current

[22] William of Saint-Cloud, *Kalendarium regine* 12 (ed. Harper 1966: 129). For William's solar observation of 1290, see the introduction to his *Almanach planetarum* 5 (ed. Pedersen 2014: 9–10). In the tables of the *Almanach* proper, the longitude of the Sun at noon of 13 March 1298 is given as 0;24°, which agrees well with a true vernal equinox at 14 p.m. of the previous day. See the two known copies of the *Almanach planetarum* in MSS Paris, BnF, lat. 16210, fol. 62v; Vatican City, BAV, lat. 4572, fol. 55r.

[23] William of Saint-Cloud, *Kalendarium regine* 6–7 (ed. Harper 1966: 101–8). The precise time of noon could be found with the help of a *Directorium*, a portable compass sundial that William had described in a separate treatise. See ibid. (ed. Harper 1966: 4, 22–3, 29–30, 56–8).

[24] Ibid. 8 (ed. Harper 1966: 109). Caesar's authorship is also asserted in the preface to the Oxford calendar for 1292, as found in MS Oxford, BodLib, Digby 149, fol. 125r (see n. 8 in this chapter).

[25] William of Saint-Cloud, *Kalendarium regine* 8 (ed. Harper 1966: 112).

day if a dot appeared before the number, but added to noon, if the dot came after (as shown in Figure 5.3).[26]

One significant factor in the growing popularity of astronomically enhanced calendars during the late medieval period was the widespread conviction that the Moon exerted a significant influence on human affairs, which made it seem necessary to track its phases and motion through the zodiac with a degree of accuracy much greater than that afforded by traditional *kalendaria*. Late thirteenth- and fourteenth-century physicians were in widespread agreement that the Moon's position in the zodiac determined the propitious or harmful times for letting blood from certain parts of the body, while its waxing and waning had implications for an individual's resistance to disease or for the efficacy of a particular cure. More mundane aspects of daily life that were believed to be influenced by the lowest of the seven planets included diet and hygiene as well as the appropriate times for travelling, getting betrothed, or cutting one's hair.[27] That knowing the Moon's sidereal longitude could be useful for medical purposes, but also for the weather predictions habitually made by farmers and sailors, was an idea expressly affirmed by William of Saint-Cloud, who showed how to derive the mean zodiacal position of the Moon from the data displayed in his *Kalendarium regine*. Step one was to multiply the current lunar age by 12 in order to yield the elongation from the Sun since the last conjunction. In a second step, the result had to be combined with the Sun's present position in the zodiac, adding one more degree for each 5-day interval. This rule of thumb was based on an understanding that the Moon will travel approximately 13;11° per day, whereas the Sun's daily motion is closer to 1°.[28] The contemporary interest in such rough-and-ready solutions to the calculation of lunar longitude is further reflected in the numerous manuscripts containing the *Kalendarium* of Peter Nightingale, which often supplemented the main calendar with a *Tabula de loco lune*, composed of 14 × 12 boxes and geared towards finding the ecliptical position of the Moon (in degrees) as a function of its age.[29]

While such calendrical schemes for tracking the Moon's sidereal motion remained on a relatively basic level, late medieval astronomers continued to invest much of their effort and ingenuity in the creation of lists of precisely timed conjunctions and

26 Ibid. 8–9 (ed. Harper 1966: 112–18). William's new moons for the years 1292–1310 occasionally appear in other calendars, for instance in MS Utrecht, Universiteitsbibliotheek, 722 (shelfmark: Eccl. 283), fols. 37r–42v, which is the one mentioned in Wijk 1936: 41. They are also found in the two leftmost columns added to the *Kalendarium* of pseudo-Grosseteste in MS Munich, BSB, Clm 5538, fols. 57r–62v. Further examples, all connected to the *Kalendarium* of Peter Nightingale, are noted in Pedersen 1983–4: 419.

27 Medieval popular and medical lunar astrology constitutes a vast field, which can be barely touched upon for the purposes of the present discussion. For orientation, see Svenberg 1936; 1963; Müller 1971; Weißer 1982. A particularly large body of literature exists for texts in Middle English. See Farnham 1923; Taavitsainen 1988; Means 1992a: 376–85; 1993. For Anglo-Norman texts, see Hunt 2013: 31–126.

28 William of Saint-Cloud, *Kalendarium regine* 10 (ed. Harper 1966: 118–24).

29 Pedersen 1983–4: 14, 30, 208–10, 326–8, 333, 360, 381–98. The table arose from an earlier tradition of zodiacal lunar tables, based on a 12 × 12 scheme, which is traceable back at least to the ninth century. See Cordoliani 1955: 308; 1960–1, pt. 1: 117 (n. 50); Chabás and Goldstein 2012a: 87; Pedersen 2002: 1610. See also the more elaborate tables described in Giles of Lessines, *Summa de temporibus* III.3.7 (ed. Steele 1926: 180–4).

oppositions, many of which took into account the true rather than the mean longitudes of Sun and Moon. The abiding interest in syzygies and eclipse calculations is well exemplified by the fourteenth-century French astronomer, mathematician, and music theorist Jean (Jehan) des Murs, who will concern us in greater detail in the following chapter (6.1). Next to a *Kalendarium solis et lune* (1321) and a number of other texts and tables devoted to the courses of Sun and Moon, Jean's many contributions to the field included the so-called *Tabule permanentes*, which sought to offer a quick and handy solution to a problem that had been vexing astronomers for centuries: how to get from calculated times of mean syzygy to the corresponding time of true syzygy. Jean, together with his collaborator Firmin de Beauval, bridged the gap by creating a double-argument table that presented the necessary time correction as a function of two components, mean solar and mean lunar anomaly.[30] This approach greatly simplified the computational procedures involved, thus recommending itself to those who sought to insert times of true conjunctions or oppositions into their calendars and ephemerides. Later authors who adapted the *Tabule permanentes* in their own work included John of Gmunden (1380/5–1442), a teacher of mathematics and astronomy at the University of Vienna, who furnished them with a new canon in *c.*1440, as well as his famous student Georg Peurbach (1423–61), who created a greatly expanded version with 32,000 entries (compared to just 1,800 entries in the original version).[31]

The idea of furnishing calendars with true instead of mean syzygy times enjoyed a particularly vigorous reception in fourteenth-century England, where Walter Elveden, a scholar at Gonville Hall, Cambridge, created a calendar for three 19-year cycles from 1330–86, in which conjunctions and oppositions are timed down to the nearest second. One special feature of this work are the appended tables and coloured drawings showing the time, duration, and magnitude of all 23 solar and 52 lunar eclipses that could be expected between 1327, the year when the calendar was made, and 1386.[32] More influential than Elveden's own composition were the continuations drawn up for the meridian of Oxford by the Franciscan friar John Somer and the Carmelite monk Nicholas of Lynn, who are both famously mentioned by Geoffrey Chaucer in his *Treatise on the Astrolabe*.[33] The earlier and more widespread of these works was the *Kalendarium* of John Somer (*c.*1340–*c.*1409), which still survives in some four-dozen copies, including six written in Middle

[30] See Porres de Mateo and Chabás 2001; Kremer 2008: 155–7.

[31] Chabás and Goldstein 1997: 99–101. On the development of computational techniques and tables for true syzygy in the late Middle Ages, see also Chabás and Goldstein 1992; 2008; 2012a: 144–50; Kremer 2003a; 2003b; 2008; 2011.

[32] MSS include Cambridge, Corpus Christi College, 37, fols. 27r–45v; London, BL, Egerton 831, fols. 1r–9v (a copy made in 1341, which omits eclipses before that date); London, BL, Sloane 286, fols. 15r–24v; Oxford, BodLib, Digby 228, fols. 50r–53r; Oxford, BodLib, Fairfax 27, fols. 71r–79r. See Thorndike 1957b: 53, 56–7; Pedersen 1983–4: 420–1; North 1988: 93–4 (n. 13); Davis and Lowne 2015: 281–5. Zinner 1925: nos. 5334–9 and p. 445, lists further examples for calendars for 1330–86, which turn out to be dependent on Elveden's work.

[33] See the overview in North 1988: 87–104. A third continuation, made in York by the Augustinian friar Richard de Thorpe, has also been preserved. It is found in MS Oxford, BodLib, Ashmole 210, fols. 2v–7v. See Mooney 1984: 407–9; Friedman 1985.

English. In a case reminiscent of William of Saint-Cloud's *Kalendarium regine*, the original was produced in 1380 for Joan, the Princess of Wales and mother of King Richard II, at the express order of the Franciscan provincial Thomas Kyngesbury. In carrying out his task, Somer retained Walter Elveden's true-syzygy approach, but computed conjunctions and oppositions for an entire 76-year period stretching from 1387 to 1462.[34] His awareness of the defects of the ecclesiastical calendar is revealed by the reported fact that he subjected it to written 'castigations' (*Calendarii castigationes*), the lost text of which started with the dramatic words: 'The calendar's corruption is horrible' (*Corruptio calendarii horribilis est*...).[35] The *Kalendarium* of Somer's contemporary Nicholas of Lynn, composed in 1386 at the request of John of Gaunt, 1st Duke of Lancaster, followed the same basic model, offering true syzygies for the years 1387 to 1462, but also added a number of new elements such as a column for the duration of twilight. Instead of filling just the habitual two facing pages (one for the *kalendarium* proper and one for lunar tables and other astronomical information), he spread each month out over three pages. The third of these was devoted to a table that allowed users to find the time of day and the Sun's current altitude from measuring the length of a man's shadow—a clever device that later served Chaucer in composing the prologue to the *Man of Law's Tale*.[36] Although originally designed for personal use by noble patrons, the works of Somer and Lynn came to dominate the English market for astronomically enhanced calendars during the period of their validity and even received further continuations, which in some cases extended to 1538. Among the features that contributed to their success were the artful diagrams for eclipses as well as the great range of auxiliary tables, for instance those devoted to mobile feast days or astrological data such ruling planets and ascendants. Another striking aspect of both works is their inclusion of astro-medical instructions, which explained how calculations pertaining to the lunar phases could be applied to phlebotomy or how the use of laxatives had to be timed in accordance with the Moon's zodiacal position. Medical 'almanacks' of this kind, in particular those connected to the calendars of John Somer and Nicholas of Lynn, were sometimes produced as folded books, which could be attached to a girdle and carried around, thus serving as a physician's or barber's *vade mecum*.[37]

By the first quarter of the fifteenth century, astro-medical compendia of the type just mentioned had also begun to mushroom in German-speaking Central Europe,

[34] The calendar is edited in Mooney 1998. See the updated list of manuscripts in O'Boyle 2005: 20–2.

[35] The only remaining witness to its existence is John Bale, *Scriptorum illustrium maioris Brytanniae, quam nunc Angliam et Scotiam vocant: Catalogus*, 2 vols. (Basel: Oporinus, 1557–59), I, p. 513 (cent. VII, no. 8). See North 1992a: 133–4.

[36] See the edition, translation, and analysis in Eisner 1980, and the update in Kuczynski 1987. A copy not noted in these publications is MS Los Angeles, J. Paul Getty Museum, Ludwig XII.7, fols. 11v–51r. Nicholas of Lynn's shadow scale and Chaucer's use of it are discussed at length in North 1988: 104–9, 116–30, 525–6.

[37] See the comprehensive catalogue of 'bat books' by Gumbert 2016: 73–206, the specimens reproduced in Murdoch 1984: 97, 318–19, and the earlier discussions in Talbot 1961; Carey 2003; 2004. The medical use of the works of John Somer and Nicholas of Lynn is stressed by O'Boyle 2005. On their long-term influence, see also Mooney 1997; Lawrence-Mathers 2010.

where they gave rise to resplendent specimens such as the richly illuminated *Tübinger Hausbuch*, which originated in the region of Württemberg in the third quarter of the fifteenth century. Like various earlier German manuscripts of the same type, which can be traced back to a prototype made in 1404, the *Tübinger Hausbuch* opens with a calendar listing true conjunction times for the 19 years from 1406 to 1424 as well as the daily longitudes of the Sun and Moon for 1405.[38] Another representative piece of southern German calendar-making is Conrad Bösner's illustrated *Passauer Kalendar* of 1445, which incorporated the hours and minutes of all 1,880 mean conjunctions and oppositions from 1444 to 1519 (see Figure 5.5).[39] The source for these syzygy tables had been produced a decade earlier by the aforementioned John of Gmunden, whose various calendars have survived in well over 150 manuscripts. The most popular among these covered a 76-year cycle from 1439 to 1514, using reformed Golden Numbers from 1 to 19 as opposed to letters from A to T. Gmunden's calendars typically placed the syzygies on a separate page facing the corresponding *kalendarium*, with the latter featuring elements such as the traditional Golden Number, the Sun's zodiacal position, the length of the day, and the times of sunset and sunrise.[40] The *Passauer Kalendar* used essentially the same layout and data, though its author decided to put the calendar's starting point five years later and subtracted 10 minutes from each syzygy to reflect the insight that Passau was located roughly 2;30° degrees west of Vienna. John of Gmunden's calendar also existed in an earlier version covering only the 19 years from 1421 to 1439. It was used in *c.*1425 by the Tyrolean poet and composer Oswald of Wolkenstein to a create a version accompanied by a mnemonic calendrical 'Cisiojanus' poem with musical notation.[41] Another important aspect of Gmunden's legacy is the existence of printed calendars issued under his name, including an early xylographic edition of which only the woodblock survives. This two-sided calendar can approximately be dated to the 1450s thanks to its inclusion of corrected Golden Numbers, which happen to match John of Gmunden's syzygy data for the 19-year cycle ranging from 1458 to 1476. Among its other features are

[38] See the calendrical section in MS Tübingen, Universitätsbibliothek, Md 2, fols. 1r–50r. On this particular codex, see Mütz 2008, and the introductions to the facsimile edition in Brinkhus, Juste, and Lengenfelder 2005. On the wider family of manuscripts of what has become known as the German '*Volkskalender*' of the fifteenth century, see Zinner 1952; Brévart 1988b; 1995–6.

[39] MS Kassel, Universitätsbibliothek, 2° Ms astron. 1, fols. 2v–14r. For a study, edition, and partial facsimile of the manuscript, see Müller 2009.

[40] See Kaltenbrunner 1876: 360–5; Klug 1912: 16–35; 1943: 71–9; Mundy 1943: 202–3; Pedersen 1983–4: 422–3; Gumbert-Hepp 1987: 22–3; Chlench 2006a: 127–8, and the repertory of Gmunden manuscripts in Chlench 2007: 200–9. Lunar columns derived from Gmunden's calendars appear in the famous Schürstab codex, MS Zurich, Zentralbibliothek, C 54, fols. 5v–16v (Nuremberg, 1459/71). This calendar features new and full moons based on mean syzygy for 38 years, from 1458 to 1495, timed to the nearest hour. See Keil 1983: 20–42, 152–3. Gmunden was also the template behind the calendar made for Count Eberhard V of Württemberg (1445–96), which features lunar tables for the years 1458–1515. See Mütz 1996.

[41] Kersken 1975: 50–61, 205–59. Contrary to what Kersken (p. 211) asserts, both Wolkenstein's and Gmunden's calendars use a noon epoch. His claim that Gmunden's calendar was a 'Weltneuheit' (p. 209 n. 4) is impossible to maintain given the existence of various thirteenth-century precedents. Cf. also Kersken 1978: 426, 436, 448.

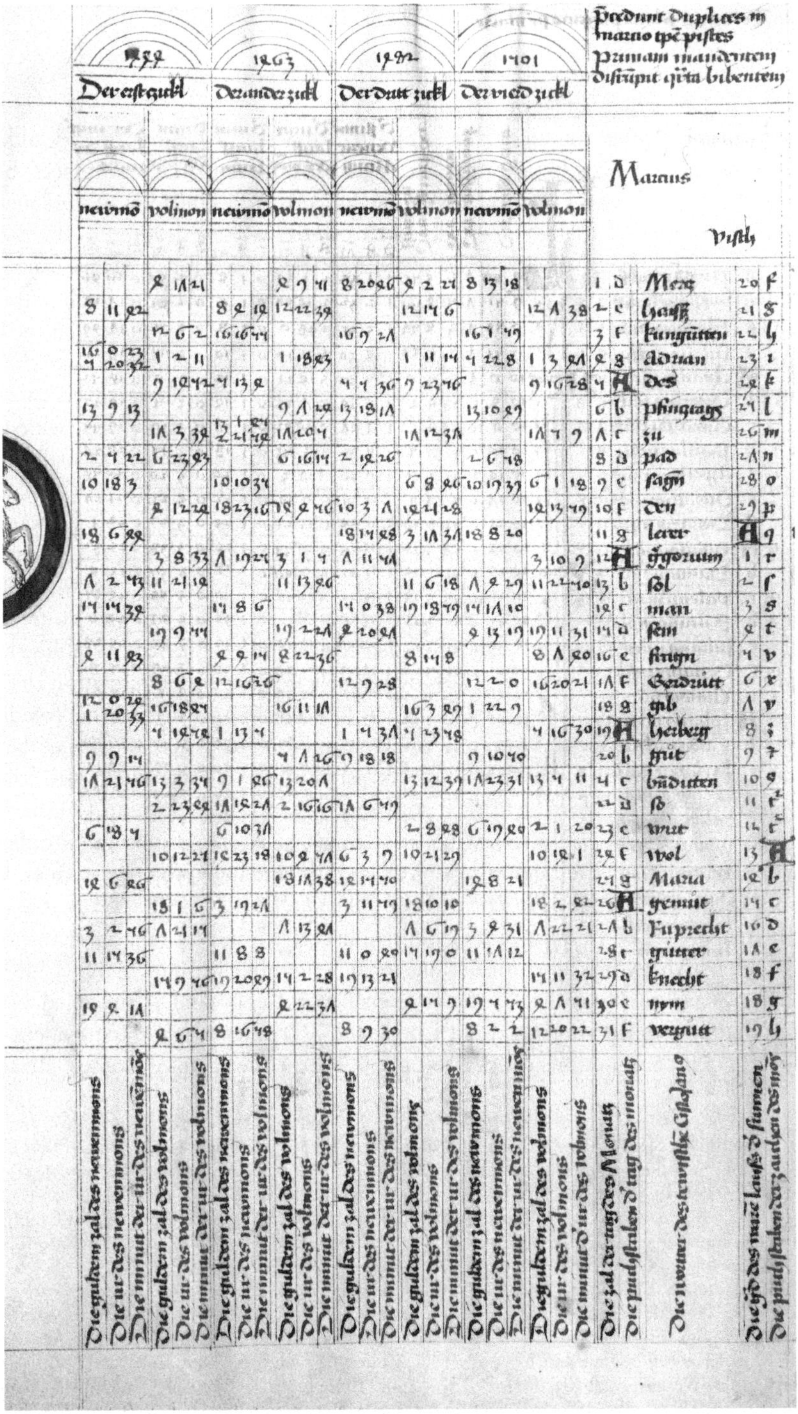

Figure 5.5. *Passauer Kalendar* (1445), syzygy times in March for 1444–1519. Kassel, Universitätsbibliothek, 2° Ms. astron. 1, fol. 4v.

Source: Universitätsbibliothek Kassel—Landesbibliothek und Murhardsche Bibliothek der Stadt Kassel

sidereal lunar letters and accurate dates for the Sun's entry into the zodiacal signs (Aries on 11 March, Cancer on 13 June, etc.).[42]

The computational practices that supported the creation of reformed Golden Numbers and improved lunar tables varied across different times and places, but the most straightforward method always remained to add multiples of the length of the mean lunation to a given epoch date or *radix*. Writers devoted to the *computus naturalis* such as Walcher of Malvern had tried to establish such a starting date by observing eclipses, but for many it remained acceptable to anchor their method in some authoritative source. One such source could be discerned in the Hebrew lunar *computus* supposedly established by Moses at Mount Sinai, whose techniques were first advertised by Reinher of Paderborn in the twelfth century and later described in great detail by Franciscan authors such as Robert of Leicester and Friar John (see Ch. 3.2 and Ch. 4.3). A long-standing tradition of using the Jewish calendar for lunar computations can be detected in Italy, which is the likely place of origin of a brief manual on *molad* reckoning, referred to in one manuscript as the *Liber erarum*. If one can judge from the appended material, the text may have gone through a stage of redaction in Cremona in *c.*1191, although the bulk of the treatise was almost certainly based on an earlier Hebrew source.[43] Besides the *Liber erarum*, Italian scribes also copied long sequences of tables that displayed the starting conjunctions of the Hebrew lunar months in relation to the Julian calendar. A particularly lavish set can be found appended to the Toledan Tables in a thirteenth-century manuscript now in Florence, where the *molad* dates from January 1248 to December 1295 are listed on six consecutive pages. They are preceded by tables for calendrical conversion identical to those in Reinher of Paderborn's *Compotus emendatus*, while another six pages of columns and numbers offer Julian dates for the beginnings of all months in the Arabic lunar calendar from 1232 to 1279.[44] The Florentine list of *moladot* for 1248–95 appears to have originated as part of an even more extensive list still found in a manuscript now in Erfurt, which records all conjunctions from September 1236 to March 1348—or Tishri 4997 to Nisan 5108 of the Jewish world era—and is preceded by a canon that advises the subtraction of 1 hour and 360 parts/*ḥalaqim* (equivalent to 20°) to obtain conjunctions times appropriate for the meridian of Rome.[45] In the

[42] Kaltenbrunner 1876: 365–7; Klug 1912: 32–3; 1943: 79–85; Lindhagen 1917; Zinner 1964: 1–2.

[43] See the introduction and edition in Nothaft 2014d: 69–127. A version of the *molad* tables survives separately in MSS Paris, BnF, lat. 7434, fol. 105r; Vatican City, BAV, lat. 3118, fol. 5r–v. The text of the *Liber erarum* was drawn upon by the Franciscan John of Presorio (see n. 193 in Ch. 4.3), who paraphrases large parts of it in the canons to his 'Hebrew' *kalendarium*. See MS London, BL, Add. 17368, fols. 37v–38r, 51v.

[44] MS Florence, BML, San Marco 185, fols. 84v–85r (Reinher), 86r–88v (Hebrew), 89r–91v (Arabic). See Nothaft 2014d: 614–17. The Julian beginnings of the Arabic lunar months, from AD 1233 to 1348, are also recorded in MS Paris, Bibliothèque de l'Arsenal, 877, fols. 5r–12r. Those for AD 1270 to 1318 can be found in MS New York, Columbia University, Rare Book and Manuscript Library, Smith Western Add. MS 06, fols. 137v–140r. Those for AD 1322 to 1333 are in MS Prague, NKCR, XIV.A.18, fol. 46r–v. An even shorter list, for January 1282 to February 1293, appears at the end of MS Paris, BnF, lat. 7409, fol. 38v.

[45] MS Erfurt, UFB, CA 8° 82, fols. 102r–111r. Further partial lists appear in MSS Oxford, BodLib, Lyell 52, fol. 75r–v (1280–7); Paris, Bibliothèque de l'Arsenal, 879, fols. 10r–12r (1248–87);

Florentine manuscript and in a related list preserved in Prague, the difference Jerusalem–Rome is instead given as 1 hour and 432 parts (or 21°).[46] That this Christian appropriation of the Jewish *molad*-system persisted well into the fifteenth century is evident from sources such as the extensive 'Table of Solomon' (*Taula de Salamon ala luna zudesca*), which is included in a notebook the Venetian mariner Michael of Rhodes began to compile in 1434. It lists the times, ferials, and Julian dates of all *moladot* in the years 1435–1530, albeit with some degree of reckoning error.[47] The popularity of this technique in quattrocento Italy is further underscored by the fact that the *Treviso Arithmetic* of 1478, which is generally regarded as the first printed mathematical textbook, included new-moon calculations based on the Hebrew value of 29 days, 12 hours, and 793 parts, and with results identical to those implicit in the Jewish calendar.[48]

North of the Alps, the idea of computing syzygies in the manner of the Jews was reinvigorated with the appearance of a *Computus Judaicus* (1342), which circulated widely enough during the fourteenth and fifteenth centuries to be preserved in more than 60 manuscript copies. Although the anonymous author adopted certain bits of Hebrew vocabulary, his interest in the subject was mostly limited to the *molad*-system, the handling of which he explained with the aid of didactic verses and other mnemonic devices, in particular the replacement of numbers with syllables to produce imaginary compound words. In nearly all preserved versions, the text opens with a poem of 20 lines, starting with the confession: 'It shames me to hear that the Jew knows such things' (*Me pudet audire Iudaeum talia scire*). It uses drastic verbal imagery to complain that the Jews—here disparagingly referred to as being 'akin to Satan' (*Sathane quos patet esse pares*)—are superior to the Christian clergy and intelligentia in such an important practice as finding the precise time of conjunction.[49] Aided perhaps by this strong anti-Jewish slant, the treatise gained considerable popularity and was even used as a teaching tool in universities such as Vienna and Kraków, giving rise to a great variety of redactions, abridgements, extensions, and satellite texts as well as to extensive commentaries. Flyleaves and marginalia in dozens of manuscripts reveal that readers actively employed its method of lunar computation, as many of them left traces of their attempts on parchment and paper.[50] For the most part, these notes indicate that users were content with reproducing *molad* times exactly as predicted by the Jewish calendar, which in theory were meant to refer to the meridian of Jerusalem. An effort to adjust these

Paris, Bibliothèque de l'Arsenal, 1128, fol. 30r–v (1267–70; on a bifolium inserted between fols. 18 and 19); Prague, NKCR, XIV.A.18, fol. 39r (1262–7).

46 MSS Florence, BML, San Marco 185, fol. 87r; Prague, NKCR, XIV.A.18, fol. 39r.

47 See the edition in Stahl 2009: 394–415 (fols. 131a–5a), and the remarks in Wallis 2009: 305–8, 314–15, which do not take into account the evident connection to the Jewish calendar.

48 See Boncompagni 1862–3: 689–842, 909–1044, and the unpaginated *Larte de labbacho* (Treviso, 1478).

49 *Computus Judaicus* (ed. Nothaft 2014d: 442). To the 59 manuscripts listed in Nothaft 2014d: 379–92, one may add MSS New Haven, Beinecke Rare Book and Manuscript Library, 1024, fols. 1r–3v; Oxford, BodLib, Digby 17, fols. 27v–31r; Prague, NKCR, I.G.6, fol. 201r–v (excerpts); Vienna, ÖNB, 4781, fols. 6r–11v.

50 See Nothaft 2014d: 400–39; 2014e, for further discussion.

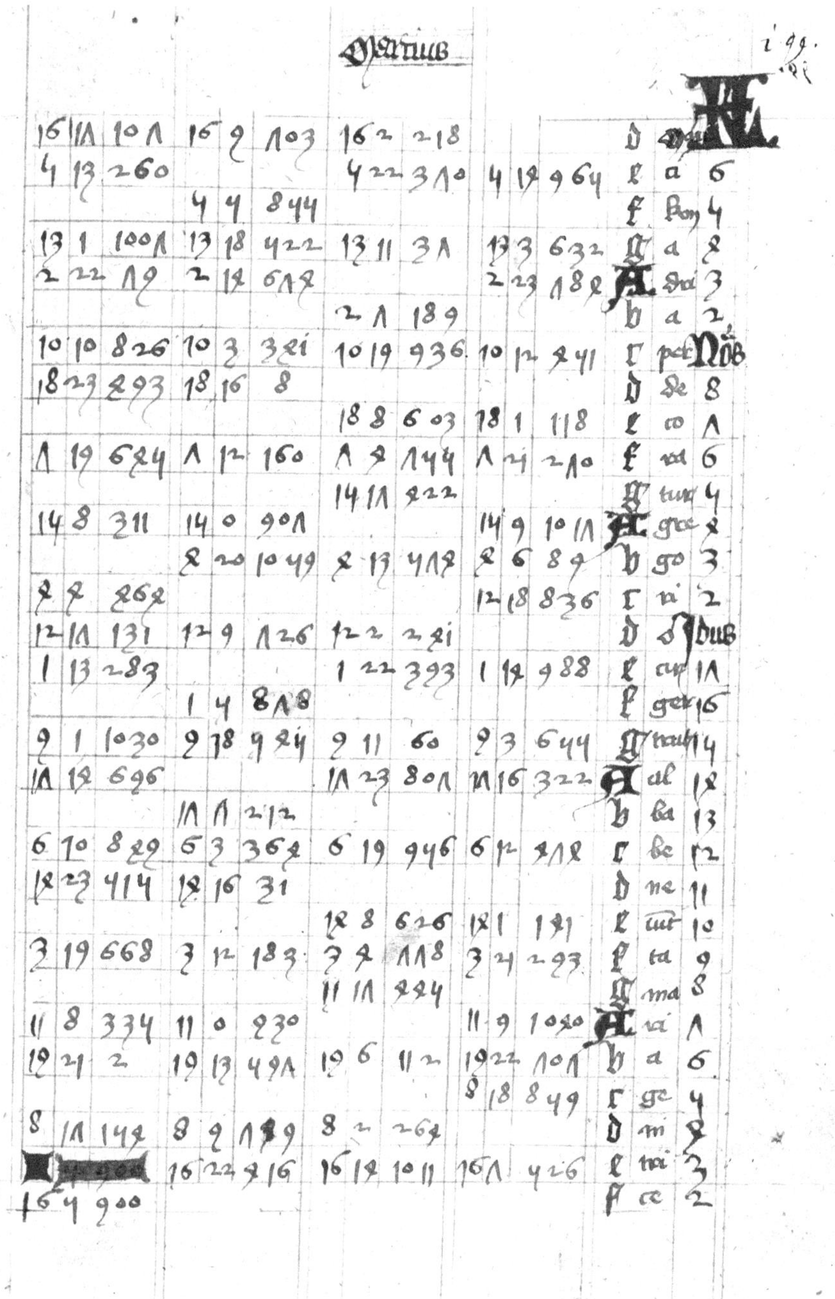

Figure 5.6. Southern German *molad* calendar, March conjunctions for 1428–1503. St Gall, Stiftsbibliothek, Cod. Sang. 827, p. 199.

Source: Stiftsbibliothek, St. Gallen

times to longitudes closer to home is attested in a brief treatise attached to the *Computus Judaicus* by a southern German author active in 1393, who promised to provide the new moon for the meridian of Constance 'in perpetuity' and 'at the blink of an eye' (*in perpetuum novilunium ad oculi conclusionem*), while also stating that the results could be made congruent with the meridian of Jerusalem by adding 3 hours and subtracting 286 parts of the hour.[51] The same doctrine regarding the time delay between Constance and Jerusalem was evidently followed by the author responsible for a 76-year lunar calendar covering the years 1428–1503, which accompanies the *Computus Judaicus* in a manuscript from the Lake Constance region (see Figure 5.6). In contrast to other tables of this type, which usually list sexagesimal minutes and seconds of the hour, this one separates its conjunctions by 29d 12h 793p and advocates the subtraction of 5h 860p to keep the calendar in tune with the Moon after their first 76 years. In line with the rule given in the treatise of 1393, the tabulated conjunctions all fall precisely 3 hours earlier and 286 parts later than the *moladot* predicted by the Jewish calendar.[52]

5.2. LATIN ASTRONOMY IN TRANSITION (*c.*1292–1317)

The nascent spread of astronomically enhanced calendars in the final decade of the thirteenth century, which can be linked to names such as Peter Nightingale, William of Saint-Cloud, and (pseudo-)Roger Bacon, coincided with an important phase of transition during which Latin astronomers, who initially had been wont simply to repeat the doctrines of their Arabic and Hebrew predecessors, progressed toward becoming more innovative and critically minded practitioners of their craft. Outward signs of this trend include a heightened interest in observational and computational instruments, which increased both in number and in type between the middle of the thirteenth and the middle of the fourteenth centuries.[53] One case in point is the proliferation of mechanical computing devices such as the volvelle-based *eclipsorium* to which the aforementioned Peter Nightingale dedicated a separate treatise.[54] Similar interests are reflected in the work of his namesake Peter of Saint-Omer, another astronomer active in Paris during the 1290s, who wrote a *Tractatus de semissis* describing an equatorium for the calculation of planetary

[51] MS Munich, BSB, Clm 14504, fol. 171ra. The same adjustment is recorded in MS Vatican City, BAV, Pal. lat. 1354, fol. 32vb. It amounts to an addition 2h 794p or *c.*41°.

[52] MS St Gall, StB, Cod. Sang. 827, pp. 196–208, where the calendar is immediately followed by a copy of the *Computus Judaicus* (pp. 209–16). The codex also contains lunar tables for Nuremberg, which use an 'Arabic' lunation of 29d 12h 44m and are calculated for the years 1406 to 1481 (pp. 1–14). The copy appears to date from 1425. Lists of *molad* times, often based on the *Computus Judaicus*, appear in numerous further manuscripts. See e.g. MS Munich, BSB, Clm 21569, fols. 9v, with a list for 1411. See also Chabás 2000: 336, who discusses lists of conjunctions from January 1355 to June 1381 and from March 1386 to December 1400 encountered in MS El Escorial, Real Biblioteca de San Lorenzo, K.II.7, fols. 32r–33v, 36r–38v.

[53] Poulle 1991; Pedersen (Olaf) 1978: 322–9; Thorndike 1950b; Lejbowicz 1997.

[54] See the editions and analyses in Pedersen (Fritz) 1978; 1983–84: 457–512. See more generally Price 1955; North 1976: ii. 249–86; King 1978; Poulle 1980b; Helfand 2002.

longitudes.[55] His other known contribution to the field is a treatise on the *quadrans novus* (1293/4), which had been invented only a few years earlier by Jacob ben Machir ibn Tibbon of Montpellier, better known as Profatius Iudaeus.[56] Next to this 'new quadrant', which ingeniously transferred features of the planispheric astrolabe to the format of the traditional quadrant, Profatius is famous for his *Almanach perpetuum* (1300), which was later used by Dante Alighieri. *Almanach*, an Arabic loan word of unclear etymology, makes one of its first known appearances in the *Liber de rationibus tabularum* by Abraham Ibn Ezra, who was probably also responsible for a short Latin text explaining a work of this type.[57] In contrast to astronomical tables in the *zīj* tradition, an almanac listed the calculated true position of the Sun, Moon, and five planets for successive days or brief intervals of days and thus relieved its audience of the arduous task of having to compute these positions from separate tables of mean motions and equations. 'Perpetual' almanacs of the type created by Profatius additionally relied on approximative periods of recurrence for planetary positions, which made it possible to keep the tables in long-term use.[58]

One of the first Latin Christian authors to create a complete almanac of his own was William of Saint-Cloud, whose *Almanach planetarum* for the years 1292–1311 offered not just planetary longitudes for every day of the year, but also data relating to planetary aspects and the conjunctions and oppositions of Sun and Moon, including the hourly solar and lunar velocities one needed to convert mean into true syzygy times.[59] William's extensive introduction to these almanac-tables is especially noteworthy for his detailed remarks on the astronomical observations he had made in preparation, some of which had furnished him with reasons to take a critical stance towards the Toledan Tables still used by most of his contemporaries. Not only did William find it necessary to apply corrections to some of the relevant epoch values (*radices*), among them a subtraction of 0;40h from all mean syzygy times, but his empirical findings made him engage critically with the access-and-recess model the standard tables employed to close the gap between sidereal and tropical longitudes. Having measured this gap himself on the basis of the vernal equinox of AD 1290, William concluded that the model commonly ascribed to

[55] Pedersen 1979; 1983–4: 649–720 (see also pp. 42–9, where Pedersen convincingly rejects the identification with Peter Nightingale upheld by earlier scholars).

[56] See Boffito and Melzi d'Eril 1922, for Profatius's work, and Pedersen 1983–4: 571–648, for Peter of Saint-Omer.

[57] Abraham Ibn Ezra, *Liber de rationibus tabularum* (ed. Millás Vallicrosa 1947: 119, l. 22). See also the *Liber cursuum planetarum* written in 1141 by Raymond of Marseille (ed. Alverny, Burnett and Poulle 2009: 152) and the mid-twelfth-century *Liber Hermetis Mercurii Triplicis de VI rerum principiis* (ed. Silverstein 1955: 284, 288) as well as the remarks in Millás Vallicrosa 1951; Chabás and Goldstein 2000: 16–19; Kremer 2017: 352. On the almanac-related text in MS London, BL, Cotton Vespasian A.II, fols. 32rb–35rb, which may be the work of Ibn Ezra, see Millás Vallicrosa 1946; Benjamin and Toomer 1971: 375; Samsó 2004b: 290–1; 2012: 191–2.

[58] Profatius's *Almanach* is edited in Boffito and Melzi d'Eril 1908, and discussed in Thorndike 1923–58: iii. 197–204; Toomer 1973; Gingerich and Welther 1977: 155–6; Boudet 1994: 52–4; De Meis 2009.

[59] The introductory canons to the *Almanach planetarum* are edited and translated in Pedersen 2014, who also offers a detailed analysis of the calculations involved. See also Harper 1966: 40–56, 65–8; Mancha 1992a; McCluskey 1998: 200–2.

Thābit ibn Qurra 'has no truth to it, or else that it is subject to some shrinkage or error'. Rather than accepting a variable precession and tropical year, William suggested returning to the linear motion proposed by Ptolemy, who had gauged the rate of precession at 1°/100y. This value had been superseded first by al-Battānī's 1°/66y and later on by the *magistri probationum*, whose rate of 1°/70y William would have been able to glean from Ibn Ezra. Instead of explaining this diversity of estimates with a change in the motion of the eighth sphere over time, William affirmed that 'the cause of this disagreement is the length of time between the observations of our predecessors; the *magistri probationum* had a longer period of time at their disposal than Ptolemy had had, and so they were in a better position to assess the truth'.[60]

With this iconoclastic stance on trepidation, the way was cleared for William to record in his *Almanach* the longitudes of planets as calculated for their tropical positions on the ninth sphere, which implied that the length of the tropical year remained constant. From his *Kalendarium regine* of 1296 it emerges that William considered this fixed length to be $365\frac{1}{4}-\frac{1}{131}$ days (see Ch. 5.1), which can be connected to a precession rate of 1°/70y through the work of Ibn Ezra and his adaptation of the tables of al-Ṣūfī.[61] Like-minded contemporaries of William of Saint-Cloud, who rejected the theory of access and recess in favour of linear precession, include John of Sicily (1290/5) and Peter of Abano (d. 1316), who commented on the issue in a separate *Tractatus de motu octave spere* (1310).[62] North of the English Channel, William's stance was shared by the Benedictine monk Walter Odington, who in *c.*1301 created a calendar and almanac mentioning the coordinates of Evesham Abbey in Worcestershire. The work came with copious supplementary material, including an adjustable star table that instructed users to increase the stellar longitudes by approximately 0;0,51° per year, or 1° for every 70 years.[63] In the accompanying text, Odington cited stellar observations going back as far as the mythical Hermes to support his argument that the motion of the fixed stars was 'always progressive and that they have no contrary motion', noting that 'the experience of instruments' favoured the ancients over Thābit and Azarquiel.[64] This

[60] See William of Saint-Cloud, *Almanach planetarum* 6 (trans. Pedersen 2014: 34), and the remarks in Duhem 1913–59: iv. 16–17; Harper 1966: 41–6; Mercier 1976–7, pt. 1: 201–4; Lejbowicz 1997: 209–10; Chabás and Goldstein 2003b: 262–4. The tables that William critiqued in his *Almanach* were a thirteenth-century adaptation of the Toledan Tables for the Christian calendar and meridian of Toulouse, for which see Poulle 1994; Pedersen 1983–4: 674–80; 1998; 2002: 1197–205, 1344–6, 1546–7, 1550, 1561–2, 1568.

[61] See Abraham Ibn Ezra, *Liber de rationibus tabularum* (ed. Millás Vallicrosa 1947: 78, 83), where al-Ṣūfī is claimed to have upheld a precession of 1°/70y and a sidereal year of 365 days + a revolution of 92;30°. Taken together, this implies a tropical year of $365\frac{1}{4}-\frac{1}{132.45}$ days, close to the $365\frac{1}{4}-\frac{1}{131}$ days used in the Pisan Tables. See Nothaft 2017c: 222–3, and n. 76 in Ch. 3.3.

[62] *Scriptum Johannis de Sicilia super canones Azarchelis de tabulis Toletanis*, ed. Pedersen (Fritz) 1986: 252–57 (J536–44); Peter of Abano, *Lucidator dubitabilium astronomiae* (diff. 2), and Peter of Abano, *De motu octave spere*, ed. Federici Vescovini 1992: 171–92, 347–65. See Nothaft 2017c: 218–19, 226–7, for further discussion.

[63] MS Cambridge, University Library, Ii.1.13, fols. 160r–170r (almanac), 170r–176r (calendar), 176v–181v (appended material). The star table appears on fol. 179v. On Walter Odington and the content of this manuscript, see North 1976: iii. 238–52.

[64] MS Cambridge, University Library, Ii.1.13, fol. 180r.

preference for a steady precession also came to the fore in a brief treatise on the age of the world (*Summa de etate mundi*) Odington wrote at some point between 1308 and 1316. One of his main sources in compiling this text was the twelfth-century *Compotus* of Master Cunestabulus, from whose discussion of the solar year (see Ch. 3.3) he lifted the dates and hours of two vernal equinoxes observed by Hipparchus in 145 BC (24 March, fourth hour) and Ptolemy in AD 140 (22 March, eighth hour). Taking the middle between the two observations, he was able to postulate that the equinox in the year of Christ's incarnation had occurred at the sixth hour of 23 March, which was 10d 6h earlier than the corresponding moment in the year 1316, which Odington expected to occur on 13 March 'in the morning' for English longitudes. Taken together, these numbers invited the conclusion that the Sun's cardinal points had regressed at a rate of *c.*0;11,13h per year or very nearly $J = 128$ years,[65] but the question remained whether it had remained constant over the past 1,316 years. Although Odington repeated Cunestabulus's statement that the access and recess of the eighth sphered caused the vernal equinox to move back and forth between 13 March and 19 April over a period of 2,520 years, he was on the whole unwilling to commit himself to the existence of a recess period, as no recorded stellar observations were available to confirm that the stars ever went the other way. What he did instead was to place the tropical year of $365\frac{1}{4} - \frac{1}{128}$ days against a sidereal year of $365\frac{1}{4} + \frac{1}{150}$ days, finding that the difference between the two was equivalent to 5;13° of a diurnal rotation of the celestial sphere or very close to $\frac{1}{69}$ of a full revolution. As Odington realized, this came reassuringly close to the precession rate of 1°/70y he had previously accepted for the purpose of his own star table.[66]

It should come as no surprise that not all astronomers in Walter Odington's day felt the same need to break free from the constraints created by the Toledan Tables and their various Latin derivatives. One contemporary voice who preferred to remain within this framework was the author of an obscure treatise starting with the words *Autores kalendarii*..., which has only come down to us in two interrelated copies from the early sixteenth century.[67] Written in 1317, *Autores kalendarii* is of some interest for being the first medieval treatise since Reinher of Paderborn's *Compotus emendatus* of 1170/1 to be concerned exclusively with the inadequacy of the 19-year cycle and its consequences for both the calculation of Easter and the historical dates of Jesus's life. Before he came to deal with these topics in depth, however, the author decided to preface his work with four chapters' worth of excerpts from Giles of Lessines, whose indictment of the *auctores kalen-*

65 Ibid., fol. 183v. The interval of 10d 6h suggests that Odington equated 'mane 3° idus Martii' to 13 March, 00:00h, but it remains unclear how Odington arrived at this prediction. The Toledan Tables would have mandated placing the vernal equinox of 1316 on 13 March, at 10:50h, whereas Odington's own almanac gives the solar longitude at the start of 13 March (midnight) in 1304 as 0;46° (fol. 171v). With a daily solar motion of 0;59,8°, the vernal equinox would have occurred roughly 18;30h earlier, on 12 March at *c.*05:30h in the morning, but by 1316 it would have already shifted by *c.*2h towards 03:30h.

66 Ibid., fol. 183v. Further details on Odington's *De etate mundi* are provided in Nothaft 2016c.

67 See the edition and analysis in Nothaft 2013d, and the discussion of authorship in Nothaft 2015d.

darii in Book III of the *Summa de temporibus* provided the backdrop for his own discussion.[68] Rather than following Giles's choice of parameters, however, the anonymous astronomer put his trust in the Tables of Toulouse, which was the same Christian adaptation of the Toledan Tables that William of Saint-Cloud had sought to improve upon when he wrote the *Almanach planetarum*. In stark contrast to William's critical stance, the author of *Autores kalendarii* appealed to these tables as being 'truer than other tables and easier to handle when it comes to finding the mean time of the conjunction of the luminaries'.[69] Using them for his own calculations, he determined that the vernal equinox of AD 1308 should have occurred on 13 March, 00:10h after noon, which is roughly 22 hours later than the correct time according to modern calculation.[70] This imperfect result may be contrasted with some of the observations made for the meridian of Paris by his contemporary Alard of Diest, who found that the vernal equinox of 1312 had occurred on 12 March, 00:50h after noon. Alard's result deviates from modern calculation by a mere hour, a success made possible by his close attention to 'the Sun's altitude at noon and its declination from the equinoctial'.[71] The author of 1317, too, claimed to have studied the true vernal equinox during the period from 1290 to 1310,[72] but instead of systematic observations, these were in fact instances of calculation using the Tables of Toulouse and their access-and-recess model, which yielded a tropical year of approximately 365d 5;58h for the period in question. The inadequacy of this estimate was noticed by a later copyist, who corrected the value in question to the 365d 5;49h endorsed by William of Saint-Cloud.[73] Better values could also be obtained from fresh observations or calculations, as was done *c.*1308 by Alard of Diest, who noted a length of 365d 5;47h.[74]

In line with the model underlying the tables for Toulouse and Toledo, the astronomer responsible for *Autores kalendarii* treated his own value as a mere snapshot of the present, claiming that the divergent estimates for the tropical year found in ancient Greek and Arabic sources had all been correct at some point in the past and would again become correct in the future due to the access and recess of the eighth sphere.[75] Rather than fretting over the error of the Julian calendar, a correction of which may have struck him as futile in light of this theory, the author

68 *Autores kalendarii* 1–4 (ed. Nothaft 2013d: 32–43). Cf. Giles of Lessines, *Summa de temporibus* III.2.18 (ed. Steele 1926: 146, ll. 10–11).

69 *Autores kalendarii* 16 (ed. Nothaft 2013d: 62).

70 Ibid. 4 (ed. Nothaft 2013d: 41).

71 See the record of observations made in Paris, as preserved in MS Oxford, Corpus Christi College, 144, fols. 97r–98v, and edited by Thorndike 1948: 203–5. The quoted remark appears on p. 204: '. . . per altitudinem solis in meridie et declinationem eius ab equinoctiali et per calculationem motus solis diversi'. The identification of the Parisian astronomer with Alard of Diest is convincingly made by Zinner 1951.

72 *Autores kalendarii* 1 (ed. Nothaft 2013d: 34).

73 See Nothaft 2015d: 221–5. The 'Toledan' solar year of 365d 5;58h is also implicit in the *Tabula solstitiorum et equinoctiorum* for 1281–84 in MS Bernkastel-Kues, Bibliothek des St. Nikolaus-Hospitals, 212, fol. 258r, for which see n. 175 in Ch. 4.3.

74 See parchment flyleaf no. 7 in MS Erfurt, UFB, CA 4° 371, and Zinner 1951: 40. Alard's tropical year is also mentioned in a marginal note in MS Berlin, Staatsbibliothek, lat. fol. 610, fol. 130r, here given more precisely as 365d 5;47,15h.

75 *Autores kalendarii* 1 (ed. Nothaft 2013d: 34).

decided to dedicate the bulk of his treatise to the Easter full moon or *terminus paschalis*, knowing that its faulty calculation created numerous instances where feasts were celebrated on the wrong date. An especially troubling recent example of this kind had been witnessed in 1291, when the Church celebrated Easter Sunday on 22 April even though astronomical reckoning showed this to be the 23rd day of the second lunar month of spring. 'And this is what the Jews ridiculed and taunted us for in that year.'[76] Similar cases, where Easter Sunday fell on the illegitimate lunar days 22, 23, or 24, had occurred in 1294, 1298, 1301, 1304, 1305, 1307, 1308, 1311, and 1314. They could also be expected for the future years 1318, 1321, 1322, 1325, 1328, and 1329, making for an error rate of 16 out of 39 years or 41 per cent.[77] His chosen path out of this quandary was to redeploy the existing Golden Numbers with the aid of the mean syzygy tables standardly included among the Tables of Toulouse, an excerpt of which accompanied the treatise.[78] He made such calculations for the 19-year cycle from 1311 to 1329, exhibiting the results in an attached *kalendarium* with 'reformed' Golden Numbers that ran between three and five days ahead of their traditional positions.[79] As a consequence of this intervention, the Golden Number for year 16 migrated from 8 to 5 March, which in turn brought the earliest Easter Sunday from 22 March up to 19 March. The author duly illustrated the new range of calendrical limits by means of a separate table, which was devoted to the 35 possible Easter dates and the corresponding *intervallum minus*, that is, the interval in weeks and days between Christmas and Quadragesima Sunday.[80] What this improvement consciously overlooked was the actual date of the vernal equinox, which at the time of writing could already be found on 12/13 March. In the author's view, it was apposite to ignore the classical rule of the equinox for fixing the *terminus paschalis*, since its implementation would have required moving the new moon of the Easter lunation to the end of February, thus violating another computistical principle according to which the Easter lunation always terminated in April. According to a rule of thumb he placed towards the end of his treatise, the Easter lunation was henceforth best defined as taking its beginning from any conjunction that fell on or after 4 March, as encapsulated by these verses:

> *A quarto Martis primam perquirito lunam,*
> *Et post bis septem lux proxima pascha celebrat.*
>
> Search for the first day of the moon on any [date] from the 4th of March,
> And the next light [of day] to come after twice-seven [days] celebrates Easter.[81]

In spite of this little excursion into calendrical mnemonics, our author found it important to emphasize that he had:

> no desire to offer any rules for the lunar age on any given day, still less for the heads of the months, concerning which the computists weaken themselves and their disciples

[76] Ibid. 15 (ed. Nothaft 2013d: 60). Passover in AD 1291 fell on 17 March and hence more than a month before Easter (22 April).

[77] Ibid. (ed. Nothaft 2013d: 60–1).

[78] Ibid. 16–18 (ed. Nothaft 2013d: 62–8).

[79] Ibid. (ed. Nothaft 2013d: 78–89).

[80] Ibid. 20–1 (ed. Nothaft 2013d: 69–73).

[81] Ibid. 22 (ed. Nothaft 2013d: 75). See also ch. 20 (p. 70). Since the Easter lunation is defined as a hollow lunation of 29 days, it needs to begin no earlier than 4 March in order to terminate on 1 April.

> through lunar regulars and epacts, not knowing that the anticipation of the conjunction makes it impossible for their rules to remain eternally fixed and infallible. This they could not know, because it is the remit of a higher science and demonstrated by a more profound and nobler craftsman, the astronomer, who is nonetheless subservient to the astrologer in his further operations. I therefore omitted the science of the *claves terminorum*, if it can be called a science, which is superfluous since all mobile feasts depend on Easter. For those rules concerning the *claves terminorum* are complicated and obscure, not sources of enlightenment (*lumina intellectus*).[82]

It would be an exaggeration to call the author of *Autores kalendarii* a man at the cutting edge of his discipline, and yet the words just cited make it clear that he regarded himself as a practitioner of a craft more refined than that of the vulgar computists, who in his view kept needlessly wrapping the ecclesiastical calendar in an intricate padding of rules that did nothing to increase its accuracy. His depreciatory remarks about the epacts and *claves terminorum* indicate a turning point in the history of the calendar problem as far as the relation between astronomy and *computus* is concerned. The idea of fusing *computus* and astronomy into a unified science of time reckoning, as pursued by Roger of Hereford in the twelfth and still upheld by authors such as Robert Grosseteste, Campanus of Novara, and Giles of Lessines in the thirteenth century, had lost much of its practical utility and persuasive force by the year 1317. In its place, the fourteenth and fifteenth centuries were to put an increasingly rigorous division of labour between computists and astronomers. The former would continue to serve as guardians of the traditional calendar by authoring texts that explained its rules, but it was the task of astronomers to test these rules, critique them in light of the actual motion of the heavens, and, where feasible, propose more adequate ones.

5.3. THE ALFONSINE TURN

To replace the ecclesiastically sanctioned Golden Number with a truer or reformed equivalent was certainly one of the most straightforward and, hence, most common ideas voiced in the context of the medieval debate surrounding the calendar problem. As seen from this and the previous two chapters, it can be traced from Roger of Hereford and John of Sacrobosco to the treatise *Autores kalendarii* of 1317 and further to the countless enhanced or 'improved' calendars that were drawn up between the thirteenth and sixteenth centuries.[83] These calendars range from widely circulated standard works such as Peter Nightingale's *Kalendarium* to unique specimens like the calendar that famously opens the *Très Riches Heures*, a richly illuminated book of hours created between *c.*1412 and 1416 for Jean, Duke of Berry. Although not a 76-year lunar almanac in the tradition of Nightingale's

[82] Ibid. 23 (ed. Nothaft 2013d: 76–7).

[83] In areas such as Sweden, where the Gregorian reformed calendar was not adopted until 1753, such corrected Golden Numbers persisted for even longer, especially in the form of runic calendars carved into wooden staffs or tablets. The best survey of this phenomenon, with a strong focus on Scandinavian rune staff calendars, is Lithberg 1953: 129–210. See also Lindhagen 1914: 13; Cucina 2011; 2013.

Kalendarium, the calendrical tables in the *Très Riches Heures* nevertheless juxtapose the traditional Golden Number with a column for the *Nombre d'or nouvel*, the dates in which can be shown to have originally been calculated for the 19-year period from 1349 to 1367 by Jean des Murs and Firmin de Beauval.[84] What set these calculations apart from those performed by Peter Nightingale or William of Saint-Cloud in the 1290s, or by Alard of Diest and Raymond Bancal soon after 1300, was their theoretical basis, which was no longer provided by the Toledan Tables or any of their Christian adaptations. Instead, Jean and Firmin are known to have availed themselves of the so-called Alfonsine Tables, which came to dominate the practice of mathematical astronomy in medieval Europe during the fourteenth century and would continue to do so far into the sixteenth century.

The watershed between these two phases in Latin computational astronomy, Toledan and Alfonsine, can be roughly located around 1320, as seen for example from the work undertaken by the Parisian astrologer Geoffrey of Meaux, who in that year produced a new set of tables with the express goal of providing relief for 'all those who desire to know the true locations of Sun and Moon, but withdraw themselves from this knowledge because of the multitude of tables and the difficulty and tedium [involved in using them]'.[85] His solution was to display in the format of a calendar the daily mean longitudes of the Sun and Moon (signs, degrees, and minutes) for an entire year as well as further values relevant to finding the Moon's mean anomaly and argument of latitude, which could be combined with one of the epoch values listed in an auxiliary table to yield results for 1321 or any other year in the near future. The calendar also featured the old Golden Number, which was after all still needed to compute the mobile feast days 'according to the Church' (*secundum ecclesiam*), but in addition it provided a separate column for the *littera primacionis* from A to T. These lunar letters made it possible to apply the tabulated longitudes in finding the precise hour and minute of any mean conjunction and opposition, and even to investigate the possibilities of eclipses. Of particular historical interest is Geoffrey's proclamation to have based the relevant calculations on the '*radices* of Azarquiel that are placed in the Toledan Tables', rather than the newly available '*radices* of Alfonso, because of certain insoluble arguments I proposed in everyone's presence when I lectured'.[86]

[84] See Nothaft 2015h, and n. 101 in Ch. 6.4.

[85] Geoffrey of Meaux, *Kalendariorum flos*, MS Madrid, Biblioteca nacional de España, 9288, fol. 1ra: 'Cunctis solis et lune scire desiderantibus vera loca, ab hoc tamen scire retrahentibus se propter pluralitatem et difficultatem et tedium tabularum, Gaufridus de Meldis presens novum opusculum composuit bipartitum.' Copies that include these tables appear in MSS Erfurt, UFB, CA 4° 369, fols. 6r–14v; Los Angeles, J. Paul Getty Museum, Ludwig XII.6, fols. 1r–11r; Madrid, Biblioteca nacional de España, 9288, fols. 1r–9v; Paris, BnF, lat. 15118, fols. 67r–75v; Toledo, Archivo y Biblioteca Capítulares, 99-5, fols. 13v–25v; Uppsala, Universitetsbibliotek, C 653, fols. 189r–197r. Incomplete copies appear in MSS Erfurt, UFB, CA 4° 381, fols 40v–41r (canons only); Paris, BnF, lat. 7281, fols. 160v–163r (calendar missing). The introduction mentions a second part of the work (*Solis et lune verum inventarium*) planned to contain an *instrumentum astrologicum* for finding true syzygies and longitudes, but none of the manuscripts I have seen includes it. On the author, see Duhem 1913–59: iv. 69–74; Weijers 1994–2012: iii. 65–6; Thorndike 1923–58: iii. 281–93, 715–17; 1950a: 208–25.

[86] Geoffrey of Meaux, *Kalendariorum flos*, MS Madrid, Biblioteca nacional de España, 9288, fol. 1rb: 'Et sciant omnes ad quos presens opus devenerit quod ego, Gaufridus de Meldis, non accepi

This resolute dismissal of the '*radices*', that is, the epoch values of 'Alfonso' constitutes one of our earliest pieces of documentation for the arrival and incipient use of the Alfonsine Tables at the University of Paris, where Geoffrey of Meaux had been teaching since the 1310s. The tables he reacted against had originally been a product of the court of King Alfonso X of Castile and León (1252–84), under whose sponsorship they were assembled between 1263 and 1272 by two Jewish astronomers, Isaac ben Sid and Judah ben Moses ha-Cohen.[87] Through undisclosed channels, the fruits of their labour eventually arrived in France, where a Latin set of tables attributed to King Alfonso emerged at the beginning of the 1320s. This Latin-Parisian recension of the Alfonsine Tables, which focuses on planetary longitudes and the motions of the fixed stars and apogees, survives in hundreds of late medieval manuscripts and a dozen printed editions, starting in 1483 with the *editio princeps* produced by Erhard Ratdolt in Venice.[88] While there is an abundance of material to document the influence of the Alfonsine Tables on the practice of astronomy during the late Middle Ages, the canons to what has been taken to be their original, but lost Castilian version survive only in a single sixteenth-century manuscript.[89] As far as can be ascertained, the tables described in these Castilian canons were meant to provide sidereal longitudes, plotted against the eighth sphere, which could then be converted into tropical longitudes with the aid of tables for the eighth sphere's access and recess.[90] They thus followed the same basic principle as the old Toledan Tables, taking the sidereal year to be fixed and the tropical year to be a variable quantity. In striking contrast to this, the Latin Alfonsine Tables marked a return to the tropical longitudes of the ninth sphere that had already characterized the works of Ptolemy and al-Battānī as well as the *Almanach planetarum* of William of Saint-Cloud.[91]

radices Alfonsi propter quasdam rationes insolubiles quas proposui coram omnibus dum legebam, sed accepi radices ab antiquis sapientibus approbatas secundum radices Azerchelis positas in tabulis Toletanis.'

[87] The standard monograph on the Alfonsine Tables is Chabás and Goldstein 2003b. On Alfonso X's scientific patronage, see also Wegener 1905; Procter 1945; Roth 1990.

[88] For the tables in the 1483 edition, see Poulle 1984. See also Wegener 1905: 165–71; Gingerich 1987; Poulle and Savoie 1988, and the 'List of incipits of Alfonsine canons and related works' in North 1977: 296–300. On the early Alfonsine astronomers active in Paris, see Duhem 1913–59: iv. 60–9, 74–90; Weijers 1994–2012: v, 110–13, 161–3; Chabás and Goldstein 2003a; 2003b: 266–90; 2004; Poulle 2005.

[89] These canons appear in MS Madrid, Biblioteca nacional de España, 3306, fols. 34v–73v, which was first edited by Rico y Sinobas 1863–7: iv. 111–83, and again in Chabás and Goldstein 2003b: 19–94.

[90] See ch. 49 of the Castilian canons in Chabás and Goldstein 2003b: 89–90, and the commentary on pp. 217–20.

[91] Owing to these and further significant differences between the Castilian and Latin versions of the Alfonsine Tables, Emmanuel Poulle came to the far-fetched conclusion that the Latin version had been elaborated independently by Parisian astronomers and that the common attribution to Alfonso X was a case of pseudepigraphy. See Poulle 1987; 1988, 2004a: 17–24; 2004b; 2005. A more convincing account is presented in Chabás and Goldstein 2003b: 243–303, who argue for a strong dependence of the Parisian astronomers on Toledan-Castilian sources. See on this intricate question also Goldstein 1980; Samsó 1994a: 565–6; Goldstein, Chabás, and Mancha 1994; North 1996; Swerdlow 2004; Husson 2011.

According to the tables adopted in Paris around 1320, the mean Sun progresses at a daily rate of 0;59,8,19,37,19,13,56°, implying a mean tropical year of $365;14,33,9,57\ldots\mathrm{d} = 365\mathrm{d}\ 5;49,15,58,49\ldots\mathrm{h} \approx 365\frac{1}{4} - \frac{1}{134}$ days.[92] The significance of this value for the history of calendar reform is best grasped by comparing it with modern estimates for the vernal equinox year—that is, the interval between two solar crossings of the vernal point—which over the past millennium has remained somewhat below 365.242375d or *c.*365d 5;49,1,12h.[93] Out of the various different values accepted by Latin astronomers during the twelfth and thirteenth centuries, only the $365\mathrm{d}\ 5;49\mathrm{h} \approx 365\frac{1}{4} - \frac{1}{131}$ days associated with al-Ṣūfī came closer to reality. The great accuracy of the Alfonsine tropical year was to some degree a matter of serendipity, given that the value in question was contingent on the long-term periods used in the Alfonsine precession model, which encompassed both a steady and a variable component (see Ch. 5.4). The steady or linear component was known as the 'mean motion of the apogees and fixed stars' (*medius motus augium et stellarum fixarum*), whose daily rate of 0;0,0,4,20,41,17,12° amounted to a full revolution of 360° after very nearly 49,000 years.[94] The rationale behind this number becomes apparent once the mean precessional term is subtracted from the Sun's daily mean motion in tropical longitude, leading to a sidereal motion of $0;59,8,19,37,19,13,56°/\mathrm{d} - 0;0,0,4,20,41,17,12°/\mathrm{d} = 0;59,8,15,16,37,56,44°/\mathrm{d}$, which is virtually the same as the mean solar motion implied by the Julian calendar (0;59,8,15,16,37,56,48°/d). It hence appears that the length of the tropical year in the Parisian Alfonsine Tables was originally found by taking the Julian year and combining it with a precessional period of 49,000 years, such that the tropical year ended up being exactly $\frac{49000}{49001}$ of $365\frac{1}{4}$ days, which leads to the aforementioned $365;14,33,9,57\ldots\mathrm{d}$ (or 365.24254607d).

A comparable instance, where a fixation on round values and cyclical periods produced a parameter of exquisite accuracy, is enshrined in a set of astronomical tables for the meridian of Cambridge that John Holbroke (d. 1437), a one-time chancellor of Cambridge University and master of Peterhouse, produced in the year 1433.[95] In tabulating planetary mean motions, Holbroke followed the structural template of the standard Latin Alfonsine Tables, but instead of simply copying their numbers he decided to re-compute all mean motion parameters on the basis of a different length of the tropical year, which Holbroke asserted to be $365\frac{1}{4} - \frac{1}{132}$ days or 365d 5;49,5,27,16,21,49,5,27,16,21,49...h. From the remarks

[92] See the *Tabula medii motus solis, Veneris et Mercurii*, as edited in Poulle 1984: 134. The corresponding Julian drift of $J = 134$ years was first mentioned in 1321 by Jean des Murs in his *Expositio intentionis regis Alfonsii circa tabulas ejus* (ed. Poulle 1980a: 251).

[93] Meeus 2002: 362.

[94] Poulle 1984: 133. The tabulated daily rate of precession was probably derived as follows: (1) $360° \div 49{,}000\mathrm{y} = 0;0,26,26,56,19,35,30\ldots°/\mathrm{y} \approx 0;0,26,26,56,20°/\mathrm{y}$; (2) $0;0,26,26,56,20°/\mathrm{y} \div 365.25\mathrm{d}/\mathrm{y} = 0;0,0,4,20,41,17,12,26,36,42\ldots°/\mathrm{d} \approx 0;0,0,4,20,41,17,12°/\mathrm{d}$.

[95] On Holbroke, see Price 1955:8, and Nothaft (forthcoming-1). His tables of 1433 survive in MS London, BL, Egerton 889, fols. 152v–160v, which is Holbroke's autograph, as well as in MSS Cambridge, University Library, Ee.3.61, fols. 63v–68v; Cambridge, Peterhouse, 267, fols. 96r–100r; Gloucester, Cathedral Library, 21, fols. 164r–171r.

Holbroke made in his preface, it appears that this alternative value was inspired by the perceived necessity of stabilizing the equinoxes and solstices in the Julian calendar. If the value of the Julian error (J) was assumed to be exactly 132 years, which was a multiple of 4, the solution to the problem reduced to the simple step of omitting the bissextile day in every 132rd year. Had this measure already been taken at the time of Julius Caesar, the current anticipation of the equinoxes and solstices would have never arisen 'nor would one see the error that happens all too often in the celebration of Easter and the other mobile feasts'.[96] Holbroke's 'true' solar year of $365\frac{1}{4}-\frac{1}{132}$ days was hence the ideal natural basis for a perfect and permanently valid solar calendar, whose structure chimed with the idea that God had created the Sun and Moon specifically to serve as 'signs, and for seasons, and for days and years' (Genesis 1:14). In 1434, just one year after Holbroke presented his new parameter values, an English chaplain named Richard Monke used the same estimate of the 'true length of the year' for his *Kalendarium verum anni mundi*, which charted daily mean and true solar longitudes for the hypothetical year of creation 4909 BC, but also for the recent year AD 1428 and every 33rd year before or after.[97] The solar data contained in this calendar could be adjusted to any other year of the world by adding the increments displayed in a set of 'true and perpetual' tables for the Sun (*Tabule solis vere atque perpetue*), in which it was tacitly assumed that the Julian bissextile day would be omitted at the end of every 132 years.[98]

One particularly striking feature of Monke's 'true and perpetual' tables is their division into four columns, each of which represents a sub-cycle of 33 years. The significance of this period lies in the fact that a multiplication of the proposed year length of $365\frac{1}{4}-\frac{1}{132}$ days ($=365.\overline{24}$ days or 365 days and $5;\overline{49,5,27,16,21}$ hours) by 33 will yield 12,053 complete days without any fractional surplus. As a result, the longitudes of the Sun and the times of its entries into the zodiacal signs are bound to repeat identically after 33 years, and hence after one-fourth of the total correctional period of 132 years.[99] From a calendrical point of view, the most important consequence of this 33-year cycle is the way it might have suggested a simple, yet highly efficient modification of the existing leap-year pattern. Instead of intercalating 8 days over the course of 32 years in accordance with the traditional rule, one could have expanded every eighth leap-year period from 4 years to 5, thus creating an average year of $365\frac{8}{33}=365\frac{1}{4}-\frac{1}{132}$ days. The resulting 33-year intercalation cycle would have carried the great advantage of being short enough perpetually to retain the vernal equinox within the same date 24-hour

[96] MS London, BL, Egerton 889, fol. 152v: 'Quod si ab ipsius tempore Iulii factum esset, non hec solstitiorum aut equinoctiorum anticipatio, non error qui in pasche ceterorumque mobilium festorum celebritate frequenter nimis accidit, videretur.'

[97] Monke's *Kalendarium* appears together with some auxiliary tables in MS Oxford, BodLib, Laud. Misc. 594, fols. 14v–21v. The material is discussed briefly, but not very correctly, by North 1983: 91–4; 1992a: 134. See now Nothaft (forthcoming-1).

[98] MS Oxford, BodLib, Laud. Misc. 594, fol. 21r.

[99] The same principle is revealed by two tables in the mid-fifteenth century MS Oxford, BodLib, Digby 97, fols. 33r and 41r.

period, as the annually accumulating difference between the Julian year and the assumed length of the tropical year would have grown to no more than 6 hours during the interval in question. That said, Richard Monke's terse remarks in the headings and margins of his tables contain no hints that he ever considered this a viable possibility, as his preserved instructions concern no more than the omission of a day after 132 years.[100] It was only 145 later, in 1579/80, that a 33-year intercalation cycle of the type just mentioned was brought up in the context of the discussions leading up to the Gregorian calendar reform. Its advocate at the time was Ignatius Naʾmatallah, an exiled former patriarch of the Syriac Orthodox Church, who became associated with Pope Gregory XIII's reform commission after his arrival in Rome and who drew his knowledge of a tropical year of $365\frac{1}{4}-\frac{1}{132}$ days from Islamic sources.[101] Rather than following Ignatius on this point, the commission in charge of planning the new calendar opted for intercalating 97 days in 400 years, which was easily achieved by omitting three bissextile days in the course of 400 years. The solar year implied by this reform hence comprised $365\frac{97}{400}$ days = 365;14,33d = 365.2425d, leading to a value that was indistinguishable from the Alfonsine length, if the latter was rounded to the second sexagesimal-fractional place.[102]

This tight fit between the Gregorian year length and the Alfonsine value for the tropical year is a telling sign of the enormous diffusion and popularity King Alfonso's tables came to enjoy during the 250 years preceding the calendar reform of 1582. One by-product of their spread from Paris to other parts of Western and Central Europe was the emergence of various local adaptations and modified versions, which were often aimed to make the existing tables more user-friendly without changing the underlying parameters in any significant way. A sophisticated example of this type of creative recasting were the *Tabule Anglicane* for the meridian of Oxford, which a certain William Batecombe calculated for a start in 1348. Following a precedent set by his Parisian predecessor Jean de Lignères, whose *Tabule magne* date from the 1320s, Batecombe used double-argument tables to offer a swift and easy way of inferring a planet's true longitude from the arguments of centre and anomaly.[103] Meanwhile, the strong grip on astronomical practice the Alfonsine Tables managed to exert over the course of the fourteenth century was only reinforced by enhanced calendars like those produced in the 1380s by Nicholas of Lynn and John Somer (see Ch. 5.1), who relied on the new lunar and solar parameters when calculating the times of true syzygies and the durations and magnitudes of eclipses. In the case of Nicholas of Lynn, his deviations from mod-

[100] This instruction is made explicit in the following remark recorded in a table heading in MS Oxford, BodLib, Laud. Misc. 594, fol. 14v: 'Et pro fixione perpetua equinoxii vernalis in kalendario veri anni mundi nulla habetur littera rubia in ultima linea transversali.' In the table in question, a red letter (*littera rubia*) signifies a bissextile year.

[101] Ziggelaar 1983: 215–17. Another early modern advocate of the 33-year cycle was the Protestant astronomer Gottfried Kirch (1639–1710), for whom see Koller 2014: 364–5, 404–5.

[102] See Swerdlow 1974, and n. 58 in Ch. 8.

[103] On Batecombe and other English adaptations of the Alfonsine Tables, see North 1977: 272–91; 1992a: 122–7; Chabás and Goldstein 2003b: 291; 2009b: 4–11; 2016; 2017. On 'user-friendly' tables more generally, see Husson 2012; Chabás and Goldstein 2013.

ern data rarely exceed 30 minutes for eclipse times and 0.10 for magnitudes and often hit much closer to their target, which is suggestive of the overall quality of results one could attain with the Alfonsine Tables.[104] Extended tabular lists of conjunctions and oppositions according to Alfonsine true motions were soon also produced on the continent, as evidenced by a calendar for the meridian of Verdun covering the years 1392–1448, which a certain Jacobus Jacobi composed in honour of Robert, Duke of Bar (1344–1411). In a manner reminiscent of Walter Elveden's calendar of 1327, it comprised only three instead of four 19-year cycles and added a list with the dates and times of eclipses expected for 1411–48.[105]

When it comes to tables of mean (rather than true) syzygy, those created in the Alfonsine mould can easily be distinguished from their predecessors by taking into consideration the time subtractions that were necessary to derive the conjunctions or oppositions in one cycle from those in the previous cycle. According to the Toledan Tables and the Jewish calendar, this value was *c.*1;27h (1h 485p) after 19 years or *c.*5;48h (5h 860p) after 76 years, whereas calendars derived from the Alfonsine Tables normally apply subtractions of 1;28h and 5;52h. This slight increase reflects the fact that King Alfonso's astronomers sped up the Moon's mean motion from the standard Toledan value of 13;10,34,52,48,47°/d to 13;10,35,1,15,11,4,35°/d.[106] Together with the Alfonsine value for the mean solar motion per day, this implied a daily elongation between Sun and Moon of 13;10,35,1,15,11,4,35°/d − 0;59,8,19,37,19,13,56°/d = 12;11,26,41,37,51,50,39°/d,[107] which could in turn be converted into a mean synodic month of 360° ÷ 12;11,26,41,37,51,50,39°/d = 29;31,50,7,37,27,8,25d = 29d 12;44,3,2,58,51,22,8h as opposed to the classic Babylonian, Ptolemaic, and Jewish value of 29;31,50,8,20d = 29d 12;44,3,20d. Part and parcel of this change in parameters was an implied increase in the error produced by the ecclesiastical lunar calendar, which now could be found to undergo a slippage of approximately $M = 310\frac{2}{3}$ years instead of $M = 314\frac{2}{3}$ years. To be overly precise, the ecclesiastical 19-year cycle takes $19 \div (6939;45\text{d} - 29;31,50,7,37,27,8,25\text{d} \times 235) = 310.713443\ldots$ years to gain a complete day relative to the Alfonsine synodic month, which is the same as 310 Julian years and 260.58 . . . days.

One of the earliest calendars to feature mean conjunctions based on the then-new Alfonsine Tables is the *Kalendarium novum* drawn up in 1329 by a Dominican friar named Isenbrand, which precedes the *Almanch perpetuum* of Profatius Iudaeus in a fourteenth-century astrological anthology of Flemish origin. From the new moons displayed in Isenbrand's calendrical tables, it can be inferred that he calculated the mean conjunctions for the four 19-year cycles from 1330 to 1405, applying the necessary subtractions of 1;28h per cycle and adjusting his epoch

[104] The accuracy of Nicholas of Lynn's data is assessed in Beard 2004. Somer's data show only minor deviations from Lynn's, which are probably due to diverging ways of calculating the equation of time. See North 1988: 98–100; Kremer 2003a: 272–5. On late medieval eclipse predictions using the Alfonsine Tables, see also Thorndike 1951; 1952; 1957b; Chabás and Goldstein 2012a: 180–2.

[105] MS Lübeck, Stadtbibliothek, Ms math. 1, fol. 110r–117r.

[106] Poulle 1984: 135. [107] Ibid.: 138.

value to a meridian 14;15° east of Toledo.[108] The difference is increased to 17;30° in a calendar for 1361 to 1436 devised by the Swabian astronomer Heinrich Selder, who in 1378 incepted as Master of Arts at the University of Paris. Its columns contain mean conjunctions timed to the nearest second of the hour as well as a separate column for the 'new Golden Number' (*novus aureus numerus*), the entries for which are moved upwards by three days compared to the traditional *aureus numerus*.[109] As Selder himself informs us, his new-moon times were all calculated for the meridian through Padua and Salzburg, although he did not forget to tell readers that they could adapt the calendar to other locations by subtracting or adding a minute for every four German miles further east or west. Neither did he omit to mention that the calendar could be maintained in use for another 76 years (1437–1512) by subtracting 5;52,13h from each conjunction. This parameter was obviously derived from the Alfonsine synodic month of 29d 12;44,3,2,58,51 . . . h, which yields a 76-year period of 27,758d 18;7,46,42h, being 5;52,13,18h short of 27,759 days. Selder was therefore correct when he revealed that, according to the Alfonsine Tables, the traditional 19-year lunar cycle was in error by 23;28,53,12h$(= 4 \times 5;52,13,18\text{h})$ after 16 iterations or 304 years.[110] The influential John of Gmunden, whose name is attached to the most popular batch of Alfonsine lunar calendars (see Ch. 5.1), followed essentially the same principles, listing the hours and minutes (but not seconds) of the mean conjunction for periods of 19 or 76 years. From the fact that his conjunction times fall 1;20h later than those derivable from Alfonsine epoch values, one may conclude that they had been computed for a longitude 20° east of Toledo, here understood to refer to Vienna, where Gmunden started teaching in 1408.[111]

One noteworthy feature of the Alfonsine Tables in their Latin-Parisian incarnation is the strict 'sexagesimalization' of all tabulated parameters, as a consequence of which motions and time intervals were consistently dissolved into base-60 multiples and fractions of days or degrees. The main advantage of this approach was that it reduced the number of tables required, while the data contained in them could be freely adapted to a variety of different epochs and calendrical schemes.[112] Later adaptations of the Alfonsine Tables, like those made by John of

[108] MS London, BL, Harley 267, fols. 175r– 178v. There is no accompanying text except for a note at the end stating: 'Explicit Kalendarium novum editum a venerabili patre fratre Isenbrando ordinis predicatorum anno domini 1329. Sed istud scriptum fuit anno domini 1343.' The syzygy times are calculated from midnight, starting with a conjunction on 20 January 1330, 03:41h.

[109] MS Rottenburg am Neckar, Diözesanbibliothek, H 15, fols. 1r–8r. Images of fols. 1r–2r and 7v–8r, are provided in Mütz 1999, whose commentary must be used with caution. For more on Heinrich Selder, see Nothaft 2016b.

[110] MS Rottenburg am Neckar, Diözesanbibliothek, H 15, fol. 8r. On the same page, the length of the lunar month is given as 29d 12;44,3,24h, which must be a scribal error for 29d 12;44,3,2h.

[111] See n. 40 in this chapter.

[112] Chabás and Goldstein 2012a: 54–7. The preserved Castilian canons to the Alfonsine Tables instead maintain the traditional use of separate tables for a planet's motion per day, month, and expanded year, the values of which are then added to an epoch date derived from collected-year cycles of 20 Julian years, but it nevertheless remains possible that the sexagesimal structure of the Latin tables was already introduced at the court of Alfonso X. See Samsó 1987a: 31–2; Swerdlow 2004: 483–4.

Gmunden,[113] forfeited this sexagesimal format and instead returned to a layout more familiar from the old Toledan Tables and their Western adaptations by listing epoch values in collected-year intervals that were counted from the beginning of the Christian era. An example that became especially widespread in the fifteenth century and still appeared in print as late as 1588 are the so-called *Tabule resolute*, which appear to have been aimed primarily at calculators of calendars, almanacs, and ephemerides. Their initial compilation was probably due to a certain Petrus Cruciferus, who produced a version for the meridian of Wrocław with an epoch date in 1424. By the middle of the fifteenth century, the *Tabule resolute* were being used as a teaching tool at the University of Kraków, where they received new canons by Marcin Król of Żurawica (d. 1459) and his disciple Andrzej Grzymała of Poznań (d. 1466) in *c*.1445 and 1448.[114]

A Mediterranean route of transmission completely independent of either Paris, Oxford, Vienna, or Kraków, appears to have been responsible for the early presence of the Alfonsine Tables in Italy, where a set of tables for Padua based on Alfonsine parameters was created by the clockmaker Jacopo de' Dondi (*c*.1298–1359) and later, in 1424, modified by Prosdocimo de' Beldomandi.[115] Another significant Italian product of Alfonsine astronomy were the tables designed *c*.1442 by Giovanni Bianchini for the longitude of Ferrara. Bianchini's tables, which received three printed editions between 1495 and 1553, greatly exceeded most other adaptations in length and featured a number of intriguing innovations relative to the mainstream tradition, among them the idea of combining the linear and variable components of the Alfonsine precession model into a single table.[116] By the end of the century, the famous Sephardic astronomer and historian Abraham Zacut was able to count Italy next to Spain, Germany, France, and England among the countries where astronomers had discarded all their former tables in favour of the Alfonsine ones—a fitting reminder that Europe had become fully united in astronomical terms.[117]

5.4. ALFONSINE WORRIES

With the benefit of scientific hindsight, it is easy to praise the Alfonsine Tables for the relative accuracy of their mean synodic month and especially for their tropical

[113] Klug 1943: 32–42; Porres de Mateo 2002; 2006; Chabás 2002: 169–71; Chlench 2006b: 214–20; 2007: 435–68.

[114] Dobrzycki 1987; Chabás 2002; Kremer 2007. See also Chabás 1998; Porres de Mateo and Chabás 1998; Chabás and Goldstein 2000: 20–1, for an adaptation of the *Tabule resolute* to the meridian of Salamanca.

[115] Chabás 2007; Chabás and Goldstein 2003b: 248, 292. An early indirect witnesses to the presence of the Alfonsine Tables in Italy is a passage in Thaddeus of Parma's commentary on the common *Theorica planetarum Gerardi*, completed on 12 July 1318. See e.g. MSS Florence, BML, Ashburnham 205, fol. 24rb; Florence, BML, Plut. 29.07, fol. 153v; Venice, Biblioteca Nazionale Marciana, Fondo antico lat. Z. 340, fol. 78v. The 'nove tabule Alfunsi' are also mentioned in the gloss to a star catalogue in MS Florence, BML, Conv. soppr. 414, fol. 60r, which gives the present year as 1304.

[116] Chabás and Goldstein 2009b.

[117] Chabás and Goldstein 2003b: 236. On Zacut's own astronomical work in the Alfonsine tradition, see Chabás and Goldstein 2000.

year, which eventually came to underpin the civil calendar still in use today. For late medieval astronomers, however, any assessment of this accuracy would have required painstaking observations, which were often enough fraught with considerable uncertainty. Perhaps the first author to put the new solar parameters to something like an empirical test was Jean des Murs, whose *Expositio intentionis regis Alfonsii circa tabulas ejus* of 1321 is a key document for the beginnings of Alfonsine astronomy in a Parisian context.[118] The text is noteworthy for containing Jean's personal recollection of having measured the Sun's meridian altitude on 13 March 1319 with the help of a large quadrant-type instrument 15 feet in radius, which he had set up in the town of Évreux. Comparing his result with an earlier observation made by William of Saint-Cloud in 1290, he found a gap between vernal equinoxes that came reasonably close to what one could expect based on an annual difference of 0;10,44h between Alfonso's tropical year (365d 5;49,16h) and the Julian calendar (365d 6h).[119] A comparison of this type, however, offered positive confirmation only if one focused solely on the time interval without also taking into account the actual solar longitudes predicted by the new tables. In Jean's estimation, these predicted longitudes were 0;20° ahead of the observations made in 1290 and 1319. Unwilling to impute this error to the illustrious King Alfonso, he concluded that the relevant epoch value for the Sun's mean motion recorded in Alfonso's tables had been disfigured by some scribal mistake.[120]

Jean's faithful attitude towards the Alfonsine Tables is one of several factors that speak against his authorship of a polemical *Expositio tabularum Alfonsii* written at the University of Paris in 1347.[121] The scholar responsible for this text, whom the manuscripts variously identify as Jean des Murs (unlikely), Henri Bate (impossible), and Geoffrey of Meaux (somewhat plausible), attacked the new tables on a number of different grounds, including their failure to agree with ancient and contemporary observations. Included among the latter was William of Saint-Cloud's vernal equinox of 1290 as well as an anonymous observation made in 1346 with the aid of a 'large instrument' (*cum instrumento magno*), which showed that the equinox in that year had happened at an hour far removed from what the tables would have predicted.[122] Further misgivings of this kind were recorded by the aforementioned Heinrich Selder, who in 1365 claimed to know, 'because ancient and modern observations have taught me so, that virtually all mean motions displayed in the Alfonsine [Tables] are diminished. And certain tables for equations, like those for Sun, Jupiter, and

[118] The *Expositio* was edited by Poulle 1980a. See the discussions in Casanovas 1987: 79–85; Goldstein 1994: 191–3; Chabás and Goldstein 2003b: 247, 251–2, 255–67, 277–81; Poulle 2005: 16–22; Husson 2011.

[119] Poulle 1980a: 265–7; 2011: 80–2. The essential data points in this discussion are as follows: William of Saint-Cloud's equinox in 1290 fell on 12 March, 16:00h from noon. On the Julian year length, the equinox in $1290+29=1319$ should have been on 12 March, 22:00h, whereas Jean des Murs believed to have found it on 12 March, 16:40h (an error for the 07:20h actually implied by Jean's own observation). The discrepancy of 5;20h is equivalent to $29\times 0;11,2$h and thus would support a year length of *c.*365d 5;49h, as upheld by William of Saint-Cloud (see n. 20 in this chapter).

[120] Poulle 1980a: 264–5.

[121] See the analysis of this text in Nothaft 2015b.

[122] MS Berlin, Staatsbibliothek, lat. fol. 192, fol. 213ra; Nothaft 2015b: 85–6.

Venus, are in excess, whereas the equations for the lunar arguments are deficient.'[123] He returned to the topic in his *Tractatus de tempore dominice annunciationis, nativitatis et passionis* (completed after 1378), in which he commented at length on the weak technical foundations of contemporary astrology. In one instance, Selder noted that the much-used tables of King Alfonso often failed to track the true position of Mars, deviating from it by up to 2°, while their solar longitudes were found to be off by 0;25°, the rough equivalent of a quarter-day.[124] In principle, the latter type of defect could simply be the result of faulty epoch values, yet in Selder's estimation 'more truthful observations' (*veriores observationes*) had shown the vernal equinox to regress at a rate of $J = 112$ years, which was a good deal swifter than the Alfonsine $J = 134$ years.[125] Empirical criticisms like the ones just cited were unlikely to appear in any of the more mundane works on calendrical reckoning, which included the *Computus novus chirometralis* written between 1375 and 1390 by the Strasbourg prebendary and chronicler Jakob Twinger von Königshofen for the benefit of his students. In spite of its textbook character, the work came with a lengthy and outspoken discussion of the defects of the ecclesiastical calendar, which focused mostly on the drift of the equinoxes and solstices. According to Twinger, none of the traditional estimates for the rate of this drift—the computistical $J = 120$y, the Ptolemaic $J = 300$y, Sacrobosco's $J = 288$y, or al-Battānī's $J = 100$y—achieved quite the congruence with experienced reality as the one derivable from:

> the tables of the illustrious King Alfonso, which the moderns now claim to be more reliable, as is shown to be true by the observational test. For I say that on the occasion of many eclipses, both of the Sun and the Moon, it has been experienced that these tables contain the precise truth where there was an error of about an hour in the Toledan Tables and other similar ones.[126]

[123] Heinrich Selder, *Canones tabularum Alphonsinarum* (2.8), MS Freiburg, Universitätsbibliothek, 28, fol. 92v: 'Scio enim quia antique observaciones et moderne me hoc docuerunt quod quasi omnes medii motus qui ponuntur in Alfoncio sunt diminuti. Et quedam tabule equacionum, sicud Solis, Iovis, Veneris, superfluunt. Equaciones vero argumentorum lune sunt deficientes.' Selder goes on to assert that the tables still yield accurate syzygy times due to the way the errors in solar and lunar parameters cancel each other out. This may help explain why he still relied on the Alfonsine Tables for his enhanced calendar for 1361–1436. See n. 110.

[124] MS Munich, BSB, Clm 18298, fol. 11ra–va. I argue for Selder's authorship of this treatise in Nothaft 2016b: 269–79.

[125] MS Munich, BSB, Clm 18298, fol. 31rb. The value $J = 112$ years comes close to one of the values for the tropical year attributed to Alard of Diest (see n. 74), which is $365\text{d}\ 5;47,15\text{h} \approx 365\frac{1}{4} - \frac{1}{113}$ days. Observations to correct astronomical tables were also carried out in fourteenth-century Italy, as seen from the remarks on Paolo dell'Abbaco (d. 1367) in Filippo Villani, *De origine civitatis Florentie et de eiusdem famosis civibus* 45.4–7 (ed. Tanturli 1997: 147–8, 406–8, 461). See Thorndike 1923–58: iii. 198–9.

[126] Jakob Twinger von Königshofen, *Computus novus chirometralis*, MS Solothurn, Zentralbibliothek, S I 167, fol. 24v: 'Nobis autem videtur quod nullus istorum predictorum concordat cum eo quod nunc verius reputatur, hoc est quod secundum tabulas illustris regis Alfonsi, quos [*sic*] moderni nunc reputant certiores, ut veraciter probat experta consideratio. Expertum est enim, inquam, pluribus tam solis quam lune eclipsibus ipsas tabulas habere precisam veritatem, ubi in tabulis Tolotanis et aliis consimilibus fere unius hore fuit error.' Twinger's text appears twice in the same manuscript. The first copy (fols. 12r–29v) was made in 1390 by his student Kunrat Justinger, while the other, expanded,

In reality, even the Alfonsine predictions could still miss the precise truth by more than a quarter-hour, as had already been noticed by the assiduous Jean des Murs on the occasion of two solar eclipses in 1333 and 1337.[127] That Jakob Twinger's expertise with regard to the Alfonsine Tables was somewhat limited is further suggested by his quoting their tropical year as $365\frac{1}{4}-\frac{1}{144}$ days instead of $365\frac{1}{4}-\frac{1}{134}$ days. More significant than this gaffe, however, is his indication that the precise rate of equinoctial retrogression thus caused was 'sometimes a little more, sometimes a little less' than the stated value of $J=144$ years, 'because of the inequality of the course of the Sun, which inequality happens because of the variation of the solar apogee'.[128] That the Alfonsine tropical year of *c.*365d 5;49,16h was merely a mean value, around which the true length could be presumed to vary, was a point already conceded by Jean des Murs in 1339, the year in which he penned new explanatory canons for King Alfonso's tables while at the Collège de Sorbonne. Against those who simply equated the annual motion implied by the tables for the mean Sun with the actual solar year, Jean insisted that 'it is necessary to treat individual years as unequal to each other, because of the motion of the eighth circle and the variable motion of the solar apogee and the [resulting] change in the [solar] equation'.[129]

A non-linearity in the change of the equation for the true Sun was indeed one of the predictions that followed from the Alfonsine precession model, the details of which posed heavy interpretative challenges even to its most committed followers. As already mentioned (Ch. 5.3), the tropical mean motion in the Alfonsine Tables was tied to a quasi-sidereal Julian year of 365;15d via a daily motion of the fixed stars and apogees of 0;0,0,4,20,41,17,12°. This suggested a rate of precession close to 1°/136y and hence nearly twice as slow as it should have been at the end of the thirteenth century (*c.*1°/71.8y). In order to produce more adequate predictions for stellar and apsidal longitudes, the Latin Alfonsine Tables supplemented this linear term with additional tables for the access and recess of the eighth sphere. These described the same variable back-and-forth motion familiar from *De motu octave spere*, but exchanged pseudo-Thābit's original parameters for a maximum 'equation' of ±9° (instead of ±10;45°) and a total access-and-recess cycle of 7,000 years (instead of *c.*4,057 years), which was exactly one-seventh of the 49,000

version (fols. 97r–110v) was added by Werner Mardersperger in Rottweil in 1394. See the detailed study of the manuscript and its scribes in Vetter 1906: 112–55.

127 See the autograph notes in MS El Escorial, Real Biblioteca de San Lorenzo, O.II.10, fols. 92v–93v. John's eclipse observations and calculations are discussed in Beaujouan 1975: 28–30; Steele 2000: 137–9; Kremer 2003a: 271–2; Poulle 2011: 85–6; Husson 2016.

128 Jakob Twinger von Königshofen, *Computus novus chirometralis*, MS Solothurn, Zentralbibliothek, S I 167, fol. 24v: 'Secundum igitur tabulas predictas Alfonsi deficit anno in quarta diei predicta 144 pars diei, aliquando parum minus, aliquando parum plus, quod contingit propter inequalitatem cursus solis que inequalitas sit propter variationem augis solis.'

129 Jean des Murs, *Canones tabularum Alfoncii*, MS Oxford, Hertford College, 4, fol. 144r: 'Non tamen ita est, sed neccessarium est propter motum octavi circuli et varium motum augis solis mutationemque equacionis annos ponere singulos inequales.' The same passage appears in MS Oxford, BodLib, Digby 97, fol. 122r. The influence of apsidal motion on the true length of the tropical year is also acknowledged in Heinrich Selder, *Canones tabularum Alphonsinarum* (3.31), MS Freiburg, Universitätsbibliothek, 28, fols. 134v–135r.

years that defined the linear term.[130] As Jakob Twinger and Jean des Murs both rightly perceived, the periodic or trepidational term represented by the 7,000-year cycle was bound to vary the motion of the solar apogee, leading to non-linear changes in the value of the solar equation for a given point in the year and hence to a lack of uniformity when it came to the Sun's period of return to the same true tropical longitude.

It is difficult to know what prompted this curious combination of precession theories, other than the fact that the traditional access-and-recess model, which had been disseminated for nearly two centuries as part of the Toledan Tables, was no longer empirically adequate. By the fourteenth century, astronomers had seen enough evidence to believe that the total distance in tropical longitude the stars had travelled since antiquity was greater than the maximum of 21;30° allowed for by pseudo-Thābit's theory.[131] It was also true that the old Toledan model had started to decline drastically in its predictive capacity owing to the way it decelerated the rate of precession towards an assumed standstill and reversal in the seventeenth century.[132] The effects of this deceleration are clearly evident from the treatise *Autores kalendarii*, whose author found a tropical year that was only two minutes shorter than the average year in the Julian calendar (see Ch. 5.2). By 1317, the year in which *Autores kalendarii* was written, the flaws of the existing model had inspired several Latin astronomers to reject the idea of trepidation entirely and instead embrace a strictly linear model of precession, as it was known from the works of Ptolemy and al-Battānī, yet such a move could always be countered by pointing to the long-term changes in precession that were suggested by the divergent observations these ancient authorities reported. The combination of a linear precession model with a trepidational component could hence appear like a reasonable way to square the circle.[133]

Although the Alfonsine Tables themselves made no clear suggestion as to how their twofold motion of the fixed stars and apogees should be imagined to work in physical terms, late medieval astronomers generally agreed that the combination of linear precession and periodic access and recess required the addition of another sphere between the eighth sphere normally thought to hold the fixed stars and the outermost *primum mobile*.[134] A detailed description of the Alfonsine model along these lines was delivered in 1450 by Nicolaus de Comitibus in a treatise *De triplici motu octave spere* and again in 1454 by Georg Peurbach in a series of lectures given in Vienna, which were printed in 1472 as the immensely influential *Theoricae*

[130] The Alfonsine precession-cum-trepidation model is discussed to varying extents in Price 1955: 104–7; Dobrzycki 1965/2010: 34–40; Swerdlow 1973: 448; North 1976: iii. 197–9; 1996: 459–64; Mercier 1976–7, pt. 2: 58–60; Samsó 1984: 87–109; 1987b; Samsó and Castelló 1988; Casanovas 1987; Gerl 1989: 273–80; Hugonnard-Roche 1992: 64–8; Chabás and Goldstein 2003b: 256–66; 2012a: 48–52; Nothaft 2015b: 80–1.

[131] See e.g. Jean de Lignères, *Theorica planetarum*, MS Paris, BnF, lat. 7281, fol. 166v, who noted an increase of 21;40° between Timocharis and the current year 1335.

[132] Gingerich and Welther 1977: 151–2, 160.

[133] See Nothaft 2017c for further examples and discussion.

[134] See the general overview in Grant 1994: 315–23.

novae planetarum.[135] Attempts to make full geometrical sense of this combination of spheres were not made easier by the way the periodic component continued to follow the model ascribed to Thābit, which had the heads of Aries and Libra of the eighth sphere revolve around the heads of Aries and Libra of the ninth sphere. In the Alfonsine model, these points on the ninth sphere underwent their own precessional motion, implying that the centres of pseudo-Thābit's small circles kept travelling along the zodiacal circle. Their resulting gradual displacement from the celestial equator carried implications for the obliquity of the ecliptic and for the true position of the equinoxes relative to the fixed stars and apogees that the existing tables did not even begin to capture.[136] The theoretical gap inherent in the Alfonsine Tables triggered a drawn-out technical discussion among early sixteenth-century astronomers, involving names such as Jacob Essler, Albert Pighius, Marco da Benevento, Johann Werner, and, finally, Nicholas Copernicus, who ended up translating the idea of a combined linear precession and trepidation into a heliocentric framework.[137]

Even those who ignored such geometrical puzzles, however, could find the coherence of the Alfonsine model undermined by its use of a tropical frame of reference, which to most readers would have implied a fixed length of the corresponding solar year (discounting the minor fluctuations of the solar equation just mentioned). This was an odd assumption considering that the Sun occupied an orb within the sphere of the fixed stars, whose motion of access and recess it therefore should have shared. Protests against the immobilization of the tropical year were already voiced in the *Expositio tabularum Alfonsii* of 1347, the author of which insisted that historical estimates showed the length of the year to have been shrinking since antiquity. A second problem the Parisian astronomer felt the need to address was the way the Alfonsine Tables altered the year and date at which the equation of the eighth sphere reached zero. The old Toledan Tables had located one such moment in AD 604, whereas the Alfonsine Tables already arrived there on 16 May AD 16, almost 1,236 years before the beginning of Alfonso's regnal era (1 June 1252). For this entire period, the total precession predicted by the Latin Alfonsine Tables was 17;07,56°, implying an excellent mean rate of 0;0,49,54° per year or *c.*1°/72y. It also necessarily implied, however, that the precession of the fixed stars had reached its maximum pace close to the beginning of the Christian era, which failed to agree with the apparent acceleration that had taken place between

[135] For the work of Nicolaus de Comitibus, see n. 81 in Ch. 3.3. For the *Theoricae novae planetarum Georgii Purbachii astronomi celebratissimi*, see the facsimile edition in Schmeidler 1972: 787–92, the translations in Aiton 1987: 36–40; Gerl 1989: 276–9, and the commentary by Albert of Brudzewo, ed. Birkenmajer 1900: 8–9.

[136] This specific problem could be avoided by assigning the access and recess to the ninth sphere and the linear precession to the eighth sphere, as was done by Johannes Tolhopf in his treatise on the *Stellarium* for King Matthias Corvinus of Hungary (1480). See MS Wolfenbüttel, Herzog-August-Bibliothek, Cod. Guelf. 84.1 Aug 2°, fols. 2v–3v, 19r. See also the anonymous *Tractatus introductorius in totam astronomiam speculativam* of *c.*1414, in MS Vienna, ÖNB, 5247, fol. 6r.

[137] The discussion engendered by the physical interpretation of the Alfonsine precession up to Copernicus is traced in Dobrzycki 1965/2010: 40–8; 1971; 1983; Lerner, Segonds, and Verdet 2015: i. 110–19. On Copernicus's role in the calendar reform debate, see Ch. 8.

Ptolemy's observations in the second century and al-Battānī observations in the ninth century, with their suggested increase from 1°/100y to 1°/66y.[138] This failure of the Alfonsine model correctly to reproduce ancient findings remained a source of discontent throughout the period of its use. It was still commented upon by the Bavarian humanist Johannes Tolhopf, who submitted opulently produced astronomical treatises to Pope Sixtus IV (1476) and the Hungarian king Matthias Corvinus (1480).[139] In the later of the two works, which accompanied an equatorium-type instrument dubbed the *Stellarium*, Tolhopf argued that the actual change of stellar longitude between Ptolemy and the present year 1480 was *c.*2° in excess of the precession one could compute from the Alfonsine Tables. He concluded that a different theory was needed in order to save past as well as present and future phenomena.[140]

Those who investigated the motion of the fixed stars during the fourteenth and fifteenth centuries repeatedly found evidence for a rate of precession significantly faster than what was predicted by the Alfonsine model. A linear precession of 0;0,54° per year, or 1° in $66\frac{2}{3}$ years, was expressly used for a star table for the year 1340 constructed by Heinrich Selder, who promised to spell out his reasons for rejecting the access-and-recess theory in a separate treatise 'On the motion and form of the eighth sphere' (*De motu et dispositione octave spere*) that, regrettably, has not come to light.[141] To more than one contemporary, Selder's adoption of a uniform parameter for precession would have appeared problematic for the way it ignored Ptolemy's discrepant findings, which trepidation models had traditionally sought to reconcile with those of al-Battānī. From remarks made by Johannes Regiomontanus in 1464 and by Albert (Wojciech) of Brudzewo in 1482, it transpires that even in the second half of the fifteenth century some still preferred to conceive of the motion of the eighth sphere as a periodic access and recess in the tradition of pseudo-Thābit, without accepting the convoluted double motion proposed by the Alfonsine Tables.[142] Johannes Regiomontanus, for his part, rejected both these models as flawed and even announced his plans to write a treatise 'On the eighth sphere against Thābit and his followers' (*De motu octavae spherae contra Tebith suosque sectatores*).[143] This did not mean, however, that he was ready simply

[138] MS Berlin, Staatsbibliothek, lat. fol. 192, fol. 212va; Nothaft 2015b: 84–5.

[139] The presentation copies of these works survive as MS Vatican City, BAV, lat. 3103 (*De motibus celestium mobilium* for Pope Sixtus IV, written 1476) and MS Wolfenbüttel, Herzog-August-Bibliothek, Cod. Guelf. 84.1 Aug 2° (*Stellarium* for Matthias Corvinus, written 1480; see n. 136).

[140] MS Wolfenbüttel, Herzog-August-Bibliothek, Cod. Guelf. 84.1 Aug 2°, fol. 19v: 'Quare motus harum stellarum alius erit assignandus ut cuncti hi sapientissimi viri salventur.' A similar point had already been made by Johannes Regiomontanus in 1471. See *Scripta clarissimi mathematici M. Ioannis Regiomontani*, ed. Johannes Schöner (Nuremberg: Montanus & Neuber, 1544), fols. 42v–43r (repr. Schmeidler 1972: 658–9).

[141] Heinrich Selder, *Canones tabularum Alphonsinarum* (3.15–16, 35), MS Freiburg, Universitätsbibliothek, 28, fols. 111r–113v, 136v. See Nothaft 2017c: 233–5.

[142] See Albert of Brudzewo, *Commentariolum super Theoricas Novas Planetarum Georgii Purbachii* (ed. Birkenmajer 1900: 148–9), and the letter Regiomontanus wrote to Giovanni Bianchini in 1464 (ed. Curtze 1902: 263).

[143] See the printed prospectus of *c.*1472 reproduced in Schmeidler 1972: 533, and edited in Malpangotto 2008: 147–54.

to ignore or discard all ancient observations that indicated the motion of the fixed stars might be variable. For all anyone knew in Regiomontanus's day, the striking discrepancies between these observations may have been the result of instrumental error, but it was equally possible that they reflected a component in the motion of the fixed stars astronomers had yet to understand. Either way, Regiomontanus and his contemporaries faced a problem of considerable difficulty—so difficult indeed that the key to it might only be found by future generations:

> For if our predecessors were deceived by instruments, we shall also necessarily be deceived, since our observations will never teach us anything, unless they are compared with the observations of the ancients. But if we shall think that that hidden motion is inherent in the stars, it will be expedient to keep a constant eye on the fixed stars and, in the same way, that those who come later weigh the observations thus recorded.[144]

Regiomontanus's words are a fitting reminder that at the end of the medieval period the nature of precession—and therefore also the length of the tropical year—was still an object of open astronomical speculation, with no definitive answer in sight. The solution the Alfonsine Tables could offer to this problem was recognized as being of temporary value at best.

[144] For this quote, which was inspired by al-Battānī, *De motu stellarum*, trans. Plato of Tivoli (Nuremberg: Petreius, 1537), fol. 81r (ch. 52), see bk. VII.6 of the *Epitome Almagesti*, MS Venice, Biblioteca Nazionale Marciana, Fondo antico lat. Z. 328, fol. 65r: 'Nam si maiores nostri suis decepti sunt instrumentis et nos decipiemur necessario, cum nostre considerationes, nisi antiquorum conferantur observacionibus, nihil umquam edocebunt. At si occultum illum motum inesse stellis estimabimus, expediet oculum ad stellas fixas habere assiduum posterosque itidem scriptas consideraciones librare.' Other manuscripts of the *Epitome* as well as the 1496 printed edition (repr. in Schmeidler 1972: 177) give the final verb as *liberare* (instead of *librare*), which alters the sense and leads to the alternative translation I gave in Nothaft 2017c: 236. Further information on Regiomontanus will be provided in Ch. 7.3.

6

The Papal Reform Project of 1344/5 and Its Protagonists

6.1. JEAN DES MURS AND CLEMENT VI'S INITIATIVE

After Roger Bacon's failed appeal to Clement IV and the long period of papal inactivity that followed, a fresh attempt to get the head of the Church interested in the correction of the calendar was made in 1318, the year when a Dominican friar named Pierre Vidal (*Petrus Vitalis*) presented his astronomically enhanced *kalendarium* for the coordinates of Montpellier to Pope John XXII (1316–34).[1] Vidal's dedicatory prologue made it clear from the outset that he intended to encourage and facilitate an 'emendation of the calendar of the most holy Church, which, having neither spot nor wrinkle, must not contain anything false or indecent'.[2] He warned that this immaculate condition was undermined by the way the date of Easter nowadays often failed to satisfy the biblical stipulations for Passover, which had to be celebrated under the first full moon in spring in commemoration of the Exodus as well as of Christ's salvific death and resurrection. In the present year AD 1318, the Sunday after this astronomical event would have been 19 March, yet the official date of Easter was 23 April, leaving a discrepancy of 35 days that exposed Christ's holy Church to 'the mockery of those who are outside of it, namely the faithless Jews, whom God has rightly rejected'.[3]

Vidal's solution to this quandary contained relatively few new elements. As with other enhanced calendars produced at the time (Ch. 5.1), the primary function of his *kalendarium* for John XXII was to demonstrate the correct date of the new moon for a period of 76 years by listing the hour and minute of mean conjunction in a layout of parallel columns, which here covered the years 1311–86. In this one

[1] The work only survives in MS Paris, BnF, lat. 7420A, fols. 101r–108v. On the author see Kaeppeli 1980: 268.

[2] MS Paris, BnF, lat. 7420A, fol. 101ra: '[N]omini eius ascribo hoc presens novum kalendarium super Montem Pessulanum ordinatum, per quod duo magna luminaria, videlicet sol et luna, in suis motibus quantum ad aliquid in presenti tempore corriguntur ad emendationem kalendarii ecclesie sacrosancte, qui, non habens maculam neque rugam, nichil debet falsum vel indecens continere.' Parts of the prologue are transcribed in Meyer 1897: 237, and (more fully) in Thomas 1921: 625–6.

[3] MS Paris, BnF, lat. 7420A, fol. 101ra: 'In hoc autem festo et aliis festis mobilibus quandoque Christi ecclesia ab institutione Domini supradicta multum deviat et discordat, in subsannationem sui ab hiis qui foris sunt, Iudeis scilicet perfidis a Deo merito reprobatis. Sic enim patuit isto anno Domini M°CCC°XVIII°, quia Pasca nostrum, quod secundum predicta esse debuit in 19ª die Marcii, celebratum fuit 23ª die Aprilis, distans a primo termino per 36 dies, quod ecciam potest distare amplius in futurum.'

respect the calendar strongly resembled the *Correctio kalendarii* devised by another mendicant friar based in southern France, the Franciscan Raymond Bancal, whose work appears in the same manuscript as Vidal's contribution.[4] While both friars evidently used the Toledan Tables to perform their calculations, Vidal decided to shift the beginning of the day from noon to the time of equinoctial sunrise (+6h) and the meridian of reference from Toledo to Montpellier. From the fact that his syzygy times tend to exceed those in Bancal's calendar by 7;7h, one may infer that he assumed Montpellier to lie 16;45° further to the East (+1;7h), which more than doubles the correct longitudinal difference. In addition to the timed conjunctions, Vidal's tables for each Julian month sport two columns that juxtapose the conventional Golden Number (*ciclus 19^{lis} ecclesie*) with a reformed equivalent. In nearly all cases, this *aureus numerus novus* precedes the old one by four days, agreeing with the author's verdict that 'today the Moon is already on its fifth [day] when the Church announces it to be on its first and, unless this is corrected, it will eventually be announced to be on its first [day] when it is [already] fully illuminated'.[5]

Attached to the main *kalendarium* were various auxiliary tables that contained the data necessary to convert equinoctial into temporal hours, find the noon altitude of the Sun, and locate the Moon's position in the zodiac.[6] A table of mobile feast days and their intervals compared the traditional gamut of dates for Septuagesima, Quadragesima, Easter, and Pentecost with a new set that placed the earliest permissible dates of each feast one week earlier. Under this new regime, Easter Sundays would have been allowed to range from 15 March to 18 April, while Septuagesima Sunday could fall as early as 11 January (as opposed to 18 January).[7] This curtailment of the period between Epiphany and the beginning of Shrovetide, which could in the future fall within the same week, was not without unwelcome consequences for the ritual life of the Church and even the laity at large. Vidal specifically mentioned the chant *Alleluia*, which had to be omitted from the Roman liturgy during Shrovetide and Lent, as well as the *tempus nuptiarum*, in allusion to the way the Church had tried to limit the periods of the year during which weddings were supposed to take place. In some regions, the prohibition of weddings during the Christmas and Easter seasons was strict enough to leave only a relatively short time window between Epiphany (6 January) or the octave of Epiphany (13 January) and Septuagesima Sunday, which was the start of Shrovetide.[8] With the anticipated set of *termini*, this window could occasionally close entirely, since the earliest date of Septuagesima Sunday, on 11 January, preceded the octave of Epiphany. Fortunately, rules pertaining to weddings and the specific elements of the liturgy were not divinely ordained like the celebration of Easter, but had been

4 Ibid., fols. 57v–63r, 71r. See Nothaft 2015e.

5 MS Paris, BnF, lat. 7420A, fol. 101ra: 'Nam hodie luna est iam 5ª quando ab ecclesia pronunciatur esse prima et, nisi in posterum corrigatur, pronunciabitur esse prima quando erit lumine tota plena.'

6 Ibid., fols. 101v, 108v.

7 Ibid., fol. 108r. The shift by seven days implies a preponement of the vernal equinox to 14 March. This is in rough agreement with the Toledan Tables, which in the years around ad 1318 would have placed the true equinox on 13/14 March.

8 Freisen 1893: 643–51; Korpiola 2008: 636–43.

instituted by the Church itself, raising strong hopes that John XXII, as the 'supreme pontiff' and 'vicar of Christ', would prescribe a remedy for any of the collateral problems a reform of Easter reckoning was apt to create.[9]

No documentation has come to light that would tell us whether John XXII, who had confirmed his predecessor Clement V's decision to move the papal curia to Avignon, ever considered such a remedy or viewed Pierre Vidal's intervention with any kind of appreciation. This continuing silence on the part of the papacy was not broken until Pierre Roger, a Benedictine canon lawyer and master of theology, acceded to the chair of St Peter under the name Clement VI (1342–52). In the autumn of 1344, the papal chancery in Avignon dispatched letters to two French members of the clergy renowned for their astronomical expertise. One of them was the highly accomplished mathematician and music theorist Jean (Jehan) des Murs, who had been practising astronomy in the newly established Alfonsine tradition since the early 1320s, as is evident from a large number of texts and tables from his pen.[10] Following a long period of loose affiliation with the Collège de Sorbonne in Paris, he had recently settled down in Mézières-en-Brenne in the diocese of Bourges as member of a collegiate church founded by Alix of Brabant. Manuscript notes from Jean's hand show that he used some of his spare time in Mézières for continued astronomical work, which included measurements of the Sun's noon altitude around the equinoxes and solstices. The latest of these recorded observations was made on 17 June 1344, only months before the papal request to appear in Avignon would have reached his doorstep.[11] The other summons was addressed to Firmin de Beauval (*Firminus de Bellavalle*), who is otherwise best known for his work on weather prognostication, *De mutatione aeris* (1338). At the time of his invitation to Avignon, Firmin resided as a canon in Amiens while simultaneously enjoying a benefice in the diocese of Cambrai that Clement VI had granted him in October 1343.[12] The two papal letters of 1344, both of them signed on 25 September, began by mentioning 'certain doubts or defects concerning the Golden Number', signalling that the lunar calendar was in need of correction 'for the benefit of the universal Church'. Having been notified of Jean's and Firmin's expertise in this area, the pope ordered them to come to the Apostolic See quickly and without delay, promising that he would ask the bishops in their respective home dioceses to

[9] MS Paris, BnF, lat. 7420A, fol. 101rb: 'In novitate autem istius kalendarii sequi videretur inconvenientia pro tempore isto, scilicet quod quia Septuagesima quandoque erit infra octavas Epifanie, quod tamen raro cotinget, tunc sesabit alleluia et tempus nupciarum tolletur, in quibus tamen summus pontifex Christi vicarius potest remedium adhibere, cum hoc sit ab i<n>stitutione ecclesie, non autem pasca, quod est ab ordinatione.'

[10] For the relevant bio-bibliographical information, see Weijers 1994–2012: v. 124–31; Lejbowicz 2004, as well as Duhem 1913–59: iv. 30–8; Thorndike 1923–58: iii. 294–324; Gushee 1969; 1998; Poulle, 2005: 5–35; Nothaft 2015d; 2015g: 284–8.

[11] The observations are recorded in MS El Escorial, Real Biblioteca de San Lorenzo, O.II.10, fols. 68v–69r. See Gushee 1969: 18–20; Beaujouan 1975: 28; Gautier Dalché 2009: 126–7; Poulle 2011: 83.

[12] See the papal letter edited in Van Isacker and Berlière 1924: 371 (no. 1061), and the information on Firmin provided by Duhem 1913–59: iv. 41–2; Thorndike 1923–58: iii. 268–80, 304–9; Gack-Scheiding 1995: 43–6; Weijers 1994–2012: ii. 89–90.

cover all expenses.[13] Clement made good on this promise on 1 October with two letters addressed to the archbishop of Bourges (Roger le Fort) and the bishop of Paris (Foulques de Chanac), in which he once again announced his intention of presiding over a reform of the Golden Number. He claimed that he had been encouraged to do so 'by the insistences of many people' and that the necessary correction was going to be carried out by 'skilled and experienced men', Jean and Firmin, whom the bishops and their clergy were ordered to provide with all the means necessary for their journeys from northern and central France to the pope's headquarters in Provence.[14]

In spite of their brevity, the four letters mark an epochal shift in the history of calendar reform. For the first time in the annals of the Latin Church, the calendar's correction was no longer an issue that occupied only scholars, but one that had officially entered the arena of ecclesiastical politics. With Clement VI, the one person in Western Christendom with the authority and leverage necessary to change the calendrical order had decided to take action. That this shift took place in or around 1344 must to a large degree be credited to the personality of Pierre Roger, whose reign as pope was characterized by a legendary degree of opulence as well as an extensive patronage of the arts and letters.[15] The circle of scholars and artists who enjoyed Clement's support during his decade-long run as pope was nothing if not diverse, ranging from Petrarch and the composer Philippe de Vitry to the Jewish scholar Levi ben Gerson (or Gersonides, 1288–1344), who was perhaps the greatest European astronomer of the fourteenth century. By the year 1336, Levi had come into contact with a circle of 'noble Christians' whom he named as the recipients of a Provençal version of his tables for the times of mean and true syzygy. The accompanying canons mention among the work's potential uses 'the movable feasts of the Christians, because the computation of the Moon is a sign of these feasts, and it is necessary for them for [determining] all their feasts insofar as all of them depend on the lunar month'.[16] More than by these tables, Levi's connections to the curia in Avignon are revealed by two Latin texts he prepared with the help of his long-time Christian collaborator and translator, an Augustinian friar named Peter of Alexandria. Next to the treatise *De sinibus, chordis et arcubus, item instrumento revelatore secretorum* (1342), which contains a dedication to Clement VI and is noteworthy for describing the Jacob's Staff (*baculus Jacob*), an instrument Levi had invented for measuring celestial altitudes and angles of separation, the pope appears to have been the intended recipient of an astrological prognostication, in which the Jewish astronomer examined the effects of a rare conjunction of the three superior planets (Saturn, Jupiter, and Mars) on 28 March 1345.[17]

[13] Déprez 1925, col. 209 (no. 1134). See also Déprez 1899, for an earlier edition and discussion of these letters.

[14] Déprez 1925, col. 211 (no. 1339): 212 (1340).

[15] See most recently Anheim 2014; Nothaft 2015g: 279–84.

[16] Mancha 1998: 291. Levi's manifold contributions to astronomy are summarized in Goldstein 1992.

[17] On the Latin translations of Levi ben Gerson's works, see Mancha 1992b; 1997; 2012. The prognostication, extant in both Hebrew and Latin, is edited and discussed in Goldstein and Pingree 1990.

Although Levi ben Gerson would have been more than qualified to assess the astronomical problems besetting the ecclesiastical calendar, it is on the whole unsurprising that the pope and his advisers relied on two Christian scholars to devise a remedy. The case for an invitation was especially strong with Jean des Murs, whose expertise in the area of calendrical astronomy had been in evidence at least since 1321, the year in which he produced several original works, including a *Kalendarium solis et lune* derived from the Latin Alfonsine Tables. As the name would suggest, the *Kalendarium* was created around an astronomically enhanced calendar, which featured the traditional ferial letters next to the true position of the Sun in the zodiac and its meridian altitude for each day of the year 1321, while the numerous appended tables included a list of all mean conjunction times for a period of 19 years.[18] Computing conjunctions, oppositions, and eclipses remained a core aspect of several of Jean's later works, which included a sophisticated set of astronomical tables for the five planets and two luminaries as well as a different work known from its incipit as *Patefit*. The latter may be interpreted as a reworked and greatly expanded edition of the *Kalendarium solis et lune*, given the way it maintained the starting epoch in 1321 and the primary focus on the courses of Sun and Moon, with extensive sets of tables devoted to both mean and true syzygies.[19] Of the handful of surviving copies of the *Patefit*, the one most sumptuous in appearance has a colophon indentifying the work as a *Kalendarius* [*sic*] *solis et lune* produced for the abbey of Le Bec-Hellouin (diocese of Rouen), where Jean had been the holder of an expectative benefice since 1329. It also specifically mentions 'the emendation of the old calendar, which the Romans put together in a less than good manner' as one of the work's chief purposes, even though this aspect is broached only briefly in the accompanying canons, which are divided into separate parts devoted to solar and lunar astronomy.[20] Of particular interest in this regard is the chapter on the length of the solar year, in which Jean compares different historical estimates, coming down in favour of the opinion of the 'astrologers of Alfonso, King of Castile', who subtracted 0;10,44h from the Julian length of $365\frac{1}{4}$ days and, by doing so, implied $J = 134$ years:

> And this value appears to be closest to the truth in our times. And I used it to put this *kalendarium* in order and it is the time in which the Sun returns to the same equinox. I am of the opinion, however, that it is impossible to have precisely one true length of the year such that an error will not eventually arise with the succession of time, even though it may remain unnoticed for a long period. Yet the founders of the ecclesiastical

[18] Chabás and Goldstein 2012b.

[19] See North 1977: 284–5; Kremer 2008: 148–55; Chabás and Goldstein 2009a: 313–17. An unpublished partial edition of the *Patefit* was prepared by Plassard 1975: 177, 181.

[20] See MS London, BL, Royal 12.C.XVII, fols. 145v–210r, which was copied in 1332 or not long thereafter. The beginning of the colophon on fol. 210rb reads: 'Explicit kalendarius solis et lune cum suis utilitatibus prenotatis, quem ego sub ingenii paucitate composui primo propter utilitatem communem, que est eligenda pocius singulari, secundo propter emendationem kalendarii antiqui, qui fuit minus bene compositus a Romanis, tertio propter reverentiam et honorem preclari domini viri religiosi reverendi patris in Christo, Gaufridi nutu divino abbatis monasterii beate Marie de Becco Helluyni, ordinis sancti Benedicti, cuius sum compatriota natione et minimus vicinorum vivens sub eo.' See the fuller citations provided in Gushee 1969: 21–2, and Poulle 2005: 27.

calendar followed Hipparchus in accepting a full quarter-day for each year, which is not true. And for this reason there is already a noticeable error in the immobile feast days of 10 days and more. And this defect might increase so much that the nativity of the Lord would be celebrated in summer and the nativity of St John in winter. Similarly, there is a defect of three days and more with respect to the new moon. As a result of this the mobile feasts are not celebrated on their due dates, which is also something that cannot be hidden from the common folk. Yet our present *kalendarium* corrects and emends all of this.[21]

Jean decided to expand on this assertion in his *Sermo de regulis computistarum* (*c.*1332), a brief text that appears as an add-on to the *Patefit* in some of the early manuscripts,[22] but later also circulated on its own.[23] The text was heavily polemical in tone. At the start of it, Jean expressed his frustration with the rules 'by which [the computists] believe to find the mobile feasts. I wish to say nothing more about them at this point, except that the computation of the computists has for a long time been running at a great distance from the truth and still does so'.[24] Alluding to the Church's ongoing complacency with this situation, he granted that:

the limbs are not allowed to correct what the head has passed over, and perhaps for good reason. But I say this one thing without risk of danger, [namely] that the length of the solar year that the computists affirm or take for granted in constructing their calendar is not precisely true, but in our times already fails us by 10 days and more, as has become manifest from what has been said before.[25]

Even the 19-year cycle that determined the recurring sequence of Golden Numbers had already incurred an error of three days or even more, creating a readily observable

[21] MS London, BL, Royal 12.C.XVII, fols. 205va–206ra: 'Astrologi autem Alfonsi regis Castelle quartam affirmant veraciter incompletam ex 10 minutis hore et 44 secundis que quidem in 134 aggregata fere faciunt diem unum. Et hec quantitas temporibus nostris apparet propior veritati. Et secundum hunc istum kalendarium ordinavi et istud est tempus quo sol redit de equinoctio ad idem. Opinor autem quod impossibile est habere precise unam veram anni quantitatem, quin finaliter successione temporis cadat error, licet permaneat longis temporibus inperceptus. Fundatores tamen kalendarii ecclesiastici Abrachim secuti sunt, quartam in anno ponentes integram, quod non est verum. Ideoque iam in eo patet error in festis immobilibus manifestus in decem diebus et ultra. Et hic defectus tantum posset augeri quod Nativitas Domini celebraretur in estate et Beati Iohannis Nativitas in hyeme. Similiter deficit in primacione lune per tres dies et plus ratione cuius festa mobilia non situ debito celebrantur, quod etiam celari non potest a vulgo populari. Set presens noster kalendarius hec omnia corrigit et emendat.'

[22] See ibid., fol. 213r–v, where the leaf containing the start of the text is missing. See also MSS Erfurt, UFB, CA 4° 360, fols. 51r–52r (*c.*1360); Erfurt, UFB, CA 4° 371, fols. 44v–45r (copied 1337), both of which join the *Sermo* to the *Patefit*. An early mention of the text appears in Schubring 1883: 520. See also Plassard 1975: 177–8, 181, whose edition remains unpublished. The date is discussed in Nothaft 2015d: 227–8.

[23] MSS Brussels, Bibliothèque Royale, 1022–47, fols. 40ra–vb, 203r–204v (two separate copies; s. $XV^{1/2}$); Milan, Biblioteca Ambrosiana, H 109 sup., fols. 119r–120r (s. XV); Paris, BnF, lat. 3123, fols. 71r–72v (s. $XV^{2/2}$).

[24] MS Erfurt, UFB, CA 4° 371, fol. 44v: 'De regulis computistarum, quia cognite sunt a multis, per quas credunt festa mobilia reperire, nihil plus volo nunc dicere, nisi quod computistarum computatio a veritate longe distans multis temporibus sic cucurrit et sic currat.'

[25] Ibid.: 'Membrorum quidem non est corrigere quod a capite pretermissum est et forte ex causa. Sed hoc unum dico tute quod anni quantitas solaris quam ponunt aut supponunt compotiste in compositione kalendarii sui non est precise vera, sed iam fallit nos temporibus nostris in 10 diebus et plus, sicut patuit in predictis.'

discrepancy of which the general public could not remain ignorant. Some modern astronomers had reacted to this embarrassing situation by discussing the shortcomings of the Golden Number in their treatises and seeking to compensate for it with their various methods of how to find the moments of mean syzygy. Such methods also underpinned the popular success of astronomically enhanced calendars, in which precisely timed conjunctions took the place of the traditional *numerus aureus*. Jean briefly mentioned that he had 'already seen five calendars that have been put together by various writers—those who have seen them know what they are worth'.[26] On the other side of the divide were the traditional tables of the computists, which raised puzzling questions none of their users seemed in a position to answer, at least according to Jean des Murs, who claimed that nobody could offer an informed opinion on the time at which the Golden Number had first been inserted into the calendar. It was similarly nebulous whether the dates to which these numbers were assigned had originally been intended to coincide with the 24-hour period during which the conjunction occurred or whether they identified as *primacio* only the day that followed. The inadequacy of the computists' tools even blinded them to the true chronological facts behind Easter, rendering them unable to establish Christ's age at the time of his crucifixion or the calendrical date on which he rose from the dead. A judicious use of syzygy tables, by contrast, made it possible to establish that these events must have occurred on 3 April (Passion) and 5 April (Resurrection) in the year AD 33, when Jesus was only 31 years and *c.*100 days of age.[27] Returning to the knowledge required to fix Easter in the here and now, Jean des Murs expressed the hope that the whole procedure could be reduced to a simple matter of finding the astronomical full moon that followed immediately after the vernal equinox, which in present times resided on 12 March:

> Yet if this is not the way it happens currently, the reason is that the Church still does not want to alter anything about the ancient doctrine of how to select the paschal *terminus*. But perhaps it will be changed in a short while. For if things continue like this, it will eventually happen that the paschal *terminus* is held on the [day of the] new moon, which must be avoided at all costs, lest there be a natural eclipse of the Sun. [For] at the time of [Christ's] Passion [the eclipse] happened through a miracle, not by nature.[28]

In order to avoid the disquieting scenario of Good Friday coinciding with a non-miraculous solar eclipse, the Church needed to avail itself of an astronomically adequate Easter calculus, which according Jean des Murs was easily obtained if the

[26] Ibid.: 'Et hoc scit etiam gens popularis. Unde quidam de modernis astronomis defectum aurei numeri in suis tractatibus ostenderunt, nec aliquem latet, quapropter vias suas dederunt de mediis coniunctionibus habendis aureo numero iam dimisso. Et ego iam vidi 5 kalendarios a diversis compositoribus ordinatos, quid et quantum valeant sciunt qui viderunt.'

[27] Ibid., fols. 44v–45r. See on the background Nothaft 2012: 178–201.

[28] MS London, BL, Royal 12.C.XVII, fol. 213rb–va: 'Si autem nunc non sic fiat, hoc est quia adhuc etiam de antiqua doctrina in sumptione termini paschalis ecclesia nihil voluit immutare. Sed forte mutabitur et in breve. Si autem sic currat contingit aliquando et in novilunio pasche terminum celebari, quod summe cavendum est ne fiat eclipsis solis per naturam que tempore passionis apparuit miraculo, non natura.'

doctrines enshrined in his own *Kalendarium* were properly heeded. A necessary first step was to shift the solstices and equinoxes back to their original dates, which could be done by desisting from the bissextile intercalation for a period of 40 years or, alternatively, by curtailing eleven months of a given year by one day each. In place of the traditional Golden Number, it was expedient to have a table of new-moon dates laid out according to four consecutive 19-year cycles, knowing that one cycle was not enough to guarantee full precision. For the *terminus paschalis*, Jean recommended that it be redefined as the 15th day of the moon after the equinox, to make sure it agreed with the lunar age on the day of Christ's death (15 Nisan). Easter had to be sought on the Sunday thereafter. What he saw no need for was to continue the use of 'as many diverse and obscure rules and speeches' as the computists were wont to employ in pursuit of their goal.[29] As with the writer responsible for the treatise *Autores kalendarii* of 1317, Jean des Murs thus pitched his reform project as a clean break with tradition, implicitly calling for a changing of the guard at the gates to the ecclesiastical calendar. The business of improving it was from now on best left to the astronomers, whose tools, approaches, and insights were in every way superior to those of the calendar-reckoners of old.

6.2. A RADICAL PROPOSAL

While Jean des Murs's authorship of the *Patefit* and *Sermo de regulis computistarum* goes a long way toward explaining the summons to Avignon he received later in life, there is evidence to show that he and Firmin de Beauval were not the only experts who went to work on the problem at the pope's behest. A third name that can be attached to the reform initiative of 1344/5 is that of the monk Johannes de Termis, whose back-story emerges from a formal supplication presented to Pope Clement's successor Innocent VI (1352–62) on 30 August 1353. According to this document, Johannes had once been a member of the Dominican order endowed with a certain degree of expertise in astronomy, which appears to have been the reason why Clement VI, at some unspecified point in time, ordered him to come to Avignon to devote himself to the 'reparation of the calendar'. The supplicant, who hoped that Innocent VI would grant him a benefice Clement had promised shortly before his death, ended up staying at the curia for two years, during which he composed a 'great and laborious work' at his personal expense. Eventually, the

[29] Ibid., fol. 213va: 'Recta quoque situatio festorum tam mobilium propter lunam quam immobilium propter solem facilis est atque brevis secundum doctrinam nostri kalendarii supradicti. Non enim oportet nisi quod solstitia et equinoctia ad statum pristinum reducantur cessando a bissexto per 40 annos, vel demendo unum diem a singulis undecim mensibus, et fiant 4 cicli 19[les] qui sunt uno precisiores, quoniam unus precise non sufficit. Et observare lunam 15 post equinoctium veris, quia ibi est terminis pasche rectus, sicut fuit dum Christus passus est, et in proxima dominica fiat pascha.... Nec video necessitatem utendi tot regulis et sermonibus variis et confusis quot fingunt in isto negocio compotiste.' Earlier on in the text, Jean placed the current vernal equinox on 12 March, which would imply a necessary correction of 9 rather than the 10 or 11 days suggested in this passage. Note that the phrase 'vel demendo unum diem a singulis undecim mensibus' does not appear in MSS Erfurt, UFB, CA 4° 360, fol. 52ra; Erfurt, UFB, CA 4° 371, fol. 45ra.

pope decided to transfer Friar Johannes to the Galician monastery of S. Salvador da Torre (diocese of Tui), turning him from a Dominican into a Benedictine, with the expectation that this move would offer him the freedom to bring his work to a conclusion.[30] Johannes de Termis's supplication does not reveal whether he ever achieved this goal, but it appears all but certain that some version of his 'great and laborious work' survives under the title *Expositio kalendarii novi et correctio veteris*. Johannes's authorship of this lengthy treatise in two books and 54 chapters, which appears anonymously in the only known manuscript,[31] is suggested by its content and confirmed beyond reasonable doubt by a note found at the end of an unrelated fourteenth-century astronomical codex. The note concerns the new moons in the '*kalendarium* of Johannes de Termis', explicitly connecting this calendar to a chapter in the *Expositio* devoted to the calculation of the so-called *clavis primacionum*.[32] It is hence clear that the *Expositio* was originally meant to be accompanied by a reformed *kalendarium*, which has not come down to us, but whose existence is suggested by the very title, *Expositio kalendarii novi*, and by various references contained in the text itself.

Johannes de Termis's written exposition of this 'new' calendar, which dates from 1345, was a complex and wide-ranging work, full of unusual and at times startlingly original ideas concerning the reform of Easter reckoning and the calendar's astronomical underpinnings. Book I concentrated on the solar component of the problem at hand, which was rooted in the fact that the average Julian year exceeded the length of the tropical year by a certain amount. Going against the Alfonsine consensus that had already begun to form in his time, Johannes gauged this gap to be exactly 12 minutes and, in doing so, reverted to the solar year of 365d 5;48h or $365\frac{1}{4}-\frac{1}{120}$ days that writers on the 'vulgar' or ecclesiastical *computus* had been championing in considerable numbers since the early thirteenth century (see Ch. 4.1). Although he freely admitted that he fell in line with the *compotiste* as far as the solar year was concerned,[33] this did not mean that he simply took their parameter on faith. Far from it—he reserved ample space for convincing his readers, and perhaps himself, that this value was borne out by observational data both past and present. Next to information drawn from Ptolemy's *Almagest*, Johannes was

[30] The supplication is printed in full in Berlière 1908: 241. The remainder of the text expresses Johannes's hope that Innocent VI would bestow upon him a benefice attached to the abbey of Saint-Germer-de-Fly (diocese of Beauvais). We also learn that he spent some of the intervening years studying theology in Paris.

[31] MS Vatican City, BAV, Ott. lat. 842. In what follows, I shall always cite this source as *Expositio*, followed by book and chapter, followed by the manuscript's folio-number in parentheses. See the preliminary study in Nothaft 2017b.

[32] MS Paris, BnF, lat. 18504, fol. 216v: 'In kalendario Io. de Termis nota circa illam partem de invencione lune primacione [*sic*] que incipit "Cum itaque diem et horam" etc. quod clavem acceptam extendere est numerum quem clavis designat in litteris ferialibus computare.... Et hoc supposito quod istud kalendarium in usu haberetur. Quia vero non adhuc habetur in usu, accipienda est semper primacio in quarta littera precedente et hoc exclusive computando.' The reference is to *Expositio* I.4 (fol. 56r). A second note on the same page of MS Paris, BnF, lat. 18504 concerns the *Tabula terminorum* and *Tabula festivitatum mobilium* appended to the *Expositio* in MS Vatican City, BAV, Ott. lat. 842, fol. 71r–v.

[33] *Expositio* I.2, 10 (fols. 1r, 12r).

able to harness the 'observed' autumn equinox of the previous year 1344, which according to his own testimony had occurred on 14 September 1344, at 09:11 p.m., as well as the calculated autumn equinox of 1289, which the Alfonsine Tables turned out to place 20,088 days and 7 hours earlier once the longitude difference between Toledo and Paris (0;48h) and the change in the solar equation wrought by the motion of the solar apogee (0;1h) were taken into account. Conveniently enough for Johannes's purposes, the interval thus established was identical to 55 mean tropical years of the precise length assumed by the computists, since $20{,}088\text{d } 7\text{h} \div 55 = 365\text{d } 5;48\text{h}$.[34] For the Sun's mean rate of tropical motion, this dictated a value of $360^\circ \div 365\text{d } 5;48\text{h} = 0;59,8,20,8,4,36^\circ/\text{d}$, which Johannes went on to accept in place of the Alfonsine 0;59,8,19,37,19,13,56°/d.[35]

In spite of his departure from the Alfonsine Tables on this specific point, other passages in the *Expositio* reveal that Johannes de Termis remained anchored to the general outlines of the solar theory the tables had imposed on late medieval astronomy. In particular, the Alfonsine theory motivated his claim that the Sun's period of return to its apogee (i.e. the anomalistic year) was exactly $365\frac{1}{4}$ days and that it was this definition of the year that Hipparchus and Ptolemy had had in mind when they proposed their seemingly exaggerated values.[36] With an assumed 12-minute excess of the Julian year relative to the tropical year, it followed that the solar apogee went full circle in relation to the equinoxes in a period that took 44,000 years (as opposed to the Alfonsine 49,000 years) to complete, since $365\text{d } 5;48\text{h} \approx 365\frac{1}{4} \times \frac{44000}{44001}$ days. Still assuming Alfonsine principles, Johannes noted that the circular motion of the apogee led to constant changes in the solar equation and, by extension, minor alterations in the duration of the solar year, if the latter was defined as the interval between two true vernal equinoxes.[37] Calendar reformers could safely ignore such complications, provided they were willing to base their calculations on the mean length of the tropical year, which comprised the aforementioned 365d 5;48h. Part and parcel of this strategy was to focus one's attention on the mean instead of the true time of the vernal equinox. This had the added benefit of lending support to a doctrine espoused by many computists, namely that Jesus Christ and his precursor John the Baptist had both been conceived at one of the equinoxes and born at the opposite solstice. Using the observations and solar tables contained in Ptolemy's *Almagest* as the basis for his discussion, Johannes established that the mean vernal equinox in AD 1 had fallen on 25 March at approximately 01:48 p.m., whereas the winter solstice arrived on 24 December at noon, just half a day before the night of Christ's birth.[38]

Strong intellectual debts to the thirteenth-century *computus* tradition can also be detected in Johannes's treatment of the historical dimensions of the topic at hand, which was at odds with the best evidence available in medieval libraries. Seemingly unaware of the Julian reform of 45 BC, he located the creation of the

[34] *Expositio* I.2 (fol. 3r–v). On the author's *modus operandi*, which very likely involved some degree of data-fudging, see Nothaft 2017b: 5–6.

[35] *Expositio* I.30 (fol. 39v).

[36] *Expositio* I.10, 17 (fols. 12r, 22v).

[37] *Expositio* I.14 (fol. 15v).

[38] *Expositio* I.14 (fols. 15v–16r).

Roman solar year in a period when the solstices and equinoxes still occupied the first day of their respective month, a date which at the assumed rate of $J = 120$ years would have been located $18 \times 120 = 2{,}160$ years in the past.[39] The Golden Number, too, could be treated as an ancient invention, as could the present rules for locating the date of Easter. If Johannes was right, Christians had been using the present solar and lunar calendar since the very beginning, even though the original boundary for Easter had been on 25 March, thus coinciding with the mean vernal equinox in the year of Christ's incarnation. The only contribution the fathers at the Council of Nicaea made in his narrative of events was to move this boundary upwards to 22 March in order to take into account the shift of nearly three days that had accrued since the year of the nativity. In a startling break with traditional interpretations, he declared that the Nicene fathers had consciously selected 22 March as the then-current date of the mean equinox, believing that Easter Sunday should always fall on or after this date. Contrary to all those who associated the equinox with 21 March and allowed this date to coincide with the 14th day of the lunar month, his *Expositio* hence asserted that it was the exact midpoint of the Easter lunation—*luna* 15—that could fall on the day of the vernal equinox, but no earlier.[40]

The Nicene intervention had led to a readjustment of Easter limits, but no steps to keep the vernal equinox from regressing had been taken, a failure that Johannes put down to the instability of the early Church and to the ignorance of the true length of the solar year that still prevailed at the time.[41] Confident that he possessed the knowledge the ancient fathers had lacked, Johannes concluded that the equinoxes and solstices could be fastened to their appropriate places in the calendar by omitting the bissextile day at the end of every 30th intercalation cycle, or once every 120 years.[42] To do nothing of the sort and simply let the paschal *termini* drift in accordance with the vernal equinox was to make the mobile feast days shed their accustomed relation to the liturgical year's numerous fixed feasts, with the inevitable result that 'all the feasts of the whole year could take place successively now in Lent (*Quadragesima*), now at the time of Easter, and thus one would have to make so many alterations and such frequent ones in the breviaries that there would be no end to these alterations'.[43] The question remained whether the modern-day Church should imitate the Church of the fourth century in moving the valid Easter limits further upwards to the current mean equinox, on 14 March, or whether a restoration of the calendar to its ancient state was in order. Johannes de Termis's discussion of this and a cluster of related problems took up 13 lengthy chapters, several of which placed a special emphasis on the theological interpretation of the Easter date as laid down in St Augustine's response to

[39] *Expositio* I.9 (fol. 11r). [40] *Expositio* I.13, 15–16 (fols. 14v–15r, 17v–20r).
[41] *Expositio* I.16 (fol. 18v). [42] *Expositio* I.3, 10, 15, 18–19 (fols. 4r, 11v, 16v–17v, 22v–6r).
[43] *Expositio* I.9 (fols. 9v–10r): 'Sed istud non valet... quia omnia festa tocius anni possent esse successive, modo in Quadragesima, modo in tempore paschali, et ita opporteret facere tot mutaciones et tociens in breviariis quod mutacionum illarum non esset finis.'

Januarius.[44] Compared to the mean equinox supposedly selected by the Council of Nicaea, the gap that had accumulated in the meantime was eight days (22 March vs. 14 March), yet Johannes did not believe that a reduction of this size was quite enough to qualify as a restoration in the proper sense. Since the Church chose to count its years from the moment of Christ's incarnation, it was mandatory to remove all eleven days that had needlessly and spuriously been intercalated since that time (11×120 = 1,320). Failure to do so would have meant that the Church continued to mislead the faithful whenever it made reference to the 'Years of the Lord', seeing how 1,345 natural or 'physical' years since the incarnation were in fact eleven shorter than the same number of years counted in the Julian calendar.[45] It was hence necessary to bring the mean equinox from 14 March back to its former location on 25 March, which also involved undoing the Nicene move of adjusting the Easter limits from 25 March–28 April to 22 March–25 April. Was it legitimate to intervene on such a drastic level? Johannes's affirmative answer to this question took recourse to historical precedent, more specifically to the apocryphal story of the Emperor Augustus, who snatched a day from February and added it to the newly baptized August in order to make sure its length was not excelled by the month that had previously been named after his great-uncle Julius Caesar (see Ch. 4.1):

> For if the emperors Julius and Augustus, whose rule was only of this world, were allowed to increase certain months...and shorten others, add to some and subtract from others, in order to make a name for themselves, and if the whole world upheld this in their favour unanimously and without scandal, it will be even more licit for the vicar of Christ, who by right of his vicariate is both the spiritual and worldly governor of the whole word, to subtract just one time single days from the months except for February—so that this way the things that pertain to the calendar, which are found to proceed in a disorderly manner as a result of the calendar's imprecision, can return to their due order and state—especially since he does not intend to do this in order to make a name for himself, but he does it for the benefit of the most holy Church, for which reason...everybody, and especially Catholics, will have to receive it most devoutly and embrace it as if with certain arms of their hearts and honour it.[46]

'The things that pertain to the calendar' naturally also included the Golden Number and the epacts and lunar regulars connected to it, which, too, had to be

[44] See *Expositio* I.5–17 (fols. 4v–22v), and Augustine, *Epistulae* 55 (ed. K. D. Daur, CCSL 31: 234–65).

[45] *Expositio* I.17 (fols. 20v–22r).

[46] *Expositio* I.5 (fol. 4v–5r): 'Quia si Iulio et Augusto imperatoribus, qui temporaliter tantummodo imperabant, licuit quosdam menses augere...et alios breviare, addere quibusdam et aliquibus subtrahere ad faciendum sibi nomen et hoc totus mundus concorditer et sine scandalo in eorum favorem sustinuit, multo magis vicario Christi, qui iure sui vicariatus et spiritualiter et temporaliter toti preeest mundo licebit a singulis mensibus preter quam a Februario singulos dies semel tantum modo subtrahere, ut sic illa ad kalendarium pertinentia que ob ipsius kalendarii imprecisionem inveniuntur deordinate incedere ad ordinem statumque debitum possint redire, presertim cum ex hoc non intendat facere sibi nomen, sed faciat hoc ad sacrosancte ecclesie utilitatem, propter quod nihil debebit istud quemquem offendere, sed debebit ab omnibus et precipue catholicis devotissime suscipi velutque quibusdam sui cordis brachiis amplecti ac venerari.'

restored to their correct position at the time of Christ's incarnation and birth. In order to fulfil this objective, Book II of the *Expositio* laid out Johannes's plan of equipping all calendars with a new batch of letters from A to T called *littere primacionum radicalium*. Like the letters included in earlier compositions such as William of Saint-Cloud's *Kalendarium regine*, the *Expositio*'s 'new-moon letters' were supposed to be accompanied by the number of hours before or after noon at which the Moon's *primacio* was supposed to occur.[47] With the help of the appropriate auxiliary tables, these times could be adjusted from one 19-year cycle to the next by subtracting the difference that separated the combined length of 235 mean synodic months from 19 Julian years. In Johannes's estimation, this gap was no more and no less than 1;28,7,6,5h, as he assumed a single month to last precisely 29d 12;44,3,2,1h (235 × 29d 12;44,3,2,1h − 19 × 365d 6h = 1;28,7,6,5h).[48] As in the case of the tropical year, he believed he could justify this value by appealing to past and contemporary observations, although the calculations he offered were not nearly as successful in supporting the conclusion in question. Next to two eclipses mentioned in the *Almagest* (491 BC and AD 125), the data he selected included a partial lunar eclipse he claimed to have observed himself in 1339 in the early morning hours of 26 January, although the times of true and mean opposition he cited were identical with those predicted by the Alfonsine Tables. The time that had elapsed since the corresponding syzygy of 1289 was 18,013d 15:51h, which implied an average lunation of 18,013d 15:51h ÷ 610 = 29d 12;44,3,2,57h. Unsurprisingly, given his use of King Alfonso's tables, the interval came very close to being a multiple of the Alfonsine value (29d 12;44,3,2,59h), but Johannes nevertheless, for reasons that are not completely clear, insisted on upholding a value of 29d 12;44,3,2,1h.[49]

With the stated result in place, he was ready to face a technical problem that considerably complicated the business of calendar improvement, but which had been passed over in silence by nearly all previous writers on the subject.[50] Jean des Murs, for instance, had rounded off his *Sermo de regulis computistarum* by asserting that the continued suppression of a bissextile day after 134 years was enough to keep the date of vernal equinox from receding. For his proposed 76-year table of revised Golden Numbers, he suggested anticipating all dates by a day after 304 years, without any overt reflection on the fact that the omission of a day every 134 years did more than enough to keep the new moons from rushing towards dates earlier than those assigned to them.[51] Since the rate of calendrical regression of the vernal equinox was more than twice as swift as that of the new and full moons, it was in

[47] *Expositio* II.3–4 (fols. 55r–57v). [48] *Expositio* II.6 (fol. 58v).

[49] *Expositio* II.1 (fols. 52v–54v). See the more detailed discussion in Nothaft 2017b: 10–11.

[50] A minor exception are the remarks made *c.*1232 by John of Sacrobosco, referenced in n. 124 of Ch. 4.3.

[51] Jean des Murs, *Sermo de regulis computistarum*, MS London, BL, Royal 12.C.XVII, fol. 213va: 'Ut autem equinoctia et solstitia immobilia perseverent secundum computationem nostram, semper in quibuslibet 134 annis excipe unum diem a kalendario Latinorum vel anno illo cessetur ab interpositione bissexti. Anteriora quoque numerum aureum quatuor ciclorum per unum diem in 304 annis aut, quod precisius est, a qualibet coniunctione sive primatione deminue unum diem ad

fact impossible for the ecclesiastical calendar to be cured of all its ailments in one fell swoop. Johannes de Termis was more perspicacious in highlighting the way his suggested suppression of a bissextile day every 120 years effectively overcompensated for the error inherent in the Golden Numbers, thus causing a shift in the opposite direction. Per 19-year cycle, this calendrical creep was going to amount to $235 \times 29\text{d}\ 12;44,3,2,1\text{h} - 19 \times 365\text{d}\ 5;48\text{h} = 2;19,52,53,55\text{h}$, meaning that it took 10.294417 cycles or approximately $195\frac{1}{2}$ years for the conjunctions to fall one day later than the dates Johannes had marked in his new calendar.

The solution to this problem was to introduce a brand new device or parameter, the so-called *clavis primacionum*, which tracked the number of days by which the *radices* inscribed in the calendar missed their targets at any given time. In an effort to simplify things somewhat, Johannes proposed to increase the *clavis* in synchronicity with the omissions of the bissextile day once every 120 years. At the beginning of the newly reformed calendar, which was meant to be located in AD 1, the *clavis* was still 0, indicating that the conjunctions of Sun and Moon fell on the dates actually occupied by the *littere primacionum radicalium*. With each subsequent interval of 120 or 240 years, the *clavis* was supposed to increase by one unit, thus indicating the number of days by which the actual conjunctions of Sun and Moon had moved away from their calendrical locations. By 1345, Johannes's year of writing, the *clavis* had reached 7, meaning that the first *primacio* of March had to be sought not on 2 March, as marked in the new calendar, but on 9 March, which was indeed the date of the mean and true conjunction in 1331 and 1350, both of them year 2 in the 19-year cycle (as was AD 1). All future increases were marked in a separate table (Figure 6.1), which suggested that the *clavis primacionum* reached a maximum of 29 in the year AD 5520 only to return to 0 after another 240 years, in AD 5760. This assumption was indeed justified, at least in an approximate sense, as the error accumulated over 5,776 years or 304 consecutive 19-year cycles happened to be nearly equal to a mean synodic month of 29d 12;44,3,2,1h years.[52] There remained a small difference of $5{,}776 - 5{,}760 = 16$ years, which, on a long-term perspective, could be taken into account if the final increase of the *clavis* was postponed from the 5,760th to the 5,880th year in certain later iterations of the proposed cycle of corrections. Johannes decided that this postponement had to be spaced *de octonis revolucionibus in septenas et de septenis in octonas*, meaning that the number of iterations had to alternate between 8 and 7. This rule respected the fact that 16 years fit 7.5 times into a 120-year interval, from which it followed that a full cycle of *claves* took not 5760 years, but 86,640 years $(= 5760 \times 6 + 5880 + 5760 \times 7 + 5880)$ to complete.[53]

propinquius veritati fractionibus insensibilibus nunc dimissis. Et sic tuum kalendarium sine errore poteris observare.'

[52] *Expositio* II.2, 4 (fols. 54v–55r, 56r). The truth of this statement follows from $24 \div 2;19,52,53,55\text{h} \times 29\text{d}\ 12;44,3,2,1\text{h} = 304.000221$.

[53] See the table in *Expositio* II.6 (fol. 60v), which lists the accumulated years of 30 sub-cycles of 5,760 or 5,880 years, going up to a grand total of 173,280 years.

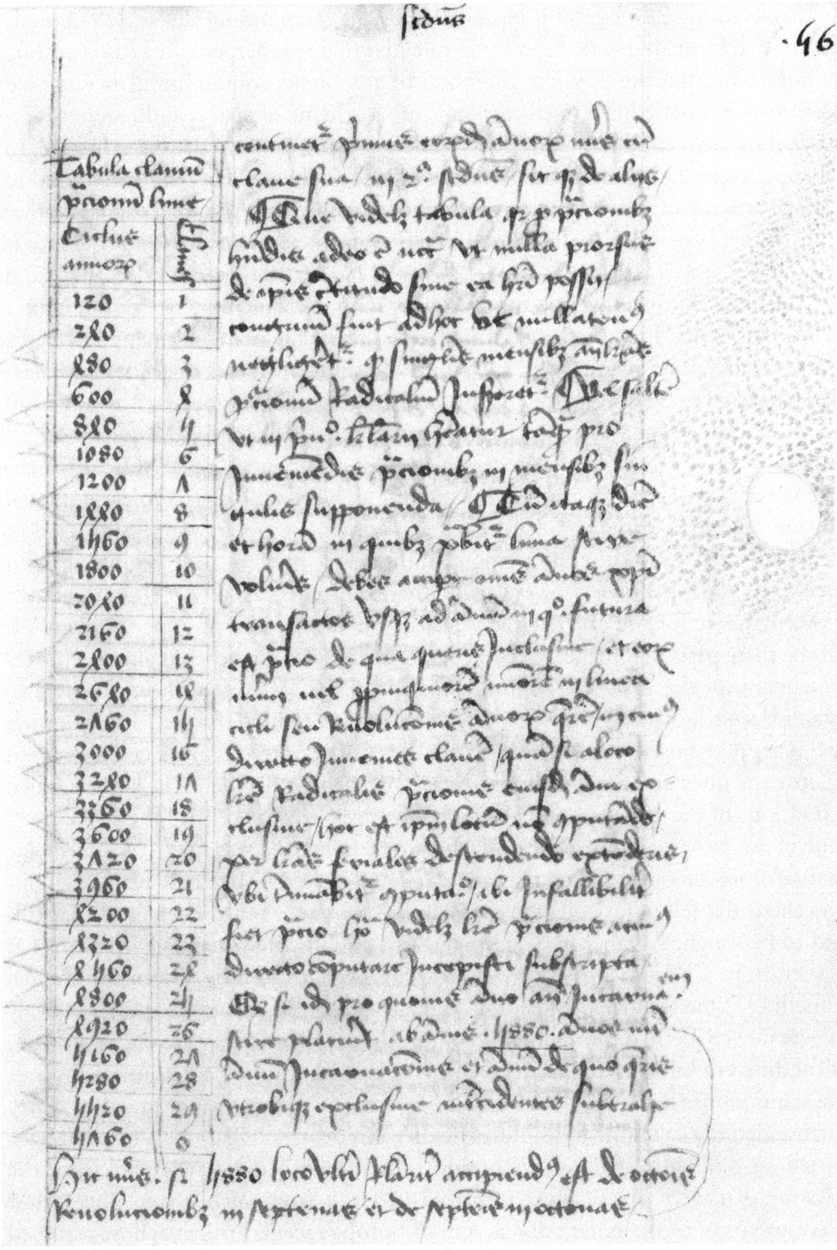

Figure 6.1. Johannes de Termis, *Expositio kalendarii novi* (1345), bk. II.4, table for the *claves primacionum*. Vatican City, Biblioteca Apostolica Vaticana, Ott. lat. 842, fol. 56r.

Source: © 2017 Biblioteca Apostolica Vaticana

Thanks to the clever gambit represented by the *clavis primacionum*, Johannes de Termis had been able to devise a lunar calendar that was perpetually self-correcting without requiring the physical alteration of any books containing it. It was even possible to translate his revised calendar into a scheme of epacts and regulars reminiscent of the techniques computists had been using by since the days of Bede. In Johannes's specific parlance, the 'epact' was the number of days one had to add to the *primaciones* of the year AD 1 in order to obtain the appropriate date in other years of the 19-year cycle. For the first 120 years after the incarnation, the epacts would have been 0, 19, 8, 27, 16, 5, 24, 13, 2, 21, 10, 28, 17, 6, 25, 14, 3, 22, and 11, yet these numbers were bound to grow with each increase of the *clavis* after 120 or 240 years. Johannes reacted to this by creating a table that mapped out the sequences of epacts adequate for each of the 30 different values the *clavis primacionum* could take (see Figure 6.2), thus anticipating an idea that was to underpin the Gregorian calendar introduced nearly two-and-a-half centuries later.[54]

For all the undeniable cleverness on display in Johannes de Termis's *Expositio kalendarii novi*, a reform proposal as radical as his was likely to rouse potential objectors. Indeed, some of Johannes's key arguments were very soon subjected to trenchant critique by the anonymous author of a brief treatise beginning with the words *Ad correctionem calendarii*... (hereafter cited as *Ad corr.*), who almost certainly wrote for the eyes, perhaps even at the instigation, of Pope Clement VI. Of the three parts that make up the structure of *Ad corr.*, the first was almost entirely dependent on the *Expositio*, reproducing bits of the text verbatim in an effort to establish ten reasons why the Julian calendar was ripe for modification. The following two parts juxtaposed these reasons with a series of counter-arguments that warned against any interference with the existing solar calendar. To these, the author added refutations of the specific points he had previously excerpted from the *Expositio*.[55] One of the principal strategies of obstruction he employed was first to acknowledge that an omission of days from the calendar could indeed be deemed justified, if this prevented the Church from transgressing its own laws, but then to add that this had to be weighed against the command to avoid any new scandals from arising, as seen from Christ's words in Matthew 18:7 ('woe to him by whom the scandal cometh').[56] One scandal an alteration to the length of the calendrical year was likely to generate concerned the veneration of the saints, whose days were either going to fall at different intervals or be erased entirely, thus constituting 'an injustice towards the saints whose feasts would fall on these days'.[57] An even greater practical worry surrounded the calendrical deadlines involved in debts, obligations, and contracts, which an omission of days was bound to throw into disarray. Apart from the lawsuits that were practically guaranteed to result from such a step, the author drew attention to the potential reaction of Europe's secular rulers, whose assumed

[54] *Expositio* II.5 (fols. 57r–58v). The Gregorian calendar will be discussed in Ch. 8.

[55] See the edition and introduction in Schabel 1998, and the previous remarks in Nothaft 2015g: 293–5.

[56] *Ad corr.* -1.2 (ed. Schabel 1998: 31). Other variations of this point are found in *Add corr.* -2, -2b, -4 (ed. Schabel 1998: 31–2).

[57] *Ad corr.* 7a (ed. Schabel 1998: 27).

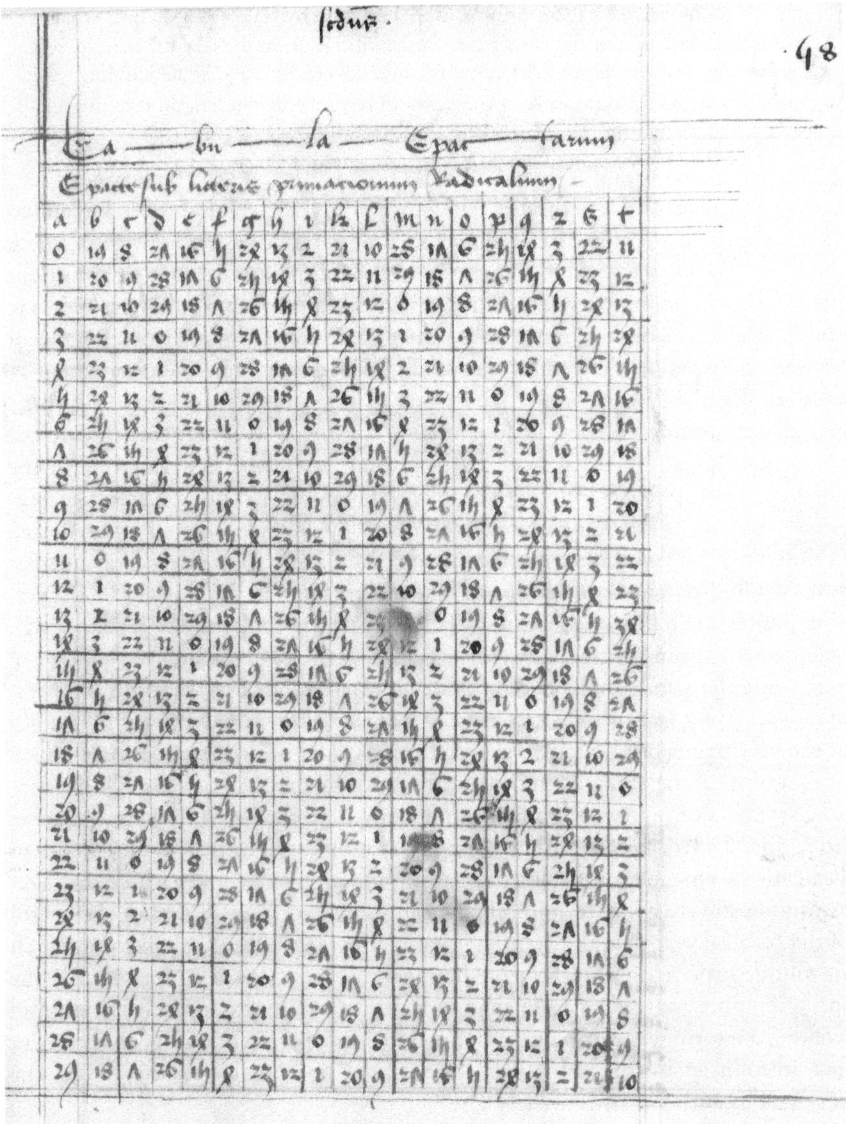

Figure 6.2. Johannes de Termis, *Expositio kalendarii novi* (1345), bk. II.5, epact table. Vatican City, Biblioteca Apostolica Vaticana, Ott. lat. 842, fol. 58r.
Source: © 2017 Biblioteca Apostolica Vaticana

authority over the civil components of the calendar might lead them to oppose the pope's attempt at reform:

> And perhaps rebellions would spring from this, such that, if some princes are unwilling to use a year different from the one they are used to in order to avoid such [problems],

> other divisions between these princes and the Church could follow. And, what is worse, the calculation of the years from Christ's incarnation the faithful were so very accustomed to would become defective, because the customary year, which until now used to be the habitual measure of time, would have a different length, resulting in a [new] calculation from the solar calendar's renovation, whereas the true and precise time of the Lord's incarnation would be abandoned, as if we created a new rite.[58]

Not all arguments he listed in the remainder of his discussion were as forceful as these appeals to the dignity of saints and the spectre of a rift between Church and state. The idea, for instance, that the commandment to celebrate Easter right after the spring equinox was not universally binding because the seasons were not experienced simultaneously across the globe ('for if Christians or Jews went beyond the equator, they could not celebrate Easter with us, because when we have summer, they have winter, and when it is springtime, they have autumn') had already been countered in Johannes de Termis's *Expositio*, which made the point that 'spring' in the Mosaic Law necessarily referred to springtime in the northern hemisphere.[59] Another complaint Johannes had done his best to preempt was that the diverging estimates espoused by past astronomers and reflected in different tables were evidence for the uncertain status of the length of the solar year. Voiding potential objections to his reform proposal, he insisted that diverging past estimates for the length of the tropical year could not be read as a sign of actual change in nature, but were instead reflective of gradual improvements in measuring the same value. He did not deem it likely, however, that this improvement was going to continue indefinitely: 'Much rather, it may be that the length of the year that is had now is so precise that no more precise one can be obtained or that, if it can be obtained in the future, the difference will be so small that very long stretches of time will pass before any noticeable change to the calendar arises from this.'[60] His critic nevertheless expressed doubts about the very feasibility of a calendrical emendation, asking once again how anybody could exclude that later astronomers were going to find different and more accurate values for the length of the tropical year than the 365d 5;48h under review. Given the uncertainty that surrounded the matter, no firm prediction could be made as to how swiftly the spring equinox was going to change its dates in the future, making it doubtful whether a reform was worth undertaking in the first place. Repeating an argument first introduced by the computists Peter and Cunestabulus in the 1170s, the anonymous author warned that:

> We do not even know whether in different times the length of the year might increase so much that the solar year would return to its pristine state by itself without a correction of the calendar, and that the equinoxes and solstices would return

[58] *Ad corr.* 1a (ed. Schabel 1998: 26).
[59] *Ad corr.* 14a (ed. Schabel 1998: 28–9). See *Expositio* I.11 (fols. 12v–13r).
[60] *Expositio* I.9 (fol. 10v): 'Quinimmo forte quantitas anni que modo habetur est sic precisa quod precisior haberi non potest vel si eam habituri sint magis precisam, differentia erit ita parva quod tempora longissima transibunt antequam ex hoc certa kalendarii appareat variacio aliqua.'

[to their ancient dates] in the same way in which all the planets sometimes go fast or slow, stationary or retrograde.[61]

When everything was said and done, the critic responsible for *Ad corr.* still remained sympathetic to a correction of the calendar's lunar component, which seemed astronomically feasible and posed a much more serious problem to the life of the Church, since 'the age of the Moon is announced incorrectly on a daily basis because of the defect of the Golden Number'.[62] Johannes de Termis had used the example set by the fathers of the Council of Nicaea as a justification for more thoroughgoing measures, claiming that the adjustment of Easter limits carried out in the fourth century constituted a mere ad hoc solution in place of a permanent one that should restore the calendar to its ancient state. The author behind *Ad corr.* strongly disagreed, insisting that the very fact that these Fathers refrained from omitting days demonstrated that no such correction was required from their modern descendants.[63] He further supported his argument with an appeal to the pseudo-Isidoran decretals, which seemed to document that the Nicene Fathers specified a range of new-moon dates (8 March to 5 April), but made no reference to the vernal equinox.[64] What this suggested was a conservative reform, which improved the computation of the lunar age, but left the *terminus paschalis* and the range of possible Easter dates in their accustomed places. To fix the equinox on whatever date it currently occupied and to move the Easter limits upwards was an option he was at least willing to consider, suggesting that the required changes to the relevant books might be a price worth paying, provided the calendar was henceforth intercalated according to the length of the solar year known from present-day experience ('The breviaries would be changed a single time and only to a minor extent, but afterwards everything would endure in a state of truth').[65] He was indeed aware that not even Johannes de Termis's proposal was immune to such inconveniences, since a reduction of the calendar to its state at the beginning of the Christian era was not feasible without a concomitant change to the Easter limits (from 22 March–25 April to 25 March–28 April). Either way, the feast was going to 'move out of the boundaries instituted by the holy Fathers, and some days that had never been within the Lenten period would be found within it and vice versa'.[66]

6.3. INTERLUDE IN CONSTANTINOPLE

For all the reticence on display in *Ad corr.*, the calendrical conservatism espoused by its anonymous author was on the whole relatively moderate, especially if compared with the prevalent view in faraway Constantinople, where the question of

[61] *Ad corr.* 11a (ed. Schabel 1998: 27). See also *Ad corr.* 5a, -1b (ed. Schabel 1998: 27, 30), and the preemptive response in *Expositio* I.9 (fols. 10v–11r).

[62] *Add corr.* 11a.1 (ed. Schabel 1998: 28). [63] *Ad corr.* 4a (ed. Schabel 1998: 27).

[64] *Ad corr.* 16a, -2 (ed. Schabel 1998: 29, 31). See the *Praefatio in Collectione canonum pseudo-Isidoriana* (PL 130: 9–10; <http://www.pseudoisidor.mgh.de/html/001.htm>), and the references to this passage in *Expositio* I.6, 13, 15 (fols. 5v, 14v, 17v).

[65] *Ad corr.* -9 (ed. Schabel 1998: 34). [66] *Ad corr.* 10a (ed. Schabel 1998: 27).

calendar reform had only recently begun to attract the attention of the learned.[67] The cultural backdrop for the debate in the Byzantine Empire was provided by the Palaeologan Renaissance, which had produced a florescence of literary, scholarly, and artistic activity in the wake of the reconquest of Constantinople from the Crusaders in 1261. Sponsored by Emperor Michael VIII Palaeologus (1258–82) and his successor Andronicus II (1282–1328), scholars during the following decades invested a great deal of effort in studying, preserving, or bringing to light classical Greek texts. Next to grammar and philosophy, this activity also extended to Ptolemaic astronomy, the practitioners of which rose in numbers during the fourteenth century, helped in no small part by introductory works such as Theodore Metochites's comprehensive *Stoicheiosis astronomike* (*c.*1316).[68] As far as can be ascertained, the first Byzantine scholar to push for an application of Ptolemaic mathematical astronomy to the calculation of Easter was Metochites's student Nicephorus Gregoras (1292/5–1358/61), who is best known for his 37-book *Roman History* (*Historia Rhomaike*) covering the events of the Eastern Empire from 1204 to 1359. His ideas regarding calendar reform are documented in a chapter in Book VIII of this *History* and in a letter he addressed to his friend Demetrios Cabasilas, both of which tell the same story of how the young astronomer raised the issue at a banquet in the presence of Emperor Andronicus II, probably in the year 1324.[69]

The specifics of Gregoras's reform proposal, as they emerge from this account, remain somewhat nebulous due to certain inconsistencies, both within the texts and between their various manuscripts.[70] For the present-day location of the vernal equinox Gregoras appears to have relied on Ptolemy's *Almagest* or *Handy Tables*, placing it on 18 March, which was six days too late compared to the best data available in the Latin West. The changed situation would have been best accommodated by subtracting three days from the calendar or by moving the earliest permissible date for Easter Sunday from 22 March to 19 March, yet Gregoras only acknowledged a total error of two days, claiming that the calendar had last been adjusted in year 6300 of the Byzantine world era (= AD 791/2).[71] As for Andronicus II's reaction, Gregoras assures us that the emperor acknowledged the validity of his astronomical arguments, but nevertheless refused to take any political measures, citing the difficulty of convincing every ecclesiastic on every Greek island to implement a reform of the existing Easter cycle. Rather than causing dissent and confusion,

[67] The status of the calendar problem in the medieval Orthodox Church is surveyed in Kuzenkov 2010. On *computus* in the Orthodox East, see Rühl 1897: 157–69; Ginzel 1906–14: iii. 288–307; Mentz 1906; Schissel and Ellend 1943/9; Zélinsky 1982; Tihon 2004; Lempire 2007; Mosshammer 2008: 278–316.

[68] The *Stoicheiosis* is studied in Bydén 2003. See also Tihon 1981; 2013; Fryde 2000: 337–73.

[69] See Nicephorus Gregoras, *Byzantina Historia* 8.13 (PG 148: 547–60), which was translated into German by Dieten 1979: 67–72, and the letter to Cabasilas, ed. in Bezdechi 1924: 330–6. See, in addition, Guilland 1926: 282–5; 1927: xiii, and the summaries in Kaltenbrunner 1876: 324–6; Bezdechi 1927: 68–74; Kuzenkov 2010: 154.

[70] See the notes on variants in Dieten 1979: 177 (nn. 145 and 147), and the detailed analysis in Tihon 2011: 395–402.

[71] Nicephorus Gregoras, *Byzantina Historia* 8.13 (PG 148: 552).

Andronicus, much to Gregoras's personal disappointment, preferred to maintain unity by perpetuating an erroneous tradition.[72]

An attitude very similar to that of the Byzantine emperor emerges from the writings of Barlaam of Seminara, a Calabrian-born abbot based in Constantinople who was connected to Nicephorus Gregoras by a long history of personal animosity and intellectual competition. Barlaam commented on the issue in a separate treatise on the date of Easter, datable to 1332/3, in which the main argument centred on the following injunction from the so-called *Canons of the Apostles*: 'If any Bishop, or Presbyter, or Deacon, celebrates the holy day of Easter before the vernal equinox with the Jews, let him be deposed.'[73] According to both Barlaam's Greek treatise and Johannes de Termis's Latin *Expositio* of 1345, the correct way to read this passage was to see it as expressing two separate prohibitions, which could be violated either individually or in conjunction. Hence, the date of Easter Sunday became illegitimate if it fell before the vernal equinox, but it was also celebrated in error whenever it coincided with the Jewish Passover, even if the latter happened after the equinox.[74] As a result, the correct dating of Easter was constrained by four independent criteria, viz.:

(1) always after the vernal equinox;
(2) never together with the Jewish Passover;
(3) always after the first full moon after the equinox;
(4) always on the first Sunday after this full moon.[75]

Barlaam freely admitted that the existing Easter cycle or 'canon' caused the new moons to anticipate their calendrical dates at a rate of $M = 304$ years, with the present computation exhibiting a delay of two days. The discrepancy allowed for cases where Easter Sunday was transposed to the second rather than the first week after the physical full moon, in violation of rule (4).[76] Even though he gauged the parallel regression of the equinoxes in accordance with the Ptolemaic $J = 300$ years,[77] Barlaam completely ignored the other main prong of the calendrical problem, which concerned rule (3): due to the regression of the vernal equinox, Easter Sunday was on occasions transposed even further, to the second rather than the first full moon after the equinox. Instead of dealing with the consequences of the calendar's defects in greater detail, the monk from Calabria insisted that a reform was futile, since it was impossible to find any other canon that would be easy

[72] Ibid. (PG 148: 557).

[73] See the Greek text and French translation in *Constitutiones apostolicae* 8.47.7 (ed. Metzger 1987: 276–7), and the quote in Barlaam of Seminara, *Traité sur la date de Pâques* 1 (ed. Tihon 2011: 368). For an earlier study of Barlaam's treatise, see Tihon 1999. For his ecclesiastical career and relation to Nicephorus Gregoras, see the biography by Cassiano 1994.

[74] Barlaam of Seminara, *Traité sur la date de Pâques* 20 (ed. Tihon 2011: 374). See the discussions of this passage in *Expositio* I.8 (fol. 9r), and *Ad corr.* 1.1, 15a (ed. Schabel 1998: 20, 29), which cite the same wording. The *Canons of the Apostles* had entered the Latin canon-law tradition in the sixth century with Dionysius Exiguus's *Collectiones canonum* (PL 67: 142B).

[75] Barlaam of Seminara, *Traité sur la date de Pâques* 22 (ed. Tihon 2011: 376).

[76] Ibid. 25–6 (ed. Tihon 2011: 376, 378).

[77] Ibid. 15–16 (ed. Tihon 2011: 372, 374).

enough to comprehend and still respect all four normative constraints for Easter Sunday. In line with the caveat expressed by Andronicus II in 1324, he also raised the extreme difficulty of persuading all monks and clerics in charge of their local Easter cycles to subtract two days, in order to emend the existing defect, and then to skip another day every 304 years.[78] Rather than reforming and then running the risk of falling into error again, Barlaam claimed it was best to tolerate the current predicament, especially since the existing canon still satisfied rules (1), (2), and (3). More than that, he pretended to see a virtue within the vice, pointing out that the discrepancy in the lunar calendar guaranteed that the Christian Easter Sunday always fell at least three days after the Jewish Passover, which he imagined to be identical with the 14th day of the Easter lunation. By occasionally celebrating Easter in the wrong week, the Church at least made sure it never violated the prohibition against celebrating with the Jews that had been passed down in the apostolic canons.[79]

Whether or not this argument was made in good faith,[80] it appears to have informed a general consensus in the Orthodox East as to why a correction of the current Easter canon was unnecessary or legally problematic. Barlaam's arguments were excerpted only a short time later, in 1335, by Matthew Blastares, who included a lengthy section on Easter reckoning in his alphabetical *Syntagma* of canon law,[81] and still echoed after the fall of Constantinople by Basilius Bessarion, who in 1470 warned Pope Paul II against a reform of the existing scheme.[82] What is perhaps more remarkable are the occasional signs of similar views being held in the Latin West. A hint to this effect appears in the treatise *Autores kalendarii* of 1317, whose author complained that the Church was so eager to avoid the charge of 'Judaizing' (*iudaisare*) that it abetted the existing error in order to ensure that Easter Sunday was never celebrated on the astronomically correct 15th day of the Moon:

> Yet, with all due obedience to the Church and [with all openness] to better judgement, the Christians would not be Judaizing simply for celebrating their Pasch on a Sunday on which the Moon is 15 days old, just as the Jews would not be Christianizing by celebrating their Pasch on a Sunday on the 15th day of the Moon.[83]

The author did not identify any members of the party he was arguing against here, but a strong worry that the Church might steer too close to the Jews in fixing the time of its own Pasch is indeed expressed in *Ad corr.*, where one passage even raised the spectre of the Antichrist, 'who will declare that the Sabbath of the Jews and the Sunday of the Christians should be observed equally, so as to cause the Christians

[78] Ibid. 30–1 (ed. Tihon 2011: 378, 380). [79] Ibid. 33 (ed. Tihon 2011: 380).

[80] Cf. Tihon (2011: 393): 'La raison de ce rejet est sans doute à chercher dans l'opposition de Barlaam à Nicéphore Grégoras.'

[81] Matthew Blastares, *Syntagma alphabeticum*, s.v. *Πάσχα* (PG 145: 65–108, esp. col. 104). Cf. Kuzenkov 2010: 151, 154–5.

[82] See the edition of his letter in Mohler 1942: 546–8.

[83] *Autores kalendarii* 15 (ed. Nothaft 2013d: 61–2). See also ibid. 14, 20 (ed. Nothaft 2013d: 60, 70) and the remarks in Nothaft 2013d: 23–4.

to Judaize'.[84] In line with this stance, *Ad corr.*'s author questioned the extent to which the Church was still bound by Mosaic Law when it came to timing the month of Easter. If Christians were not obligated to eat unleavened bread for seven days during the first spring month, why should they be expected to celebrate Easter at the beginning of spring in exactly the same way the ancient Israelites had been ordered to do? Upon closer inspection, Jews and Christians used completely different calendars for timing their feasts, which raised the question whether the lunations of Easter and Passover should really be treated as equivalent.[85] Simply to hold Easter within a fixed date rage, with no particular regard for the rule of the equinox, was not an absurd idea from this vantage point.

6.4. A LETTER FOR THE POPE

Barlaam of Seminara has been worth invoking not just for the comparative perspective his work provides, but for his personal ties to the curia of Clement VI, who received the Byzantine cleric into the Roman Church and made him bishop of Gerace (Calabria) in October 1342. Although it remains unknown whether Barlaam had any kind of influence on the calendrical discussions that unfolded in Avignon in 1344/5, his presence at the Papal Palace, which is attested at several instances between 1339 and 1344,[86] could have served as a reminder to Clement VI and his advisers that the problem they tried to solve was not exclusive to the Church of Rome. One brief sign of such awareness occurs in the *Expositio kalendarii novi* by Johannes de Termis, who wondered aloud 'what the schismatic Christians would say' if the Latin Church corrected its calendar without restoring the vernal equinox to its ancient date. Surely, 'they would think of us as incompetent and could subject us not undeservedly to mockery'.[87] Predictably, the critic behind *Ad corr.* remained unconvinced by this argument:

> They would have no rational basis for ridiculing us, because they know well that our calendar is in need of correction as far as the Moon is concerned and that it can be emended without scandal, whereas this is not the case at all with the solar calendar. They accordingly know well that it is worse not to avoid the great scandals that would follow from [a correction of the solar calendar] than to deviate to an extent, as we already have, from the rite of the Jews, whose ceremonies have expired.[88]

In the end, the arguments and caveats recorded in *Ad corr.* were too weighty to leave the pope and his advisers entirely unaffected. An executive decision against reforming the Julian calendar may in fact already have been made in 1344, if one

[84] *Ad corr.* -3b (ed. Schabel 1998: 32).

[85] *Ad corr.* -1, -3, -3b, -3.1 (ed. Schabel 1998: 29–30, 32).

[86] A payment to Barlaam 'in curia legenti grecum' is recorded for 7 August 1342. See Schäfer 1914: 198.

[87] *Expositio* I.16 (fol. 20r): 'Item . . . quid dicerent scismatici Christiani, quorum aliqui perfectissime sciunt textum Biblie . . . ? Ipsi reputarent nos insciolos et possent de nobis non immerito trufare.' This passage is repeated in *Ad corr.* 3.2 (ed. Schabel 1998: 23).

[88] *Ad corr.* -3.2 (ed. Schabel 1998: 32).

can judge from the fact that the four letters the apostolic chancery dispatched on 25 September and 1 October that year only ever mentioned the Golden Number as the object of potential correction (see Ch. 6.1). When Jean des Murs and Firmin de Beauval responded to the pope's mandate, they did so in the form of an obsequiously penned *Epistola super reformatione antiqui kalendarii*, to which they attached a treatise in four parts.[89] All of the seven extant manuscript witnesses to this text subjoin a brief postscript or 'user's manual', which confirms the time and place of writing by referring back to the 'little book on the correction on the Golden Number, which we put together in 1345 in the city of Avignon under the orders of our master, Pope Clement VI'.[90] Although the first part of this 'little book' (*libellus*) placed its focus on the solar year, the authors made it clear from the outset that correcting the Julian calendar was not part of their assignment, even citing a key passage from *Ad corr.* that warned against the political upheaval that might arise if days were dropped.[91] Jean des Murs's influence on this part of the work is especially conspicuous from a brief discussion of previous estimates of the solar year, which invoked the opinions of Hipparchus, Ptolemy, al-Battānī, Thābit, and King Alfonso, but was in fact derived wholesale from Jean's *Patefit.*[92] In line with this and other astronomical texts from Jean's pen, the authors of the *Epistola* followed the Alfonsine Tables in gauging the vernal equinox's drift rate as $J = 134$ years, claiming that the tables in question were the ones most favoured 'by many sensory experiences had in Paris and elsewhere'.[93] Looking back at the alternative values proposed by ancient astronomers, Jean and Firmin granted the possibility 'that they all made good observations' and that the actual length of the solar year may be subject to the same kind of inequality that also affected the planets with their stations and retrogradations. Since they immediately went on to mention the existing 'doctrine concerning the equation of the solar year', it appears that the two astronomers were here thinking of the kind of minor fluctuations that would have been caused by the irregular motion of the solar apogee rather than the more dramatic changes described in the classical theory of access and recess, which would have even allowed the vernal equinox to go backwards in the calendar.[94]

The second part of the *Epistola* began with the authors citing the lengths of the common and embolismic lunar years (354d 8;48,36h and 383d 21;32,39h),

[89] A first summary of this source was published by Kaltenbrunner 1876: 315–22. In what follows, I shall cite the whole text as 'Jean/Firmin, *Epistola*', following the critical edition by Schabel 1996. The earlier edition by Gack-Scheiding 1995, although based on one more manuscript, is on the whole less reliable. Gack-Scheiding's extended introduction to the text can only be used with the greatest caution.

[90] Jean/Firmin, *Epistola* (ed. Schabel 1996: 212). On the 'user's manual', see also Schabel's remarks on p. 190. In the introductory letter (p. 196), Jean and Firmin allude to their work having been directed by the Cardinal of Rodez (Bernard d'Albi), of whose involvement in the calendar reform project nothing else has come to light. See the biographical summary in Lützelschwab 2007: 434–7.

[91] Jean/Firmin, *Epistola* 1.3 (ed. Schabel 1996: 199). See also 1.4 (ed. Schabel 1996: 199–200), tacitly quoting *Ad corr.* 1a (ed. Schabel 1998: 26).

[92] See the texual comparison in Nothaft 2015d: 220–1.

[93] Jean/Firmin, *Epistola* 1.3 (ed. Schabel 1996: 198).

[94] Jean/Firmin, *Epistola* 1.1 (ed. Schabel 1996: 198).

which were both based on a simplified mean synodic month length of 29d 12;44,3h.[95] That they had obtained this value by rounding down from the Alfonsine 29;31,50,7,37,27,8,25d = 29d 12;44,3,2,58,51,22,8h (see Ch. 5.3) becomes clear from their parallel assertion that the new moons will anticipate the dates marked by the Golden Numbers at a rate of $M = 310$ years and 260 days. This estimate was a rare example of extreme mathematical diligence within a corpus of literature that usually only offered multiples of 19 when speaking of the calendrical slip of the new moon, as was the case with the ubiquitous $M = 304(= 16 \times 19)$. To obtain Jean's and Firmin's more precise number, it was necessary to follow the Alfonsine value down to the fifth sexagesimal-fractional place, as $19 \div (6939;45\text{d} - 29;31,50,7,37,27\text{d} \times 235) = 310.713227\ldots$ or 310 Julian years and *c.*260.5 days. How they had managed to arrive at their approximation to this result, the two astronomers did not reveal, apart from a general appeal to the *calculus numerorum*.[96]

Rather than on a technical analysis of the error, Jean and Firmin concentrated on its various unwelcome consequences, which included the undeniable and embarrassing fact that the Church committed a manifest blunder whenever it announced the age of the Moon based on its existing tables. They highlighted the importance of calculating the lunar age of Easter and other feasts days in accordance with the truth, warning that 'foreigners or adversaries' of the Church were well versed in this sort of knowledge and that toleration of the current defect was likely to stir up murmurs among the common people.[97] Ridicule was to be feared not just from the laity, but from the 'Jews and gentiles', who could easily perceive that Easter Sunday was no longer accurately keyed to the full moon of Passover, which compromised the harmony between Old and New Testament as manifested by the symbolic relation between the slaughter of the Passover lamb and Christ's self-sacrifice on Good Friday.[98] Repeating a point already made by Jean in his *Sermo de regulis computistarum*, the two astronomers warned that a continuing elongation between the calendrical and observable full moon was eventually going to land Good Friday on the day of an ecliptic conjunction, which might lead some to question the miraculous status of the darkness seen on the day of Christ's crucifixion. 'It is hence good to bring the Golden Number back to the new moons so that the holy Church of God is able to avoid the inconveniences mentioned above'.[99] Having demonstrated the reform's necessity, they dedicated the third part of their work to discussing four different methods of how the ecclesiastical Moon might be brought back into agreement with the phenomena. One well-established approach, which had already been put into practice in various calendars drawn up in the late thirteenth and early fourteenth centuries, was to supplement or replace the traditional Golden

[95] Jean/Firmin, *Epistola* 2.1 (ed. Schabel 1996: 200).
[96] Jean/Firmin, *Epistola* 2.2 (ed. Schabel 1996: 202).
[97] Jean/Firmin, *Epistola* 2.2 (ed. Schabel 1996: 203).
[98] Jean/Firmin, *Epistola* 2.2 (ed. Schabel 1996: 203).
[99] Jean/Firmin, *Epistola* 2.2 (ed. Schabel 1996: 204).

Number with precise conjunction times calculated with the aid of astronomical tables. Jean and Firmin testified to having:

> already seen many calendars and many tables composed on this basis. But since anyone who makes such tables places in them the hours and parts of the hours at which the Sun and Moon are in conjunction with respect to his own meridian, we shall put down several versions of this approach, which allows one to obtain the true day of the mean conjunction, using Jerusalem for the meridian while leaving out the hours and minutes, because the Church, requiring the simplest [possible] rules, has always been wont to make calculations based on complete days.[100]

The most straightforward approach was hence to redistribute the Golden Numbers according to the dates the conjunctions were going to occupy in the forthcoming 19-year cycle, which was to begin in 1349. A table attached to the *Epistola* showed all 235 of the conjunction dates in question, which were evidently calculated from the Alfonsine Tables using a midnight epoch and the notional meridian of Paris (+12°) instead of the Jerusalem one alluded to in the text (Figure 6.3).[101] Jean and Firmin were aware that their various predecessors, such as Peter Nightingale, had been more fastidious in creating tables for 76 years, taking into account the 18-hour surplus contained in 19 Julian years (19 × 365d 6h = 6,939d 18h), which was bound to make some of the conjunction dates shift from one lunar cycle to the next. They mapped out such a solution for the month of January, using letters from A to T in place of Golden Numbers, but were quick to suggest a third option, which relied on a single column of new-moon dates for 19 years (see Figure 6.3). The dates inscribed into this column remained valid even after the first 19-year cycle from 1349 to 1367, provided readers shifted the day-epoch from midnight to sunrise (6 a.m.). As with the other two options, long-term validity was only guaranteed if the new Golden Numbers were collectively shifted to an earlier date at the end of every 310 or 311 years, as laid out in a separate wheel diagram.[102]

One drawback all of these proposals shared was that they were going to render useless the innumerable *kalendaria* currently in the service of the Church or, at any rate, require a great amount of material to be added to each of them. In principle, such excess could be avoided if the Church was prepared to omit a certain number of days from the calendar, for instance by suppressing the bissextile day in a series of leap years—enough of them to reunite the Golden Numbers with the actual dates of the new moon. According to Jean and Firmin, a reform of this sort had already been 'well discussed in a certain book' by the head of a monastery in Grasse whom the manuscripts variously identify as *Alnadus*, *Alva(l)dus*, or *Arnaldus de*

[100] Jean/Firmin, *Epistola* 3.1 (ed. Schabel 1996: 204).

[101] The table is reproduced and analysed in Nothaft 2015h, where it is argued that Jean's and Firmin's reformed Golden Numbers underlie the *Nombre d'or nouvel* that adorns the calendar in the *Très Riches Heures du Duc de Berry*.

[102] Jean/Firmin, *Epistola* 3.1–2 (ed. Schabel 1996: 204–6).

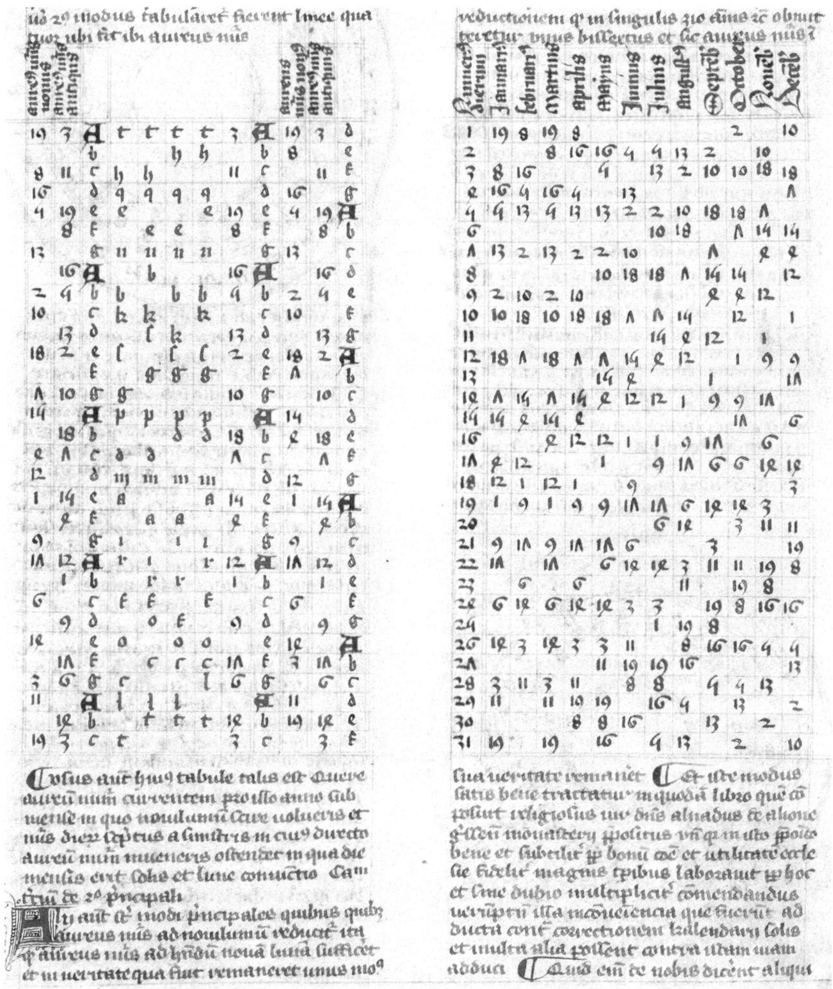

Figure 6.3. Jean des Murs and Firmin de Beauval, *Epistola super reformatione antiqui kalendarii* (1345), tables of reformed Golden Numbers. Oxford, Bodleian Library, Canon. Misc. 248, fol. 25r.

Alione.[103] No further information about this man and his work appear to have survived, apart from the claim, made in the *Epistola*, that he dated the beginning

[103] Jean/Firmin, *Epistola* 3.3 (ed. Schabel 1996: 206). The two earliest manuscripts, MSS Oxford, BodLib, Canon. Misc. 248, fols. 25rb, 25vb, and Paris, BnF, lat. 15104, fol. 119r–v, have 'Alnadus de Alione' and 'Alnaldus'. The attempt in Gack-Scheiding 1995: 9, 50, 134, to identify this person with the twelfth-century computist Arnoul d'Amboise is without foundation. Schabel 1996: 206 (n. 22), suggests Alard of Diest, but Alard's *Kalendarium*, mentioned in Ch. 5.1, bears no resemblance to the work described here.

of the Church's use of the Golden Number to the year AD 209.[104] Jean and Firmin made no attempt to deny the merit of their predecessor's work. Indeed:

> since he spent a great amount of time working on this proposal in a good and subtle manner, for the common good and utility of the Church, there are without doubt many reasons to recommend it. And yet the same inconveniences that were adduced against a correction of the solar calendar, and many others, could be adduced against this approach. For what would the schismatic Christians, who are wont to celebrate Christ's nativity on the same day as we do, say about us? And after 12 years they would see how we, who used to be steady [in intercalation], differ [from them] by three days in all our immovable feasts without any reason, because the astronomers of this sect and all other sects would recognize that this has not [yet] led to a correction of the solar calendar. But perhaps they do not know the Golden Number and instead make use of a different, more precise way of finding the new moon. And if they know about the Golden Number, they might perceive that it could be rescued by some other, better, way than by omitting bissextile days, wondering not without justification why we are willing to make such a great and sudden change to our most sacred rite.[105]

Jean and Firmin moved on to propose an alternative in the shape of a *numerus elongationis*, which is superficially reminiscent of the *clavis primacionum* suggested in the same year by Johannes de Termis. Unlike the latter, however, which implied an addition to the *primaciones* found in the calendar, the elongation number indicated how many days one needed to subtract from the dates occupied by the conventional Golden Number in order to yield the proper day of the new moon. For dates in the present, this elongation number was 4, but it was set to increase by another unit with every 310 or 311 years, coming back to zero after 9,582 years. The two astronomers expressed a personal preference for this method of correction, emphasizing its 'brevity, precision, durability, and the paucity of inconveniences' that followed from it.[106] They nevertheless suggested one more alternative, which was to alter the Golden Number itself by adding a constant they referred to as the *corrector*. In present times, this *corrector* was to be 14, thus indicating that the Golden Number for the first year of the 19-year cycle was $1+14=15$, with the first new moon of the year on 19 January instead of 23 January. After 310 years, the correct date should have shifted to 18 January, yet this was not a date occupied by any Golden Number in the old calendar, forcing Jean and Firmin to retain the previous *corrector* for the entire period from 1349 to $1970(=1{,}349+310+311)$.[107] The *numerus elongationis* method, too, was somewhat compromised by the imprecisions inherent in the existing lunar calendar, which it simply duplicated in structural terms. In addition, neither approach had the capacity to stop the calendrical drift of the vernal equinox, which happened at a rate twice as fast as the slippage of the new moons. As long as the Church stuck to its traditional date-range from 22 March to 25 April even a reformed lunar cycle could not prevent situations in

104 Jean/Firmin, *Epistola* 3.4 (ed. Schabel 1996: 207).
105 Jean/Firmin, *Epistola* 3.3 (ed. Schabel 1996: 206–7).
106 Jean/Firmin, *Epistola* 3.4 (ed. Schabel 1996: 208).
107 Jean/Firmin, *Epistola* 3.5 (ed. Schabel 1996: 208–9).

which the feast was celebrated in the second, rather than the first, month of spring. Jean and Firmin were fully cognizant of this dilemma and addressed it in the fourth and final part of their treatise, in which they suggested that the Church might simply alter the existing rules and let the earliest Easter Sunday fall on the day after the currently observed equinox, which would have potentially put it on 13 March. As they themselves realized, however, this gambit had a significant downside, given that:

> all mobile feasts would leave their accustomed boundary-dates, such that festivities that have never taken place within Lent would end up [in this period], and ones that have only ever taken place in Lent would be found outside it. And this way it would be necessary to alter the divine service to an extraordinary degree. And since the equinox would not remain fixed where it is, it would be necessary to make this correction many more times, and this way a great disturbance would follow in the Church, which one should seek to avoid.[108]

Sufficiently alarmed by this prospect, the two astronomers proposed a pragmatic alternative, which was to conceive of the astronomical criteria for Easter as a convention that merely reflected the situation in late antiquity, when the present set of rules had been first introduced:

> The second mode is to continue doing what we are used to do, [staying] within the accustomed boundary-dates, the only exception being that we operate with a corrected new-moon [date] the way we have operated with the Golden Number, while caring neither about the Passover of the Jews nor about the true place of the equinoxes. And then we would be able to interpret the pronouncements of the saints in the following way: when they speak of the first month or the vernal equinox that [something happens] 'from the 14th day of the Moon of the first month until the 21st', this refers to the 14th day of the Moon of those days of our year that used to fall within the first month of the Hebrews at the time when these rules were put into place. Or, whenever it is said '[after] the vernal equinox etc.', this is after the day of our year on which the equinox fell at the aforementioned time, which was 21 March. This way, all the rules concerning Easter would remain true and infallible from now on.[109]

Jean and Firmin ended the main part of their treatise at this point, leaving it up to the pope to decide whether he wanted to steer his Church towards the Scylla of altering the breviaries or the Charybdis of divorcing Easter from the vernal equinox, to which it had been tied by several layers of authoritative texts. The unsavoury nature of either of these options probably goes a long way towards explaining why no actual reform ever issued forth from Avignon, whether in 1345 or any of the following years. In addition to the unresolved practical issues the submitted reform proposals had raised, one should not underestimate the devastating influence of the Black Death, which swept over Provence and the rest of Europe in 1347 and 1348 and may have kept both Jean des Murs and Firmin de Beauval

[108] Jean/Firmin, *Epistola* 4.1 (ed. Schabel 1996: 210).
[109] Jean/Firmin, *Epistola* 4.1 (ed. Schabel 1996: 210).

from making any further contributions to calendar improvement.[110] Apart from the pope himself, the only other participant of Clement VI's reform project known to have survived the plague years was Johannes de Termis, who in 1354 made an ineffective attempt to revive the debate by submitting a *Tractatus de tempore celebrationis paschalis* to Clement's successor, Innocent VI.[111]

Dependent in part on arguments and quotations first presented in his *Expositio kalendarii novi*, Johannes's *Tractatus* directed its focus towards the theological and exegetical foundations of the date of Easter, but added little that could be used to bring the calendar problem to a satisfactory conclusion. On a technical level, Johannes maintained his preference for mean over true equinoxes and solstices, claiming that the vernal equinox had been displaced by three days, from 25 March to 22 March, between Christ's incarnation and the first 'reform' of Easter reckoning at the Council of Nicaea. Nowadays it resided on 14 March, calling for a long-overdue adjustment of the paschal *termini* to the new situation. In the *Expositio*, the Nicene example had still served to bolster the idea of restoring the equinoxes and solstices to their former dates, but the criticism this proposal had received in 1345 motivated the Benedictine monk to reconsider his stance in the meantime. The colophon to his *Tractatus* promised 'a brief, clear, and easy way of correcting' the reckoning of Easter 'without any change to the old calendar, but rather with its perpetuation'.[112] In line with these goals, Johannes showed himself content with nothing more than periodic shifts of the Church's mobile feast days and boundary-dates to match the continuously drifting date of the mean vernal equinox. What this proposal left entirely unaddressed was how such adjustments could be tolerated in the light of the changes to the existing breviaries and the reshuffled order between mobile and immobile feast days that would have followed from it. The lack of any recorded response to Johannes de Termis's second intervention confirms the impression that the reformist spirit that had prevailed at the court of Clement VI in 1344/5 had almost entirely vanished only ten years later.

[110] See Gushee 1998: 362–3; Poulle 2005: 9. None of Jean des Murs's works can be securely dated to later than 1345, notwithstanding the fact that MS London, BL, Sloane 3124, fols. 2r–8r, contains a calendar of conjunctions from 1356 to 1374 that was allegedly cast in 1346/7 by Jean des Murs and other astrologers (fol. 8r: '[E]t fuit composita et ordinata Parisius per magistrum Johannem de Muris et plures alios magistros in astrologia expertos ad mendatum [*sic*] domini Clementis pape sexti, anno quinto'). Attention to this calendar was first drawn by Thorndike 1923–58: iii. 317–18, who was doubtful about its attribution to Jean, and rightly so given the decree of incompetence on display. The calendar gives conjunction times in a format familiar from the Jewish *molad*-system, with hours and 'points' (1h = 1,080p), but erroneously assumes their cyclical repetition after 19 years. The same 19-year cycle of pseudo-*moladot* survives in MS Vatican City, BAV, lat. 696, fols. 136r–142v (*c.*1410), where the canon mentioning Jean des Murs is translated into Castilian, and in MS Bamberg, Staatsbibliothek, R. B. Msc. 176, fols. 6v–12r, 14v (*c.*1392), where it appears in a column of 'lunationes secundum equactionem Judaycam' as part of the *Kalendarium* of Peter Nightingale. It also appears in a Book of Hours from Modena, discussed by Neugebauer 1989: 400–2.

[111] See MS Vienna, ÖNB, 5292 (s. XVI$^{1/4}$), fols. 231r–242v, which was the exemplar for MS Vienna, ÖNB, 5273, fols. 122v–138v (copied in 1527), and the summaries in Kaltenbrunner 1876: 322–3; Nothaft 2015g: 291–6.

[112] MS Vienna, ÖNB, 5292, fol. 242v: 'Explicit Tractatus de tempore celebrationis paschalis . . . qui modum omnia et singula premissa corrigendi brevem, clarum et facilem et absque calendarii immutatione antiqui, ymmo et cum perpetuatione ipsius ubi quibus et quotiens oportunum fuerit paratus est demonstrare.'

7

Church Councils and the Question of Easter in the Fifteenth Century

7.1. PIERRE D'AILLY AND THE *VIA CONCILII*

For the entire second half of the fourteenth century the issue of calendar reform continued to lie dormant and unattended by the papacy and the higher clergy, quite as if the failed reform of 1344/5, together with the pressing crises that swept over Europe—from the Black Death to the Hundred Years War—had dissuaded ecclesiastical authorities from trying to deal with the problem. A further obstacle to progress in this area was created by the turmoil surrounding the papal election of 1378, which had the unprecedented outcome of the same collegium of cardinals creating two different popes, one Italian (Urban VI), the other French (Clement VII). The resulting Great Western Schism of 1378–1417 paralysed the papacy's ability to pursue another initiative of the kind Clement VI had launched in 1344/5. As long as the kingdoms, territories, bishoprics, and universities of Western Europe were torn between opposing papal obediences, one centred in Rome, the other in Avignon, an actual reform of the Church's calendar was even less of a political possibility than it had been during previous centuries.[1] The loss of practical authority the papacy inevitably suffered as a result of this split benefited certain avenues of ecclesiological thought in which the sovereignty of the universal Church was placed above that of the pope as its solitary head. Followers of such theories upheld 'that under certain circumstances the general council representing the Church—acting even apart from or in opposition to the pope—could exercise a jurisdictional or governmental authority superior to his, and, by so doing, impose constitutional limits on the exercise of his prerogatives or serve as a control function to prevent their abuse'.[2] An awareness that conciliarism might offer the solution not just to the schism but to the still-unattended calendar problem may be detected in some remarks attached to the widely diffused German calendar of 1404, whose author endorsed the traditional computistical view according to which $J = 120$ years, causing Christmas to migrate from winter to summer in 21,930 years:

> But if we wished to overcome these aberrations, we would have to remove and omit a leap year once every 120 years. This way, the feasts would remain in the year unchanged.

[1] For an overview of the history of the Great Schism, see the essays in Rollo-Koster and Izbicki 2009.
[2] Oakley 2003: 15. See also Black 1988.

> Yet this is prohibited, unless it happens with the common counsel of the clergy and all of Christendom. And this is the reason why we have to endure such errors.[3]

A less ambiguous expression of the idea that a general council was the right place to address the problem can be detected in the writings of the French theologian Pierre d'Ailly (1350–1420), whose early career steps included a six-year run as chancellor of the University of Paris (1389–95).[4] D'Ailly's open zeal for ecclesiastical reform and his determination to solve the schism by the *via concilii* did not keep him from temporarily aligning himself with the French Pope Benedict XII (1394–1417), who made him bishop of Le Puy-en-Velay (1395) and bishop of Cambrai (1397–1411), as a result of which he rose to the position of an imperial prince. Originally loyal to Benedict during the French withdrawal of obedience from 1398 to 1403, d'Ailly finally broke with the Avignonese pope in 1408/9 and rejoiced when a large group of dissident cardinals from both sides of the schism convened for the Council of Pisa, which opened in March 1409. Failing to secure the cooperation of Benedict XII and his Italian opponent Gregory XII (1406–15), the prelates assembled in Pisa declared the existing claimants to be obdurate heretics and elected in their stead a new pope, Alexander V (June 1409–May 1410). Since neither of the two schismatic popes recognized this decision, the Pisan intervention meant that Western Christendom was now faced with the choice between three conflicting obediences, one based in Rome, one in Avignon, and a third in Pisa.[5] When the newly created pope died after less than a year, the cardinals of the Pisan obedience chose as his successor the former *condottiere* Baldassare Cossa, who become known as Pope John XXIII (1410–15). On 29 April 1411, John XXIII made good on one of his predecessor's promises by convoking a general council that was meant to open in Rome in the forthcoming year 1412. Encouraged by this prospect as well as by his own elevation to the rank of cardinal (on 6 June 1411), d'Ailly presented John XXIII with a recently composed *Exhortatio ad concilium generale super kalendarii correctione* in six chapters, which argued for the necessity of correcting the calendar as part of the imminent reform and reunion of the Latin Church.[6] The treatise, which had been written in the early months of 1411 for future use at the general

[3] MS Tübingen, Universitätsbibliothek, Md 2, fol. 31rb–va: 'Wolten aber wir der jrronge uber werden so mustent wir ye in 120 jaren eyn schaltjare hant abe tun und under wegen lassen. So blieben die hochgezite unferwandelt in dem Jare. Aber das ist verboten, es geschee dan mit gemeynem rade der Paffheit und der Cristenheit. Und darombe so müszen wir semlich yrronge liden.' Köll 1983: 48–9 quotes the same passage from MS Innsbruck, Universitätsbibliothek, 750, fols. 10vb–11rb, which is a calendar manuscript from *c.*1440 owned by the Cistercian monastery of Stams. Müller 2010: 266–73 treats this passage in the *Calendarium Stamsense* as an original contribution by its supposed author Vitus de Augusta rather than the copy of a much earlier work (*c.*1404; see n. 38 in Ch. 5.1). The more recent piece by Bohnert and Müller 2012 fails to correct this gaffe.

[4] On d'Ailly's life and career, see Salembier 1932; Guenée 1991: 102–258; Millet and Maillard-Luypaert 2015. On his political and reformist thought, see Oakley 1964; Pascoe 2005.

[5] On the historical background and the decrees of the Pisan Council, see Girgensohn 2007.

[6] A critical edition with French translation appears in Solan 2016: 200–29. For previous discussions of this text, see Kaltenbrunner 1876: 326–36; Duhem 1913–59: iv. 174–82; Tschackert 1877: 331–3; Stegemann 1955: xxxi–xxxiv; North 1983: 87–8; Gumbert-Hepp 1987: 20–1; Sudmann 2005: 264–5; Müller 2010: 153–6; Steinmetz 2011: 56–7; Solan 2016: 181–91. The recent publications by Lebigue 2014: 372–8, and Ribordy 2017, must be used with caution.

council, began with a passionate round of applause for the diligence of past Church authorities, who had invested time and effort into properly ordering the ecclesiastical calendar and its mobile feast days. What made men like the second-century Pope Victor I, as well as more recent writers such as John of Sacrobosco and Robert Grosseteste, praiseworthy was the way they devoted themselves 'to calculating days and moments more so than to counting denars and money'.[7] In doing so, they offered an embarrassing contrast to the inertia of today's churchmen, who neglected their duties and allowed the celebration of Easter and other feasts to stray from their proper dates. Determined to fight this scandal, d'Ailly felt encouraged to put together a compilation based on previous works on the same topic, depicting himself as the hunting-dog whose barking was supposed to spur the universal assembly of bishops into action—to chase off 'the scandalous error and the erroneous scandal' and to completely extirpate 'the wicked beast and horrible monster' from the Church's vineyard. The hounds were released not a day too early:

> For 'Behold, now is the acceptable time! Behold, now is the day of salvation!' [2 Corinthians 6:2]. Clearly, [it is] a day highly useful for the salvation of the Church and acceptable to God and all the Christian populace, on which according to the decree of the last council, [held in] Pisa, a universal synod must be celebrated. This [synod] was specially ordained to reform this Church, which is now—Oh pity!—so extraordinarily and miserably deformed.[8]

Most of the intellectual groundwork for the chapters that follow these exclamations appears to have been laid during the year 1410, which had seen d'Ailly withdraw from the turmoil of domestic and ecclesiastical politics to instead focus on his literary activities.[9] One source he studied intensively during this retreat to Cambrai was Book IV of Roger Bacon's *Opus maius*, whose calendrical appeal to Pope Clement IV provided him with most of the facts and arguments he deployed in the *Exhortatio*. As a result, the English Franciscan's defence of the legality of a calendar reform, his various cries of indignation at the scandalous state of the calendar, his warnings against the ridicule of unbelievers, and his praise for the astronomical abilities of the ancient Hebrews are all echoed at some length in d'Ailly's own treatise, about 35 per cent of which paraphrases or reproduces passages in the *Opus maius*.[10] Calendars aside, another theme in Bacon's work that visibly impressed the French churchman was the reliability, scope, and legitimacy of astrology, in particular the prediction of political and religious developments on the basis of planetary conjunctions. This kind of historical astrology, which had been carried to the Latin West by the works of Arabic writers such as Abū Maʿshar (Albumazar), became an important preoccupation for Pierre d'Ailly during the final decade of his life (1410–20). From this period dates a series of widely read treatises in which the cardinal defended the reconcilability and mutual harmony between astrology,

[7] D'Ailly, *Exhortatio*, prologue (ed. Solan 2016: 202).
[8] Ibid. (ed. Solan 2016: 204). [9] Guenée 1991: 224–32.
[10] See Solan 2016: 182–6, and the source apparatus on pp. 633–5. The dependence on Bacon was already demonstrated by Kaltenbrunner 1876: 329–34. For an assessment of Bacon's influence on d'Ailly, see Smoller 1994: 54–60.

history, and Christian theology, as expressed in work titles such as *Vigintiloquium de concordia astronomice veritatis cum theologia* and *Tractatus de concordantia astronomice veritatis et narrationis hystorice* (both from 1414). Roger Bacon's influence is most clearly apparent in d'Ailly's *De legibus et sectis contra superstitiosos astronomos*, finished in December of 1410, which expounded the Franciscan scholar's doctrine according to which the rise of six major religions over the course of history was connected to conjunctions of Jupiter with each of the other six planets. The ability of astrologers to predict religious change held a special interest in light of the ongoing schism within Western Christianity, which in the eyes of many, d'Ailly included, indicated that the Church might have entered the end times. Bacon's and Abū Maʿshar's historical astrology promised a way of assuaging such anxieties. In d'Ailly's case, it showed that a political solution to the schism could be sought, as the advent of the Antichrist was still centuries away (d'Ailly eventually dated it to the distant year 1789).[11]

The derivative—some might say plagiarizing—nature of *De legibus et sectis* extends from the parts inspired by Roger Bacon to the discussion of astronomical parameters in chapters 8–10, which conclude the treatise as a whole. Large parts of the wording in these chapters were borrowed, with little modification, from a *Tractatus de commensurabilitate et incommensurabilitate motuum celi* by the Parisian mathematician and philosopher Nicole Oresme (d. 1382), who had been grand master of the Collège de Navarre from 1356 to *c.*1362, only a few years before Pierre d'Ailly himself enrolled there, in *c.*1365.[12] Oresme's *Tractatus* was a sophisticated mathematical and kinematic exploration of the consequences of circular motions that were either commensurable or incommensurable in relation to each other. In the third and final book, which was the only one utilized by d'Ailly, he presented his readers with a fictional debate between the personifications of Arithmetic and Geometry, who took opposing positions on whether the motions of the celestial spheres were commensurable (Arithmetic), meaning that specific conjunctions and constellations were bound to recur with the passage of time, or incommensurable (Geometry), meaning that they happened only once (or never). Although Oresme freely admitted that no definitive proof could be given either way, he argued in favour of incommensurability on probabilistic grounds, stating that, 'for any proposed set of unknown magnitudes, it is more probable that they are incommensurable rather than commensurable, just as it is more probable for any proposed unknown magnitude that it is a non-perfect rather than a perfect number'.[13] For practising astrologers, such scepticism about the cyclical nature of celestial motions posed a serious problem, as it threatened to undermine the very

[11] Thorndike 1923–58: iv. 101–13; Smoller 1994. On the wider context, see North 1980; Boudet 1995.

[12] Pierre d'Ailly, 'De legibus et sectis contra superstitiosos astronomos' (8–10), in *Tractatus de ymagine mundi*, etc. (Leuven: Johannes de Westphalia, 1480/3), sigs. f6v–g4v. The borrowings from Oresme are listed in Grant 1971: 131 (n. 121). See also Caroti 1979: 629–43.

[13] Nicole Oresme, *De commensurabilitate* 3.459–63 (ed. Grant 1971: 320). Oresme had already argued for this conclusion in his earlier treatise *De proportionibus proportionum* 3.10 (ed. Grant 1966: 246–55). See also Small 1991: 122–7.

notion that celestial influences were predictable on account of their continuous recurrence. The corrosive consequences of celestial incommensurability had already been noticed by the Parisian author responsible for the *Expositio tabularum Alfonsii* of 1347, who protested that only perfect commensurability was congruent with the dignity and perfection of the celestial bodies and treated it as a necessary prerequisite for a harmonic musical relation between the spheres.[14] Nicole Oresme, for his part, was a noted opponent of judicial astrology and used his mathematical conclusions to undermine the validity of astrological concepts such as the Great Year championed in ancient sources.[15] This position comes to the fore with particular clarity in the imaginary orations that appear at the end of Oresme's *Tractatus de commensurabilitate*. Eager to safeguard the orderliness of the heavens, Arithmetic protests that unless the motions of the celestial spheres are truly commensurable:

> no one could ever have foreknowledge of aspects, predict conjunctions, or foresee [their] effects. Instead, astronomy will lie forever hidden, be unknown and even unknowable... Also, the celestial bodies would never return to a state similar to the disposition in which they are now after the completion of that Great Year of all circular [motions], whose existence has been affirmed by many philosophers.[16]

Arithmetic's opponent Geometry shows herself unimpressed by these objections and instead underlines the inherent limits of man's ability to determine astronomical parameters down to the very last second, inch, or fraction. To attain such exactitude was not even desirable, because:

> if men knew all motions with absolute precision, it would no longer be necessary to make further observations or note the revolutions of the heavens with watchful attention. Accordingly, it was better that of things as excellent as these something would be known and something would always remain unknown and open to further enquiry, so it would draw noble minds, who have tasted its sweetness in advance, away from earthly things and perpetually keep them occupied, with aroused desire, in the venerable exercise of such high-minded an endeavour.[17]

In another salient passage Oresme has Apollo, the judge in the dispute between Arithmetic and Geometry, declare that the conflict between the two parties is beyond full resolution. Since whether or not a ratio of motions is commensurable can depend on tiny, and even on wholly imperceptible, differences, the senses are not sufficient to provide a definitive answer to the question, leading Apollo to sneer at the 'presumptuous individuals who boast that they have passed on a perpetual almanac and a true calendar, when it is perhaps not even possible to find such a thing. And what is surely impossible is for them to know that they have found

[14] See Nothaft 2015b: 78–81.

[15] See, in particular, Oresme's treatise *Ad pauca respicientes* (ed. Grant 1966: 61–5, 382–429), and the discussions in Caroti 1979: 584–7; Callataÿ 1996: 189–204.

[16] Nicole Oresme, *De commensurabilitate* 3.263–5, 275–8 (ed. Grant 1971: 304–6). This passage is paraphrased in d'Ailly, 'De legibus et sectis' (8), sig. f8ra–v.

[17] Nicole Oresme, *De commensurabilitate* 3.442–8 (ed. Grant 1971: 318–20). The second sentence appears in d'Ailly, 'De legibus et sectis' (10), sig. g4v.

anything of this kind.'[18] Pierre d'Ailly, who drew upon these passages in his *De legibus et sectis*, realized that he could use Oresme's stance to validate his own position, according to which very specific astrological predictions had a tendency to fail, not because of a lack of any causal connection between planetary conjunctions and processes in the sub-lunar sphere, but because of the difficulty involved in gaining precise knowledge of heavenly motions, which forced practitioners to work with rough approximations and flawed computations. At the same time, he expressly connected the issue of commensurability and the limitations imposed on the human senses to the uncertainties surrounding the length of the solar year, which had long been troubling potential calendar reformers. A step in this direction had already been taken by Oresme himself, who in Book II of his *Tractatus* pointed out that the Sun's diurnal and annual revolutions might be mutually incommensurable, 'in which case it is impossible to discover a precise number for the length of the year or put together a perpetual almanac or find a true calendar'.[19] In seizing upon this line of criticism, Pierre d'Ailly appealed to the wisdom of Robert Grosseteste, whose discussion of the solar year in the opening chapter of his *Compotus* suggested very strongly that, in spite of the best efforts of the ancient Greeks and Arabs, no certainty had been reached in these matters. Next to the Ptolemaic $J = 300\text{y}$ and the value supposedly found by al-Battānī, which d'Ailly (here following Jean des Murs) gave as $J = 112\text{y}$, more recent estimates of the error inherent in the Julian calendar included $J = 120\text{y}$ (the computists), $J = 125\text{y}$ (Roger Bacon), as well as the 11 minutes per year used in William of Saint-Cloud's *Kalendarium regine*, which implied $J = 131\text{y}$. In addition, there was the widely accepted parameter of King Alfonso, according to whom 134 years were enough to cause the equinoxes and solstices to recede by 23;58,18,36,51h or almost an entire day.[20] Popular as the Alfonsine Tables still were, however, some continued to object to their treating the tropical year as invariable and permanent, which was in violation of the theory of access and recess associated with Thābit, a version of which featured in Alfonso's own precession model.[21]

The complications wrought by the solar year were a theme the learned cardinal was to return to in his *Elucidarium astronomice concordie cum theologica et hystorica veritate*, which he wrote in September 1414. D'Ailly here cited at length from the tenth chapter of Campanus of Novara's *Compotus maior* (see Ch. 4.2), which offered a complex discussion of the shift of the equinoxes and solstices and the

[18] Nicole Oresme, *De commensurabilitate* 3.29–31 (ed. Grant 1971: 286).

[19] Ibid. 2.389–93 (ed. Grant 1971: 278). Oresme makes the same point in his *Ad pauca respicientes* 2.16 (ed. Grant 1966: 420–3), and in the *Questiones super geometriam Euclidis* 9 (ed. Busard 2010: 132), where Oresme also addresses the possible incommensurability of lunar and solar revolutions.

[20] See d'Ailly, 'De legibus et sectis' (10), sigs. g2v–3v, large parts of which are taken from Robert Grosseteste, *Compotus correctorius* 1 (ed. Steele 1926: 213–17). Note that d'Ailly here corrects Grosseteste's simplified value for the solar year according to al-Battānī ($365\frac{1}{4}-\frac{1}{100}$ days) by replacing it with the 365d 5;47,9h or $365\frac{1}{4}-\frac{1}{112}$ days used by Jean des Murs and Firmin of Beauval in their *Epistola* to Clement VI (ed. Schabel 1996: 197), and in d'Ailly, *Exhortatio* 2 (ed. Solan 2016: 206).

[21] D'Ailly, 'De legibus et sectis' (10), sig. g3v. Cf. the corresponding complaint in the *Expositio tabularum Alfonsii* of 1347, discussed in Nothaft 2015b: 85.

changes in the length of the tropical year predicted by pseudo-Thābit's theory.[22] As before, d'Ailly pointed out that some experts were sceptical of the tropical year length used in the Alfonsine Tables, not least because the doctrine of a fixed tropical year was contrary to the findings of astronomers in different centuries. He confirmed this point by excerpting bits from the Parisian *Expositio tabularum Alfonsii* of 1347, whose author had noted discrepancies between the Alfonsine Tables and equinoxes observed in 1290 and 1346.[23] A further testimony to d'Ailly's extensive reading in Parisian astronomy is the way he followed this up with a reference to Jean des Murs's canons on the Alfonsine Tables, which the latter had written at the Collège de Sorbonne in 1339. According to Jean, the precise length of the solar year in these tables, as derived from the daily rate of motion, was 365d 5;49,15,58,49,46h, but even this number could only count as an average, since the actual length varied from year to year as a result of the unsteady motion of the Sun's apogee.[24]

The relevance of all these different data-points to Pierre d'Ailly's reform project becomes clearly manifest from a look into his *Exhortatio ad concilium generale super kalendarii correctione*, the third chapter of which updated Roger Bacon's thirteenth-century discussion of the regression of the equinoxes and solstices by confronting it with the more recent Alfonsine Tables.[25] When it came to the error of the 19-year cycle, d'Ailly was content with accepting the rough estimate of $M = 304$ years that had been used by previous authors such as Bacon and Grosseteste. If one assumed that the Golden Numbers had already been placed in the calendar by Julius Caesar, whose reform d'Ailly dated to 48/7 BC, it followed from this continuous slippage that the gap between calendrical and astronomical new moons was going to reach five days by AD $1473 (1{,}473 + 47 = 1{,}520 = 5 \times 304)$.[26] A number of possible reactions suggested themselves, the most obvious of which was to move the Golden Numbers towards the currently valid dates of the new moons or *primaciones*. The overestimation of the mean lunation by the 19-year cycle made it necessary after 304 years either to move them upwards by another day or to change the definition of the Golden Numbers and treat them as signifying the second day of the Moon.[27] Another possible alternative was to omit a bissextile day every 304 years, but Pierre d'Ailly was well aware that this solution clashed with the simultaneous need to forestall the drift of the equinoxes and solstices, thus revealing a basic dilemma

[22] D'Ailly, 'Elucidarium astronomice concordie cum theologica et hystorica veritate' (35–8), in *Tractatus de ymagine mundi* etc. (n. 12 above), sigs. gg3r–5v. The copies of the *Elucidarium* in MSS London, BL, Harley 3742, fols. 203v–204v; London, BL, Add. 29969, fol. 57v–58v, add a diagram of pseudo-Thābit's trepidation model with an explanation.

[23] See d'Ailly, 'Elucidarium', sigs. gg4v–5r, where he tacitly quotes the *Expositio tabularum Alfonsii*, as found in MS Berlin, Staatsbibliothek, lat. fol. 192, fol. 213ra. See also Duhem 1913–59: iv. 172–3; Nothaft 2015b: 86.

[24] See the 'Sermo de quantitate anni' contained in Jean des Murs's *Canones tabularum Alfoncii* (extant in MSS Oxford, Hertford College, 4, fols. 143v–144r; Oxford, BodLib, Digby 97, fol. 122r), as well as the corresponding reference in d'Ailly, 'Elucidarium', sig. gg5r.

[25] D'Ailly, *Exhortatio* 3 (ed. Solan 2016: 208–14).

[26] Ibid. 2, 4 (ed. Solan 2016: 206–8, 214–18).

[27] Ibid. 4 (ed. Solan 2016: 218).

every aspiring calendar reformer had to face. According to the values d'Ailly himself accepted, $J \approx 134$ years and $M \approx 304$ years, the vernal equinox drifted through the Julian calendar about 2.3 times as swiftly as the concomitant slippage of the new moons. It followed that an omission of the bissextile day adequate to M would not have been enough to stop the vernal equinox from drifting towards the beginning of the year. Conversely, a reformed intercalation pattern that was sufficient to fix the vernal equinox would have overcompensated for the deficiency of the 19-year cycle, to the extent that the new moons would again have started to slip away from the dates assigned to them, but in the opposite direction, from earlier towards later dates.[28]

A well-established way of avoiding this dilemma was to abolish the Golden Numbers altogether and in their stead to calculate the date of Easter with the aid of astronomical tables, or by using the lunar calendars of the Arabs or Hebrews, as had already been suggested by Robert Grosseteste and Roger Bacon. D'Ailly recycled passages from both these authors and duly highlighted the greater reliability and precision such an approach promised, but not without also noting that it was going to look 'more alien to the common use and the custom of the Church. For it destroys the Golden Number, which has been in use for so long, and introduces new and unaccustomed tables.'[29] One advantage of a purely astronomical solution was its potential of circumventing some of the problems caused by the apparent variability of the tropical year and the difficulties of establishing its actual length. Rather than trying to determine a single value for this parameter, one could simply locate the changing position of the vernal equinox on a year-to-year basis, using 'instruments or accurate astronomical tables'.[30] Without urging the adoption of any particular option, the bishop of Cambrai ended his treatise by insisting that the Church should at the very least adjust the calendrical equinoxes and solstices to their actual position and determine the paschal *termini* with the aid of a corrected sequence of Golden Numbers. Over time, further adjustments to these new-moon dates could be made to prevent the present error from arising again, thus shielding the Church 'against the derision of the infidels, and especially of the Jews'.[31] Imprecise as these suggestions were, there are signs that the *Exhortatio* was intended to be no more than the starting point for a more refined investigation. In his dedicatory letter to John XXIII, Pierre d'Ailly expressed his hope that the pope was going to commission experienced astronomers for the task of examining his treatise and the suggestions found therein. Once these astronomers had worked out the pertinent details, the next step was to have the matter examined by the prelates assembled at the approaching Council of Rome, who could then join the pope in sanctioning a

[28] Ibid. 6 (ed. Solan 2016: 224–6).

[29] Ibid. (ed. Solan 2016: 226). Pierre d'Ailly shed more light on the option of using the Arabic 30-year lunar cycle in his treatise *De vero cyclo lunari* (1411), which serves as an appendix to the *Exhortatio* in the 'master copy' of the text (see n. 42 in this chapter) and in the printed anthology *Tractatus de ymagine mundi* etc. (n. 12 in this chapter), sigs. h2v–5v. See Nothaft (forthcoming-2), for an edition and discussion of this text.

[30] D'Ailly, *Exhortatio* 6 (ed. Solan 2016: 226), quoting Robert Grosseteste, *Compotus correctorius* 10 (ed. Steele 1926: 259, ll. 12–15).

[31] D'Ailly, *Exhortatio* 6 (ed. Solan 2016: 228).

calendar reform, which was meant to take effect only once unity within the Church had been restored.[32] Eager to facilitate the steps towards this goal, d'Ailly and some of his supporters went on to petition John XXIII with the draft text for a papal decree, which appears as an appendix to his *Exhortatio* in about half of the preserved manuscripts.[33] It shows that d'Ailly wished for John XXIII to sign off on the following declaration:

> As is incumbent upon our office, we have convened a universal synod for the reformation of the Church. With its approbation, we have decided to reform that which is relevant to the divine worship as well as several other things. This is why we have entrusted our beloved sons, etc., diligently to concentrate on the examination of the aforementioned errors and their correction with the counsel of experienced astronomers, and to administer [the necessary steps] to bring the Roman calendar in line with the truth. Yet since this undertaking needs a great amount of time for discussion and will not come into common use without difficulty, and in order to have the error in determining Easter corrected more quickly and more easily, we have ordained, after due deliberation with experts and the holy council, that the following mode should be observed, namely that the vernal equinox and the new moons, which change their locations in the calendar over time by ascending continuously, are brought to their correct positions.[34]

The portion of the decree text that follows gauges the Julian solar drift at $J = 134$ and the lunar slippage at $M = 304$, claiming that the vernal equinox had reached 12 March already some time before the present year AD 1412.[35] As a means of counteracting this error, the text describes the same simplified solution that d'Ailly had already adumbrated at the end of the *Exhortatio ad concilium generale* and spelled out in more detail in the guise of two wheel diagrams attached to a companion treatise, *De vero cyclo lunari* (1411).[36] This solution amounted to a new list of Easter full-moon dates or *termini paschales*, which exhibited an upwards shift of three days compared to the lunar ages in the traditional 19-year cycle while taking 12 March instead of 21 March as the new equinoctial limit. The new rules placed the earliest possible *terminus* on 13 March, in year 11 of the 19-year cycle (in which the *terminus* used to occur on 15 April), and the latest possible such date on 10 April in year 3 (down from 13 April).[37] After a brief remark suggesting the construction

[32] D'Ailly, *Epistola domino Johanni pape super correctione kalendarii* (ed. Solan 2016: 200). See also d'Ailly, *Exhortatio* 6 (ed. Solan 2016: 224): 'Licet autem superius aliqualiter apertus sit correctionis modus, tamen plenior et precisior hujus rei certitudo debet per astronomos declarari.'

[33] See Solan 2016: 16, 187–90, who edits the decree and the preceding supplication on pp. 230–5. The supplication was also printed in Finke 1896–1928: i. 164–5 (no. 39).

[34] See the Latin and French text in Solan 2016: 230–1.

[35] The chosen equinoctial date on 12 March was correct, both according to Alfonsine and modern parameters, for the years 1409–11 and 1413–15, but not for the present year 1412, which was bissextile and had the equinox on 11 March.

[36] See *Tractatus de ymagine mundi*, etc. (n. 12 in this chapter), sig. h6r, and the 'master copy' in MS Cambrai, Bibliothèque municipale, 927, fol. 105v.

[37] See Nothaft 2015h: 122–3, and the tables in Solan 2016: 234–5. Note that the suggested shift of paschal full moons by three days implies a more conservative estimate of the total slippage since antiquity than the close to five days that d'Ailly had alleged in c. 4 of his *Exhortatio* (ed. Solan 2016: 218). It is possible that d'Ailly's decision of a 3-day shift was inspired by the lunar ages obtained from

of revised Easter tables 'for the instruction of the simpleminded' (*ad instructionem simplicium*), the final portion of the decree postpones the implementation of the reform thus outlined 'until the current schism has been put to rest and God has given his Church the union that we strive for with all our might'.[38]

As with previous attempts to get the papacy involved in a calendar-reform project, nothing is known about John XXIII's reaction to d'Ailly's proposal or whether the reform of the calendar was even brought up at the Council of Rome (14 April 1412–3 March 1413), which was poorly attended and remained without any lasting impact. Faced with this disappointing outcome, d'Ailly's hopes for a solution to the calendar problem would have naturally focused on the next general council, which met in Constance in 1414. The Council of Constance quickly became the greatest ecclesiastical gathering of the Middle Ages, and generated high hopes that it would not just end the schism but satisfy the urgent demands for a thorough reform of the entire Church *in capite et membris*, in addition to quenching the new heresies associated John Wycliffe and Jan Hus. Its high points included the formal deposition of John XXIII (29 May 1415), who had fled town two months earlier, the abdication of the Roman pope Gregory XII (4 July 1415), and, finally, the deposition of the Avignonese pope Benedict XIII (26 July 1417). Between 9 and 11 November 1417 an electoral college of cardinals and deputies gathered in conclave to decide upon a new pope, who took the name Martin V (1417–31). In the months prior to this election, the council's fathers had already agreed on the convocation of future councils, as stipulated in the decree *Frequens* (9 October 1417), which were hoped to balance out papal claims to supremacy and further the cause of ecclesiastical reform.[39]

Pierre d'Ailly came to play a key role in many of the council's proceedings, yet the evidence to suggest that his calendrical ideas came to the attention of the other prelates assembled in Constance is remarkably thin. All we have is the colophon added to one of the many manuscript copies of the *Exhortatio*, which claims the text was 'written in Constance, at the holy council, in the year 1417, in the month of March, and recited at [the church of] St Paul in the said town'.[40] More than a year prior to this alleged public lecture, the issue of calendar reform had already been mentioned in a memorandum drawn up by Jacobus Arrigoni, the bishop of Lodi, which is dated 16 February 1416. The text correctly predicted that several burning issues were not going to receive sufficient treatment during the present council and were thus best deferred to a future one. Among these matters, Bishop Arrigoni counted the union with the Greek Church, the correction of heresies, the correction of the calendar, 'in which there are many notorious and scandalous errors', and even 'the verification of the [number of] years from the beginning of the world'. He considered it expedient to entrust these subject matters to individual cardinals, who were supposed to consult with other prelates and doctors and then

the 30-year cycle of the Arabs, which is the topic of his treatise *De vero cyclo lunari* (see n. 29 in this chapter).

[38] Solan 2016: 232.

[39] See Stump 2009, with references to earlier literature.

[40] MS Vienna, ÖNB, 5113, fol. 214v: 'Scriptum Constantia in sacro concilio anno 1417, mense martii, etc. et pronunciatum ad Sanctum Paulum in dicta civitate.' See Solan 2016: 30.

report their findings to the next general council. Ideally, the Council of Constance was also going to contact various universities, 'whether in the form of open letters or through the delegates of these universities who are present here', to ask them to nominate *commissarii* charged with preparing reports on the individual issues as well as to collect the relevant literature, or to solicit the production of new texts.[41] None of these steps appears to have been taken at the time and the contributions made by Europe's universities to the medieval calendar-reform debate remained on the whole very meagre.

Even though it ultimately failed to lift the calendar onto the agenda of the Councils of Rome and Constance, Pierre d'Ailly's *Exhortatio ad concilium generale super kalendarii correctione* of 1411 became a canonical entry in the medieval literature on calendar improvement, owing in particular to its frequent inclusion in larger anthologies of the cardinal's 'scientific' (i.e. cosmographical, astrological, and calendrical) writings, which enjoyed considerable popularity during the fifteenth and sixteenth centuries.[42] At some point between 1480 and 1483, the German printer Johannes de Westphalia, who was based in Leuven, issued a book containing a dozen or so of the cardinal's scientifica, making the *Exhortatio* and its associated texts (i.e. the treatise *De vero cyclo lunari* and the decree of 1412) the first work(s) on calendar reform ever to appear in print. This collection is now chiefly remembered for its flagship text, the cosmographical compendium *Ymago mundi*, which d'Ailly had completed half a year prior to the *Exhortatio*, on 12 August 1410. In the eighth chapter of this work, d'Ailly reproduced at length Roger Bacon's discussion of the circumference and habitability of the Earth in the *Opus maius*, and in doing so made the Franciscan's views available to a wider audience.[43] Among the owners of Westphalia's famous incunabulum was the Genoese mariner and merchant Christopher Columbus, who adorned his copy with nearly 900 marginal annotations, more than 54 of which pertain to the pages containing the *Exhortatio* and *De vero cyclo lunari*.[44] In one of the more detailed of these marginalia, Columbus used d'Ailly's stated time of the vernal equinox in 1411 and the annual drift of 0;10,44h

[41] See the text edited in Finke 1896–1928: iv. 712–13, and the discussion in Solan 2016: 27–9. The original document is contained in MS Paris, BnF, lat. 1485, fol. 232v.

[42] The *Exhortatio*'s 'master copy' is MS Cambrai, Bibliothèque municipale, 927, which contains d'Ailly's scientific works up to 1411 and was perhaps dictated by the author himself. See fols. 91r (dedicatory letter to John XXIII), 91v–100r (*Exhortatio*), 100v–105v (*De vero cyclo lunari*), 106r–107v (supplication and draft decree). It was one of the exemplars for a more luxurious and comprehensive collection up to 1415, now MS Cambrai, Bibliothèque municipale, 954. See fols. 38r (letter to John XXIII), 38v–43r (*Exhortatio*), 43v–46r (*De vero cyclo lunari*), 46v–47r (supplication and draft decree). On both these manuscripts, see Muzerelle 2000: 110, 114; Boudet 2002: 136–7; Lebigue 2014: 376–8, 385–7. See also Solan 2016: 15–16, 191–7, who mentions a total of 16 extant manuscripts of the *Exhortatio*, to which one should add MS Lincoln, Cathedral Chapter Library, 232 (B.6.10), fols. 114r–122r (*Exhortatio*), 122v–125v (*De vero cyclo lunari*).

[43] The text with Columbus's marginalia is found in Buron 1930. The entire volume owned by Columbus, now in the Biblioteca Colombina de Sevilla, was issued in facsimile in 1990 and accompanied by a Castilian translation by Ramírez de Verger 1990. On d'Ailly's text, see most recently Bouloux 2011.

[44] These annotations are reproduced in facsimile and edited in Lollis 1892: 84–6 and tab. lxxxiv–lxxxvi. See also Buron 1930: iii. 734–5. On the authorship of these postills, see Perez de Tudela 1991: 207–47.

to infer the corresponding time in the 'present year' 1491. The passage thus presents precious evidence that the explorer had a chance to imbibe the cosmographical information presented in *Ymago mundi* before his departure to find Asia on a western route.[45]

Aside from d'Ailly's own astrological writings, the Leuven edition of *Ymago mundi* also included texts on the same topic by his former student Jean Gerson (1363–1429), who took a somewhat less enthusiastic stance on the possibility of reconciling the science of the stars with theology. Born in Champagne, Gerson succeeded his teacher and long-time benefactor as chancellor of the University of Paris in 1395, and later came to play a prominent and vocal role at the Council of Constance.[46] After the council, Gerson spent the final decade of his life, from 1419 to 12 July 1429, at Lyons. On 4 April 1428, one year before his death, he wrote a letter to an unnamed correspondent (perhaps Oswald of Corda, vicar of the Grande Chartreuse) that decried the 'inattention, if not to say laziness, of those presiding over the Church', which had allowed the dates of the new moons and equinoxes to incur an all too manifest error.[47] Unsurprisingly, Gerson's assessment of the situation shared a good deal in common with that of his former teacher 'of famous memory', who had already attempted to effect a correction 'by imploring the Apostolic See and the Supreme Pontiff presiding over it' as well as by petitioning the general council of Constance, 'but so far no [provision] has been made because of the great number of constantly fluctuating worries or for God knows what other reason'.[48] Gerson was also aware of the opinion of another 'son of the Church', who had recently recognized the scandalous consequences of this error, which invited the laughter not just of Christian experts, but, worse even, of those who remained outside the fold of the Church—Jews, Saracens, and pagans, who only had to observe the heavens to notice the ignorance and neglect that caused the Church to declare the first day of the Moon when in reality it was the fourth or fifth. Without a correction, it was only going to take until the year AD 4565 for the ecclesiastical new moon to occupy the date of the full moon (*luna* 15), and vice versa.[49]

According to Gerson's own testimony, the claim involving the year AD 4565 came from a treatise *De astrologia* supposedly by Ramon Llull, which is preserved as an appendix to his letter. Far from being a lost work by the Catalan philosopher, however, the text is easily identified as the first two chapters of the *Summa astrologiae* by the Franciscan friar John, who in *c.*1276 had analysed the failures of the Julian calendar in a somewhat imprecise fashion, basing himself on the parameters of the

[45] Lollis 1892: 84 (no. 621); Buron 1930: iii. 735.

[46] McGuire 2005; Oakley 2006. For Gerson's attitude towards astrology, see Thorndike 1923–58: iv. 114–31.

[47] Glorieux 1960: 314–15. The letter previously already appeared in Glorieux 1951: 282–7. See Solan 2016: 31–7.

[48] Glorieux 1960: 315. Gerson expressly locates two copies of d'Ailly's *Exhortatio*, one of them kept in Cambrai by the archdeacon of Hainault (p. 316).

[49] Glorieux 1960: 315.

Jewish calendar (see Ch. 4.3).[50] For Jean Gerson, of course, the point of his letter was not to provide a sophisticated astronomical analysis, but to underline the necessity of a reform. He therefore articulated his hopes that someone among the cardinals of the Church would soon follow d'Ailly's example and once again promote the calendar's cause—a task he characterized as 'special and difficult', 'excellent and worthy of a great man', and above all, 'necessary to the entire Catholic commonwealth, both the spiritual and the secular one'.[51] Compared with much of the synodal legislation of his day, the correction of the calendar not only seemed more useful, but also in many ways more practical, less expensive, and less disruptive in its execution. At the same time, Gerson saw no point in waiting for the next general council to examine the task, but believed that it was incumbent upon the pope to decree the needed reform by his own authority and direct the various ecclesiastical institutions in its implementation. He was confident that this step was going to bestow the utmost glory not only on the pope willing to authorize it, but also on whomever the pontiff was going to choose as the project's main 'promoter and executor' (*promotor hujus negotii et executor*). Eager to see this scenario become reality ('May it be done! May it be done!'—*fiat, fiat*), Gerson finished his letter with four lines of verse:

Vera kalendarii fiat <correction> tandem
ne sit ridiculus calculus Ecclesiae.
Compotus insinuat errorem qualiter illum
emendare potest summus in Ecclesia.[52]

May a true correction of the calendar finally happen
so the calculation of the Church can cease to be the object of ridicule. May the *computus* point out the error and show
how the highest man in the Church can correct it.

7.2. ACCOMPLISHMENT AND FAILURE AT THE COUNCIL OF BASEL

Contrary to what the aging Jean Gerson may have hoped, it was not the pope who soon brought the undertaking of a calendar reform close to the brink of success, but the very next general council, which—in accordance with the decree *Frequens* passed in Constance in 1417—convened in the town of Basel in 1431. In its initial stages the council was primarily focused on finding a solution to the Hussite problem next to addressing the most pressing issues of ecclesiastical reform, but these individual projects were soon overshadowed by questions of authority and representation within the Church, which led the council onto a confrontation course with Pope Eugene IV (1431–47). One key factor in the escalating conflict

50 Glorieux 1960: 316–19. On p. 317, read 'ccc' for 'xxx' and 'xviii^{a} die' for 'xiiij^{a} die'. Pereira 1976: 198–9 was already sceptical of Llull's authorship, chiefly because Llull uses the annunciation style, whereas the author begins the year from January.

51 Glorieux 1960: 315.

52 Glorieux 1960: 316. See also Thorndike 1955: 173.

between the papacy and the general council was the ongoing negotiations towards a renewed union with the Greek Orthodox Church, which was regarded as a precondition for Western European military aid to an ailing Byzantine Empire, whose survival was acutely threatened by the advance of the Ottoman Turks. The council's diplomatic efforts to have the Greek party come to Basel or another town within its sphere of influence were effectively thwarted in May 1437, when the delegates of a pro-papal minority faction left Basel and united with the pope in Italy. On 18 September 1437, Eugene IV issued the Bull *Doctoris gentium*, which transferred the general council to Ferrara. Soon afterwards, a papal delegation arrived in Constantinople and successfully convinced the Byzantine emperor, John VIII Palaeologus, to journey to the new site together with the foremost prelates of his Church. The Union Council of Ferrara began in April 1438, but had to be transferred to Florence in January 1439 on account of plague and financial difficulties. After fifteen months of arduous negotiations, both parties came together on 6 July 1439 for the solemn reading of the bull *Laetentur coeli*, which contained a formal declaration of papal primacy, but also expressed compromises on the various issues that divided East and West. Although the Greek delegation was ultimately unable to convince the Church at home to assent to the terms of this union, the gathering of Ferrara–Florence proved to be an important victory for Pope Eugene IV in his political struggle against the conciliarists in Basel, who ended up taking the drastic measure of declaring the existing pope contumacious and electing Duke Amadeus VIII of Savoy in his place. As anti-pope Felix V (1439–49), Amadeus spent the next decade in an unsuccessful bid to obtain the support of Europe's secular rulers as the council's political momentum petered out, ending with its relocation from Basel to Lausanne in 1448 and formal dissolution on 25 April 1449.[53]

One of the most significant novelties the Council of Basel introduced while it was still in full vigour was the organization of its delegates and most of its decision-making processes on the basis of deputations rather than nations, as had still been the case in Constance. There were four such deputations, each of them dedicated to a particular thematic area: 'faith' (*deputatio fidei*), 'peace' (*deputatio pacis*), 'reform' (*deputatio reformatorii*), and common affairs (*deputatio pro communibus*). An egalitarian principle was maintained by distributing members of different clerical ranks equally across the four deputations. By giving each participant an equal vote, the council created an unprecedented opportunity for clerics below the rank of bishop as well as university scholars to influence the course of events.[54] Among the numerous members of the lower clergy who soon came to fill the council's ranks was Hermann Zoest (*c.*1380–1445), a Cistercian monk from Marienfeld Abbey in Westphalia, who defended the extension of voting rights in a treatise *De vocibus definitivis in conciliis generalibus* (1438). In another work from his pen, *De potestate ecclesiastica*

[53] On the history of the Council of Basel and its struggle with Eugene IV, see Stieber 1978; Helmrath 1987; Sudmann 2005; Decaluwé 2009, and the articles assembled in Decaluwé, Izbicki, and Christianson 2017. Specifically on the calendar reform effort and its sources, see now Solan 2016: 39–104.

[54] On the council's organization and rules of procedure, see Lazarus 1912; Gilomen 2017.

et papali (1436), he subscribed to a radical view of ecclesiastical hierarchy, according to which the pope merely functioned as an executive servant or administrator (*villicus*), who had to answer to the Church-in-council as the actual seat of ecclesiastical authority.[55] While these treatises did not remain unread, the original reason for Hermann Zoest's delegation to Basel by the head of his order was not his political thought but his recognized expertise in matters relating to chronology and astronomy.[56] That Easter reckoning had been on Zoest's mind since his early days as a monk in Marienfeld is seen most clearly from his *Tractatus phase*, a text of *c.*6,000 words and four chapters written in 1424.[57] More so than previous works on calendar reform, this *Tractatus* reflected its author's preoccupation over the reputation and education of his fellow clergymen, whose ignorance of matters calendrical could easily become a source of embarrassment in their interactions with laypeople. In line with this worry, Zoest claimed that one of his motivations for writing a treatise on the Easter problem was to make sure that:

> clerks are not ignorant of things that laypeople know...and have a basis from which they can respond to them. For it is a disgrace and a confusion on the part of the clerk, if he does not know how to respond to the layperson even though he is required to possess this kind of knowledge according to Saint Augustine. For he says that 'Priests are required to know the *computus*, otherwise they are hardly worthy of the name "priest"'.[58]

Clerical ignorance and negligence was, according to Hermann Zoest, a major cause why the reckoning of Easter had fallen into such disarray. Yet even those well educated in astronomy, such as Zoest himself, often failed to apply their expertise in the right direction, 'which is why I did not pay attention to the issue before I heard about it from others—for the laypeople have already become our masters'.[59] The widespread awareness of the calendrical error among the lay public was a major point of concern for Hermann Zoest, who warned that it was bound increasingly to scandalize the common folk:

> and most especially peasants and tenants, who know from ancient custom and the tradition of their fathers that Easter is always celebrated on the next Sunday after the 14th [day of the] Moon of the vernal equinox, that is, on the third Sunday of the aforementioned lunar month. For this purpose they also have this rule in their vernacular: *Na der Epyphanien dre mane nighe, darna de derde sindach is de hilghe Paschedach*. And in case this work should come into the hands of someone who is not German, he may not understand this vernacular rule, which is why it strikes me as good to translate it. So, the rule of the laypeople goes as follows: one has to find three

[55] On these and other works from Hermann's pen, see Wattenbach 1884: 102–6; Zurbonsen 1884: 21–31; 1899: 162–72; Tönsmeyer 1930: 128–89. On the reception of Hermann's political ideas, see Schröer 1951: 309–10. Further literature on Hermann Zoest is cited in Nothaft 2014a; 2014d: 478–503; Solan 2016: 283–4.

[56] This, at least, is suggested by the Marienfeld chronicle, cited in Zurbonsen 1899: 157.

[57] See the edition, French translation, and introduction in Solan 2016: 285–353.

[58] Hermann Zoest, *Tractatus phase*, prefacio (ed. Solan 2016: 298–9). For the quote attributed to Augustine, see n. 5 in the Introduction.

[59] Hermann Zoest, *Tractatus phase* 2 (ed. Solan 2016: 308).

new lunations after the feast of Epiphany [6 January], then the third Sunday after the [start of the] third lunation is the day of Easter.[60]

The distressing fact was that this peasant's rule of thumb or *regula rusticorum*, which can be traced back at least to the year 1390,[61] all too often came into conflict with the date of Easter as actually celebrated in the Church. In order to highlight the full extent of the problem to his readers, Zoest had to do no more than point to the present year 1424, in which the Church celebrated Easter on 23 April. This was indeed the third new lunation since Epiphany, as demanded by the *regula*, but it also qualified as the second rather than the first month of spring, since the lunation whose 14th day fell on or after the vernal equinox had already started on 1 March. Eager not to overstate his criticism, Zoest took into consideration a stipulation in the Book of Numbers (9:10–11), which permitted Israelites who had become unclean or were away on a journey to celebrate Passover in the second month.[62] Viewed from this vantage point, the late Easter of 1424 might have seemed tolerable, yet the problem remained that, if the age of the Moon was calculated astronomically, the official Easter date corresponded to *luna* 24, making it the fourth rather than the third Sunday of the lunar month in question.[63] According to the monk from Marienfeld, this was an unacceptable state of affairs for at least three reasons. One was the already mentioned way in which the error threatened to scandalize the common people, whom Zoest identified with the *pusilli* or 'little ones' whose peace of mind had already been defended by Jesus Christ in the Gospels of Matthew (18:6) and Luke (17:2). Another obvious problem was that celebrating Easter in the wrong week caused Christ's people to consume flesh and engage in carnal pleasures during what should normally be Lent, and to fast and perform Passion plays when it was really the Easter Week. Finally, and most conventionally, Zoest warned that this error gave a pretext to Jews to blaspheme against Christ and his Church.[64]

Far from merely railing against the Jews and their supposed calumnies against the Church, Hermann Zoest's perception of the subject was in fact heavily coloured by his familiarity with the *Compotus emendatus* of his twelfth-century Westphalian compatriot Reinher of Paderborn (see Ch. 3.2). Like Reinher before him, the Cistercian monk was ready to blame the disarray in which the Christian lunar *computus* found itself on Dionysius Exiguus, whose Easter table strayed from the ancient intercalation cycle of the Jewish calendar and assumed an incorrect relation between the Easter lunation and the date of the equinox.[65] From contemporary

[60] Ibid. (ed. Solan 2016: 310–12). See also Hermann Zoest, *Phaselexis (2nd version)* 5 (ed. Solan 2016: 432).

[61] Nothaft 2015c: 121–2. A Latin version is documented in Mooney 1983.

[62] Hermann Zoest scrutinized this issue in a separate treatise, *Quare interdum secundo mense pascha celebratur*, preserved in MS Copenhagen, Det Kongelige Bibliotek, Thott 825 4°, fols. 177ra–178rb.

[63] Hermann Zoest, *Tractatus phase* 2 (ed. Solan 2016: 310).

[64] Ibid. (ed. Solan 2016: 312). On the danger of being ridiculed by Jews, see also the preface to the *Tractatus phase* (ed. Solan 2016: 298–300).

[65] Hermann Zoest, *Tractatus phase* 1, 4 (ed. Solan 2016: 304, 322–6). See also the indictments of Dionysius Exiguus in Zoest, *Phaselexis (2nd version)* 7 (ed. Solan 2016: 446–8, 452); Zoest,

Jewish practice, he knew that the Passover was sometimes eaten as early as 15 March, citing 'a certain rather understanding Jew' as his source. This was the case in 1424, when the full moon on 15 March (14 Nisan) should have been followed by Easter Sunday on 19 March, if the date of Easter was correctly defined as the Sunday after Passover. Instead, the ecclesiastical date was 23 April.[66] In contrast to Reinher of Paderborn, who had advocated the wholesale adoption of the Jewish *molad* reckoning, Hermann Zoest sought to avoid future mistakes of this kind by completely recasting the Golden Number inscribed into the calendar, following the supposed ancient precedent of Julius Caesar. Zoest's new sequence of Golden Numbers, as displayed in a calendar attached to his *Tractatus*, tracked the dates of the mean conjunctions as predicted by the Alfonsine Tables, with times adjusted to the meridian of Münster. The sequence started with the imminent year 1425, the first year of the next 19-year cycle, in which the first conjunction was going to fall on 19 January, 2 p.m. By placing the Golden Number 1 on this new date and by generating the 234 remaining Golden Numbers in an analogous fashion, it was possible to forestall most of the errors presently encountered, at least temporarily. In order to fall completely in line with the Nicene stipulations, however, the Church also had to move the limits for its paschal *termini* in accordance with the date of the true vernal equinox. The latter, as Hermann Zoest did not fail to point out, now fell on 11 March and hence ten days earlier than traditionally assumed.[67]

Following the initial salvo fired in the *Tractatus phase*, Hermann Zoest renewed his agitation in 1432 with a *Tractatulus exhortatorius*, whose existence is known only from a brief reference in his later writings.[68] From the fragmentary evidence at hand, it may be concluded that this lost treatise, or some other text based on it, played a crucial role in lifting the issue of calendar reform onto the agenda of the Council of Basel. According to the brief summary of events provided in John of Segovia's monumental *Historia* of the council, the subject was first introduced on 18 June 1434, which was also the day on which Zoest was formally admitted to the council as a member of the *Deputatio pacis*. On the very same day, the notary of Zoest's deputation appeared before the general assembly to read out a written plea (*cedula cum exhortacione*), which demanded in no unclear words that the council should attend to the concerns of the lower clergy and the general population, to whom the inaccuracies of the calendar had become a source of increasing grief. As the text went on to complain in a manner highly reminiscent of Zoest's *Tractatus phase*, the frequency of years in which the mobile feast days were celebrated incorrectly had turned the holy Church into an object of derision by infidels and compelled many believers to correct their own calendars.[69]

Compendium paschale 8 (ed. Solan 2016: 558–64). Zoest's use of the *Compotus emendatus* and attitude towards the Jewish calendar are discussed in Nothaft 2014a.

66 Hermann Zoest, *Tractatus phase* 3 (ed. Solan 2016: 320).

67 See ibid. 4 (ed. Solan 2016: 324–6), and the calendar edited on pp. 330–53.

68 Hermann Zoest, *Phaselexis (2nd version)* 5 (ed. Solan 2016: 428).

69 John of Segovia, *Historia gestorum generalis synodi Basiliensis* 8.19 (ed. Birk 1873: 709). The following account of the calendrical proceedings at Basel rests on the documentation assembled and

Hermann Zoest's arguments for the urgency of this issue appear to have been persuasive. The council reacted by putting Antonio Correr (1369–1445), at the time cardinal bishop of Ostia, in charge of preparing a reform of the ecclesiastical calendar. After Correr's premature departure from the council in September 1434, the matter passed into the hands of eight theologians and canon lawyers, two from each deputation, who were tasked with considering a range of different solutions to the calendar problem in order to pick the one that seemed most suitable.[70] Although there is evidence that the council made efforts to support the work of this newly formed commission with reports from universities and other external advisers,[71] the project's main astronomical navigator appears to have been none other than Hermann Zoest himself, who had previously already been given a role as an adjunct to Antonio Correr.[72] By the spring of 1435, the delegates had found enough common ground to allow for the drafting of a decree authorizing a calendar reform slated for February 1436. Credit for preserving this decree text is due to the Polish canon lawyer Thomas Strzempiński (1398–1460), who joined the commission as a member of the *Deptutatio fidei* and may have presided over the initial discussions of the year 1435. A formal report (*Relacio*) on these discussions, signed by Strzempiński, prefaces the draft decree in the only known copy (see Figure 7.1).[73] Our understanding of the tersely worded stipulations contained in this document would remain incomplete, however, were it not for the help provided by Hermann Zoest, who completely rewrote his *Tractatus phase* and turned it into a new treatise of six chapters entitled *Phaselexis*—a portmanteau of *phase*, the Old Vulgate's rendering of the Hebrew *pesaḥ*, and the Greek *lexis* ('word').[74] Its main purpose was to supplement the decree text with its proper elucidation and defence, while further sections were supposed to fill in the historical background of the Easter problem or provide computists with the information necessary to correct their existing books and tables. That the Westphalian monk's enthusiasm for the reform project was unbroken can be seen from the prologue to his work, which treated the anticipated reform of 1436 as a fait accompli:

> Rejoice and be happy, O most holy [and] excellent Council of Basel, because you have become a companion of the sacrosanct councils of old!... For that which the supreme pontiffs and these councils have often attempted and set out to do you have brought to the desired and auspicious end. Indeed, you have carefully emended the calendar,

discussed in greater detail in Nothaft 2015i; Solan 2016: 48–104. See, in addition, Kaltenbrunner 1876: 336–42; Wattenbach 1884: 96–100; Zurbonsen 1899: 156–60; Honecker 1940; Stegemann 1955: xxxiv–lii; Sudmann 2005: 260–72.

[70] See the protocols edited in Haller et al. 1896–1936: iii. 126 (18 June 1434), 203 (15 Sept. 1434), 358 (4 Apr. 1435), 365 (8 Apr. 1435); v. 94 (18 June 1434), and Wattenbach 1884: 109. See also Hermann Zoest's own retrospective account in *Phaselexis (2nd version)* 8 (ed. Solan 2016: 460).

[71] Haller et al. 1896–1938: iii. 289 (19 Jan. 1435).

[72] This role is documented in a marginal gloss in MS Munich, BSB, Clm 18470, fol. 13r. See Nothaft 2015i: 175–7.

[73] The *Relacio* and decree text are jointly edited in Nothaft 2015i: 225–35, and Solan 2016: 239–63.

[74] See the edition in Nothaft 2015i: 239–90.

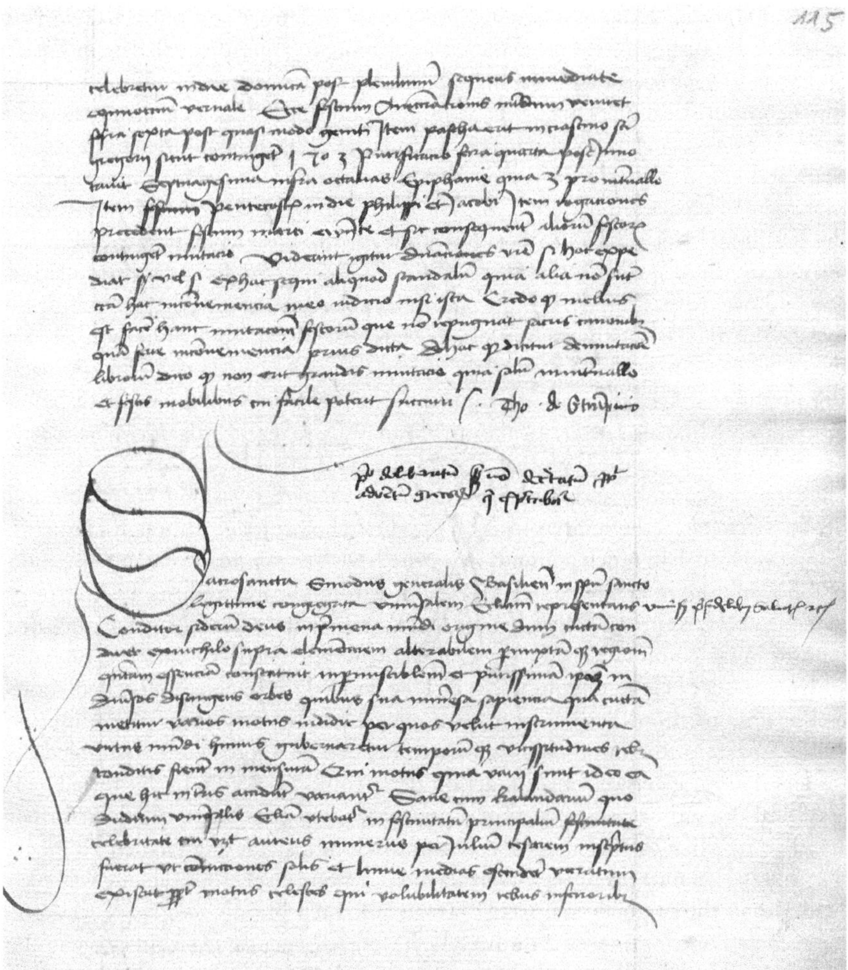

Figure 7.1. *Relacio* and draft decree of 1435, with signature by Thomas Strzempiński. Kraków, Biblioteka Jagiellońska, 4164, fol. 115r.

Source: Biblioteka Jagiellońska

> [once] full of vice and corrupt, and expelled the defect in the celebration of the most honoured feast of Easter beyond the doorstep of the House of God.[75]

Midway into the treatise, Zoest indicated his awareness of the long line of distinguished scholars who had laboured on the topic before him, ranging from Jean des Murs, Reinher of Paderborn, Roger Bacon, and Pierre d'Ailly to his own 'lowly' self (*et ego tamquam scobs et fex omnium ultimus*).[76] He also repeated his charges

[75] Hermann Zoest, *Phaselexis (1st version)*, prefacio (ed. Nothaft 2015i: 239).
[76] Ibid. 3 (ed. Nothaft 2015i: 255).

against Dionysius Exiguus, detailing how the sixth-century computist had wished to relieve the bishop of Rome of the task of having to announce the date of Easter on a year-to-year basis by furnishing the Church with a perpetual table. The result was a computistical mess, incorrect astronomically and out of step with the calendar of the Jews, causing Christians to intercalate embolismic months in the wrong years and celebrate Easter in Iyyar rather than Nisan.[77] In order to return to the legitimate intercalation pattern of the ancient Hebrews, it was necessary to weaken the traditional rule of the equinox and allow for the Easter lunation to fall in the second month of spring in certain embolismic years. A corollary of this stipulation was to move the paschal *termini* upwards from their accustomed dates, but only by five days: from 21 March–18 April to 16 March–13 April. To do otherwise and tie the earliest *terminus* to the date of the vernal equinox (12 March)—an approach still presupposed by Thomas Strzempiński in his report—would have been tantamount to celebrating Easter one month ahead of Passover on at least two occasions per 19-year cycle. In the view that Zoest articulated in 1435, the Jewish calendar's importance as a guide and model for the Easter *computus* was non-negotiable and hence necessitated the reinterpretation of entrenched rules.[78] Although Hermann Zoest evidently had much admiration for the Jewish *computus*, he had no intention of following the Jews into some of the more astronomically sophisticated territory that the *molad*-system of their lunar calendar entailed. Instead of re-creating this type of lunar computation, Zoest equipped the reformed ecclesiastical calendar with a new set of Golden Numbers, which he derived from meticulous calculations of the time of mean conjunction for a period of 76 years, using the Alfonsine Tables. Aware of the imprecisions inherent in such an approach, he reminded his readers that no given set of conjunction dates could be deemed valid for all geographical longitudes. Under these circumstances, the Westphalian reformer regarded it as licit to take a *grosso modo* approach to distributing Golden Numbers over the twelve months of the Julian calendar. In one respect his scheme was even cruder than the original one, seeing how it consisted of only two sequences, one for even and one for uneven months, which were to be repeated six times each. The advantage thus gained was one of simplicity, as the whole scheme could be summarized in a few lines of mnemonic verse from Hermann Zoest's pen, which appear at the very end of the decree text.[79]

When it came to the all-important paschal lunation, whose new moons all fell in March, Zoest's main strategy was to select the appropriate Golden Number from the latest possible mean conjunction date that could be expected for the fifteenth century. In Zoest's own parlance, this measure was guided by the necessity 'not to anticipate the Jews' (*ne Iudeos preveniremus*),[80] whom he and his colleagues perceived as celebrating Passover in strict correlation with the astronomical full moon. Since this full-moon date could on many occasions be identical with the

[77] Ibid. 4 (ed. Nothaft 2015i: 261–3).
[78] See ibid. 6 (ed. Nothaft 2015i: 271–5), and the discussion on pp. 190–4.
[79] See §14 of the decree text (ed. Nothaft 2015i: 235), and the discussion on pp. 200–8.
[80] Hermann Zoest, *Phaselexis (1st version)* 5 (Nothaft 2015i: 267).

16th day inclusive from the last conjunction, the Basel reformers discerned the need to exclude the day of conjunction from the ecclesiastical count of days of the Moon. According to Thomas Strzempiński, whose use of sources reveals his dependence on Hermann Zoest's *Tractatus phase*, this exclusive count was sanctioned by the authority of Augustine, Bede, Theophilus of Caesarea, and Isidore of Seville, who all testified to the importance of not anticipating the day of the full moon when celebrating Easter.[81] As an illustrative example he brought up the conjunction of 1439, which in the newly reformed calendar was found to occur on Sunday, 16 March. According to the traditional lunar limits of *luna* 15–21, this would have mandated placing Easter Sunday 14 days later, on 30 March. Yet the Moon could be expected to reach a mean or true elongation of 180° only on the following Monday. Thomas warned that if the Moon was eclipsed on such an occasion everybody could see that the Church had celebrated its version of Passover before the opposition of Sun and Moon and, hence, before the Jews.[82]

In looking after the solar component of the calendar, the commission took the convenient step of relying on the tropical year enshrined in the Alfonsine Tables, which Hermann Zoest cited as 365d 5;49,16h.[83] In principle, the difference of 0;10,44h between this value and the Julian year called for an omission of a day every 134 years, but the decree text increased this span to 136 years. The obvious reason, as Zoest himself took care to explain, was that 136 was a multiple of 4, making it possible perpetually to restrict such excisions to the bissextile day in February, whose intercalation was to be suppressed in the appropriate years.[84] According to the decree text, the first such intervention was to come in February 1436, the targeted year of reform, when the omission of the *bissextus* could be expected to bring about a new set of canonical dates for the equinoxes and solstices: 12 March, 14 June, 15 September, and 13 December.[85] This was a straightforward series of steps as far as the Julian calendar was concerned, but far from ideal if the goal was simultaneously to cure the defect of the ecclesiastical lunar reckoning, which fell out of synchronicity with the observable phases of the Moon at a rate less than half as great as the recession of the equinoxes. The members of the calendar commission sought to counteract the looming imbalance by decreeing that every second suppression of the leap day, slated to take place every $2\times136=272$ years, was to be accompanied by a downward shift of all Golden Numbers by one day. The first such shift was scheduled for the year $1708(=1,436+272)$, when the earliest possible Easter full moon (*luna* 15), in the final year of the 19-year cycle, was supposed to move from 16 March down to 17 March. Analogous steps were to be taken in AD 1980, 2252, and 2524, by which year the earliest paschal full moon would have arrived at 20 March and hence very close to its traditional date.[86] Beyond the year

[81] Thomas Strzempiński, *Relacio* §9 (Nothaft 2015i: 231–2).

[82] See ibid. §6 (ed. Nothaft 2015i: 229–30), and the discussion on pp. 195–200.

[83] Hermann Zoest, *Phaselexis (1st version)* 4 (ed. Nothaft 2015i: 260).

[84] Ibid. 5 (ed. Nothaft 2015i: 270).

[85] See §13 of the decree text (ed. Nothaft 2015i: 234). The dates conform to the Alfonsine true solar longitudes for 1436.

[86] See §12 of the decree text (ed. Nothaft 2015i: 234).

2800, the reform proposal of 1435 did not even dare to look, owing to three reasons Zoest spelled out when he attached a table elucidating these corrections to the end of his *Phaselexis*:

> No continuation beyond these years was made, partly because it is believed the world will not last for this long, partly because it is possible that our successors will discover certain celestial motions yet to be found, most of all because the true length of the year has not yet been found. For this same reason I do not promise that this table will last for that many years, as this would be rather presumptuous.[87]

The passage suggests that Hermann Zoest, and presumably also some of the delegates he assisted as an adviser, shared the same uncertainty as to the actual length and nature of the tropical year that had been highlighted by Pierre d'Ailly two decades earlier (see Ch. 7.1). He was thus willing to concede that the 'opinion of Alfonso' was currently 'reported to be closer to the truth' than other estimates, but in the absence of a generally agreed model for the motion of the apogees and eighth sphere it appeared wise to reserve definitive judgement.[88] Long-term changes in the length of the tropical year still seemed possible, as did future improvements in the accuracy of measurement, both of which would have entailed alterations to the correction mechanisms proposed in the decree and the accompanying *Phaselexis*. For the meantime, the Basel reform proposal effectively assumed a solar year that lasted $365\frac{1}{4}-\frac{1}{136}\,\text{days}=365.242647\text{d}$ and a lunar calendar that deteriorated at a rate of $M=272$ years, which implied a mean synodic month of 29.5305538d.[89] This came rather close to $29\text{d}\ 12;44\text{h}=29.530555\text{d}$ and hence to the average lunation in the Arabic calendar, whose use had been endorsed by previous authorities such as Robert Grosseteste and Pierre d'Ailly.

The astronomical underpinnings of the reform proposal of 1435 offered relatively little contact surface for criticism, but there was more to calendar reform than just choosing the right parameter value. A clear hint that the plan provoked fierce debates within the commission itself comes from a later version of Hermann Zoest's *Phaselexis*, in which he recalls how his initial proposal was met with considerable resistance 'on account of the first month and the paschal *termini*, for which reason the pursuit of our task was much impeded'.[90] In a similar vein, Thomas Strzempiński admitted to the worries generated by Easter dates that preceded the traditional paschal boundaries by several days. In the extreme case of Easter Sunday falling on 13 March—which would have been allowed for by the projected date of the vernal equinox, but not by the precise rules laid down in the decree, where Easter does not precede 17 March—Septuagesima Sunday would have ended up falling within the week after Epiphany, and Candlemas on the Wednesday after Quadragesima Sunday. Similarly, Rogation Sunday was allowed to precede St Mark's Day (24 March), with Pentecost falling as early as 1 May, the

[87] See the canon and table edited in Nothaft 2015i: 290, and the discussion on pp. 209–12.
[88] Hermann Zoest, *Phaselexis (1st version)* 4 (ed. Nothaft 2015i: 260).
[89] $365\frac{1}{4}-\frac{1}{272}\,\text{days}\times 19\div 235=29.530553817\ldots\text{days}$.
[90] Hermann Zoest, *Phaselexis (2nd version)* 8 (ed. Solan 2016: 460).

feast of Saint Philip and Saint James.[91] A related problem concerned the inevitable changes that the reform proposal of 1435 spelled for the existing breviaries, *computus* manuals, and calendars, which in the anticipated future had to be brought into agreement with a different set of Easter limits as well as new locations for the Golden Numbers. The provision at the end of the decree text of a simple set of mnemonics, which codified the complete set of reformed Golden Numbers, gives some indication of how Hermann Zoest and the members of the calendar commission tried to overcome the logistical challenge inherent in their project. Existing calendars were not meant to be discarded and replaced by new ones, but gently emended by a few additional entries in the margins. Hermann Zoest accordingly voiced his expectation that those accepting the reform would adorn the calendar page for February with a reminder as to the years in which the bissextile day needed to be omitted against the standard rule.[92] A reformed calendar attached to Zoest's *Phaselexis* demonstrated in full detail how these annotations together with the new dates of the equinoxes and solstices and the new set of Golden Numbers and paschal *termini* were supposed to be inscribed by future users (see Figure 7.2).[93]

Thomas Strzempiński closed his report on the matter by insisting that the alteration of existing books that arose from the planned reform was not particularly worrisome, 'because it only affects the interval and the mobile feast days, which can easily be provided for'.[94] And yet, the surviving documentation shows that an even more conservative option was available to the delegates in 1435. When summarizing the discussions that took place in the early months of that year, Strzempiński briefly mentioned an alternative reform scheme, which involved no more than the omission of a complete week from 1436 together with a simultaneous change of the Golden Number valid for the present year from XII to IX.[95] Strzempiński's report has nothing to say about the precise purpose of this combination of steps, nor does it identify the individual(s) who first introduced it into the discussion. It is not unlikely, however, that its original architects included the German canon lawyer Nicholas of Cusa, also known as Nicolaus Cusanus (1401–64), who had joined the Council of Basel in 1432 as the legal representative of Count Ulrich of Manderscheid, one of the rival claimants to the archbishopric of Trier. Cusanus—like Thomas Strzempiński a member of the *Deptutatio fidei*—soon became one of the council's chief theorists, as manifested by works such as *De maioritate auctoritatis sacrorum conciliorum supra auctoritatem papae* (1433) and *De concordantia catholica* (1433/34). Having spent most of 1435 and parts of 1436 on legal and diplomatic missions in Germany, Cusanus returned to Basel in August of 1436 and subsequently presented to the calendar commission a treatise *De correctione kalendarii*, which carefully elaborated on the alternative proposal that had already

91 Thomas Strzempiński, *Relacio* §10 (ed. Nothaft 2015i: 232).

92 Hermann Zoest, *Phaselexis (1st version)* 5 (ed. Nothaft 2015i: 270).

93 See the two preserved copies of Zoest's reformed calendar in MSS Gniezno, Archiwum Archidiecezjalne, 17, fols. 353v–356r; Uppsala, Universitetsbibliotek, C 15, fols. 273r–273ev.

94 Thomas Strzempiński, *Relacio* §10 (ed. Nothaft 2015i: 232).

95 Ibid. §8 (ed. Nothaft 2015i: 230–1).

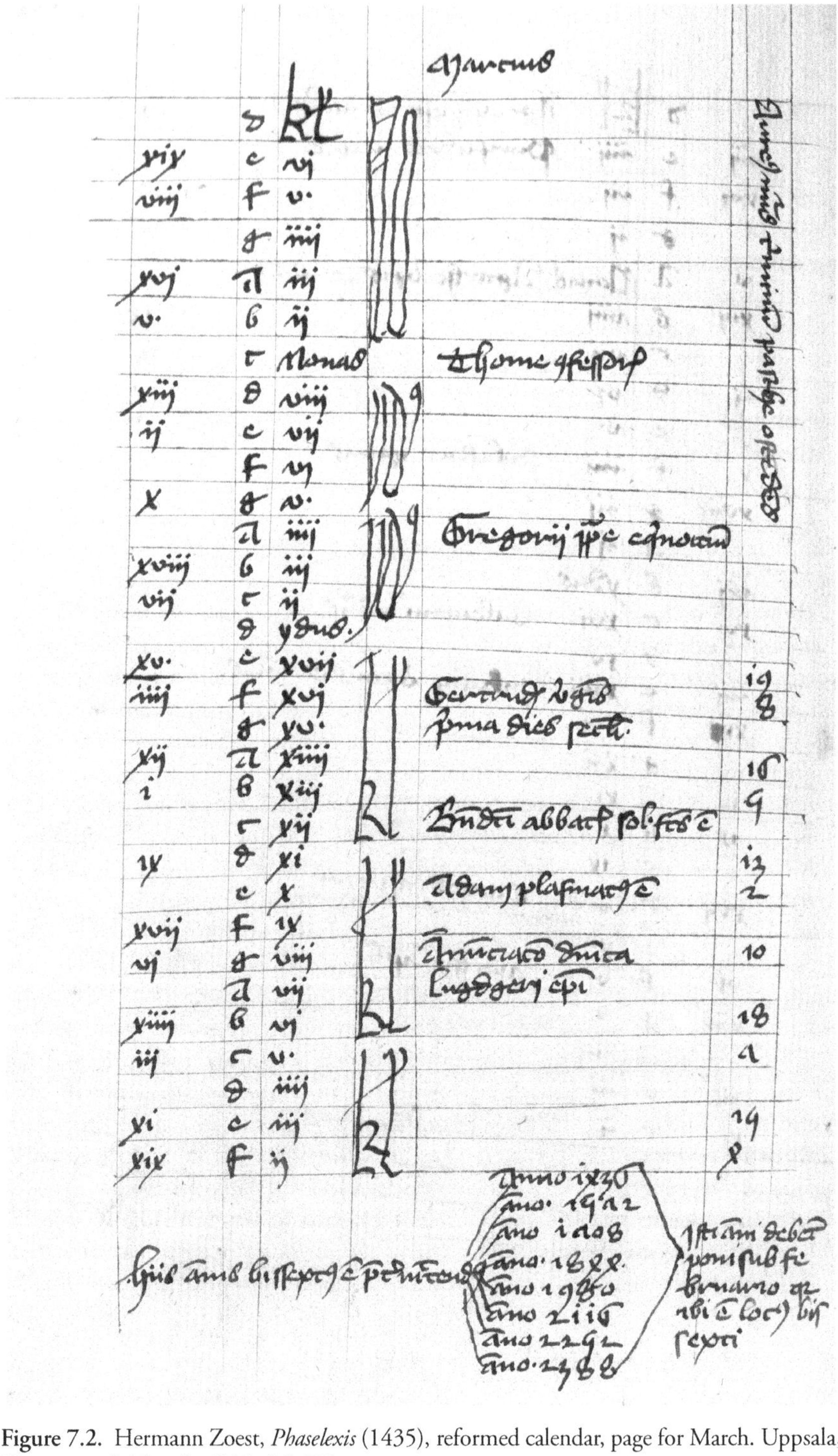

Figure 7.2. Hermann Zoest, *Phaselexis* (1435), reformed calendar, page for March. Uppsala, Universitetsbibliotek, C 15, fol. 273v.

Source: Uppsala universitetsbibliotek

been sketched in Strzempiński's *Relacio*.[96] At the core of Cusanus's plan was an attempt to restore the calendar to its state at the time of the Council of Nicaea, whose stipulations with regard to the reckoning of Easter had to be maintained at all costs. In order to secure this goal, he proposed the simple formula of 'converting the Golden Number into the *cyclus lunaris* while omitting one week' (*quod obmissa una septimana aureus numerus in cyclum lunarem commutetur*).[97] Computists in Cusanus's day would have readily understood the *cyclus lunaris* to be a variant of the common 19-year cycle, which differed from the latter in assuming a start in year 4 of the conventional cycle, when the Moon on 1 January was 12 days old.[98] The formula proposed in *De correctione kalendarii* was hence structurally identical to the idea already on the table in 1435; but whereas the original version opted for the erasure of 7 days from the start of 1436, the passage of time forced Cusanus to postpone the envisioned reform to 1439. At the same time, he moved the targeted week from the period preceding Lent to one that was less prone to interfere with the calculation of mobile feast days: the week between Pentecost or Whit Sunday (24 May) and the beginning of June. The end result of this twofold correction was that the calendrical new moons were moved upward by 3–4 days compared to their previous location, while the equinox was simultaneously brought one week closer to 21 March, all without causing any changes to the traditional Easter limits and paschal *termini*.

For a better understanding of this reform and its underlying rationale, one may turn to Table 7.1, where column I shows the 19 paschal new moons as marked by the traditional Golden Numbers. Column II juxtaposes the corresponding dates that will result if three years are subtracted from the 19-year cycle, such that years of the *cyclus decemnovenalis* are counted as if they started three years later, in synchrony with the so-called *cyclus lunaris*. Column III indicates the shift these new-moon dates will undergo relative to the old calendar once 7 days are subtracted, while column IV records the difference between the traditional and reformed new-moon dates, that is, between the dates shown in columns I and III.[99] The astronomical success of this reform can be gauged from a look at the remaining two columns, which show the mean conjunctions for the 19-year cycles from 1425 to 1443 (V) and from 1444 to 1462 (VI), as calculated by the Alfonsine Tables (= AT). Note that because 19 years in the Julian calendar can comprise either 6,939 or 6,940 days (depending on whether the 19 years in question contain four or five bissextile years), the dates in column VI sometimes fall on a later date than those in column V, as is the case in years 4, 8, 12, and 16. More often than that, however, the conjunctions in column VI are found to fall a day earlier (years 1, 2, 3, 7, 9, 13, 14, 18), as one would expect from the decrease by $6h + 1;28h = 7;28h$ that the Alfonsine

[96] See the editions of *De correctione calendarii* in Stegemann 1955 and Müller 2010: 287–319. For the work's genesis and content, see Honecker 1940; Solan 2016: 74–87; Nothaft 2017d.

[97] Nicholas of Cusa, *De correctione calendarii* 8 (ed. Stegemann 1955: 68, ll. 20–1).

[98] See Mosshammer 2008: 85–95.

[99] In years 8 and 19, the relevant comparison is between the reformed new-moon dates on 2 March and 1 March and the new moons of March in the old 19-year cycle, which are 6 March and 5 March respectively.

Table 7.1. The paschal new moons in Cusanus's *cyclus lunaris* compared with the old ecclesiastical calendar and the conjunctions in the Alfonsine Tables (AD 1425–62)

I		II		III	IV	V	VI
Year	GN	Year	GN	GN new	Diff.	AT 1425–43	AT 1444–62
1	23 Mar.	17 em.	27 Mar.	20 Mar.	3d	19 Mar. 18:34h	19 Mar. 11:06h
2	12 Mar.	18	16 Mar.	09 Mar.	3d	09 Mar. 03:22h	08 Mar. 19:54h
3 em.	31 Mar.	19 em.	04 Apr.	28 Mar.	3d	28 Mar. 00:55h	27 Mar. 17:27h
4	20 Mar.	1	23 Mar.	16 Mar.	4d	16 Mar. 09:43h	17 Mar. 02:15h
5	09 Mar.	2	12 Mar.	05 Mar.	4d	05 Mar. 18:32h	05 Mar. 11:04h
6 em.	28 Mar.	3 em.	31 Mar.	24 Mar.	4d	24 Mar. 16:05h	24 Mar. 08:37h
7	17 Mar.	4	20 Mar.	13 Mar.	4d	14 Mar. 00:53h	13 Mar. 17:25h
8 em.	05 Apr.	5	09 Mar.	02 Mar.	4d	02 Mar. 09:42h	03 Mar. 02:14h
9	25 Mar.	6 em.	28 Mar.	21 Mar.	4d	21 Mar. 07:15h	20 Mar. 23:47h
10	14 Mar.	7	17 Mar.	10 Mar.	4d	10 Mar. 16:03h	10 Mar. 08:35h
11 em.	02 Apr.	8 em.	05 Apr.	29 Mar.	4d	29 Mar. 13:36h	29 Mar. 06:08h
12	22 Mar.	9	25 Mar.	18 Mar.	4d	17 Mar. 22:24h	18 Mar. 14:56
13	11 Mar.	10	14 Mar.	07 Mar.	4d	07 Mar. 07:13h	06 Mar. 23:45h
14 em.	30 Mar.	11 em.	02 Apr.	26 Mar.	4d	26 Mar. 04:46h	25 Mar. 21:18h
15	19 Mar.	12	22 Mar.	15 Mar.	4d	15 Mar. 13:34h	15 Mar. 06:06h
16	08 Mar.	13	11 Mar.	04 Mar.	4d	03 Mar. 22:23h	04 Mar. 14:55h
17 em.	27 Mar.	14 em.	30 Mar.	23 Mar.	4d	22 Mar. 19:56h	22 Mar. 12:28h
18	16 Mar.	15	19 Mar.	12 Mar.	4d	12 Mar. 04:44h	11 Mar. 21:16h
19 em.	04 Apr.	16	08 Mar.	01 Mar.	4d	01 Mar. 13:33h	01 Mar. 06:05h

Tables predict for the times of two mean conjunctions that are separated by 19 Julian years. Both sequences of conjunctions are calculated for the longitude of Jerusalem (+55;30°), which is the only meridian Cusanus ever mentions in connection with the Alfonsine Tables. Times are counted, in conventional fashion, from midnight of the date in question.[100]

From protocol entries for 15 February and 22 March 1437 it can be inferred that the four deputations found the new plan unobjectionable and recommended that it be passed on to council's general assembly.[101] John of Segovia, in his *Historia* of the council, names March as the month in which Nicholas of Cusa reported the commission's findings to the 'assembly of fathers sitting in the refectory of the Franciscans'.[102] The report in question appears to have survived in the form of an *Avisamentum de correctione kalendarii*, which follows upon Cusanus's treatise in the two earliest preserved manuscripts and is largely derivative of its contents.[103] This

[100] See Nicholas of Cusa, *De correctione calendarii* 9 (ed. Stegemann 1955: 80, ll. 5–7). The meridian of Jerusalem is also expressly used for calculations in Hermann Zoest, *Compendium paschale* 8 (ed. Solan 2016: 560–2). Its difference of 55;30° relative to Toledo (at 11°) is confirmed in Hermann Zoest, *De fermento et azymo*, MS Munich, BSB, Clm 3564, fol. 147vb, and by the table of coordinates in MS London, BL, Harley 3072, fol. 48r, which codex was once owned by Cusanus.

[101] Haller et al. 1896–1936: vi. 15 (15 Feb. 1437), 36 (22 Mar. 1437).

[102] John of Segovia, *Historia* 8.19 (Birk 1873: 709). Cusanus's appearance before the council is discussed in Solan 2016: 87–92.

[103] MSS Vienna, ÖNB, 5266, fols. 283va–4rb; Wolfenbüttel, Herzog-August-Bibliothek, Cod. Guelf. 354 Helmst., fols. 126r–127v. The Viennese manuscript was used for the edition in Solan 2016: 274–9.

turn of events strongly suggests that by March 1437 Nicholas of Cusa had succeeded Thomas Strzempiński as the main spokesman of the calendar commission, never mind the fact that he left Basel soon thereafter, on 20 May 1437, to join the side of Pope Eugene IV. Although his involvement in the attempted calendar reform thus came to an abrupt end, the outlines of his proposal were carried over into the draft text of an official decree, the original version of which has been preserved in a manuscript in Brixen, the seat of a prince bishopric later to be occupied by Cusanus (1450–64).[104] The text in question begins by acknowledging the special care the holy Fathers of old had taken when setting down the rules for the celebration for Easter in a series of decrees passed by general councils and the Holy See. These rules had been observed without change for many centuries, but with the passing of time the calendrical regression of the equinoxes, solstices, and lunisolar conjunctions had led to serious defects in the celebration of mobile feast days, with the result that, on an increasing number of occasions, Easter was fixed not on the Sunday closest to the 14th day of the Moon of the first spring lunation but on some later date. Not only was this state of affairs in violation of the decrees of the holy Fathers, but over time the error in the calculation of Easter threatened to scandalize the Christian public, 'if no provisions are made':

> In order, therefore, to compensate for this defect and to reduce what the ancient Church once decreed back to its due order, we have enjoined a great number of prelates and doctors to consult on this matter with experienced men and examine all difficulties, so they can seek the most convenient way in which this can be provided for. From their report it has been ascertained that there is one way that is the most convenient for repairing the aforementioned defect, namely that in the coming year of the Lord 1439 the final seven days of May are omitted from the computation of the calendar, such that the month of June begins on the day that falls after the feast of Pentecost of this year and the feast of St Barnabas [11 June] is assigned to the date of the feast of Corpus Christi.... The number that is commonly named the 'Golden' will be converted into the ancient lunar cycle, such that [the number] that used to be 15 will become 12 and it will show the date of the conjunction of Sun and Moon without any change to the position this Golden Number had in the calendar until now. Also, the first Easter in the calendar will be assigned to 22 March and the last one to 25 April, as used to be the case.... Accepting this method as the easiest and most convenient one, the holy synod has decreed that the universal Church shall observe it with diligence, ordering all believers in Christ to assign the beginning of the month of June in the year 1439 to the day after Pentecost and to make sure they celebrate the feast of St Barnabas and the nativity of John the Baptist and the other [feasts] that follow according to the order just mentioned. Moreover, to prevent any error in these things from arising in the future, this holy synod has decided that the bissextile day shall be omitted once after every 300 years. And to make sure that no one can pretend to have been ignorant of this solemn decree, we instruct and order all archbishops and bishops to publish this decree in all cathedral and parish churches of their dioceses in time for Easter of the aforementioned year 1439 and to continue this publication

[104] MS Brixen, Bibliothek der Philosophisch-Theologischen Hochschule, C 16, fols. 81v–82v. See Nothaft 2015i: 292–303.

> in all of the aforementioned churches on all the Sundays that follow from Easter to Pentecost, to ensure that on the day following Pentecost and thereafter this salutary ordinance will be unanimously observed by everybody.[105]

The decree of 1437 represents an undeniable high point in the history of the calendar debate prior to the Gregorian reform. No other calendar-reform project attempted during the Middle Ages came closer to fruition and none displayed greater seriousness in advancing a solution that would not just address the astronomical problem, but appear feasible on a practical level. That the logistics of reform received close attention from the members of the commission can be seen from the way the draft decree gave explicit instructions to archbishops and bishops on how to publicize the reform in advance. More than that, the decree's authors made sure to add a paragraph that addressed the legal ramifications that might arise from the omission of 7 days, thereby placating some of the worries about deadlines and contracts that had first been raised at the court of Clement VI in 1344/5:

> Finally, in order to forestall various legal quarrels, which might be occasioned by this reform, this holy synod declares that each and every one who is obliged, by contract or otherwise, to pay back or give or do something on a certain day or within a certain deadline must in this one year not be in any way required or compelled to [make this payment or perform this deed] until the eighth day after the stipulated deadline has come.[106]

This concern for the practicalities involved was mirrored by the technical elements of the reform itself, which was designed to retain the traditional Golden Numbers as well as the customary paschal *termini* and Easter limits. By doing so, Cusanus and his colleagues were able to save existing breviaries and calendars, as well as all other books dealing with the *computus* and the divine services, from the threat of obsolescence. The centrality of this consideration to the reform project of 1437 is amply confirmed by Hermann Zoest, who went ahead and reworked his *Phaselexis* into an even more elaborate treatise, now consisting of ten chapters, which was geared towards defending the various measures the reform commission had come to endorse.[107] One of several new elements Zoest introduced in this final version of his treatise, which was copied widely during the fifteenth century,[108] was a narrative structure that followed the monastic division of the night into four 'watches' or *vigiliae*, called *conticinium* ('still of night'), *intempestum* ('dead of night'), *gallicinium* or *gallicantus* ('cock's crow'), and *antelucanum* ('before daybreak'). By linking successive parts of his treatise to these and other parts of the 24-hour day, the Cistercian calendar expert suggested that his work's long chain of arguments and historical excursions was akin to a march from the darkest night into the light

[105] This is a free translation of the Latin text edited in Nothaft 2015i: 300–3.

[106] Nothaft 2015i: 303.

[107] See the introduction, edition, and French translation in Solan 2016: 357–523.

[108] A total of 15 surviving manuscripts are listed and discussed in Nothaft 2015i: 294–5; Solan 2016: 362–87. Eight of these present Hermann Zoest's treatise together with the decree text of 1437.

of dawn. Broad daylight was finally reached by the time Zoest arrived at chapters 9 and 10, which contained the principal justification for the reform outlined in the decree of 1437. Among many other points made in this part of the treatise, Zoest underlined the care the commission had taken in devising a reform that would be easy enough to understand and implement, but without necessitating revisions to the countless numbers of books that were in daily ecclesiastical use. The removal of exactly one week from the calendar seemed ideal in light of this goal, as it caused no interruption in the sequence of seven ferial letters (A to G) and even left the 28-year cycle of dominical letters intact for the time being. The proposal of 1435, which had scheduled the omission of a bissextile day for 1436, would have disrupted this solar cycle right away. In the proposal of 1437, by contrast, the 28-year cycle could be retained in its old form until the Julian leap day was omitted for the first time in 1740, thus guaranteeing a continuity of computistical methods for the next three centuries.[109] The decision to suppress the *bissextus* once every 300 years instead of 304 years, as had initially been envisioned by Cusanus, was again due to considerations of simplicity and intelligibility, which in Hermann Zoest's opinion increased the likelihood that the necessary omission would be correctly implemented by future generations.[110]

Zoest was well aware that the simplicity and conservatism achieved by this reform was possible only because its advocates had been willing to accept certain compromises. Although the new calendar was successful at restoring Easter Sunday to the astronomically correct date in the vast majority of years, the omission of 7 days only moved the true vernal equinox back to 18/19 March. In years where the full moon fell between 19 March and the traditional equinoctial limit on 21 March, the question inevitably arose whether the Church celebrated Easter in the wrong month (in the lunation of April rather than March) or whether one should follow Nicholas of Cusa and Hermann Zoest, who both argued that the embolismic years affected by this scenario were exempted from the 'rule of the equinox'.[111] Even then, however, critics could object that a solar calendar that only corrected for an error of $J = 300$ years represented a sub-optimal solution in light of all the evidence that pointed to a much swifter retrogression for the equinoxes and solstices. While the omission of a bissextile day every three centuries was by and large sufficient to halt the calendar's 'lunar slip', it left unresolved the much greater discrepancy between the tropical year and the average Julian year, which according to the Alfonsine Tables amounted to $J = 134$ years. That these problems were not hidden to contemporary observers is clear from the sceptical comments made by John of Segovia, who underscored his point by raising the existence of a calendar of feast days 'made in pagan times' (*kalendario tempore facto gentilium*), where the vernal equinox and summer solstice were allegedly annotated for 28 March and 26 June. For modern times, the combined testimony of wise men, the general

[109] Hermann Zoest, *Phaselexis (2nd version)* 9 (ed. Solan 2016: 480–94). See also Nicholas of Cusa, *De correctione calendarii* 8 (ed. Stegemann 1955: 64, 68–72).

[110] Hermann Zoest, *Phaselexis (2nd version)* 10 (ed. Solan 2016: 496–8).

[111] Nicholas of Cusa, *De correctione calendarii* 10 (ed. Stegemann 1955: 84–6); Hermann Zoest, *Phaselexis (2nd version)* 8 (ed. Solan 2016: 460–70).

populace, and experience itself showed the dates to be 10/11 March or June, indicating a much greater calendrical gap than that covered by the reform plan of 1437.[112] Eager to forestall criticisms of this kind, Nicholas of Cusa and Hermann Zoest did their best to portray the astronomical shortcomings of the new reform proposal as an intentional act of caution in light of the uncertainty that surrounded the true length of the solar year, which included the possibility of long-term fluctuations. Rehearsing a line of argumentation familiar from Nicole Oresme and Pierre d'Ailly (see Ch. 7.1), Cusanus linked the confusing variety of available estimates to the notion that celestial motions might be incommensurable, making it impossible to find one true parameter.[113] Hermann Zoest insisted on the 'uncertainty of the length of the year', the correct measure of which 'nobody has found as yet',[114] and followed Cusanus in casting strategic doubt on the Alfonsine Tables, which placed the true vernal equinox about two days earlier than the adopted reform plan would have demanded. While Cusanus limited himself to noting past instances where the Alfonsine predictions had failed to conform to observations of the vernal equinox, for instance those made by William of Saint-Cloud and the Parisian astronomer of 1347, Zoest alluded to his own empirical labour: 'And I myself have experienced it with regard to the summer solstice, using a large quadrant.'[115]

A second trump card the two reform advocates tried to play in their respective works was the testimony of Isidore, abbot of the Constantinopolitan monastery of St Demetrios, who had come to Basel in July 1434 as the head of a delegation representing the Byzantine emperor. As a noted bibliophile and scholar, Isidore was an ideal contact for members of the calendrical reform commission, from whom they could not only receive information on the Easter reckoning employed in the Greek Church, but also learn about some of the new influences that had enriched Byzantine astronomy since the beginning of the fourteenth century. Both Zoest and Cusanus associate Isidore with a set of Persian astronomical tables, written in Greek, which can be identified as some version of the widely circulated *Persian Syntaxis* created *c.*1347 by George Chrysokokkes, a physician in Trebizond. The Persian–Greek vernal equinox, as reported by Zoest and Cusanus on the basis of Isidore's tables, fell on 13/14 March, which would have lent a certain degree of justification to the plan of only omitting 7 days from 1439.[116] Strangely enough, neither author admitted that this result was derived from mean motion tables and

112 John of Segovia, *Historia* 8.19 (ed. Birk 1873: 710). John here confuses omission of a bissextile day every 304 years with intercalating one day every 304 years. He also wavers as to whether the omission of seven days was supposed to restore the equinoctial dates at the time of Nicaea (as was the intention) or the time of Christ's birth.

113 Nicholas of Cusa, *De correctione calendarii* 2 (ed. Stegemann 1955: 18, ll. 10–28). The passage is discussed in Nothaft 2017d: 91–9.

114 Hermann Zoest, *Phaselexis (2nd version)* 7 (ed. Solan 2016: 454).

115 Ibid. 8 (ed. Solan 2016: 470).

116 Ibid. 10 (ed. Solan 2016: 502); Nicholas of Cusa, *De correctione calendarii* 6, 9 (ed. Stegemann 1955: 54, 72–6). Isidore's role is discussed in more detail in Nothaft 2015i: 217–19; 2017d: 84–90. For the origin and history of the Persian tables available in Byzantium, see Leurquin 1985; Tihon 1987; Nicolaïdis 1995; Bydén 2003: 241–57; Ragep 2014.

hence not really at odds with the Alfonsine true equinox, as a competent application of the solar equation would have easily shown.

Another item Cusanus had spotted in one of Isidore's books was an Easter table that contained the Greek equivalents to the Roman *termini paschales*. For reasons that remain unclear, the German canon lawyer went away convinced that the Byzantine Church began not just its 19-year cycle but also its sequence of common and embolismic years three years later than the Church of Rome. In this case, the planned 3-year shift to reduce the *cyclus decemnovenalis* to the *cyclus lunaris* would have had the added advantage of reuniting the lunar cycle of the Latin Church with the cycles used by the Greeks and Hebrews, thus ending a computistical divergence that, according to Cusanus and his colleague Hermann Zoest, had arisen only when the Latin Church adopted the cycle of Dionysius Exiguus.[117] Misguided as this idea certainly was, it may reveal the psychological impact of the ongoing three-way negotiations between the Council of Basel, Pope Eugene IV, and John VIII Palaeologus about the terms of a reunification between the Greek and Latin Churches, whose leaders had been going separate ways since the eleventh century.[118] A unilateral reform of the calendar on the part of the Council of Basel would have inevitably compromised the goal of a Church union, causing East and West to celebrate Easter on different dates in a great number of years. Anxieties of this sort may explain why the only preserved copy of the first draft decree of 1435 is preceded by the following brief note in what appears to be Thomas Strzempiński's hand (see Figure 7.1): 'first deliberated, but not decreed, because of the arrival of the Greeks, which was expected [at the time]' (*Primo deliberatum sed non decretatum propter adventum Grecorum qui expectabatur*).[119]

In spite of Cusanus's and Zoest's joint insistence that the planned reform was apt to bring Latins and Greeks closer together, 1439, the targeted year of reform, came and went without any further results. Signs of resistance to the reform are already visible in the conciliar protocols for 2–4 September 1437, when the three deputations for faith, peace, and reform agreed to have the submitted decree text examined by a commission of four, which was to comprise Abbot Alexander of Vézelay, Master Thomas of Courcelles, Abbot Thomas Livingstone of Dundrennan, and Nicholas, the prior of Heilsbronn Abbey in Franconia. On 6 September 1437 this motion was presented to the general assembly, but no further steps appear to have been taken. Instead, a motion dated 29 May 1439 indicates that the deputations saw the need to repeat their earlier decision of September 1437.[120] The delays that the reform project faced during these years also left their mark on the transmission of the draft decree, which in its most widely circulated version prescribed the omission

[117] Nicholas of Cusa, *De correctione calendarii* 6–8 (ed. Stegemann 1955: 50–60, 70–2); Hermann Zoest, *Phaselexis (2nd version)* 9 (ed. Solan 2016: 486–8, 492).

[118] See now Mariano 2017.

[119] See MS Kraków, Biblioteka Jagiellońska, 4164, fol. 115r, which is reproduced in Fig. 7.1, and the more detailed discussion in Nothaft 2015i: 217–23.

[120] See Haller et al. 1896–1936: vi. 103 (6 Sept. 1437), 462 (29 May 1439), 713 (16 Nov. 1439), and the excerpts that follow the decree text in MS Berlin, Staatsbibliothek, lat. fol. 246, fol. 101va, edited in Wattenbach 1884: 108–9. For discussion, see Solan 2016: 96–100.

of seven days not from May 1439, but from October 1440, demanding that the 20th of the month be followed immediately by the feast of Saints Simon and Jude on 28 October.[121] This wording was evidently known to the Bavarian theologian Johannes Keck (1400–50), who took into account the planned reform when drawing up a *Kalendarium* for the diocese of Freising during the first three days of March 1440. The calendar itself featured no major technical innovations, thus highlighting the reform's success at leaving the Golden Numbers in their accustomed places. In the accompanying wheel diagrams and computistical instructions, Keck proceeded to spell out the consequences the decree of the Council of Basel was going have for the computation of the lunar and solar cycles, which likewise only suffered minimal disruptions. The 28-year cycle of concurrents and dominical letters was going to continue as usual for the next three centuries until 1740, when the omission of the bissextile day forced the computist to jump back 12 years and treat the year in question as common or non-intercalary.[122] Keck's work reveals that he had seen a copy of the decree and expected the reform to take effect later in the same year, yet on 12 September the deputation for common affairs—which is the only deputation for which substantial protocols have been preserved—was still unwilling to let the revised decree pass the final hurdles of legislation. In reaction to a prompt received from the council's steering committee (also known as the *domini duodecim* or Committee of Twelve), the deputies decided that the council, before taking a final decision on this matter, should first seek the written advice of various universities.[123] On 12 December 1440, the same deputies followed the steering committee's recommendation in voting to refrain from any further decision concerning the calendar 'in light of the current circumstances' (*attenta disposicione temporis*).[124] This vote came to pass a year after the Council of Basel had deposed Eugene IV and elected its own pope in Felix V, thus creating a set of circumstances that was indeed everything but conducive to a reform project that was meant to affect the Church in its entirety. That the calendar reform fell prey to the renewed schism is amply confirmed by the council's historian John of Segovia, who wrote that:

> as long as, after the suspension of Eugene IV from administering the papacy, some were obedient to the council, and some to him, it seemed better not to insist on [the reform of the calendar], in order to keep a new cause of division, relating to the divine service in the Church, from arising, such that those who obeyed the pope would end up observing Lent and celebrating Easter and the other mobile feasts at one time, and those who obeyed the council at another. And since the party loyal to the holy synod was smaller, it seemed better to the fathers not to conclude upon the proposed correction [of the calendar].[125]

[121] This version appears in seven manuscripts alongside Hermann Zoest's *Phaselexis* (see Nothaft 2015i: 294–6) and in MS Vatican City, BAV, Barb. lat. 445, fols. 444r–445v, amidst other decrees of the council. It is also worth noting that five copies of Zoest's work change the year and month of reform from May 1439 to February 1441. See Solan 2016: 383–7, for an overview of the text's recensions.

[122] See the introduction, edition, and French translation of Keck's calendar in Solan 2016: 577–631.

[123] Haller et al. 1896–1936: vii. 245 (12 Sept. 1440).

[124] Haller et al. 1896–1936: vii. 293 (12 Dec. 1440).

[125] John of Segovia, *Historia* 8.19 (ed. Birk 1873: 709). See also Solan 2016: 102.

The failure of the Council of Basel to bring the reform of the calendar to fruition will have troubled few protagonists more than Hermann Zoest, who makes his final recorded appearance in the council's protocols on 16 December 1440, just four days after the project was officially swept aside.[126] He is nevertheless known to have stayed in Basel for three more years, until 1443, when the anticipation of the equinox and the error in the lunar *computus* once again caused Easter Sunday to be celebrated 35 days after the legitimate date, on 21 April instead of 17 March. Zoest addressed this scandalous error in his *Compendium paschale* (1443), which was a passionate final attempt to alert the council's remaining fathers to the necessity of a calendar reform.[127] Instead of acknowledging the adverse political situation, the work's preface spoke of the delays the reform project had incurred in the intervening years with thinly veiled bitterness:

> This holy synod of Basel did well to decide upon the eradication and reformation of this error both in matter and form. But—Oh pity!—many mock, slight, and ridicule this reform of Easter and have impeded, and still impede, its publication. The cause of this, however, is ignorance. For knowledge has no enemy except the ignorant person.[128]

Zoest's *Compendium paschale* was hence in part aimed against persistent nay-sayers and adversaries of the reform project, with whose objections he had been wrestling ever since composing his first treatise on the subject in 1424.[129] A vivid glimpse of the sort of doubts the reform initiative could engender among ordinary members of the council can be had from John of Segovia, who concluded his account of the rise and fall of the Basel 'calendar project' with a reference to the Book of Job (3:1–3), whose fate-stricken protagonist cursed the day of his birth, wishing for it to perish from the days of the year. Given that Job's intent had been unsuccessful, was it not downright presumptuous to try to remove not just one but seven days from the existing calendar?[130] Hermann Zoest's *Compendium paschale* confronted such doubters with an extensive *florilegium* of source quotations, spread out over six chapters. These showed in sequence that a failure to correct the error was contrary to (1) the mandates of Sacred Scripture, (2) the decrees of past Church councils, (3) the ordinances of past popes, (4) the writings of the Church Fathers, and (5) the standard glosses to the Bible. In addition, such inactivity was an insult to (6) the 'writings and labours of great men', which included past advocates of a calendar reform such as Reinher of Paderborn, Roger Bacon, Robert Grosseteste, Jean des Murs, Firmin de Beauval, and Pierre d'Ailly. In the seventh and penultimate chapter of his *Compendium*, Zoest returned to the error itself, showing that the incorrect lunar calculation was the current calendar's most deplorable sin, worse even than the quartodeciman practice once observed in the early Church, as it violated the biblical precepts of observing the seven days of unleavened bread. In order to remove any second thoughts as to the legitimacy of his demands, Hermann

[126] Haller et al. 1896–1936: vii. 294 (16 Dec. 1440).
[127] See the introduction, edition, and French translation in Solan 2016: 527–73.
[128] Hermann Zoest, *Compendium paschale* (ed. Solan 2016: 534).
[129] See the objections and rejoinders in Hermann Zoest, *Tractatus phase* 3 (ed. Solan 2016: 314–20).
[130] John of Segovia, *Historia* 8.19 (ed. Birk 1873: 710).

Zoest began the eighth and final chapter with a potted history of the ancient Easter controversy and the attempts of successive councils to establish a universally observed set of rules. Consistent with his earlier writings, he identified Dionysius Exiguus as the main culprit responsible for the present predicament, since it was Dionysius who had replaced the 'tables and calculations of the Hebrews' once followed by the Alexandrian patriarchs with own erroneous 532-year table.[131] When it came to the most adequate way to banish this error, the monk from Marienfeld still insisted on the one enshrined in the decree of 1437. Referring back to this proposal, Zoest reminisced how he and the members of Basel's calendrical 'task force' had invested three years of diligent labour in their project, sending letters 'to astronomers of various nations in various parts of the world, not only to Christians, but also Jews'.[132] He went on to repeat many of the arguments already familiar from his second *Phaselexis* as well as from Nicholas of Cusa's *De correctione kalendarii*, concluding that the plan adopted in 1437 was the only 'feasible or thinkable' (*dabilis vel cogitabilis*) way to salvage all the decrees of past Church councils.[133] In a final bid to animate the council's fathers and convince them to take action, Hermann Zoest finished his treatise by reminding them of their duties towards their predecessors:

> For if you want that your decrees are followed, it is proper that you also maintain and preserve the decrees of the old councils, which must be upheld and venerated like the four holy Gospels, for what you do not want to be done to you, do not do to others. After all, there could be no greater way of keeping [this] council in memory than by [carrying out] this aforementioned correction, to which future calculators of time would always pay attention and take recourse.[134]

In spite of Hermann Zoest's best efforts to turn the tide, the political will to push the matter forward seems to have completely dissipated by 1443. The only sign that the Council of Basel still dealt with the calendar problem after 1440 comes in the form of a brief document appended to the *Compendium paschale* in the only known manuscript, which used to belong to Zoest's home monastery of Marienfeld. It purports to be a communiqué or *rescriptum* the council sent out in 1444 to advise the clergy of Cologne against celebrating Easter on the astronomically correct date, which would have been 5 April (following an opposition of Sun and Moon on Thursday, 2 April). The council's *rescriptum* acknowledged the need for a correction, but also made it clear that, for the time being, all Christian were called upon to follow the accustomed *computus* of the Church, which placed Easter in 1444 one week later, on 12 April.[135] The diocese of Cologne, where the Easter date of 1444 appears to have generated a major controversy (*magna valde discensio*), was at the time one of the last in Europe to support the Basel's general council and

[131] Hermann Zoest, *Compendium paschale* 8 (ed. Solan 2016: 558).
[132] Ibid. (ed. Solan 2016: 562). [133] Ibid. (ed. Solan 2016: 566).
[134] Ibid. (ed. Solan 2016: 570).
[135] MS Copenhagen, Det Kongelige Bibliotek, Thott 825 4°, fol. 186r–v: *Rescriptum conilii Basiliensis ad clerum Coloniensem de celebratione Pasche anno 1444º quia tunc fuit valde magna discensio pascalis festi.* The text is edited in Nothaft 2015c: 146–7, and again in Solan 2016: 570–3 (with French translation).

its pope Felix V, whose power-base was rapidly diminishing. The same held true for the diocese of Freising in Bavaria, where Cardinal-Bishop Johannes Grünwalder saw the need to ask the aforementioned Johannes Keck to adjudicate in a similar dispute in the very same *annus confusionis*.[136]

Perhaps related to these events, a public disputation held at the University of Erfurt in the summer of 1444 gave the Arts master Heinrich of Runen an opportunity to determine a quodlibetal *quaestio* on whether or not the calendar should be brought back into harmony with the rules of the ancient Church. Although the preserved text nowhere mentions the reform efforts made in Basel, some of Heinrich's remarks went to the heart of the dilemma faced by Hermann Zoest and his colleagues during the preceding years, showing more clearly than other documents why the solution they ended up proposing was viewed with scepticism by some of their contemporaries.[137] As Heinrich of Runen explained in the course of his carefully worded argument, a *correctio reformativa*, which sought to restore the calendar to its proper state, could proceed in one of two ways: *per subtractionem* or *ablacionem*, that is, by erasing or adding days to move the phenomena back to their old dates, or *per transpositionem*, which was to adjust the possible dates of Easter to the current true date of the vernal equinox (or the Golden Numbers to the actual new-moon dates). It is the former approach that had characterized the decree of 1437, which prescribed the erasure of seven days in an attempt to bring both the lunar phases and the spring equinox back on track. The main problem with this approach was that it did not harmonize with the Alfonsine Tables, which placed the present equinox on 11 March and hence 10 days earlier than where it was supposed to be according to the traditional rules. Heinrich of Runen's argument accordingly assumed that 10 days was the necessary interval for correcting the ecclesiastical calendar *per ablacionem*, but he was also quick to point out that this measure conflicted with the lunar component of this same calendar, which lagged behind the actual new and full moons by only 3 days. To eject 10 days from the calendar would therefore have caused its new-moon dates to stray in the opposite direction by 7 days. Heinrich's conclusion was that this mode of correction had no potential: 'It therefore clearly must be rejected' (*Patet ergo repugnandum*).[138] Like the correction *per subtractionem*, the alternative approach *per transpositionem* had been considered and tried out by the delegates in Basel, as revealed by the decree of 1435. The latter envisaged a *transpositio* of both the Golden Numbers and the *termini paschales* to new dates, followed by periodic omissions of days to keep them in their correct places—what Heinrich of Runen called a *correctio conservativa*. A *transpositio* was in many ways the preferable approach, as the Erfurt master happily

[136] Johannes Keck's response letter to Grünwalder, which shows signs of being influenced by Hermann Zoest's *Phaselexis*, is edited and discussed in Nothaft 2015c. The background to the dispute of 1444 is further elucidated by an anonymous text *De intervallo notabile et ciclo clavium* in MS Wolfenbüttel, Herzog-August-Bibliothek, Cod. Guelf. 36.21 Aug. 2°, fol. 358r.

[137] See Solan 1999, for a detailed analysis, followed by an edition of the text.

[138] Solan 1999: 219. How an omission of 10 days could be reconciled with the correction of the 19-year cycle was shown in 1488 by the Leuven professor Peter de Rivo, whose *Reformacio kalendarii Romani* is discussed in Ch. 8.

acknowledged, but it nevertheless ran into an impasse caused by the way the current practice of dating Easter was legally protected by the common custom (*approbata conswetudo*) of the Church. The only institution capable of changing this custom was the Church itself, but only if it acted in its entirety, lest it be hurled into a schism.[139] Nothing in Heinrich of Runen's disputation revealed that a new schism had already arisen and that the Council of Basel and its calendar project were on the losing side of it.

7.3. CALENDRICAL ASTRONOMY AND THE PRINTING-PRESS IN THE LATER FIFTEENTH CENTURY

With the slow demise of the Council of Basel, which continued until its dissolution in April 1449, advocates of a calendar reform, especial those less attached to the conciliar cause than Hermann Zoest, would have naturally looked towards Pope Eugene IV for political and institutional support. One of the few pieces of evidence to suggest that the pope was indeed petitioned in this way, and that the calendar reform was discussed outside the conciliar setting, occurs in an obscure treatise on the construction of a sundial by the Italian mathematician Leonardo of Cremona (*fl.* 1404/38), who alludes to his having sent a calendar-reform proposal not just to the Council of Basel, 'when it was still vigour', but also to Eugene IV and to 'England and Paris'. Its tenor, he claimed, was even known in Spain.[140] Another attempt to get the current pope interested in the issue was apparently made by Johannes de Vesalia, doctor of medicine at the University of Leuven and great-grandfather of the anatomist Andreas Vesalius, who addressed to Eugene IV an *Exhortatio super calendarii correctione*. This, at least, was reported in the sixteenth century by the Flemish theologian Jan Vermeulen (Johannes Molanus, 1533–85), according to whom a copy of this treatise, once in the library of Groenendael Priory, had already perished in his own day.[141] An additional sign that the calendar issue may have been discussed in the presence of Eugene IV can be spotted in a *Defensorium* of three propositions that the Castilian theologian Alfonso de Madrigal (or Alfonso Tostado, 1410–55) felt the need to write down in 1443, after his participation in a disputation at the papal court in Siena had caused offence and even accusations of heresy.[142] The *Defensorium* concluded with a series of chapters on the calendar in which Madrigal decried the numerous errors encountered in contemporary *kalendaria*, which often gave incorrect dates for the solstices and equinoxes as well as false information on the positions of the Sun in the

[139] Solan 1999: 219.

[140] Leonardo of Cremona, *Ars instrumenti horologici pro tempore sereno*, MS Paris, BnF, lat. 7192, fols. 83r–96r, at fol. 83r: 'Mea siquidem vi modica correctionis huius exemplaria transmisi ad concilium Basiliense, iam diu dum vigeret, et ad dominum papam Eugenium, ad Angliam et Parisium. Cuius tenor scio quod viget in Hispania.' See Favaro 1904; Marzi 1906: 640; Solan 2016: 46, 58–9.

[141] Molanus 1861: 560. See also O'Malley 1964: 422 (n. 11).

[142] For the details of this affair, see Nothaft 2012: 203–12.

zodiac, daytime lengths, or the historical date of creation.[143] Apart from the complaints voiced in his *Defensorium*, the Spaniard is also supposed to have written a complete treatise on the *Error del calendario*, which was listed among his lost works by his sixteenth-century bibliographer Rainerus Bovosius.[144] Some idea of its content may possibly be gained from chapter 86 of his massive five-volume Castilian commentary on the chronicle of Eusebius (*Tostado sobre el Eusebio*), where he gives the length of the true solar year in accordance with the Alfonsine Tables (365d 5;49,16h), but ends up rounding the implied Julian slip from $J = 134$ years to $J = 140$ years. On this estimate the winter solstice in the year of Christ's birth should have fallen roughly 10 days later than was the case in Madrigal's own time, yet his Eusebius commentary completely ignored this implication and instead assumed a shift of 12 or 13 days, based on a comparison between the canonical date of Jesus's birth (25 December) and the presently valid date of the solstice (12/13 December).[145]

A relatively weak grasp of the astronomical subtleties involved in the calendar problem was also displayed a quarter of a century later by one of Madrigal's pupils, the theologian Pedro Martínez de Osma (*c.*1424–80), who in 1468 delivered a twofold disputation on the date of Christ's death and the correction of the calendar at the University of Salamanca.[146] Like Heinrich of Runen in his Erfurt disputation, Osma gave the error of the Julian calendar as $J = 100$ years, but instead of tracing this value back to the solar year determined by al-Battānī—as Grosseteste and Campanus of Novara had done two centuries prior and as Osma's student Antonio de Nebrija was to do some decades later[147]—the Castilian theologian confronted readers with the baseless historical assumption that the Roman calendar was first constructed at a time when the beginnings of January, April, July, and October had still coincided exactly with the cardinal points of the solar year. The winter solstice, which was now found on 12 December, had once occurred on 1 January, while the date of the summer solstice had regressed from 1 July to 11 June. On the assumption of a solar year of $365\frac{1}{4} - \frac{1}{100}$ days, the 20-day shift implicit in these dates would have taken no more and no less than two millennia to arise, but this did not keep Osma from employing an inclusive count of 21 days, asserting that the calendar had been constructed a little more than 2,100 years before his own time.[148] His anachronistic view of the history of the Roman calendar can be

143 See Alfonso de Madrigal, 'Defensorium trium propositionum' (chs. 86–97), in *Opera omnia*, vol. 25 (Venice: Balleon, 1728), cols. 151–64

144 See the list of 'Opusculorum, quae hactenus inventa non sunt', in Alfonso de Madrigal, *Commentaria in Genesim* (Venice: Sessa, 1596).

145 Alfonso de Madrigal, *Tostado sobre el Eusebio*, vol. 1 (Salamanca: Hans Gysser, 1506), fol. 122v.

146 This text, which appears to have originated as part of a commentary on Peter Lombard's *Sentences*, is preserved under the title *Disputatio de anno in quo possimus dicere Dominum fuisse passum et de quibusdam erratis in kalendario*. It was edited by Labajos Alonso 2010: 354–83, with a Castilian translation by Pabo García Castillo. For corrections and a new transcription of the second part (dealing with calendar), see Nothaft 2013a: 544–9. The existence of Osma's treatise was first noted by Marzi 1906: 640–1. The present discussion relies on the more detailed analysis in Nothaft 2013a.

147 For Antonio de Nebrija's views on the issue, see the brief text *De anni correctione*, which is discussed, edited, and translated by Martín Baños 2015: 252–3, 256–7.

148 Pedro de Osma, *Disputatio* 2.3 (ed. Nothaft 2013a: 547).

regarded as a throwback to the thirteenth century, when some computists repeatedly treated 1 January as the original date of the winter solstice, ignoring the well-attested fact that the Roman calendar had become a solar calendar only with Julius Caesar's reform. The last advocate of a reform that was supposed to restore the solstice to the start of the year had been the anonymous author of 1276 (see Ch. 4.3), who wished the pope would omit 16 days from the calendar. Pedro de Osma, given his diverging assumptions, even called for an omission of 21 days, although he found dropping 14 inclusive days a reasonable compromise, accepting that this would be enough to move the vernal equinox and winter solstice back to the historical dates of Christ's conception and birth. Similar to the proposal of 1276, he even seriously considered the possibility of moving all immobile feast days along with this correction, having All Saints Day fall on 14 November instead of 1 November.[149]

Osma's remarks about the Julian calendar reflect an overall lack of concern for the Alexandrian rule of the equinox, which implied that a successful reform of the calendar should fix the vernal equinox not on 1 April or 25 March, but on 21 March. His reasons for ignoring this principle were located in his parallel plans for the lunar calendar, which followed an idea familiar since the days of Reinher of Paderborn, namely to tie the paschal lunation to the month of Nisan in the Jewish calendar. For the mid- to late fifteenth century, when 1 Nisan could fall as early as 1 March, this proposal implied a shift of Easter limits by up to 7 days, which was still less than the gap between the present and canonical equinox.[150] The question remained whether or not the Church could be persuaded to abandon long-standing customs, for instance when it came to counting common and embolismic years. Pedro de Osma was conscious that 'it is hard to abandon what one is used to, and perhaps it is hard for the Church to accept these [suggestions]', but he also insisted, backed by a quotation from Pope Gregory IX, that custom was no excuse for error.[151] A conservative option would have been for the Church to maintain its own cycle but bring it closer to the Jewish calendar, and to reality, by reducing the new-moon dates by 4 days. An additional 1-day adjustment after every 360 years was enough, in Osma's schematic understanding of the matter, to keep the Golden Numbers in their right places.[152] At the same time, he reminded listeners that a more rigorous submission to Jewish calendrical principles was possible and perhaps even preferable, going so far as to suggest that one should 'insert the months of the Jews and their names into our calendar, just as ours are found in their [calendar]'.[153]

Had Pedro de Osma sought a wider reception for these ideas, he might have done for his chronological *Disputatio* what he did *c.*1472 for his commentary on

[149] Ibid. [150] Ibid. 2.1 (ed. Nothaft 2013a: 545).

[151] Ibid. 2.2 (ed. Nothaft 2013a: 547), quoting Gregorius IX, *Decretales* lib. I, tit. 4, ch. 11 (ed. Friedberg 1879–81: ii. 41).

[152] See the discussion of this estimate in Nothaft 2013a: 532–4.

[153] Pedro de Osma, *Disputatio* 2.4 (ed. Nothaft 2013a: 548). Material on the Jewish calendar was also included in the report on calendar reform the University of Salamanca sent to Rome in 1515. See the edition in Carabias Torres 2012: 309–16, and the remarks in Nothaft 2014d: 622–5.

the pseudo-Athanasian Creed (*Quicumque vult*), which became one of the first books ever printed in Spain, thanks to the theologian's collaboration with the German printer Johannes Parix.[154] Quite apart from Osma's personal involvement in the early history of the printing-press in Europe, the rise of this revolutionary technology in the second half of the fifteenth century created an attractive new range of opportunities for the dissemination of calendars and related astronomical material, which was greatly helped by the relatively small size of the relevant texts and the large audience they were able to attract. That the calendar was already present at the very birth-pangs of European printing with movable type can be seen, for instance, from the so-called *Türkenkalender*, which was issued in 1454 by Johann Gutenberg's press in Mainz. Although at its core a political pamphlet calling European rulers to military action against the Ottomans, the work was nevertheless structured according to the 12 months of the Julian calendar and furnished with computistical data and the new-moon times for the year 1455. It thus became the first among a handful of printed German 'lunation tracts', in which verse poems (mostly with political invective) were placed in a calendrical and prognostic setting.[155] A related, but vastly more popular genre was the annual broadside almanac, which typically offered astronomical data such as timed syzygies together with computistical information and rules for phlebotomy for a particular year. More than 450 titles of this genre are known to have been produced in the Holy Roman Empire alone between 1457 and 1501, making the broadside almanac one of the dominant printed formats of the time.[156] On the Iberian peninsula, the mass market that existed even for sophisticated forms of calendrical astronomy is best exemplified by the *Lunari* first issued in 1485 by Bernat de Granollachs (d. 1487), who was a member of Barcelona's lesser nobility. At its core the *Lunari* was a lunar almanac listing the precise times (down to the minute of the hour) of all 1,632 conjunctions and oppositions from 1485 to 1550, based on a 31-year cycle of true syzygies that had originally been devised in the fourteenth century by the Jewish astronomer Jacob ben David Bonjorn. Each page contained the lunar data for one year, supplemented by predicted eclipses, the Golden Number, the dominical letter, the interval from Christmas to Lent, and the dates of the mobile feast days.[157] Granollach's *Lunari* quickly became one of the best-selling scientific books of the incunabula era, with 42 editions printed before 1500. Aside from the initial Catalan and Latin editions, these included translations into Castilian, Portuguese, and Italian as well as a version of Granollach's lunar tables incorporated into the more voluminous *Reportorio de los tiempos* by the contemporary astrologer Andrés de Li, which was first printed in Zaragoza in 1492. All in all, at least 90 editions of

[154] Reinhardt 1977; Marcos Rodríguez 1979–80. [155] Geldner 1975; Simon 1988.

[156] Eisermann 2004: i. 75–252, 257–63, 275, 345–6, 393–6, 423–41, 447, 567–8, 572–4, 578–84; ii. 30–1, 34–5, 44, 58, 118–20, 227–9, 232, 412, 431–5, 452, 539, 541–6, 576–7, 589–94, 623–9. See also Heitz 1905; Zinner 1964: 93–135; Brévart 1988b: 338–42; Green 2012; Hayton 2015: 118–44; Kremer 2017.

[157] See the study and facsimile of the first edition in Chabás and Roca 1985, and the astronomical analysis in Chabás and Roca 1998: 127. Chabás 2000: 337–8 mentions two further fifteenth-century Iberian lists of true conjunctions calculated with Bonjorn's method.

the *Lunari* were produced until the end of the sixteenth century, with a total print-run that is estimated to have reached 50,000 copies.[158]

North of the Alps a role similar to that of the Iberian *Lunari* was played by printed calendars like the one published in 1477/8 by Jakob Pflaum in the town of Ulm, which offered true syzygies and eclipse times with a 76-year range from 1477 to 1552.[159] An important provider of the numerical raw material and computational procedures used in publications of this type was the astronomer Johannes Müller, who for a brief period during the 1470s ran what may qualify as Europe's first scientific printing house at his Nuremberg home.[160] Born in 1436 in the Franconian town of Königsberg, from which later writers would derive the Latin name 'Regiomontanus', Müller spent his formative years at the University of Vienna (1450–61), where he became a student of Georg Peurbach. In 1460, Regiomontanus made his first acquaintance with Cardinal Basilius Bessarion, who had come to Vienna as a papal legate to secure the support of Emperor Frederick III in a planned crusade against the Turks. At this occasion, Bessarion encouraged Peurbach to produce an updated abridgement of Ptolemy's *Almagest*, which was meant to clarify the sometimes obscure Latin text and correct long-standing mistakes. Peurbach's premature death in April 1461 left Regiomontanus with the task of following Bessarion to Italy to finish the work, using Greek manuscripts to improve upon Gerard of Cremona's translation from Arabic. The resulting *Epitome of the Almagest* (1462/3) received its first printed edition in 1496 and went on to become a major source of advanced astronomy in the sixteenth century.[161] By the end of his Italian sojourn, which lasted a total of five years, Regiomontanus also had developed acute doubts regarding the Alfonsine Tables and their various derivatives, whose predictions regularly failed to agree with his own observations of planetary positions and eclipses carried out since the late 1450s. In a letter sent in 1464 to his colleague Giovanni Bianchini, he voiced his misgivings about various aspects of the received tables, stating that the Alfonsine model for precession and trepidation—just like the older model attributed to Thābit—failed to agree with observed and historical values for the obliquity of the ecliptic. Problems were also encountered in the Alfonsine predictions for lunar eclipses, which could be observed up to an hour earlier than announced by the calculators.[162]

Regiomontanus's dissatisfaction with the existing computational tools eventually led him to investigate the possibility of alternative, homocentric planetary models

[158] The editions and print-run are documented in Chabàs and Roca 1985: 39–56, 91–104. On Andrés de Li's *Repertorio*, see Lluís Martos 2014, and the edition in Delbrugge 1999. See also Zironda 1985, for a facsimile edition of the Italian *Summario de la luna*.

[159] Jakob Pflaum, *Kalender* (Ulm: Johann Zainer der Ältere, *c.*1478). See Zinner 1964: 11–23; Brévart 1988a; Kremer 2012.

[160] The standard account of Regiomontanus's life and works is Zinner 1968. See also Shank 2017 and the essays edited in Hamann 1980. Specifically on his printing business, see Wingen-Trennhaus 1991; Shank 2012.

[161] Zinner 1968: 79–144; Rose 1975: 90–117; Mett 1989; Rigo 1991; 1994; Malpangotto 2008: 33–6; Shank 2017: 87–93.

[162] The pertinent passage is edited in Curtze 1902: 263–6. See also the English translation and commentary in Swerdlow 1990: 170–90, and the German translation and commentary in Gerl 1989: 102–7, 269–314. See also Curtze 1902: 326, for the letter Regiomontanus sent to Christian Roder in 1471.

and to engage in the systematic collection of empirical data, an activity that intensified following his relocation to the city of Nuremberg in 1471. The other prong of his plan to reform contemporary astronomy consisted in the production of reliable printed editions of the most important works on mathematics and the science of the stars. Among the eight publications that eventually left Regiomontanus's printing house were his own 900-page *Ephemerides*, first issued in 1474, which showed the true longitudes of the Sun, Moon, and five planets for each day of the year between 1475 and 1506.[163] Despite Regiomontanus's reformist aspirations, his *Ephemerides* were based entirely on the Alfonsine Tables, such that corrections were made only for local longitude. The same holds true for Regiomontanus's more modestly sized *Kalendarium*, which first appeared on the market in 1474 in both Latin and German versions and went on to spawn nearly 20 editions until the end of the century, with numerous further calendars and almanacs copying or adopting its principles.[164] As so often, one of the *Kalendarium*'s main attractions was a series of tables showing the times of all conjunctions and oppositions for a given month of the calendar, which in Regiomontanus's case covered the 57-year period from 1475 to 1531 (see Figure 7.3). Unlike previous authors such as John of Gmunden, whose calendrical tables offered mean syzygy times with a steady interval between them, Regiomontanus calculated true (or 'equated') times from noon for the meridian of Nuremberg and adjusted the result to the apparent solar time (i.e. by adding the equation of time). The remaining pages of his work included eclipse diagrams, tables, and instructions for the calculation of the mobile feast days, as well as construction plans for sundials and a volvelle-type equatorium for the course of the Moon. Towards the end, the Latin version added two pages on the errors that currently affected the calculation of Easter Sunday. Regiomontanus's discussion of this long-standing problem culminated in a table covering the years 1477–1531, which listed and compared all instances where the date of Easter 'according to the practice of the Church' (*iuxta usum ecclesie*) clashed with the legitimate date 'according to the decrees of the fathers' (*iuxta decreta patrum*), which was based on the true date of the vernal equinox and the opposition of Sun and Moon. The table listed no fewer than 30 such problematic or paradoxical cases for a period of just 55 years, with a separate column signalling that the discrepancy amounted to a week in 18 of these years, to 28 days in another five, and to 35 days in seven more instances (see Figure 7.4). No wonder, then, that:

> there frequently arise certain small questions concerning the paschal Moon such as the one brought to Cardinal Bessarion when he exercised the office of Legate in Venice. Such [questions] have also been thrown before me on many occasions by many people and by some Jews.[165]

Regiomontanus did not offer further specifics regarding his discussions with Jews, but it is striking to also find them alluded to in a calendrical *consilium* the Viennese

[163] Zinner 1968: 185–98; De Meis 2006: 401–6.

[164] Kaltenbrunner 1876: 367–74; Pfaff 1944: 109–15; Zinner 1968: 198–206, 347–51; Reich 1990: 352–5; Kremer 2012: 325–6, 328–9; 2013.

[165] Johannes Regiomontanus, *Kalendarium* (Nuremberg, 1474), fol. 30r (not paginated). See Mütz 1996: 79, who discusses a similar passage in a contemporary calendar in the German vernacular.

	·1475·						·1494·						·1513·					
	CON			OPPO			CON			OPPO			CON			OPPO		
1	8	23	46	4	2	25	8	8	58							4	8	30
2				12	21	15				12	5	12	8	1	11	12	18	44
3							16	18	31				16	2	7	1	23	9
4	16	6	48							1	15	19	5	20	58			
5	5	18	17	1	1	1	5	11	18							9	18	14
6				9	10	48				9	6	25						
7	13	1	45	17	17	41	13	2	41	17	18	15	13	0	33	17	17	50
8	2	10	13				2	9	40				2	10	17			
9	10	21	46	6	3	31	10	17	9	6	1	13	10	17	41	6	2	10
10				14	7	24				14	16	58				14	9	42
11	18	4	54				18	13	2				18	3	7	3	21	21
12	7	23	7	3	2	12				3	9	39	7	17	22			
13				11	19	16	7	7	7							11	12	40
14	15	19	27				15	14	33	11	4	36						
15				19	11	26				19	8	53	15	1	55	19	22	46
16	4	3	2	8	18	26	4	2	18	8	18	55	4	18	54			
17	12	9	56				12	10	54				12	8	26	8	14	9
18				16	5	54	1	18	13	16	1	47	1	17	51	16	2	27
19	1	0	53	5	20	45				5	11	25				5	9	22
20	9	17	27				9	5	28				9	1	15			
21										13	14	38	17	20	25	13	0	29
22				13	6	42	17	12	6							2	17	5
23	17	3	6	2	22	35				2	9	26	6	14	18			
24	6	17	5	10	10	41	6	6	25							10	11	46
25	14	3	8	18	18	37				10	2	42	14	22	8			
26	3	10	22				14	3	27	18	19	33				18	16	37
27	11	19	37	7	2	38	3	11	9	7	19	25	3	10	10			
28							11	18	2	15	13	32	11	18	57	7	2	54
29	19	22	6	15	4	0	19	8	20							15	9	49
30				4	19	29				4	4	5	19	2	6	4	19	7
31	8	16	54				8	0	36	12	21	45	8	13	1			

Figure 7.3. Johannes Regiomontanus, *Kalendarium* (Nuremberg, 1474), syzygy times in March for 1475–1531. Oxford, Bodleian Library, Auct. VII. Q. VII. 50, fol. 3v.

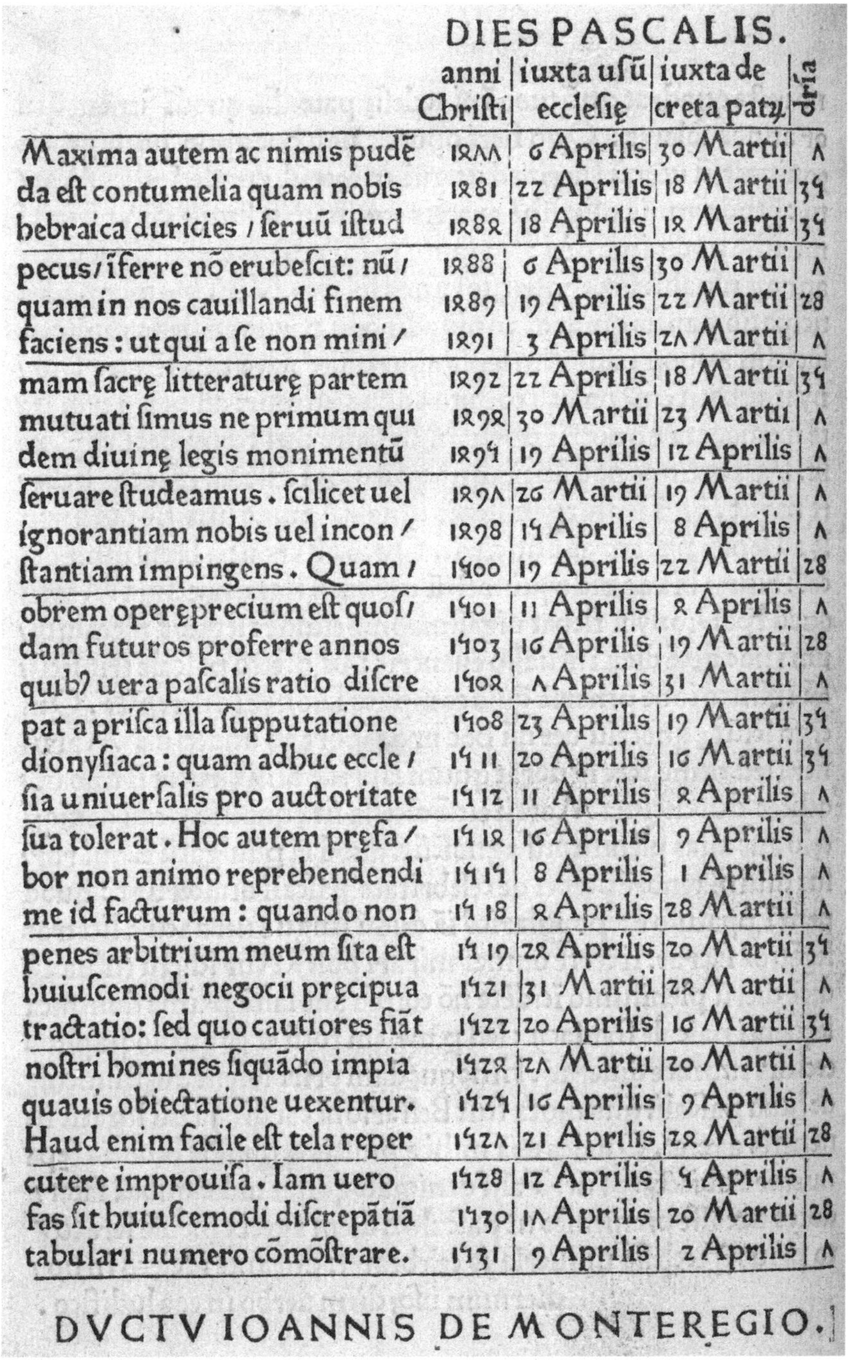

DIES PASCALIS.

	anni Chriſti	iuxta uſũ eccleſię	iuxta decreta patꝝ	dñica
Maxima autem ac nimis pudẽ	1477	6 Aprilis	30 Martii	7
da eſt contumelia quam nobis	1481	22 Aprilis	18 Martii	35
hebraica duricies / ſeruũ iſtud	1484	18 Aprilis	14 Martii	35
pecus / ĩferre nõ erubeſcit: nũ /	1488	6 Aprilis	30 Martii	7
quam in nos cauillandi finem	1489	19 Aprilis	22 Martii	28
faciens: ut qui a ſe non mini /	1491	3 Aprilis	27 Martii	7
mam ſacrę litteraturę partem	1492	22 Aprilis	18 Martii	35
mutuati ſimus ne primum qui	1494	30 Martii	23 Martii	7
dem diuinę legis monimentũ	1495	19 Aprilis	12 Aprilis	7
ſeruare ſtudeamus. ſcilicet uel	1497	26 Martii	19 Martii	7
ignorantiam nobis uel incon /	1498	15 Aprilis	8 Aprilis	7
ſtantiam impingens. Quam /	1500	19 Aprilis	22 Martii	28
obrem operępręcium eſt quoſ /	1501	11 Aprilis	4 Aprilis	7
dam futuros proferre annos	1503	16 Aprilis	19 Martii	28
quib⁊ uera paſcalis ratio diſcre	1504	7 Aprilis	31 Martii	7
pat a priſca illa ſupputatione	1508	23 Aprilis	19 Martii	35
dionyſiaca: quam adhuc eccle /	1511	20 Aprilis	16 Martii	35
ſia uniuerſalis pro auctoritate	1512	11 Aprilis	4 Aprilis	7
ſua tolerat. Hoc autem pręfa /	1514	16 Aprilis	9 Aprilis	7
bor non animo reprehendendi	1515	8 Aprilis	1 Aprilis	7
me id facturum: quando non	1518	4 Aprilis	28 Martii	7
penes arbitrium meum ſita eſt	1519	24 Aprilis	20 Martii	35
huiuſcemodi negocii pręcipua	1521	31 Martii	24 Martii	7
tractatio: ſed quo cautiores fiãt	1522	20 Aprilis	16 Martii	35
noſtri homines ſiquãdo impia	1524	27 Martii	20 Martii	7
quauis obiectatione uexentur.	1525	16 Aprilis	9 Aprilis	7
Haud enim facile eſt tela reper	1527	21 Aprilis	24 Martii	28
cutere improuiſa. Iam uero	1528	12 Aprilis	5 Aprilis	7
fas ſit huiuſcemodi diſcrepãtiã	1530	17 Aprilis	20 Martii	28
tabulari numero cõmõſtrare.	1531	9 Aprilis	2 Aprilis	7

DVCTV IOANNIS DE MONTEREGIO.

Figure 7.4. Johannes Regiomontanus, *Kalendarium* (Nuremberg, 1474), comparative table of 'paradoxical' Easter dates. Oxford, Bodleian Library, Auct. VII. Q. VII. 50, fol. 30v.

professors Andreas Stiborius (Stöberl) and Georg Tannstetter wrote in 1514 at the behest of Emperor Maximilian I. Like Regiomontanus, the two astronomers strongly advocated the use of true syzygies and equinoxes for computistical purposes, warning that the use of mean longitudes could cause the calculated Easter to be separated from the true one by a whole month, since the true and mean opposition could fall on different sides of the vernal equinox. Stiborius and Tannstetter announced that more about these subtleties could be gleaned 'from the writings of the great and never sufficiently praised astronomer Johannes Regiomontanus', in particular his 'table on the celebration of Easter at the end of his *Kalendarium*',[166] and went on to mention the year 1503 as an instance where Regiomontanus used the difference between true and mean longitudes to convict the Jews of erroneously celebrating their Passover in the second spring month. In the Jewish calendar, 15 Nisan of AD 1503 fell on Tuesday 11 April, which agreed with the way the mean opposition of March (12 March) had preceded the mean equinox (13 March) in the year in question. The Church observed the same postponement, celebrating Easter on 16 April, the Sunday after Passover. Yet a look at the true syzygy times and solar longitudes revealed that both feasts should have occurred a month earlier. If Vienna was used as a reference meridian, the true vernal equinox had occurred at noon on Saturday, 11 March, followed by a true opposition *c.*11 hours later, at the start of Sunday, 12 March. The Jews, or so the argument went, therefore should have celebrated Passover on the latter date, with Christian Easter following a week later, on 19 March. The divergent Easter dates are indeed recorded in the table at the end of Regiomontanus's *Kalendarium*, but, contrary to what the remarks of Stiborius and Tannstetter suggest, there was no mention here of any error in the Jewish celebration of Passover. The *consilium* by the two Viennese professors nevertheless reports that:

> when it was pointed out to the Jews that they were going to err in this year, they rose up against Johannes Regiomontanus in a surprising way and ridiculed him in a most shameful manner, asserting that this master and great Christian astronomer had erred. Yet Johannes offered to prove that the Jews were in error and to demonstrate this before suitable judges.[167]

Whether the argument was ever settled in this way we do not know, and the reliability of Stiborius's and Tannstetter's report—which may be no more than a creatively amplified retelling of the remarks already contained in Regiomontanus's *Kalendarium*—is certainly up for debate. Even if there is more to the story, it is not especially likely that Regiomontanus would have had much success convicting the Jews of any violation of their own calendrical rules. After all, the claim that Passover in 1503 fell a month too late depended on the notion that Passover had to follow as soon as possible after the vernal equinox. As Hermann Zoest could have told the

[166] Andreas Stiborius and Georg Tannstetter, *De Romani calendarii correctione consilium* (Vienna: Singrenius [*c.*1515]), sig. A4r. See also the references to Regiomontanus on sigs. A2r, A4r–v, and the summaries of the work in Kaltenbrunner 1876: 386–90; Marzi 1896: 96–100; Graf-Stuhlhofer 1996: 125–8; Steinmetz 2011: 66–7; Hayton 2015: 122–6.

[167] Stiborius and Tannstetter, *De Romani calendarii correctione consilium*, sig. A4v.

astronomer, this was not a principle the present-day Jewish calendar observed in any strict sense. According to Regiomontanus's own words, doubts concerning the date of Easter had also been brought before Basilius Bessarion, which may be an allusion to the letter his erstwhile patron sent to Pope Paul II in 1470.[168] In this letter, the Greek-born cardinal insisted that a change to the existing paschal tables was neither practical nor desirable and that the existing error could be tolerated, since it ensured Passover and Easter were perpetually kept apart. Bessarion's attitude to the calendar problem, which was characterized by the same Byzantine type of conservatism already expressed by Barlaam of Seminara in the fourteenth century (see Ch. 6.3), hence differed to an extent from that displayed by his one-time protégé Regiomontanus, who reminded readers of his *Kalendarium* of a famous story according to which the clergy of Bremen once 'observed the paschal day nearly an entire month before the universal Church celebrated it, such that a proverb that makes fun of them has lasted down to the present day'.[169] While this fifteenth-century anecdote about the feckless 'asses from Bremen' existed primarily as a means of emphasizing the importance of Christian unity in the celebration of Easter—which could be endangered by computistical incompetence as much as by heretical intent—there were also reasons to be less complacent than Bessarion about the way the Catholic Church currently fixed its feast days. Chief among those was the:

> shameful indignity that the obdurateness of the Jews, that servile flock, does not blush to inflict upon us, never ceasing to mock us, [claiming that] we, who have borrowed not the least part of Sacred Scripture from them, do not even devote ourselves to keeping the fundamental principles of the divine law, that is to say, they accuse us of either ignorance or inconsistency. For this reason it is worthwhile to set out certain future years in which the correct determination of the Pasch disagrees with that old Dionysiac computation that the universal Church still endures as its authority. But I say beforehand that I will not do this with the intention of assigning blame, since it is not within my authority to offer a special analysis of this matter, but so that our men will be more prudent if they are ever troubled by any irreverent charges brought against them. For it is not easy to repel unforeseen weapons.[170]

It is interesting to note that none of this material appears in the German version of the *Kalendarium*, first printed in 1474, which suggests that Regiomontanus did not deem his admissions to the faults of the ecclesiastical calendar suitable for a general lay audience—an attitude reminiscent of Hermann Zoest's worries that the situation might scandalize the common people.[171] Further evidence for Regiomontanus's personal interest in the issue of calendar improvement comes from a broadside prospectus printed by his Nuremberg press in 1474, in which he provided an extensive list of all the astronomical, mathematical, and cartographic

[168] The text of Bessarion's letter was last edited by Mohler 1942: 546–8. See also n. 82 in Ch. 6.3.

[169] Regiomontanus, *Kalendarium*, fol. 30r. Regiomontanus here refers to the rhyme starting *Bremenses azini*, for which see n. 7 in the Introduction to this book.

[170] Regiomontanus, *Kalendarium*, fol. 30v.

[171] A facsimile print of the 1474 German edition appears in Zinner 1937.

works he hoped to publish in the coming years. Among the titles mentioned, but never brought to fruition, was a treatise *De motu octavae spherae contra Tebith suosque sectatores*, which presumably took up some of the issues Regiomontanus had already discussed with Bianchini in 1464, as well as a treatise *De instauratione kalendarii ecclesiae*.[172] Neither of these works has ever surfaced, although the ambiguous statements made by Stiborius and Tannstetter in 1503 have led modern scholars to conclude that *De instauratione kalendarii ecclesiae* was indeed completed.[173] Its one-time existence might explain a puzzling reference to Regiomontanus's *Consilium de reformatione calendarii* by the Lutheran reformer Martin Chemnitz (1522–86), who remembered seeing this text 'in Biblioteca Prutenica' at some time before 1584.[174] Another curious document that has been connected to the German astronomer appears on three pages in an early sixteenth-century manuscript once owned by the humanist Willibald Pirckheimer (1470–1530). They contain 18 propositions on the correction of the Roman calendar, which appear to represent no more than the rough outline of a more extensive treatise. The author of this text based his calculations on the coordinates of Nuremberg, which he claims to have a northern latitude of 49;24,30° (prop. 3) and lie the equivalent of 1;52h west of Alexandria (prop. 5). Observations made for the meridian of this city convinced him that the best way to keep the Julian calendar in line with the solar year was to omit three bissextile days over the course of 400 years, thus implying a year length of $365\frac{97}{400}$ days (props. 10, 11). The close resemblance between this solution and the correction later implemented as part of the Gregorian reform of 1582 is further underscored by the author's recommendation that 10 days should be omitted from a single year in order to restore the vernal equinox to its location at the time of the Council of Nicaea (props. 12, 13).[175]

The evidence provided by this anonymous and undated text is tantalizing, but one cannot go further and assert with any confidence that it offers a genuine glimpse into Regiomontanus's lost work on calendar reform. Indeed, it is conspicuous that Regiomontanus's *Kalendarium* resorted to true syzygies in order to find the legitimate date of Easter, whereas the author of the 18 propositions believed the calendar's lunar component should be corrected using tables for mean syzygy.[176] What does seem to be incontrovertible is that Regiomontanus's reputation as an astronomer and his engagement with the Easter question eventually earned him a

[172] The advert is reproduced in Schmeidler 1972: 533, and edited in Malpangotto 2008: 147–54. See also Zinner 1938; 1964: 3–11; 1968: 178–80; Malpangotto 2008: 52–4, 101–7; Wingen-Trennhaus 1991: 27–32.

[173] See Kaltenbrunner 1876: 371; 1877: 523, and Zinner 1968: 201; 1988: 125, who claims that Johannes Eck attributed to Regiomontanus some Easter tables for the years 1514–52, which are not contained in the *Kalendarium*. In actual fact, Eck simply extended Regiomontanus's existing table for 1475–1531 to 1552, as he expressly states in *De vera Paschae celebratione* (Augsburg: Miller, 1515), sig. b4r.

[174] Martin Chemnitz, *Bericht vom newen Bäbstischen Gregoriano Calendario, an den Landgraffen zu Hessen* (S.l., 1584), sig. A2v.

[175] MS Nuremberg, Stadtbibliothek, Pirckheimer-Papiere 356, fols. 15r–16r. Zinner 1968: 200–1 finds it likely that Regiomontanus took a copy of this text to Rome. Likewise Rigo 1994: 113, 116–17 (n. 84).

[176] MS Nuremberg, Stadtbibliothek, Pirckheimer-Papiere 356, fol. 15v (props. 15 and 16).

summons to Rome, where he was supposed to assist Pope Sixtus IV (1471–84) in a reform of the ecclesiastical calendar. A claim to this effect first appears in a biographical note included in Giacomo Filippo Foresti's widely read *Supplementum chronicarum* (first edition 1483), who adds that the reform came to nothing because Regiomontanus died soon after his arrival.[177] Foresti believed the German astronomer to have died of the plague before completing his task, but later authors would spread the rumour that he was murdered, perhaps even poisoned by the relatives of his astronomical opponent George of Trebizond.[178] That the journey to Rome was more than just a historiographical rumour is indicated by the fact that Sixtus IV, on 16 October 1475, conferred on Regiomontanus a canonry and benefice in the Neumünster of Würzburg, which was the seat of his home diocese. From the sudden end of his recorded observations it appears that Regiomontanus had left Nuremberg after 28 July 1475, but before 2 August 1475, when the record was continued by his student Bernhard Walther.[179] And yet, no concrete traces of the reform efforts that were undertaken at Sixtus IV's curia have been found, nor is it known how or when the pope and his advisers decided to select Regiomontanus. As far as the history of calendar reform goes, the end of the Middle Ages is lost in a historical mist.[180]

[177] See Malpangotto 2008: 22–3, who also quotes biographical notes by Hartmann Schedel (1493) and Johannes Trithemius (1494). Regiomontanus's summons to Rome is also alluded to in the colophon to Hans Sporer's xylographic edition of his German *Kalendar*, first issued in 1476–7, where the author is referred to as an astronomer to the pope and the Hungarian king (see Kremer 2013: 161–2, 176, 187), and in an appended note in the University of Vienna's list of scholars that dates from *c.*1500 (Mett 1989: 17).

[178] See the remarks in Zinner 1968: 237–9, and the edition of Georg Tannstetter, *Viri mathematici quos inclytum Viennense gymnasium ordine celebres habuit*, in Graf-Stuhlhofer 1996: 160.

[179] Mett 1989: 20.

[180] A work on calendar reform reportedly penned by the Northern German humanist Alexander Hegius (d. 1498) has yet to surface. See the references in Eck, *De vera Paschae celebratione* (n. 173), sigs. b3v, c2r. Other relevant texts produced during the period between Regiomontanus and the Fifth Lateran Council are covered in Marzi 1896: 9–38.

8

The Harvest of Medieval Calendar Reform

As the fifteenth century turned into the sixteenth, the geographical centre of learned activities surrounding the calendar began to shift southward. The main harbinger of this change was a man from Zeeland named Paul of Middelburg (1446–1534), who moved to Padua in the late 1470s to take up a position as lecturer in astrology. By 1481, he had swapped his lectureship for a place in the entourage of Federico da Montefeltro, duke of Urbino, whom he served as a personal physician and court astrologer. Paul's loyalty to Federico and his sons was rewarded with a handsome ecclesiastical career, which saw him become bishop of Fossombrone in 1494 and even receive a cardinal's hat a few days before his death on 14 December 1534.[1] Paul of Middelburg's earliest publications had been astrological in nature, starting with a prognostic for the year 1479, but the first signs of his interest in the calendar problem can already be detected in an *Epistola apologetica* dated 26 February 1488 and addressed to the doctors of his *alma mater*, the University of Leuven. Although primarily a defence of his controversial views on the chronology of Jesus Christ's last days, the *Epistola* also offered Paul the opportunity to voice his concerns about the ever-growing discrepancy between the astronomically legitimate and the calendrically mandated Easter date. Echoing some of the common tropes that had been cultivated by earlier writers on the matter, Paul decried the 'indignity that these perfidious and obstinate Jews and the people of other sects inflict upon us Christians' for erroneously celebrating Easter in the second month of spring. 'Pity us!', he exclaimed, 'this matter has now been drawn into such error that it is not just the learned who have taken note, but the error does not even escape the notice of sailors and—for shame!—common working girls' (*nautas et prochpudor mulierculas non latet*).[2]

For all the harsh rhetoric, the reformist sentiment expressed in his *Epistola* remained on a fairly conventional level, at least if compared to the ensuing treatment of the date of Christ's crucifixion, which according to Paul of Middelburg's reading of the source data should have occurred on 22 March, followed by the

[1] The principal source on Paul's life is a biography written in 1582–1600 by Bernardino Baldi. It was edited by Marzi 1896: 233–50, and again by Nenci 1998: 355–97 (with critical notes and additions). See in addition Marzi 1896: 12–18, 39–53; 1906: 641–2; Struik 1925; Castelli 1983; Sambin 1976–7; Welker 2005; Boorman 2006: 230–2; Hallyn 2008; Heilen 2015, and the literature cited in Nothaft 2012: 226 (n. 45).

[2] Paul of Middelburg, *Epistola apologetica ad doctores Lovanienses* (Leuven: Johannes de Westphalia, 1488), sig. A3v. On the motif of Jewish mockery and its influence on the calendar-reform debate, see Nothaft 2016a: 229–35.

resurrection a full three days later, on 25 March. In AD 34, the putative year of the Passion, these dates fell on Monday and Thursday respectively, leading Paul to cast doubt on the feasibility of demonstrating that Jesus died on a Friday and rose again on the following Sunday. The inevitable response to Paul's subversive claims came in the form of an *Opus responsivum* written by Peter de Rivo (*c.*1420–99), a distinguished master of arts and doctor of theology as well as erstwhile rector of the University of Leuven, who on behalf of his institution asserted the orthodox opinion according to which Christ died on Friday, 25 March.[3] While this printed work focused more or less exclusively on the historical-chronological issues raised in Paul's *Epistola*, Peter found the opportunity to comment on the calendar problem in a separate treatise devoted to the 'reformation of the Roman calendar and its reduction to the original state' (*Reformacio kalendarii Romani eiusque reductio ad statum pristinum*), dated 1488, whose only known copy survives as a handwritten appendix to the *Opus responsivum* in a codex once kept at Bethleem, an Augustinian Priory in Herent near Leuven.[4] The reform proposal sketched in this brief treatise is best understood as a conscious improvement on the plan submitted to the Council of Basel in 1437, the specifics of which were known from Hermann Zoest's *Phaselexis* (second version) and from Nicholas of Cusa's *De correctione kalendarii* (see Ch. 7.2). Peter openly acknowledged the existence of the latter treatise as well as Pierre d'Ailly's *Exhortatio*, but instead of following Cusanus's lead in recommending the erasure of merely a week from the calendar, he saw the advantage of restoring the spring equinox to its canonical date on 21 March, which necessitated an omission of 10 days. In Peter's proposal the targeted 10 days were those between 10 March and 21 March of AD 1501, which was the first year of the forthcoming 19-year cycle. Their removal was guaranteed to shift the Sun's entry into Aries—as calculated according to the Alfonsine Tables for the meridian of Rome—from 11 March to 21 March, but also had the effect of cancelling important saint days, a prospect that Peter hoped to prevent by postponing St Gregory's Day from 12 March to 21 March and the feast of Gertrude of Nivelles, who was especially venerated in his Leuven environment, from 17 March to 23 March.[5]

When the reformers at the Council of Basel opted for the cancellation of a week instead of 10 days, they had done so in part because of their desire to keep the 28-year solar cycle intact, but also because the combination of this step with the

[3] Peter de Rivo, *Opus responsivum ad Epistolam apologeticam Magistri Pauli de Middelburgo de anno, die et feria dominice passionis* (Leuven: Ludovicus de Ravescot, 1488). See also Peter de Rivo, *Tercius tractatus de anno, die et feria dominice passionis atque resurrectionis* (Leuven: Johannes de Westphalia, 1492). For bio-bibliographical information on the author, see the doctoral thesis by Masolini 2016.

[4] The work's full title is *Reformacio kalendarii Romani Eiusque reductio ad statum pristinum in quo fu<i>t temporibus patrum qui decreta et ordinaciones ediderunt ut paschalis festivitas a fidelibus non nisi die dominico celebretur.* It fills four leaves in Cambridge, University Library, Inc 3.F.2.9 [3294], which once formed a single volume with MS Brussels, Bibliothèque Royale, 11750–1. For more information on the text and the codex that contains it, see ch. 5 in Champion 2017. The *Reformacio kalendarii* represents an improved and fleshed-out version of the reform plan first adumbrated by Peter de Rivo in his *Dyalogus de temporibus domini nostri Ihesu Christi* (1471), as found in MS New York, Columbia University, Rare Book and Manuscript Library, Western MS 31, pp. 313–429, at pp. 344–6.

[5] Peter de Rivo, *Reformacio kalendarii Romani* (c. 5–6), MS Cambridge, University Library, Inc 3.F.2.9 [3294], fol. 2r–v.

subtraction of 3 years from the 19-year lunar cycle opened up a convenient way of restoring the new moons to the dates already marked in the existing calendars. Peter de Rivo had come to the conclusion that an analogous solution was available to those who omitted 10 days from the calendar, provided they jumped over 13 years in the conventional *cyclus decemnovenalis*, applying to 1501 not the Golden Number I, but XIV. Under normal circumstances, this would have been tantamount to moving the calendrical new moon from 23 March to 30 March, but the removal of 10 days effectively placed the Golden Number XIV on 20 March and hence one day after a mean conjunction the Alfonsine Tables signalled for 19 March.[6] Peter was forthright about his belief that the day of the calendrical new moon always had to begin on the evening following a conjunction, but his short treatise offered no further analysis as to whether this goal was met in the other 18 years of his reformed cycle. A more meticulous discussion of the matter would have shown that Peter's intervention made the conjunction and the corresponding Golden Number of the paschal lunation coincide in about half of all cases, as the total shift of new-moon dates produced by his proposed scheme amounted to 3 days in the first 6 years of the 19-year cycle, but to 4 days in the remaining 13. Referring to the precedent set by Pierre d'Ailly and Nicholas of Cusa, Peter accepted $J = 134$ years and $M = 304$ years as the numbers that defined the calendar's deterioration over time, but the difficulty of reconciling these values made him follow the path previously taken in Basel, which was to accept a correction based entirely on the latter value. The new calendar was hence supposed to lose a day every 304 years in order to fix the new moons in their places, whereas the vernal equinox would keep on drifting, albeit at a diminished rate.[7]

One positive trait that Peter's plan shared with the Basel proposal was its practical feasibility, which was given at least in so far as the suggested scheme made it possible to keep existing calendars in use. Only the rules for calculating years in the lunar and solar cycles were going to be affected by some easily described alterations. Peter also highlighted the advantage of scheduling the reform for 1501, which was the year immediately following the next Jubilee celebrated by the Roman Church. Peter suggested that this Jubilee, which foreseeably was going to bring streams of pilgrims to Rome, could be used to publicize the reform. Moreover, 'should it turn out that some doubt arises in this matter, it could be explained to those who flow together in the city to receive the indulgences of the Jubilee, so that once they return to their own places the paschal feast will be celebrated by everyone on the same day in the most accurate manner'.[8] These words notwithstanding, there are

[6] In the earlier version of his reform proposal, as contained in his *Dyalogus* of 1471 (see n. 4), Peter had still targeted the mean, rather than the true, date of the vernal equinox, suggesting the removal of 8 days from the year 1476 and a forward jump of 5 years in the count of the 19-year cycle.

[7] Peter de Rivo, *Reformacio kalendarii Romani* (c. 3, 7), MS Cambridge, University Library, Inc 3.F.2.9 [3294], fols. 1v, 3r.

[8] Ibid. (c. 4), fol. 2r: 'Interea quidem ab hiis qui huius rei periti sunt indagari poterit modus reformandi convenientissimus. Quatinus inventus una cum iubileo per orbem terrarum fidelibus publicetur. Sique quid dubietatis in eo suboriri contigerit, id declaretur illis qui ad urbem confluent pro indulgenciis iubilei. Quatinus illis ad propria redeuntibus paschalis festivitas ab omnibus eodem die ritissime celebretur.'

no signs that Peter de Rivo made any serious efforts to have his ideas relayed to the current pope, Innocent VIII, whose predecessor Sixtus IV (1471–84) had condemned some of Peter's views on the problem of future contingents in 1473 and 1474, forcing the Leuven professor to recant.[9] Things were quite different for Peter's opponent Paul of Middelburg, who in 1491 sent to Innocent VIII a 'little book' (*libellus*) on the reckoning of Easter and the correction of the calendar, which was accompanied by an epistolary *Exhortatio pro calendarii emendatione*. Another copy of the same work was communicated to the Apostolic Penitentiary and cardinal Guiliano della Rovere, who later became Pope Julius II (1503–13) and whom Paul tried to win as an intercessor with the current pope.[10] The letter to Innocent VIII reiterated some of Paul's earlier statements about how the frequent failure to celebrate Easter in the first month of spring and no later than the 21st day of the Moon exposed the Church to accusations and calumnies from the Jews, but on the whole refrained from offering any concrete suggestions of how the necessary emendation might be achieved. The only exception were Paul's remarks on the possibility of dispensing with astronomical calculation altogether and instead going back to an empirical method of fixing the date of the new moon, as it had been practised by the ancient Hebrews before the introduction of their fixed calendar. In those days:

> scouts used to climb on mountain tops and announce the new moon and Passover to the people. If Your Holiness, too, commands for this to happen, the holy Easter will be observed much more accurately by sight than it is by customary rule. For even common working girls will be able to spot the new moon by their own eyes and sight and celebrate Easter on the full moon according to the decrees of the Fathers.[11]

Rather than dwelling on the technicalities of a future reform, Paul put most of his emphasis on the question of legality, arguing that the pope had every right to go ahead and alter the existing rules for fixing the dates of the mobile feast days. He conceded that in the days of the Council of Nicaea, when the sees of Alexandria, Constantinople, Jerusalem, and Antioch were still influential contenders for authority within the Church, issues such as the calculation of Easter could only be resolved in the context of a general council. Yet now that all remaining sites of this one-time pentarchy had been conquered by Islamic forces and thus 'removed from the middle by the most frightful enemies of Christ's name, only the Roman Church, which is left standing without contradiction as the head of all churches, is

[9] Masolini 2016: 29–30.

[10] The two letters survive in MS Vatican City, BAV, lat. 3684, fols. 2r–8v (to Innocent VIII), 9r–12r (to Giuliano della Rovere, cardinal-bishop of Ostia). Their full headings read: *Ad summum pontificem pro calendarii emendatione exhortatio ut legittima pasche observatio errore calendarii iam abolita in usum revocetur* (fol. 2r) and *Paulus de Middelburgo Zelandie reverendissimo in Christo patri ac Domino D. Iuliano Episcopo Ostiensis, Cardinali S. Petri ad Vincula ac summo penitentiario salutes* (fol. 9r).

[11] Ibid., lat. 3684, fol. 4r: 'Speculatores namque montis vertice ascendentes novam lunam et pascha populo denunciabant. Quod si et tua sanctitas fieri mandaverit rectius multo per visum quam per canonem consuetum pascha sanctum observabitur. Nam et muliercule suis luminibus et visu novilunium discernere poterunt ac pascha secundum patrum decreta in plenilunio celebrare.' Sources attesting to this empirical phase in the history of the Jewish calendar are discussed in Stern 2001: 157–64. On Paul of Middelburg's use of Hebrew texts, see Nothaft 2012: 227–30.

seen to possess the remaining powers of a general council'.[12] In a strikingly biased interpretation of recent history, Paul depicted the Councils of Constance and Basel as occasions where reform attempts made by Innocent's predecessors, Martin V and Eugene IV, had been thwarted by the factionalism of the conciliarists, who neglected the calendar over their legislative battles against the papacy. Instead of repeating the mistakes of Constance of Basel, Innocent VIII was best served if he relied on his College of Cardinals, which already included some of Christendom's most learned and able men. That the pope possessed a personal mandate to administer and, if necessary, change the ecclesiastical calendar was confirmed by several strands of evidence, which included the actions of various late antique popes, but also the calendar reforms carried out in pre-Christian times by Julius Caesar and Octavian Augustus, whose role as *pontifex maximus* lived on in the office of the Roman bishop.[13]

No copies of the larger work Paul of Middelburg sent to the curia in 1491 are known to survive, but one may suspect that parts of it were carried over into the massive *Paulina de recta Paschae celebratione*, which appeared in print in 1513. Their 800 pages offered a cornucopia of information on the history of the ecclesiastical calendar and the various ways to rectify it, followed by an astonishingly detailed analysis of the chronology of Jesus Christ.[14] Among the work's multiple addressees was the newly elected pope Leo X, who reacted in February 1514 by summoning Paul of Middelburg to Rome and making him president of a reform commission that was to take up work in the context of the Fifth Lateran Council (1512–17).[15] Eager to have the project rest on a sound scientific consensus, Leo X and his advisers proceeded in July 1514 to send out briefs to Europe's university rectors and heads of state, who were asked to send their best theologians and astronomers to the Lateran Council or at least have them submit their reports in writing. The strongest response came from within the Holy Roman Empire, where the papal request triggered an epistolary chain of communication that went from the Emperor Maximilian I to various universities (Vienna, Ingolstadt, Tübingen, Leuven), as well as to individual scholars, and back to Rome.[16] In general, however, Europe's learned were slow to react to the invitation, so slow indeed that the council's tenth session had to be postponed from December 1514 to May 1515

[12] MS Vatican City, BAV, lat. 3684, fol. 5v: 'Nunc vero reliquis omnibus ecclesiis per immanissimos Christi nominis hostes e medio sublatis sola Romana ecclesia superstes sine contradictione caput omnium ecclesiarum relicta generalis concilii vires preoccupare videtur.'

[13] Ibid., fols. 6r–7r.

[14] Paul of Middelburg, *Paulina de recta Paschae celebratione: et de die passionis Domini nostri Iesu Christi* (Fossombrone: Petrucci, 1513). For a detailed bibliographical description and a list of extant copies, see Boorman 2006: 724–40. See also the summaries in Kaltenbrunner 1876: 375–85; Marzi 1896: 53–72; Steinmetz 2011: 64–6.

[15] The main study of the calendar-reform project at the Fifth Lateran Council is Marzi 1896. See, in addition, Kaltenbrunner 1876: 375–97; O'Connell 1975: 192–8; North 1983: 95–100; Galle 2006.

[16] The letter addressed to Maximilian I is printed in Marzi 1896: 78–9. See also pp. 84–6, for the accompanying *Summarium* of 14 propositions, probably authored by Paul of Middelburg. The ensuing chain of communication in the Holy Roman Empire is elucidated in Galle 2006: 148–53.

and the curia saw the need to send out a second round of letters to bishops, rulers, and universities, this time dated 1 June 1515.[17]

One year later, the commission headed by Paul of Middelburg finally came together to discuss the matter in greater depth. The tenor of these debates is known from Paul's *Secundum compendium correctionis calendarii*, addressed to Leo X and printed only a few days after the meeting in question, on 4 June 1516.[18] If we can believe this summary, nearly all who submitted their expertise to the council were in agreement that the current rules for the calculation of Easter were in dire need of an overhaul and that it was the pope's prerogative to take the steps required. Opinions strongly diverged, however, when it came to the specific mode of correction to be followed, as some were willing to cut Easter loose from the vagaries of the Moon's course and simply assign to it a fixed date, or fixed Sunday, in the Julian year. More in keeping with established traditions were the various proposals on how to reform the lunar *computus*, where the basic choice was between adjusting the existing cycle with its Golden Numbers or switching to a more properly astronomical mode of calculation. Among those who agreed that the mobile feasts should be relocated in accordance with the current date of the vernal equinox, some believed it best to subject the immobile and fixed feasts of the liturgical year to the same type of calendrical migration. These groups were opposed by those who preferred to keep the old limits and instead adjust the date of the vernal equinox, which raised the additional question of whether this should be achieved by omitting days *en bloc* or by suppressing the bissextile day for a series of years. Paul of Middelburg, for his part, conceded that the doctrines of the Church Fathers made it mandatory to celebrate Easter after the actual full moon, but also praised the advantages of having the feast fall on the Sunday immediately after the current date of the vernal equinox, which necessarily entailed moving the paschal *termini*. This way, one could ensure that the period of Quadragesima or Lent began as early as possible in winter, which made fasting more agreeable due to shorter daytime lengths while the colder weather helped preserve the salted herrings that those living north of the Alps used as their staple Lenten diet.[19]

The publication of the *Secundum compendium* was quickly followed by a much briefer compendium outlining 13 numbered propositions, which the curia sent out together with a third round of briefs on 8 and 10 July 1516. Once again, Europe's dioceses, universities, and heads of state were asked to commission and submit learned responses, this time with a deadline marked by a council session on 30 November 1516, which was meant to bring the matter to its final conclusion.[20] In essence, the new compendium offered two possible scenarios for the calendar's successful emendation. For one thing, the Roman pontifex could imitate the

[17] The text is printed in Marzi 1896: 167–8.

[18] Paul of Middelburg, *Secundum compendium correctionis calendarii continens et exponens diversos modos corrigendi calendarium pro recta pasche celebratione* (Rome: Silber alias Franck, 1516). See the summary in Marzi 1896: 172–84. A *Primum compendium* had already been published in 1514 (pp. 91–6).

[19] Paul of Middelburg, *Secundum compendium*, sigs. c2v–4r.

[20] The relevant texts are printed in Marzi 1896: 185–6, 188–9, 191–5.

drastic steps taken by his predecessor Julius Caesar and relieve the calendar of 11 days, whether by skipping 21–31 October in a given year or by ceasing intercalation for 44 years. Just like Paul of Middelburg in some of his previous publications, the document made the inaccurate claim that an omission of this number of days had already been suggested by 'Cardinal Cusanus' in Basel.[21] Alternatively, the Church could follow the advice given by Pierre d'Ailly at the Council of Constance and do what the fathers of the Council of Nicaea had done, which was to establish a new equinoctial limit for the calculation of Easter (10 March instead of 21 March).[22] Familiar objections to the first option were that it was going to scandalize the general populace, displace important saint days, confound the established order of dominical letters, and carry corruption into all astronomical tables based on the old calendar. In response, the papal document referred to the testimony of Macrobius's *Saturnalia*, which showed how the ancients repeatedly intervened in their calendars without incurring public scandal. Astronomers, moreover, could easily avoid confusion by furnishing their tables with new sets of *radices*—an idea that had already been voiced in the works of Nicholas of Cusa and Hermann Zoest.[23] The second option had the drawback of throwing the fixed and mobile feasts out of their familiar alignment, causing the number of Sundays between Epiphany and Easter to shrink and the number between Pentecost and Advent to grow. Among the resulting inconveniences was the omission of the *Alleluia* when Septuagesima Sunday fell within the octave of Epiphany as well as the reduction of the traditional period for weddings between Christmas and Shrovetide. The compendium noted that these problems could be neutralized by slight adjustments to the existing rules. In the case of weddings, for instance, it was conceivable that the Holy See could permit their celebration for the entirety of Shrovetide, provided these celebrations were carried out 'without pomp' (*absque pompa*).[24] In either case, it seemed important to preserve the new order and affix the vernal equinox to its present date by omitting a bissextile day every 134 years. Once this step was taken, however, it was no longer possible to offer a simple emendation for the 19-year lunar cycle, whose error grew at less than half this rate. A compromise that was suggested in the document's final proposition was to reset the Golden Numbers only for the next 134 years and let future generations make further adjustment when the necessity arose. As a means of increasing accuracy in the calculation of Easter, one could use this reformed cycle in parallel with astronomical tables or lists of astronomically computed conjunctions. In cases where the cyclical and the astronomical Moon were found to diverge, the Church had to ensure that the latter was followed.[25]

[21] Proposition VII, ed. Marzi 1896: 193; Paul of Middelburg, *Paulina* (pt. 1, 7.2), sig. k4r; Paul of Middelburg, *Secundum compendium*, sig. f2v.

[22] Proposition VIII, ed. Marzi 1896: 193.

[23] Proposition IX, ed. Marzi 1896: 194. See Nicholas of Cusa, *De correctione calendarii* 10 (ed. Stegemann 1955: 86–8), and Hermann Zoest, *Phaselexis (2nd version)* 10 (ed. Solan 2016: 512), who proposed the introduction of a new 'era of the Council of Basel', to be used in astronomical tables.

[24] Proposition XI, ed. Marzi 1896: 194. See already Paul of Middelburg, *Paulina* (pt. 1, 14.2), sig. t9r.

[25] Proposition XIII, ed. Marzi 1896: 195.

A belated, but very detailed response to these suggestions was signed on 3 December 1516 by the Polish astronomer Marcin Biem of Olkusz (*c.*1470–1540), who wrote to Rome at the behest of the University of Kraków, but only managed to complete two of the three parts his treatise was supposed to comprise. His submission is noteworthy for the way it repeatedly referred back to the ideas of Nicholas of Cusa, whose *De correctione* (or *De reparatione*) *calendarii* was one of the few medieval works on calendar improvement already available in print.[26] Cusanus had advocated the omission of 7 days from the Julian calendar, whereas Biem considered it better to double this number. This way, it became possible to move the vernal equinox from its current position on 10/11 March to 24/5 March, making the earliest Easter Sunday coincide with the assumed date of Christ's incarnation and crucifixion (25 March). He favoured this solution partly on historical-theological grounds, arguing, incorrectly, that the vernal equinox in the year of Christ's Passion had coincided with 24 March, when the Jews sacrificed the Passover lamb.[27] Matters were greatly complicated, however, by the Alfonsine precession model, a proper geometrical interpretation of which urged the conclusion that the moment of the Sun's entry into Aries obtained from the tables was not always, and not usually, the moment when the Sun actually crossed the celestial equator. Biem's sophisticated analysis of this conundrum, which was based on Georg Peurbach's *Theoricae novae planetarum*, showed that the real Alfonsine equinox was currently 4;46° or roughly five days ahead of the conventional entry into Aries: 6 March as opposed to 11 March.[28] Rather than calling for an even more generous omission of days from the Julian year, Biem found it advisable to stay content with the commonly accepted or 'vulgar' equinox for calendrical-computistical purposes, as the astronomically accurate version was simply too volatile in its movement to be easily stabilized on a given date.[29] Observational data, which would have favoured the 'vulgar' equinox in 1516 and beyond, played no role in this argument.

Biem's constructive, if prolix, assessment of the calendar problem stands in stark contrast to the intervention published in 1515 by the Dominican theologian Cipriano Benet, whose *Tractatus de non mutando Paschate* was an elaborate attempt to convince Pope Leo X and the Lateran Council of the futility of the whole undertaking. A substantial part of Benet's attack was informed by his sweeping hostility towards any form of judicial astrology, which moved him to highlight the multifarious disagreements between astronomical authorities and cast doubt on the validity of their calculations. Rather than mobilizing technical arguments against the specific reform proposals the council had been discussing, Benet stuck to general theological and political considerations, among them the proposition that the Church as a whole could never be in error. If this was true, the mere suggestion

[26] See the references to Cusanus in Marcin Biem, *Nova calendarii Romani reformatio* 1.6, 2.27, 31, 34, 36–7 (ed. Birkenmajer 1918: 8, 47–8, 54–8, 63, 66–7, 75, 77–8). On Biem, see Markowski 1973.

[27] Biem, *Nova calendarii Romani reformation* 2.10–19, 25 (ed. Birkenmajer 1918: 23–33, 42–4).

[28] Ibid. 2.21–2 (ed. Birkenmajer 1918: 34–8). See also Dobrzycki 1965/2010: 42–3, and Ch. 5.4 above.

[29] Biem, *Nova calendarii Romani reformatio* 2.24 (ed. Birkenmajer 1918: 40–1).

that a set of computational rules that had been in ecclesiastical use for more than a millennium might be in need of correction could be interpreted as an act of impiety.[30] Paul of Middelburg went on the counter-attack in the *Secundum Compendium* as well as in a letter to the pope published in November 1516 as *Parabola Christi de correctione calendarii*,[31] but in the end his hopes to have his plans realized were to be disappointed: the council closed on 16 March 1517 without reaching any decision on the calendar.

One reason why sceptics such as Benet were successful in bringing the reform to a halt was the continuing uncertainty with regard to the nature and rate of precession and the length of the tropical year. Worries of this sort had already occupied the minds 80 years earlier at the Council of Basel and remained difficult to dispel in 1516/17, making it tempting to argue that attempts to reform the calendar were premature given the incomplete state of current astronomical knowledge. An often-cited hint in this very direction comes from the preface to Nicholas Copernicus's *De revolutionibus orbium coelestium* (1543), which testified that:

> not so long ago under Leo X the Lateran Council considered the problem of reforming the ecclesiastical calendar. The issue remained undecided then only because of the lengths of the year and month and the motions of the Sun and Moon were regarded as not yet adequately measured. From that time on, at the suggestion of that most distinguished man, Paul, bishop of Fossombrone, who was then in charge of this matter, I have directed my attention to a more precise study of these topics.[32]

This passage later gave rise to the persistent legend, spread most notably by Galileo Galilei, that Copernicus had been personally summoned to Rome to assist in the planned reform.[33] While this is probably false, his more remote involvement in the events of 1514–16 receives confirmation from Paul of Middelburg's *Secundum compendium*, where *Nicolaus Copernicus Warmiensis* is listed among the experts whose written reports had reached the Holy See.[34] No version of Copernicus's report has ever surfaced, leaving us with several unanswered questions regarding his work on the calendar problem and its significance in relation to his heliocentric and geo-kinetic planetary theory. According to Copernicus's own testimony, in Book III of *De revolutionibus*, 1515 was the year in which he paid particular attention to the eccentricity of the Earth's orbit and the motion of the corresponding apsidal line,[35] starting an investigation that eventually led him to conclude that both of these parameters shared the fate of the rate of precession and the obliquity

[30] Cipriano Benet, *Tractatus de non mutando Paschate: et contra servile pecus Judaeorum aculeus* (Rome: Silber alias Franck, 1515). See Marzi 1906: 644–9.

[31] Paul of Middelburg, *Parabola Christi de correctione calendarii* (Fossombrone: Petrucci, 1516). See Marzi 1896: 206–9; Welker 2005: 112; Boorman 2006: 237–8, 822–3.

[32] I here follow the English translation in Rosen 1977: 148. For the original Latin, see Nicholas Copernicus, *De revolutionibus orbium coelestium*, ed. Lerner, Segonds, and Vernet 2015: ii. 9–10.

[33] Rosen 1958: 320–2; O'Connell 1975: 196–8; Casanovas 1996: 99–100.

[34] Paul of Middelburg, *Secundum compendium*, sig. br. On Copernicus's involvement in the reform project pursued at the Fifth Lateran Council, see the divergent views expressed in Galle 2006: 153–4; Hallyn 2008: 370–1; Lerner, Segonds, and Verdet 2015: i. 101–6.

[35] Copernicus, *De revolutionibus* (3.16), ed. Lerner, Segonds, and Verdet 2015: ii: 223–4. See Rosen 1977: 148–9.

of the ecliptic in being variable and having undergone significant change since antiquity. To account for the lack of uniformity traditionally explained with the help of the 'access and recess' of the eighth sphere, Copernicus invented an ingenious new model for the Earth's motion around the Sun, which involved small oscillations of the Earth's rotational axis referred to as 'librations'. One of the effects of these librations was to make the rate of precession fluctuate over a period of 1,717 Egyptian years, leading to a tropical year that varied between a maximum of 365d 5;55,37,40h $\approx 365\frac{1}{4}-\frac{1}{329.4}$ days and a minimum of 365d 5;42,55,7h $\approx 365\frac{1}{4}-\frac{1}{84.3}$ days. It so happens that the mean value implied by these extremes was 365d 5h 49m 16.4s $\approx 365\frac{1}{4}-\frac{1}{134.2}$ days and hence barely distinguishable from the Alfonsine mean tropical year that had been so frequently invoked in calendrical-astronomical writings since the fourteenth century.[36] Yet, while these findings were certainly relevant to an assessment of the Easter problem, one must be cautious not to exaggerate the link between the calendar-reform attempt of 1514–16 and Copernicus's more far-reaching efforts to reform astronomy as a whole. It is generally agreed that the first formulation of his heliocentric hypothesis, as found in the so-called *Commentariolus*, pre-dates the year 1514 and hence the period when Paul of Middelburg's calendar commission was active. His involvement in the reform project and any solar observations he made in 1515 can therefore only have served to supplement, round off, and fine-tune an already existing theory of planetary motion.[37]

If Copernicus himself created the impression of a tight link between his own project and that pursued at the Fifth Lateran Council, this may instead tell us something about the overall rhetorical strategy he pursued in the preface to *De revolutionibus*. One plausible motivation for bringing up Leo X, Paul of Middelburg, and their failed reform effort would have been to secure support and patronage from the preface's illustrious addressee, Paul Pope III (1534–49), who took a marked interest in the calendar question and even solicited advice from astronomers such as Gaspare Insoni and Luca Gaurico.[38] Another author worth mentioning in this context is Pietro Pitati of Verona, who dedicated astronomical-chronological books to Paul III and his successor Paul IV, both of which came with exhortations to reform the existing calendar and Easter *computus*.[39] In the second of these works,

[36] The maximum, minimum, and mean values of the tropical year mentioned here are those given by Christopher Clavius, *Novi calendarii Romani apologia, adversus Michaelem Maestlinum* (Rome: Sanctius, 1588), 37, 50–3 (hereafter cited as Clavius, *Apologia*). They are repeated by North (1983: 79) and differ by only a few seconds from the modern calculation provided by Swerdlow (1980: 221). Copernicus's models and parameters for the Earth–Sun system are discussed in greater detail by Dobrzycki 1965/2010: 48–57; Swerdlow 1973: 448–54; 1975; 1980; Proverbio 1983; Swerdlow and Neugebauer 1984: i. 42–3, 61, 72–4, 127–72; Lerner, Segonds, and Verdet 2015: i. 361–4; iii. 481–92.

[37] The question is discussed at length by Krafft 1994. A view to the contrary, according to which the length of the tropical year and theory of precession played a significant role in Copernicus's overall project, is asserted in Mehl 2017.

[38] Granada and Tessicini 2005: 464–70. See also Galle 2006: 154–6.

[39] Pietro Pitati, *Paschales atque noviluniorum mensurni canones* (Venice: Giunti, 1537); Pitati, *Compendium super annua solaris, atque lunaris anni quantitate, Paschalis item solennitatis iuxta veteres Ecclesiae canones recognitione, Romanique Calendarii instauratione, deque vero passionis dominicae die* (Verona: Ravagnanus, 1560). The *Compendium* was printed again in Venice in 1564 and in Basel in 1568.

a *Compendium* on the length of the solar and lunar years first printed in 1560, Pitati outlined a proposal to turn all 31-day-long months into 30-day-long ones for two consecutive years. The net effect would have been the piecemeal removal of 14 days from the calendar that left the old solar cycle intact while bringing the vernal equinox back to 24/5 March, its original location at the time of Caesar's reform. In order to keep the equinox in its ancient place, the Veronese astronomer made the innovative suggestion of suppressing the bissextile in three out of four centennial years, that is, in years evenly divisible by 100. According to this plan, years 1600, 1700, and 1800 would have all been reduced to a 'common' length of 365 days, whereas 1900 would have been bissextile, making for an average solar year of $365\frac{97}{400}$ days = 365;14,33d = 365.2425d. If this was accepted as a *mean* value, the accuracy of which was vouchsafed not just by the Alfonsine Tables, but by more recent astronomical authorities such as Abraham Zacut, Nicholas Copernicus, and Erasmus Reinhold, then the complications wrought by the irregular motion of the eighth sphere could be safely ignored.[40]

In spite of Pitati's best efforts, truly decisive steps towards the introduction of this new leap-year rule were only taken a decade later during the pontificate of Pope Gregory XIII (1572–85). Encouraged by the wide-ranging liturgical reforms initiated at the Council of Trent (1545–63), which had led to the publication of a revised Breviary (1568) and Missal (1570) under his predecessor Pope Pius IV,[41] the new pontiff decided to convene a special commission or *congregatio*, whose task it was to work out a plan for a restored and emended ecclesiastical calendar. Of the dozen or so men who served on the commission during the 1570s, one of the more active members was the Calabrian physician Antonio Lilio, who communicated to the pope a reform proposal originally drawn up by his brother Luigi, a lecturer in medicine at the University of Perugia (1510–74/6). By the year 1577, Luigi Lilio's reform scheme had found the general approval of the other commission members, who proceeded to prepare a printed summary, entitled *Compendium novae rationis restituendi kalendarium*, and have it sent out to princes and universities throughout Europe.[42] In spite of the many alternative suggestions that were included in the dozens of reports received during the ensuing years 1578–80, the commission's final report, dated 14 September 1580, opted for measures that differed only in minor details from those outlined in the *Compendium*.[43] The nine signatories,

[40] See 'propositio II' in Pitati, *Compendium*, fols. 4v–6r; Kaltenbrunner 1876: 403–6; Swerdlow 1986: 109–10.

[41] On calendar reform proposals during the period of the Council of Trent, see Schmid 1882–4, pt. 3: 53–7; Steinmetz 2011: 70–1.

[42] On the work carried out by the Gregorian commission and the reports submitted by external scholars and institutions, see Kaltenbrunner 1877: 485–503; 1880a; Schmid 1882–4, pt. 1: 388–411; pt. 3: 57–74; Marzi 1897; Moyer 1983; Ziggelaar 1983; Maiello 1989: 114–31; Mezzi and Vizza 2010: 47–60; Steinmetz 2011: 72–132. The *Compendium novae rationis restituendi kalendarium* (1577) was reprinted in Christopher Clavius, *Romani calendarii a Gregorio XIII. P.M. restituti explicatio* (Rome: Zannetti, 1603), 3–12 (hereafter cited as Clavius, *Explicatio*). Facsimile reproductions of the 1577 edition are included in Schiavo 1993; Mezzi and Vizza 2010: 202–19.

[43] *Ratio corrigendi Fastos confirmata et nomine omnium, qui ad Calendarii correctionem delecti sunt, oblato Sanctissimo Domino nostro Gregorio XIII.* The text was edited in Kaltenbrunner 1880a: 48–54; Mezzi and Vizza 2010: 226–31.

headed by the influential Cardinal Guglielmo Sirleto, stressed that the Church had been striving towards the goal of a corrected calendar for a long time. 'There have been nearly 165 years since Pierre d'Ailly, cardinal of Cambrai, a man of great skill in things both divine and human, tackled this sort of correction at the Council of Constance.' They also remembered Nicholas of Cusa and the attempt made at the Lateran Council, where a great effort had failed at last minute 'because of the dissent of a few, as some were overly petty and feeble in desiring to call the Church back to the calculations of the astrologers, while others deemed it necessary to cling tenaciously to the cycles once accepted by the Church, however crippled they might be'. A different situation, they claimed, was enjoyed in the present year 1580, when the *Compendium* on the calendar had received an overwhelming response from Europe's learned and the planned reform was backed by the emperor, the foremost kings, and nearly all universities. The right moment had finally come.[44]

Pope Gregory's correction of the ecclesiastical calendar was promulgated on 24 February 1582 with the signing of the bull *Inter gravissimas*,[45] which ordered the adoption of a modified leap-year rule and the introduction of a new epact cycle for the calculation of Easter. In addition, the document informed all Christians that the 4th of October of the current year should be followed immediately by the 15th, which amounted to the removal of 10 days from the calendar.[46] A detailed explanation of these measures (*Liber novae rationis restituendi calendarii*) was promised at the time, but never saw the light of day, making it easy for Protestant detractors such as Joseph Justus Scaliger, Michael Maestlin, and François Viète to find fault with the new calendar on various technical grounds.[47] It was left to the Jesuit mathematician Christopher Clavius, one of the few commission members to survive the 1580s, to publicly elucidate and defend the papal reform in a series of imposing books, starting in 1588 with a 500-page *Apologia* directed against Maestlin and culminating in 1603 with the 700-page *Romani calendarii a Gregorio XIII P.M. restituti explicatio*, which became the long-awaited definitive book on the Gregorian calendar and its technical underpinnings.[48] One of the more common complaints against the final reform was its failure to restore the date of the vernal equinox to 25 March, where it was thought to have been when the Julian calendar was first introduced or in the years of Christ's birth and crucifixion. Even though this was a step favoured by many sixteenth-century writers on calendar reform,

[44] Kaltenbrunner 1880a: 49; Mezzi and Vizza 2010: 227.

[45] The text is printed in Clavius, *Explicatio*, 13–15. See also the edition in Müller 2010: 321–5, and the facsimile with Italian translation in Mezzi and Vizza 2010: 232–8.

[46] According to Clavius, *Explicatio*, 74–5, the decision for October as the month of reform was a pragmatic one, motivated by the fact that it contained relatively few important saint days that would have been affected by the omission of 10 days. A similar motif was probably behind the revised wording of the second reform decree drafted at the Council of Basel (omission of 21–7 October; see n. 121 in Ch. 7.2) and the propositions sent out by Pope Leo X on 8/10 July 1516 (omission of 21–31 October; see n. 21 in this chapter).

[47] Kaltenbrunner 1877: 536–73; Nobis 1983; Hamel 2002; Bien 2007; Mezzi and Vizza 2010: 157–66; Steinmetz 2011: 215–360; Koller 2014: 30–45; Armogathe 2017: 224–6.

[48] Clavius's calendrical writings are described in Steinmetz 2011: 314–34. On the author, see Baldini 1983; Lattis 1994.

including the aforementioned Marcin Biem and Pietro Pitati,[49] Luigi Lilio and the Gregorian commission decided against it for reasons already appreciated at the Council of Basel: any change of Easter limits entailed not only a violation of the Nicene rules, but also an unacceptable change in the order of mobile and immobile feast days. Hence, an omission of 10 days to move the equinox from its current location back to 21 March seemed like the only viable option. This could be achieved by dropping 10 days all at once, as was done in the final reform, or by omitting bissextile days for a period of 40 years.[50] With the hindsight of two decades, which had seen the death of five popes and the succession of as many new ones (Sixtus V to Clement VIII), Clavius felt vindicated in having selected the former method:

> This restitution was made through the removal of 10 days at once from a single month, namely from October... because much could have happened in 40 years that would have impeded the correction, especially given the frequent change of Supreme Pontiffs that takes place in this space of years, and also because the cycle of dominical letters would have had to be interrupted several times, which would have been inconvenient, whereas in our restitution the [same cycle] was interrupted only once, but in a way that was necessary.[51]

The 'restitution' of the equinox, although of key importance to the reform as a whole, was a relatively trivial task compared to what is generally considered to have been Luigi Lilio's main contribution to the undertaking: the creation of an astronomically accurate lunar calendar that was both truly perpetual and freely adaptable to different patterns of intercalation.[52] To this end, Lilio scrapped the familiar calendrical column for the Golden Numbers in favour of a continuous count of 30 'epacts', from 0 to 29, which ran through the calendar in reverse order. In the old ecclesiastical *computus*, the epact used to represent the lunar age at a particular reference date (22 March), which could be added to the so-called lunar regulars to obtain the age of the Moon at the start of a given Julian month. The Lilian epact, by contrast, had more in common with the early medieval *litterae punctatae* (see Ch. 2.2), in that it identified the lunar age for each day of the year without the need for further addition. Each year of the 19-year cycle had a different epact-number assigned to it, which represented the age of the Moon on the final day of the previous year and simultaneously indicated all dates of the current year that were host to the start of a new lunar month. Like the old lunar epacts, the Lilian number for the annual epact was supposed to repeat after a 19-year cycle was completed. Unlike the old system, however, the correlation between years of the cycle and epact-numbers was no longer set in stone. Two calendrical interventions,

[49] Further examples are mentioned in Kaltenbrunner 1876: 376, 402, 404; 1880a: 24–30, 44–5; Schmid 1882–4, pt. 1: 396; pt. 3: 64; Marzi 1896: 20, 98, 129; Ziggelaar 1983: 214, 226–7; Maiello 1989: 123; Milz 2001: 70; Steinmetz 2011: 110–11.

[50] *Compendium novae rationis restituendi kalendarium* (ed. Mezzi and Vizza 2010: 203–6).

[51] Clavius, *Explicatio*, 74. An earlier version of the same passage appears in Clavius, *Apologia*, 20–1.

[52] For detailed explanations and mathematical appraisals of the Lilian-Gregorian 'epact' system, see Mayr 1932; 1955: 309–15, 320–4; Lichtenberg 2003; Roegel 2004: 2–7; Bien 2007: 45–50; Mezzi and Vizza 2010: 97–149; Görke 2011: 85–108.

known in the parlance of the Gregorian calendar as 'equations', caused the epact-number valid for a given year of the cycle to change on a century-to-century basis. One of these equations, the so-called 'solar equation', was owing to the correction of the Julian calendar, for which Luigi Lilio's proposal envisaged a modified leap-year rule only marginally different from the one suggested by Pietro Pitati in 1560. Exceptions to the standard rule were introduced for all centennial years not evenly divisible by 4, such that 1700, 1800, and 1900 were all stipulated to lose their bissextile day. Lilio's solar equation protected the lunar count against these centennial omissions by lowering the epact of the year by one unit, which was tantamount to decreasing the calendrical age of the Moon and, hence, postponing the beginning of the next month. His 'lunar equation' had the opposite effect, as it was meant to compensate for the overestimation of the length of the synodic month in the 19-year cycle. This could be, and was, done by raising the annual epacts by one unit.

As with the modified Julian leap-year rule, Lilio made sure to restrict the lunar equation's implementation to centennial years, which conveniently allowed for instances where both equations coincided and thus cancelled each other out. However, whereas the solar equation was triggered three times in 400 years, the lunar equation had to be applied much less frequently. Lilio's scheme placed it at seven consecutive 300-year intervals, followed by a 400-year interval every eighth time, leading to a cycle of 2,500 years that compensated for an error of exactly $M = 312\frac{1}{2}$ years. The Alfonsine Tables, which are cited in the *Compendium* as a source for the tropical year, would have only provided him with a value of $M = 310\frac{2}{3}$ years, amounting to 8 days in $2485\frac{1}{3}$ years. It is possible that Lilio's desire to limit interventions to centennial years made him simply round from there to 2,500 years while ignoring the slight discrepancy in implied month lengths. A different line of reasoning was suggested by Clavius, who in his 1588 defence of the Gregorian calendar showed that the value of $M = 312\frac{1}{2}$ years was in nearly exact agreement with Erasmus Reinhold's *Prutenic Tables*, published in 1551 on the basis of the models and parameters in Copernicus's *De revolutionibus orbium coelestium*. They gave a mean synodic month of 29;31,50,7,57d = 29d 12;44,3,10,48h, such that $M = 312;31,30\text{y} \approx 312\frac{1}{2}$ years. Even if Lilio did not note this correspondence in his original proposal, Clavius and the other experts serving on Pope Gregory's commission would have almost certainly taken account of it in their assessment. The 'Copernican Revolution' thus contributed to the ecclesiastical calendar at least in this, admittedly minor, way.[53]

If the Lilian lunar equation is applied consistently according to the prescribed rhythm, the total pattern of corrections returns to start only after $4 \times 2{,}500 = 10{,}000$ years. At this rate, it takes $30 \times 19 \times 10{,}000 = 5{,}700{,}000$ years for the annual epacts to recur in the exact same order, which also constitutes the total length of the Gregorian Easter cycle. What made the new calendar truly perpetual in the eyes of its makers, however, was not the enormous length of this cycle, but the fact that the epact system was freely adaptable to more than one intercalation pattern. Since

[53] Clavius, *Apologia*, 84–5. More precise values are cited in Clavius, *Explicatio*, 102–4. For discussion, see O'Connell 1975: 200–1; Casanovas 1983: 338–9; Swerdlow 1986: 115–16.

the Lilian scheme neatly separated the solar and lunar equations and thereby kept the correction of the Julian calendar from interfering with its lunar counterpart, future generations were free to abandon the Alfonsine year length in favour of a different one or try to create an unsteady leap-year pattern that took into account periodic fluctuations. A pattern of this kind, based on Copernicus's model for precession, was indeed supplied in the *Compendium* of 1577 as an alternative to the standard rule of omitting 3 days over 400 years.[54] Christopher Clavius later recorded that this had been done 'on account of the uncertain and non-uniform length of the year', noting that 'the excellence and perpetuity of the Gregorian Calendar consists in this, that it can be adapted to any form of intercalating, that is, to whatever length of the year you wish, while no change of the cycle of epacts in the calendar is made'.[55] The same idea, namely that the Gregorian calendar was a creation adjustable to different year lengths, was expressed in the bull *Inter gravissimas*,[56] but not in the commission report of 1580, which simply cited the Alfonsine value as a reliable estimate of the *mean* tropical year and derived from the implied $J = 134$ years a correction of 3d/400y.[57] After the Gregorian calendar had been viciously attacked on astronomical grounds by Michael Maestlin, Clavius made sure to cover all bases, pointing out that the average length of the tropical year implied in Copernicus's precession model was practically the same as the standard Alfonsine length.[58] In general, however, his guiding principle was to avoid tying the new calendar's fate to any particular set of tables, which seemed unwise in light of the prevailing uncertainties surrounding the nature and rate of precession and the resulting length of the tropical year:

> Who will be so impudent and audacious as to wish to bind the Church of God to such uncertain and fallacious theories and tables of astronomers? Surely, I believe, no one, unless perhaps he wishes to say that the Church ought to embrace now these tables, now those, according as at this time or that he observes them to agree with the motions of the heavens.[59]

The same danger of submitting holy Easter to the fickleness of astronomers and the vagaries of their theories made it mandatory to disappoint all those who wanted to

[54] *Compendium novae rationis restituendi kalendarium* (ed. Mezzi and Vizza 2010: 218–19). The congruence with Copernicus's model is demonstrated in Swerdlow 1986: 111–15. It goes without saying that the use of Copernicus's calculations for the tropical year did not automatically entail an acceptance of heliocentrism. See on this point Duhem 1911; Baldini 1983: 148–53; Lattis 1994: 160–73.

[55] Clavius, *Apologia*, 43, repeated in Clavius, *Explicatio*, 85–6. I partly follow the translation in Swerdlow 1986: 115.

[56] See the two occurrences of the phrase 'ad quamcumque anni solaris magnitudinem accommodatum' in §§5 and 11 of the text as edited in Müller 2010: 321–3.

[57] See the editions of the commission report in Kaltenbrunner 1880a: 52–3, and Mezzi and Vizza 2010: 230. That Lilio based himself on the Alfonsine year, 'que inter varias media est, atque ideo errori minus obnoxia', is also stated in the *Compendium* of 1577 (ed. Mezzi and Vizza 2010: 205).

[58] Clavius, *Apologia*, 36–82; Clavius, *Explicatio*, 81–102. The Copernican, Prutenic, and Alfonsine mean tropical years were all identical up to the second sexagesimal place (365;14,33d = 365d 5;49,12h), thereby justifying the newly introduced leap-year rule of omitting 3d/400y ($365;15\text{d} - \frac{3}{400} = 365;14,33\text{d}$). See Swerdlow 1974; 1986: 110; Casanovas 1983: 337.

[59] Clavius, *Apologia*, 29, repeated in Clavius, *Explicatio*, 76. I follow the translation in Swerdlow 1986: 114.

replace the *computus ecclesiasticus* with a purely astronomical calculation, based on true equinoxes and syzygies. Demands for such a solution had been voiced most audibly in the context of the Fifth Lateran Council, for instance by Andreas Stiborius and Georg Tannstetter, who wrote on behalf of the University of Vienna (see Ch. 7.3), and by Johannes Stöffler, who taught mathematics at the University of Tübingen. Stöffler's report to the Lateran Council, receipt of which was acknowledged by Paul of Middelburg,[60] is no longer extant, but the author later cited some of its core propositions in his magnificently produced *Calendarium Romanum magnum* of 1518, which contained extended chapters on the history of the late antique Easter controversy and the calendar problem. As for the method of finding the *terminus paschalis*, Stöffler believed that only a computation of the true moment of the opposition of Sun and Moon was fully adequate to this highly important task. Like Stiborius and Tannstetter before him, Stöffler found it apposite to base future calculations of the full moon on the meridian of Rome, the centre of Latin Christianity. Not to pick a reference meridian for the new calendar was to risk that Christians in far-flung places, such as the Holy Land or the newly discovered West Indies, were going to celebrate the feast on different dates.[61]

Christopher Clavius had shown similar proclivities for an exact astronomical calculation during the negotiations of 1577–80, but in the end he saw himself forced to agree with his colleagues that the Church was best served with a cyclical approach. Not only did such an approach obviate the need to decide on a particular meridian, but it helped keep the calendar's principles intelligible to non-astronomers, who made up the majority of its intended users. Equally important, however, was the way the use of cycles made it possible to maintain a fundamental continuity with the methods and principles that had been observed by the universal Church since the days of the Council of Nicaea, whose decrees the Gregorian reform intended to honour.[62] A manifest signal of this continuity was the *Computus ecclesiasticus* attached to Clavius's *Apologia* and *Explicatio*, which adapted the new calendar to the traditional tools of *argumenta*, finger-reckoning rules, and Easter tables.[63] Of the minor adjustments the commission made to Lilio's original epact system, most were indeed geared towards increasing the new calendar's conformity with traditional modes of Easter reckoning. One of these, which entailed a distinction between epacts XXV and 25, was deemed necessary to ensure that the Easter lunation would always be 'hollow' and Easter Sunday was never going to fall on 26 April.[64] Another significant decision was to lower the set of epacts valid for the

[60] Paul of Middelburg, *Secundum compendium*, sig. br.

[61] See the 39th proposition in Johannes Stöffler, *Calendarium Romanum magnum* (Oppenheim: Jacobus Koebel, 1518), fols. 68r–72v. Stöffler's contribution is discussed in more detail in Kaltenbrunner 1876: 390–5; Marzi 1896: 108–12; Steinmetz 2011: 67–9.

[62] Clavius, *Apologia*, 28–34; Clavius, *Explicatio*, 75–9; Schmid 1882–4, pt. 3: 68; Baldini 1983: 146–7; Ziggelaar 1983: 218; Casanovas 1996: 101–2.

[63] See Clavius, *Explicatio*, 635–80, which reproduces Clavius's two volumes of *Computus ecclesiasticus*, published separately in 1597. An earlier version was included in Clavius, *Apologia*, 325–98. See Steinmetz 2011: 325–6.

[64] Clavius, *Apologia*, 108–12; Clavius, *Explicatio*, 120–3; Kaltenbrunner 1877: 499–503; 1880a: 16–18; Mayr 1955: 312; Ziggelaar 1983: 220–4.

current century by one unit or day, which pushed the new moons they referred to closer towards the evening of first visibility. As Clavius explained some years later, this made it more likely that *luna* 15 in the calendar fell on or after the astronomical full moon, as marked by the mean opposition, lest Easter be celebrated too early in the month.[65]

Even with these adjustments in place, the Lilian scheme could not guarantee that Easter Sunday always obeyed the astronomical criteria it was keyed to. A comparison with modern calculations reveals that the new calendar has been producing 'paradox' Easter dates—where the calendrical date of Easter Sunday fails properly to respond to the full moon, the vernal equinox, or both—at a rate close to 10 per cent, with 418 such cases between 1583 and 2000.[66] Clavius was on the whole aware of these shortcomings, but considered them a price worth paying for a fully cyclical, fully predictable solution that respected the old rules as much as possible. Just like the ages of the Moon announced in Churches and marked in calendars for centuries prior to the reform, the Moon in the new Gregorian calendar remained a legal and ecclesiastical construct that was only indirectly tied to astronomical reality. The deviations from this reality had of course become much subtler and less troubling than ever before.[67]

The calendrical correction promulgated in 1582 was hence in nearly every respect a conservative one. Even the titles of the two major explanatory works associated with the project, the *Compendium* of 1577 and the *Explicatio* of 1603, were consistent in referring to the reform as a 'restitution' (*restitutio*) of the Roman calendar to its proper state.[68] It is therefore no surprise that the new calendar, instead of breaking with all pre-Copernican conventions, was for the most part founded on ideas that had already been tried out during the three preceding centuries. A particularly striking case in point is Luigi Lilio's *Tabula Epactarum expansa*, which was reproduced in both of the works just mentioned (see Figure 8.1). The table's core consists of 30 lines and 19 columns, representing the 30 different 19-year sequences of epacts that became theoretically possible in Lilio's system. Its first line shows the epacts of the traditional 19-year cycle supposedly introduced at the Council of Nicaea. In the third year of this cycle, which is placed first in the sequence, the Lilian epact is ω or *, representing the epactal value $0 = 30$, which indicates that 1 January will be *luna* 1. As with the old calendar, the epact will increase by 11 points from year to year, except between year 19 and year 1, where the *saltus lunae* makes the epactal difference between the solar and lunar year grow to 12. Each line is preceded by an upper- or lower-case letter of the alphabet, which an auxiliary table on the facing page will assign to one or more centuries before and after the reform as a way of keeping track of the effects of the solar and lunar equations over time.

[65] Clavius, *Apologia*, 140–2, 245–74; Clavius, *Explicatio*, 128–30, 380–90. The declared rationale was similar to that followed by Hermann Zoest and Thomas Strzempiński in Basel 150 years earlier. See Nothaft 2015i: 195–200.

[66] Lange 1928. Other idiosyncratic features of the epact system are discussed in Roegel 2004; Bien 2007: 56–9.

[67] Clavius, *Explicatio*, 390–400, 562–5.

[68] See n. 42 in this chapter.

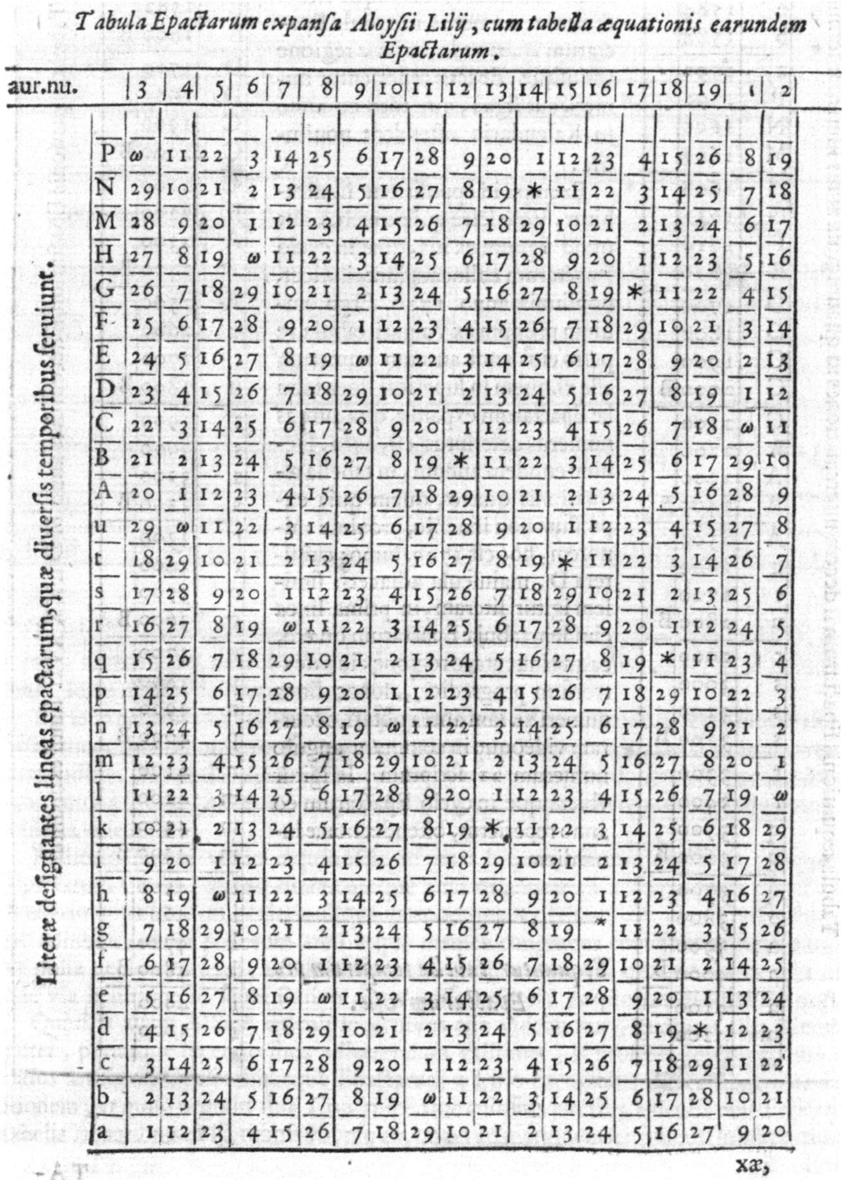

Tabula Epactarum expansa Aloysii Lilij, cum tabella æquationis earundem Epactarum.

Literæ designantes lineas epactarum, quæ diversis temporibus seruiunt.

aur.nu.	3	4	5	6	7	8	9	10	11	12	13	14	15	16	17	18	19	1	2
P	ω	11	22	3	14	25	6	17	28	9	20	1	12	23	4	15	26	8	19
N	29	10	21	2	13	24	5	16	27	8	19	*	11	22	3	14	25	7	18
M	28	9	20	1	12	23	4	15	26	7	18	29	10	21	2	13	24	6	17
H	27	8	19	ω	11	22	3	14	25	6	17	28	9	20	1	12	23	5	16
G	26	7	18	29	10	21	2	13	24	5	16	27	8	19	*	11	22	4	15
F	25	6	17	28	9	20	1	12	23	4	15	26	7	18	29	10	21	3	14
E	24	5	16	27	8	19	ω	11	22	3	14	25	6	17	28	9	20	2	13
D	23	4	15	26	7	18	29	10	21	2	13	24	5	16	27	8	19	1	12
C	22	3	14	25	6	17	28	9	20	1	12	23	4	15	26	7	18	ω	11
B	21	2	13	24	5	16	27	8	19	*	11	22	3	14	25	6	17	29	10
A	20	1	12	23	4	15	26	7	18	29	10	21	2	13	24	5	16	28	9
u	29	ω	11	22	3	14	25	6	17	28	9	20	1	12	23	4	15	27	8
t	18	29	10	21	2	13	24	5	16	27	8	19	*	11	22	3	14	26	7
s	17	28	9	20	1	12	23	4	15	26	7	18	29	10	21	2	13	25	6
r	16	27	8	19	ω	11	22	3	14	25	6	17	28	9	20	1	12	24	5
q	15	26	7	18	29	10	21	2	13	24	5	16	27	8	19	*	11	23	4
p	14	25	6	17	28	9	20	1	12	23	4	15	26	7	18	29	10	22	3
n	13	24	5	16	27	8	19	ω	11	22	3	14	25	6	17	28	9	21	2
m	12	23	4	15	26	7	18	29	10	21	2	13	24	5	16	27	8	20	1
l	11	22	3	14	25	6	17	28	9	20	1	12	23	4	15	26	7	19	ω
k	10	21	2	13	24	5	16	27	8	19	*	11	22	3	14	25	6	18	29
i	9	20	1	12	23	4	15	26	7	18	29	10	21	2	13	24	5	17	28
h	8	19	ω	11	22	3	14	25	6	17	28	9	20	1	12	23	4	16	27
g	7	18	29	10	21	2	13	24	5	16	27	8	19	*	11	22	3	15	26
f	6	17	28	9	20	1	12	23	4	15	26	7	18	29	10	21	2	14	25
e	5	16	27	8	19	ω	11	22	3	14	25	6	17	28	9	20	1	13	24
d	4	15	26	7	18	29	10	21	2	13	24	5	16	27	8	19	*	12	23
c	3	14	25	6	17	28	9	20	1	12	23	4	15	26	7	18	29	11	22
b	2	13	24	5	16	27	8	19	ω	11	22	3	14	25	6	17	28	10	21
a	1	12	23	4	15	26	7	18	29	10	21	2	13	24	5	16	27	9	20

xx,

Figure 8.1. Luigi Lilio's table of epacts, in Christopher Clavius, *Romani calendarii a Gregorio XIII. P.M. restituti explicatio S.D.N. Clementis VIII. P.M. iussu edita* (Rome: Zannetti, 1603), p. 9. Oxford, All Souls College Library, 2: Sr.41.a.3.

A *Tabula epactarum* of nearly identical structure and similar use already appeared 230 years earlier in Johannes de Termis's *Expositio kalendarii novi*, which played an important role in the calendrical discussions at the curia of Pope Clement VI (see Figure 6.2). Again, the purpose was to display 30×19 different configurations of epacts, which in the *Expositio* are shown in ascending rather than descending order. At the start of each line appears the so-called *clavis primacionum*, from 0 to 29, which was supposed to change in intervals of 120 or 240 years, thus indicating when the epacts had to be increased in order to compensate for the omission of a bissextile day from the calendar. In contrast to the Lilian epact, which served to locate the lunar age in the calendar directly, the epact in the *Expositio* had to be added to the monthly lunar regulars before it could be employed for this purpose. Also, while Lilio's scheme carefully distinguished a solar from a lunar equation, the *Expositio* employed only a single type of correction, that represented by the *clavis primacionum*. For all the visual-structural parallels just mentioned, the proposals that underlie the two tables are hence sufficiently different—and the *Expositio* sufficiently obscure—to render unnecessary the supposition that Luigi Lilio may have been directly inspired by this medieval precedent. It is nevertheless true that texts prompted by Clement VI's reform initiative were still being read 200 years later, as seen from the fact that two of these texts, Johannes de Termis's *Tractatus de tempore celebrationis paschalis* and the anonymous *Ad correctionem calendarii*, survive only in two sixteenth-century manuscripts, one copied from the other. In both instances, the works appear alongside the *Epistola super reformatione antiqui kalendarii* by Jean des Murs and Firmin de Beauval (see Ch. 6.4) and the obscure *Autores kalendarii* of 1317 (see Ch. 5.1), thereby forming a whole dossier of fourteenth-century treatises on the Easter problem. The earlier copy of this dossier appears to have been assembled in Vienna, perhaps at the time of the Fifth Lateran Council,[69] which makes it interesting to note that the two Viennese astronomers who prepared a report for the council, Stiborius and Tannsetter, were still in a position to cite the precedent of Jean des Murs, who wanted to adjust the lunar calendar with the help 'of a type of *corrector* number written in some little wheel below the Golden Number'.[70] A considerable range of medieval sources on the calendar was also available to Paul of Middelburg, whose various works feature names such as Gerland the Lotharingian, Roger Bacon, Robert Grosseteste, Jean des Murs, and Johannes Regiomontanus next to the very frequently cited Nicholas of Cusa and Pierre d'Ailly.[71] One name conspicuously absent from most such lists is Hermann Zoest, whose *Phaselexis*, though relatively popular during the fifteenth century, did not make the transition from manuscript to print. A copy of the text was nevertheless

[69] See MS Vienna, ÖNB, 5292, fols. 199r–242v, which was the exemplar of MS Vienna, ÖNB, 5273, fols. 91r–138v (dated 1527). See also Nothaft 2013d: 2–3, 29–31; 2015g: 302; 2017b: 16–17.

[70] Andreas Stiborius and Georg Tannstetter, *De Romani calendarii correctione consilium* (Vienna: Singrenius [*c.*1515]), sig. A6v. Such a wheel diagram was indeed included in the *Epistola* of Jean des Murs and Firmin de Beauval. See MS Vienna, ÖNB, 5292, fols. 227v–228v, the edition in Schabel 1996: 208–9, 212, and Ch. 6.4 above.

[71] See e.g. Paul of Middelburg, *Secundum compendium*, sigs. a2v, a4r–v; Paul of Middelburg, *Parabola Christi*, sig. B2r.

still accessible to the aforementioned Johannes Stöffler, who quoted not just from Hermann's work, but from the 1437 reform decree that was often found attached to it.[72] At the beginning of the seventeenth century, Christopher Clavius still counted Cusanus, Pierre d'Ailly, and Regiomontanus, but also Bede, Campanus of Novara, and John of Sacrobosco among the pioneers whose work had paved the way for the Gregorian calendar and his own explication of it.[73]

What set the Gregorian achievement apart from these medieval predecessor was not so much the scientific content of the reform as the ability of Pope Gregory's Church to implement it for all of Catholic Europe. The logistics of a universal calendar update in a world where each document had to be written out by hand must have had a stifling impact on the reform debate at the Council of Basel 150 years earlier, as seen from Hermann Zoest's and Nicholas of Cusa's willingness to defend an astronomically compromised reform proposal as long as this made it possible to leave existing calendars unchanged. These limitations had largely disappeared by the time Gregory XIII signed *Inter gravissimas*, which was meant to be distributed across the continent alongside new calendrical tables and instructions issued by a papally licensed Roman printing-press.[74] Even with the advantage of print at the Church's disposal, however, the reform's implementation proved difficult to achieve and caused a long-lasting calendrical schism, not just between Europe's Catholics and Protestants, but also between the Catholic Church and the various Churches of the East, whose leaders steadfastly refused to follow Rome's path.[75] In the West of Europe, inter-confessional polemic gave way to pragmatic action only at the end of the seventeenth century, when the Protestant estates of imperial Germany planned a calendar reform of their own, indistinguishable from the Gregorian one on a civil level, but with a different mode of Easter reckoning grounded in astronomical calculation. This 'improved calendar' (*Verbesserter Kalender*) took effect in 1700, but lasted only until 1776, when an imperial decree made sure that the Gregorian calendar, now re-labelled as *Allgemeiner Reichskalender*, was henceforth universally followed.[76]

In the longer term, the repeated calendar reforms of early modern Europe and the various quarrels they generated contributed to an increasingly secular perception of calendrical time. Once stripped of its former lustre as a divinely ordained

[72] Stöffler, *Calendarium Romanum magnum*, fols. 66r. See also fols. 64v, 71r, where Zoest's name is mentioned, and fols. 35r, 60v, where Stöffler tacitly utilizes passages from the *Phaselexis (2nd version)*, chs. 6–7 (ed. Solan 2016: 438–40, 448).

[73] Clavius, *Explicatio*, 1.

[74] On the logistics of the Gregorian reform, see Steinmetz 2011: 121–38.

[75] On the Protestant reaction to the Gregorian reform, see the classic study by Stieve 1880, as well as Kaltenbrunner 1877: 503–30; Hamel 1999; 2017; Smolinsky 2000; Methuen 2001; Scheutz 2005; Koller 2014: 45–277. On the attitude of the Eastern Church, see Schmid 1882–4, pt. 2; pt. 3: 76–81; Peri 1967; Ziggelaar 1983: 228–32.

[76] Steinmetz 2011: 361–91. On the learned discussion in the seventeenth century, see Gaab 2011: 402–43; Koller 2014: 278–408. Note that the *Verbesserter Kalender* did not yet affect the date of Easter in 1700, which coincided with the Gregorian Easter in spite of the astronomical data. The *Verbesserter Kalender* was also adopted in Denmark and most Protestant cantons of Switzerland. In Sweden and Finland, the astronomical Easter *calculus* was only abandoned in 1844 and 1867 respectively. See Blackburn and Holford-Strevens 1999: 798–9, 868–9; Steinmetz 2011: 392–400, 406–10.

and therefore unchanging representation of sacred time, the calendar became a malleable tool of controlling and organizing social life, fit to change along with the political and confessional tides. Nowhere is this better reflected than in the highly artificial calendar introduced in Republican France (1793–1805), which championed 'rational' principles in purposeful opposition to the calendar of the *ancien régime* and its Church.[77] By contrast, more recent calls for a new calendar reform, as they were sounded before the League of Nations and United Nations in the course of the twentieth century, have been driven less by political ideologies and more by factors such as globalization, economic rationalization, and a desire to key the calendar to scientific progress.[78]

On the other side of the spectrum, an ancient calendrical regime continues to regulate religious life in most Eastern and Oriental Orthodox Churches, where the date of Easter is still determined according to the same 19-year cycle that had been the foundation of the *computus* of medieval Latin Europe, notwithstanding the fact that *luna* 1 in this cycle is now the equivalent of *luna* 6 or 7 on an astronomical reckoning. A contemporary of Bede in the eighth century, or of John of Brunswick in the thirteenth century, would have been able to predict that Easter in the year 2016 was going to fall relatively early, on 18 April. Christians from Bucharest to Bethlehem celebrated it on 1 May, not because of any change in method, but because the discrepancy between the Julian and Gregorian calendars has now grown to 13 days. Some Churches, like the Orthodox Church of Greece and the Patriarchate of Constantinople, take this difference into account for the fixed days of their liturgical year, using a reformed calendar first adopted at the Pan-Orthodox Congress of Constantinople in 1923. Others, like the Russian Orthodox Church or the Coptic Orthodox Church of Alexandria, remain steadfast in their adherence to the Julian calendar, where the vernal equinox now resides on 7/8 March.[79]

Efforts to overcome these divisions and have Christians all over the world observe one and the same Easter date continue to play a significant part in the ecumenical debate of the twenty-first century. A summit held in Aleppo in 1997 under the aegis of the World Council of Churches (WCC) and the Middle East Council of Churches (MECC) followed medieval precedent in stressing the importance of respecting the decrees of the Council of Nicaea, thereby ignoring the fact that no decree relating to the date of Easter has survived from this ecclesiastical gathering. Accepting the pseudo-Nicene rule of the equinox, the assembled representatives agreed that 'the most likely way to succeed in achieving a common date for Easter in our own day' would be to calculate the date of the vernal equinox and paschal full moon 'by the most accurate possible scientific means', using Jerusalem

77 On the French Republican calendar, see most recently Shaw 2011; Perovic 2012.

78 On calendar reform in the modern era, see Achelis 1955; Russo 1983; Blackburn and Holford-Strevens 1999: 689–92; Steinmetz 2011: 416–24; Šuráň 1997, and the web resource <http://www.theworldcalendar.org>.

79 On Easter reckoning in the orthodox East, see Milanković 1924; Lange 1928: 52–65. For past efforts towards restoring a common Orthodox Easter, see Synodica 1981.

as the appropriate reference meridian.[80] The need to achieve an agreement on the date of Easter was again emphasized in June 2015 by the Catholic Pope Francis, who signalled the Roman Church's willingness to give up its current Easter calculation for the sake of Christian unity.[81] Rather than abiding by the 1997 Aleppo Statement, future unification efforts may well aim towards a fixed date for Easter Sunday. A solution of this type, which would redefine Easter as the second or third Sunday in April, was favoured as recently as January 2016 by the Coptic Pope Tawadros II and by Justin Welby, Archbishop of Canterbury, the latter of whom expressed his hope of seeing this goal realized within the next ten years.[82] Whatever the outcome of these initiatives may be, they remind us that, even in the realm of calendars and Easter reckoning, 'the past is never dead', and the medieval debate continues.

[80] 'Towards a Common Date for Easter', World Council of Churches, 10 March 1997, <https://www.oikoumene.org/en/resources/documents/commissions/faith-and-order/i-unity-the-church-and-its-mission/towards-a-common-date-for-easter/index>.

[81] 'Will Pope Francis change the date of Easter?', Catholic News Agency, 19 June 2016, <http://www.catholicnewsagency.com/news/will-pope-francis-change-the-date-of-easter-87684>.

[82] 'Primates support ecumenical moves over Easter', Anglican Communion News Service, 18 Jan. 2016, <http://www.anglicannews.org/news/2016/01/primates-support-ecumenical-moves-over-easter.aspx>.

References

Achelis, Elisabeth. 1955. *Of Time and the Calendar.* New York: Hermitage House.

Ackermann, Silke. 2001. 'Bartholomew of Parma, Michael Scot and the Set of New Constellations in Bartholomew's *Breviloquium de fructu tocius astronomie*'. In *Seventh Centenary of the Teaching of Astronomy in Bologna, 1297–1997*, ed. Pierluigi Battistini et al., 77–98. Bologna: CLUEB.

Ackermann, Silke. 2009. *Sternstunden am Kaiserhof: Michael Scotus und sein 'Buch von den Bildern und Zeichen des Himmels'*. Frankfurt/Main: Lang.

Albertini, Giovanni. 1967. 'Il solstizio invernale nei proverbi'. *L'Universo* 47: 533–54.

Albertini, Giovanni. 1981. 'Indagine sui rapporti tra l'errore del calendario giuliano e la "centesma" Dantesca'. *Atti dell'Accademia roveretana degli Agiati*, 6th ser., 20A: 3–41.

Allard, André. 1991. 'The Arabic Origins and Development of Latin Algorisms in the Twelfth Century'. *Arabic Sciences and Philosophy* 1: 233–83.

Allard, André, ed. 1992. *Muhammad Ibn Mūsā al-Khwārizmī: Le Calcul Indien (Algorismus).* Paris: Blanchard.

Alverny, Marie-Thérèse d'. 1982. 'Translations and Translators'. In *Renaissance and Renewal in the Twelfth Century*, ed. Robert L. Benson and Giles Constable, 421–62. Cambridge, Mass.: Harvard University Press.

Alverny, Marie-Thérèse d', Charles Burnett, and Emmanuel Poulle, eds. 2009. *Raymond de Marseille: Opera omnia*, vol. 1, *Traité de l'astrolabe, Liber cursuum planetarum.* Paris: CNRS.

Andriani, Beniamino. 1971. 'La centesima negletta'. *Studi danteschi* 48: 83–103.

Anheim, Étienne. 2014. *Clément VI au travail: lire, écrire, prêcher au XIVe siècle.* Paris: Publications de la Sorbonne.

Archer, Peter. 1941. *The Christian Calendar and the Gregorian Reform.* New York: Fordham University Press.

Armogathe, Jean-Robert. 2017. 'La réforme grégorienne du calendrier'. In *Le temps des astronomes*, ed. Edouard Mehl and Nicolas Roudet, 219–27. Paris: Les Belles Lettres.

Arrighi, Gino, ed. 1991. *M^o Bono da Lucca: Computus Lunaris (1254).* Lucca: S. Marco.

Arweiler, Alexander. 2000. 'Zu Text und Überlieferung einer gekürzten Fassung von Macrobius *Saturnalia* 1,12,2–1,15,20'. *ZPE* 131: 45–57.

Aufgebauer, Peter. 2006. '"Ein vielfaches Ärgernis in der Kirche"—Probleme der christlichen Zeitrechnung im Mittelalter'. In *Herrschaftspraxis und soziale Ordnungen im Mittelalter und in der frühen Neuzeit*, ed. Peter Aufgebauer and Christine van den Heuvel, 73–85. Hannover: Hahn.

Avi-Yonah, Reuven S. 1985. 'Ptolemy vs al-Bitruji: A Study of Scientific Decision-Making in the Middle Ages'. *AIHS* 35: 124–47.

Baker, Peter S. 1981. 'Byrhtferth's *Enchiridion* and the Computus in Oxford, St John's College 17'. *Anglo-Saxon England* 10: 123–42.

Baker, Peter S., and Michael Lapidge, eds. 1995. *Byrhtferth's Enchiridion.* Oxford: Oxford Univeristy Press.

Baldini, Ugo. 1983. 'Christoph Clavius and the Scientific Scene in Rome'. In Coyne, Hoskin, and Pedersen 1983: 137–69.

Barlow, Claude W., ed. 1970. *Martini Episcopi Bracarensis Opera Omnia*. New Haven: Yale University Press.

Bartlett, Robert. 2008. *The Natural and the Supernatural in the Middle Ages*. Cambridge: Cambridge University Press.

Bauer, Ulrike. 1983. *Der Liber Introductorius des Michael Scotus in der Abschrift Clm 10268 der Bayerischen Staatsbibliothek München: Ein illustrierter astronomisch-astrologischer Codex aus Padua, 14. Jahrhundert*. Munich: tuduv.

Baumgartner, Frederic J. 2003. 'Popes, Astrologers, and Early Modern Calendar Reform'. In *History Has Many Voices*, ed. Lee Palmer Wandel, 41–56. Kirksville, Mo.: Truman State University Press.

Baur, Ludwig. 1912. *Die philosophischen Werke des Robert Grosseteste, Bischofs von Lincoln*. Münster: Aschendorff.

Baur, Ludwig. 1917. *Die Philosophie des Robert Grosseteste, Bischofs von Lincoln († 1253)*. Münster: Aschendorff.

Bayer, Karl. 2002. 'Antike Zeitmessung: Die Kalenderreform des Caius Iulius Caesar'. In Herzog 2002: 35–64.

Beard, Darren. 2004. 'Eclipse Predictions in the Kalendarium of Nicholas of Lynn'. *Journal of the British Astronomical Association* 114: 343–8.

Beaujouan, Guy. 1975. 'Observations et calculs astronomiques de Jean de Murs (1321–1344)'. In *Proceedings of the XIVth International Congress of the History of Science*, vol. 2, 27–30. Tokyo: Science Council of Japan.

Beaujouan, Guy. 1982. 'The Transformation of the Quadrivium'. In *Renaissance and Renewal in the Twelfth Century*, ed. Robert L. Benson and Giles Constable, 463–87. Cambridge, Mass.: Harvard University Press.

Benjamin, Francis S. Jr, and G. J. Toomer, eds. 1971. *Campanus of Novara and Medieval Planetary Theory: 'Theorica planetarum'*. Madison, Wisc.: University of Wisconsin Press.

Bennett, Chris. 2003a. 'The Early Augustan Calendars in Rome and Egypt'. *ZPE* 142: 221–40. With addenda and corrigenda in *ZPE* 147 (2004): 165–8.

Bennett, Chris. 2003b. 'The Imperial Nundinal Cycle'. *ZPE* 147: 175–9.

Bennett, Chris. 2005. 'Evidence for the Regulation of Intercalation under the Lex Acilia'. *ZPE* 151: 167–84.

Berggren, J. Lennart. 2002. 'Medieval Arithmetic: Arabic Texts and European Motivations'. In *Word, Image, Number*, ed. John J. Contreni and Santa Casciani, 351–65. Florence: SISMEL.

Bergmann, Werner. 1985. *Innovationen im Quadrivium des 10. und 11. Jahrhunderts: Studien zur Einführung von Astrolab und Abakus im lateinischen Mittelalter*. Stuttgart: Steiner.

Berlière, Ursmer. 1908. 'La réforme du calendrier sous Clément VI'. *Revue Bénédictine* 25: 240–1.

Berndt, Guido M., ed. 2009. *Vita Meinwerci episcopi Patherbrunnensis: Das Leben Bischof Meinwerks von Paderborn*. Munich: Fink.

Bezdechi, Ștefan. 1924. 'Nicephori Gregorae Epistolae XC'. *Ephemeris Dacoromana* 2: 239–377.

Bezdechi, Ștefan. 1927. 'Un projet de réforme du calendrier par Nicéphore Grégoras'. In *Mélanges d'Histoire Générale*, ed. Constantin Marinescu, 68–74. Cluj: Cartea Românească.

Bien, Reinhold. 2007. 'Viète's Controversy with Clavius over the Truly Gregorian Calendar'. *AHES* 61: 39–66.

Birk, Ernst Ritter von, ed. 1873. *Monumenta conciliorum generalium seculi decimi quinti*. Vol. 2. Vienna: Typ. Aulae et Status.

Birkenmajer, Ludwik, ed. 1900. *Commentariolum super Theoricas Novas Planetarum Georgii Purbachii in studio generali Cracoviensi per Mag. Albertum de Brudzewo diligenter corrogatum* [sic] *AD MCCCCLXXXII*. Kraków: Typis et Sumptibus Universitatis Jagellonicae.

Birkenmajer, Ludwik, ed. 1918. *Martini Biem de Ilkusz Poloni Nova Calendarii Romani Reformatio*. Kraków: Gebethner.

Bisagni, Jacopo, and Immo Warntjes. 2007. 'Latin and Old Irish in the Munich Computus: A Reassessment and Further Evidence'. *Ériu* 57: 1–33.

Bisagni, Jacopo, and Immo Warntjes. 2008. 'The Early Old Irish Material in the Newly Discovered *Computus Einsidlensis* (*c*. AD 700)'. *Ériu* 58: 77–105.

Bischoff, Bernhard. 1940. 'Ostertagtexte und Intervalltafeln'. *Historisches Jahrbuch* 60: 549–80.

Bischoff, Bernhard. 1967. *Mittelalterliche Studien: Ausgewählte Aufsätze zur Schriftkunde und Literaturgeschichte*. Vol. 2. Stuttgart: Hiersemann.

Black, A. J. 1988. 'The Conciliar Movement'. In *The Cambridge History of Medieval Political Thought c. 350–c. 1450*, ed. J. H. Burns, 573–87. Cambridge: Cambridge University Press.

Blackburn, Bonnie J., and Leofranc Holford-Strevens. 1999. *The Oxford Companion to the Year: An Exploration of Calendar Customs and Time-Reckoning*. Oxford: Oxford University Press.

Blake, Martin, ed. and trans. 2009. *Aelfric's 'De temporibus anni'*. Cambridge: D. S. Brewer.

Blume, Dieter, Mechthild Haffner, and Wolfgang Metzger. 2012. *Sternbilder des Mittelalters: Der gemalte Himmel zwischen Wissenschaft und Phantasie*. Vol. 1, *800–1200; Teilband I.1: Text und Katalog der Handschriften*. Berlin: Akademie Verlag.

Boffito, Giuseppe, and Camillo Melzi d'Eril, eds. 1908. *Almanach Dantis Aligherii sive Profhacii Judaei Montispessulani Almanach perpetuum ad annum 1300 inchoatum, nunc primum editum ad fidem Codicis Laurentiani (Pl. XVIII sin. N. 1)*. Florence: Olschki.

Boffito, Giuseppe, and Camillo Melzi d'Eril, eds. 1922. *Il quadrante d'Israele di Jacob ben Machir ben Tibbon (Profacio)*. Florence: Libreria internazionale.

Bohnert, Niels, and Tom Müller. 2012. '"*Nun wil ich etwas schreiben von der natur und aigenschafft der zwelf zaichen des gürtels*": Astronomisches und Astrologisches aus dem *Calendarium Stamsense* des Vitus de Augusta'. In *Johannes von Gmunden—zwischen Astronomie und Astrologie*, ed. Rudolf Simek and Manuela Klein, 23–53. Vienna: Fassbaender.

Boncompagni, Baldassarre. 1862–63. 'Intorno ad un trattato d'arithmetica stampato nel 1478'. *Atti dell'Accademia Pontificia de' Nuovi Lincei* 16: 1–64, 101–228, 301–64, 389–452, 503–630, 683–842, 909–1044.

Boorman, Stanley. 2006. *Ottaviano Petrucci: Catalogue Raisonne*. Oxford: Oxford University Press.

Borrelli, Arianna. 2008. *Aspects of the Astrolabe: 'architectonica ratio' in Tenth- and Eleventh-Century Europe*. Stuttgart: Steiner.

Borst, Arno. 1984. 'Ein Forschungsbericht Hermanns des Lahmen'. *Deutsches Archiv für Erforschung des Mittelalters* 40: 379–477.

Borst, Arno. 1990. *Computus: Zeit und Zahl in der Geschichte Europas*. Berlin: Wagenbach.

Borst, Arno. 1993. 'Alkuin und die Enzyklopädie von 809'. In *Science in Western and Eastern Civilization in Carolingian Times*, ed. Paul Leo Butzer and Dietrich Lohrmann, 53–78. Basel: Birkhäuser.

Borst, Arno. 1994. *Das Buch der Naturgeschichte: Plinius und seine Leser im Zeitalter des Pergaments*. Heidelberg: Winter.

Borst, Arno. 1998. *Die karolingische Kalenderreform*. Hannover: Hahn.

Borst, Arno. 2001. *Der karolingische Reichskalender und seine Überlieferung bis ins 12. Jahrhundert*. 3 vols. Hannover: Hahn.

Borst, Arno. 2004. *Der Streit um den karolingischen Kalender*. Hannover: Hahn.

Borst, Arno. 2006. *Schriften zur Komputistik im Frankenreich von 721 bis 818*. 3 vols. Hannover: Hahn.

Boudet, Jean-Patrice. 1994. *Lire dans le ciel: la bibliothèque de Simon de Phares astrologue du XVe siècle*. Brussels: Centre d'Étude des Manuscrits.

Boudet, Jean-Patrice. 1995. 'Prévision de l'avenir et connaissance du passé: les relations entre astrologie et histoire à la fin du Moyen Âge'. In *Pratiques de la culture écrite en France au XVe siècle*, ed. Monique Ornato and Nicole Pons, 299–312. Louvain-la-Neuve: Fédération Internationale des Instituts d'Études Médiévales.

Boudet, Jean-Patrice. 2002. 'Un prélat et son équipe de travail à la fin du Moyen Âge: remarques sur l'oeuvre scientifique de Pierre d'Ailly'. In *Humanisme et culture géographique à l'époque du Concile de Constance*, ed. Didier Marcotte, 127–50. Turnhout: Brepols.

Bouloux, Nathalie. 2011. 'Pour une nouvelle lecture de l'*Imago mundi* de Pierre d'Ailly'. In *Représentations et conceptions de l'espace dans la culture médiévale*, ed. Tiziana Suarez-Nani and Martin Rohde, 237–64. Berlin: de Gruyter.

Bowen, Alan C., and Bernard R. Goldstein. 1988. 'Meton of Athens and Astronomy in the Late Fifth Century B.C.'. In *A Scientific Humanist*, ed. Erle Leichty, Maria deJ. Ellis, and Pamela Gerardi, 38–81. Philadelphia: University Museum.

Brévart, Francis B. 1987. 'Die "Mainauer Naturlehre": Ein astronomisch-diätetisch-komputistisches Lehrbuch aus dem 14. Jahrhundert'. *Sudhoffs Archiv* 71: 157–79.

Brévart, Francis B. 1988a. 'Johann Blaubirers Kalender von 1481 und 1483: Traditionsgebundenheit und experimentelle Innovation'. *Gutenberg-Jahrbuch* 63: 74–83.

Brévart, Francis B. 1988b. 'The German *Volkskalender* of the Fifteenth Century'. *Speculum* 63: 312–42.

Brévart, Francis B. 1995–96. 'Chronology and Cosmology: A German *Volkskalender* of the Fifteenth Century'. *Princeton University Library Chronicle* 57: 225–65.

Brewer, John Sherren, ed. 1859. *Fr. Rogeri Bacon Opera quaedam hactenus inedita*. Vol. 1. London: Longman, Green, Longman, and Roberts.

Bridges, John Henry, ed. 1897. *The 'Opus Majus' of Roger Bacon*. Vol. 1. Oxford: Clarendon Press.

Brind'Amour, Pierre. 1983. *Le calendrier romain: recherches chronologiques*. Ottawa: Études de l'Université d'Ottawa.

Brinkhus, Gerd, David Juste, and Helga Lengenfelder, eds. 2005. *Iatromathematisches Kalenderbuch/Die Kunst der Astronomie und Geomantie: Farbmikrofiche-Edition der Handschrift Tübingen, Universitätsbibliothek, Md 2*. Munich: Lengenfelder.

Britton, John Phillips. 1992. *Models and Precision: The Quality of Ptolemy's Observations and Parameters*. New York: Garland.

Bubnov, Nicolaus, ed. 1899. *Gerberti Opera Mathematica (972–1003)*. Berlin. Reprint, Hildesheim: Olms, 1963.

Bullough, Donald A. 2003. 'York, Bede's Calendar and a Pre-Bedan English Martyrology'. *Analecta Bollandiana* 121: 329–55.

Bullough, Donald A. 2004. *Alcuin: Achievement and Reputation*. Leiden: Brill.

Bultot, Robert. 1978. '"Quadrivium", "Natura" et "Ingenium naturale" chez Guillaume d'Hirsau (†1091)'. *Rivista di filosofia neo-scolastica* 70: 11–27.

Burnett, Charles. 1978. 'Arabic Into Latin in Twelfth-Century Spain: The Works of Hermann of Carinthia'. *Mittellateinisches Jahrbuch* 13: 100–34.

Burnett, Charles. 1992. 'The Translating Activity in Medieval Spain'. In *The Legacy of Muslim Spain*, ed. Salma Khadra Jayyusi, 1036–58. Leiden: Brill.

Burnett, Charles. 1995. 'The Institutional Context of Arabic-Latin Translations of the Middle Ages: A Reassessment of the "School of Toledo"'. In *Vocabulary of Teaching and Research between Middle Ages and Renaissance*, ed. Olga Weijers, 214–35. Turnhout: Brepols.

Burnett, Charles. 1997. *The Introduction of Arabic Learning into England*. London: The British Library.

Burnett, Charles. 1998. 'King Ptolemy and Alchandreus the Philosopher: The Earliest Texts on the Astrolabe and Arabic Astrology at Fleury, Micy and Chartres'. *Annals of Science* 55: 329–68. With addendum in *ibid.* 57 (2000): 187. Reprinted as chapter 1 in Burnett 2009.

Burnett, Charles. 1999. '"Abd al-Masīḥ of Winchester". In *Between Demonstration and Imagination*, ed. Lodi Nauta and Arjo Vanderjagt, 159–69. Leiden: Brill.

Burnett, Charles. 2000. 'Antioch as a Link between Arabic and Latin Culture in the Twelfth and Thirteenth Centuries'. In *Occident et Proche-Orient*, ed. Isabelle Draelants, Anne Tihon, and Baudouin van den Abeele, 1–78. Turnhout: Brepols. Reprinted as chapter 4 in Burnett 2009.

Burnett, Charles. 2003. 'The Transmission of Arabic Astronomy via Antioch and Pisa in the Second Quarter of the Twelfth Century'. In *The Enterprise of Science in Islam*, ed. Jan P. Hogendijk and Abdelhamid I. Sabra, 23–51. Cambridge, Mass.: MIT Press.

Burnett, Charles. 2006. 'Stephen, the Discipline of Philosophy, and the Exchange of Medieval Learning in Antioch'. *Crusades* 5: 113–29.

Burnett, Charles. 2009. *Arabic into Latin in the Middle Ages: The Translators and Their Intellectual and Social Context*. Farnham: Ashgate.

Burnett, Charles. 2010a. *Numerals and Arithmetic in the Middle Ages*. Farnham: Ashgate.

Burnett, Charles. 2010b. '"Ptolemaeus in Almagesto dixit": The Transformation of Ptolemy's *Almagest* in Its Transmission via Arabic into Latin'. In *Transformationen antiker Wissenschaften*, ed. Georg Toepfer and Hartmut Böhme, 115–40. Berlin: de Gruyter.

Burnett, Charles. 2011. 'Two Approaches to Natural Science in Toledo of the Twelfth Century'. In *Christlicher Norden—Muslimischer Süden*, ed. Matthias M. Tischler und Alexander Fidora, 69–80. Münster: Aschendorff.

Burnett, Charles. 2013. 'The Geometry of the *Liber Ysagogarum Alchorismi*'. *Sudhoffs Archiv* 97: 143–73.

Buron, Edmond, ed. 1930. *Ymago mundi de Pierre d'Ailly*. 3 vols. Paris: Maisonneuve Frères.

Busard, H. L. L., ed. 2005. *Campanus of Novara and Euclid's 'Elements'*. Vol. 1. Stuttgart: Steiner.

Busard, H. L. L., ed. 2010. *Nicole Oresme: Questiones super geometriam Euclidis*. Stuttgart: Steiner.

Bydén, Börje. 2003. *Theodore Metochites' 'Stoicheiosis Astronomike' and the Study of Natural Philosophy and Mathematics in Early Palaiologan Byzantium*. Göteborg: Acta Universitatis Gothoburgensis.

Caiazzo, Irène. 2002. *Lectures médiévales de Macrobe: les 'Glosae Colonienses super Macrobium'*. Paris: Vrin.

Caiazzo, Irène. 2006. 'Abbon de Fleury et l'héritage platonicien'. In Obrist 2006: 11–39.

Caiazzo, Irène. 2012. 'Sur l'astrolabe et le comput au XIIe siècle: nouveaux textes inédits'. *Sudhoffs Archiv* 96: 28–38.

Callataÿ, Godefroid de. 1996. *Annus Platonicus: A Study of World Cycles in Greek, Latin and Arabic Sources*. Louvain-la-Neuve: Université Catholique de Louvain, Institut Orientaliste.

Cameron, Alan. 2011. *The Last Pagans of Rome*. Oxford: Oxford University Press.

Campani, Romeo, ed. 1910. *Alfragano (al-Farghānī): Il 'Libro dell'aggregazione delle stelle' (Dante,* Convivio, *II, vi-134) secondo il Codice Mediceo-Laurenziano, Pl. 29, cod. 9 contemporaneo a Dante*. Città di Castello: Lapi.

Carabias Torres, Ana María. 2012. *Salamanca y la medida del tiempo*. Salamanca: Ediciones Universidad de Salamanca.

Carabias Torres, Ana María, and Bernardo Gómez Alfonso. 2014. 'Francisco de Salinas y el calendario gregoriano'. In *Francisco de Salinas*, ed. Amaya García Pérez and Paloma Otaola González, 117–45. Salamanca: Ediciones Universidad de Salamanca.

Carey, Hilary M. 2003. 'What is the Folded Almanac? The Form and Function of Key Manuscript Sources for Astro-Medical Practice in Later Medieval England'. *Social History of Medicine* 16: 481–509.

Carey, Hilary M. 2004. 'Astrological Medicine and the Medieval English Folded Almanac'. *Social History of Medicine* 17: 345–63.

Carmody, Francis J., ed. 1943. *Al Farghani Differentie*. Berkeley: University of California Press.

Carmody, Francis J., ed. 1952. *Al-Biṭrûjî: De motibus celorum*. Berkeley: University of California Press.

Carmody, Francis J., ed. 1960. *The Astronomical Works of Thabit b. Qurra*. Berkeley: University of California Press.

Caroti, Stefano. 1979. 'La critica contro l'astrologia di Nicole Oresme e la sua influenza nel Medioevo e nel Rinascimento'. *Atti della Accademia Nazionale dei Lincei, Classe di Scienze morali, storiche e filologiche*, Memorie, 8th ser., 23: 545–685.

Casanovas, Juan. 1983. 'Il ruolo delle Tavole Alfonsine e Pruteniche nella riforma gregoriana del calendario'. *Giornale di astronomia* 9: 333–41.

Casanovas, Juan. 1987. 'On the Precession Problem in the Alfonsine Tables'. In *De astronomia Alphonsi regis*, ed. Mercè Comes, Roser Puig, and Julio Samsó, 79–87. Barcelona: Instituto 'Millás Vallicrosa' de Historia de la Ciencia Árabe.

Casanovas, Juan. 1996. 'Copernicus and the Gregorian Calendar Reform'. In *Copernico e la questione copernicana in Italia dal XVI al XIX secolo*, ed. Luigi Pepe, 97–107. Florence: Olschki.

Cassiano, Domenico. 1994. *Barlaam Calabro: l'humana avventura di un Greco d'Occidente*. Lungro: Marco.

Castelli, Patrizia. 1983. 'Gli astri e i Montefeltri'. *Res Publica Litterarum* 6: 75–89.

Casulleras, Josep. 1996. 'Las *Tablas astronómicas* de Pedro Alfonso'. In *Estudios sobre Pedro Alfonso de Huesca*, ed. María Jesús Lacarra, 349–66. Huesca: Instituto de Estudios Altoaragonenses.

Chabás, José. 1998. 'Astronomy in Salamanca in the Mid-Fifteenth Century: The *Tabulae Resolutae*'. *JHA* 29: 167–75.

Chabás, José. 2000. ' "Lo compta de la Luna": lunarios medievales'. In *Actes de les V Trobades d'història de la ciència i de la tècnica (Roquetes, 11–13 desembre 1998)*, ed. Josep Batlló Ortiz, Pere de la Fuente Collell and Roser Puig Aguillar, 335–41. Barcelona: Societat Catalana d'Història de la Ciència i de la Tècnica.

Chabás, José. 2002. 'The Diffusion of the Alfonsine Tables: The Case of the *Tabulae Resolutae*'. *Perspectives on Science* 10: 168–78.

Chabás, José. 2007. 'From Toledo to Venice: The Alfonsine Tables of Prosdocimo de' Beldomandi of Padua (1424)'. *JHA* 38: 269–81.

Chabás, José. 2014. 'Aspects of Arabic Influence on Astronomical Tables in Medieval Europe'. *Suhayl* 13: 23–40.

Chabás, José. 2017. 'An Early Witness of Alfonsine Astronomy: The London Tables of 1336'. *JHA* 48: 324–8.
Chabás, José, and Bernard R. Goldstein. 1992. 'Nicholaus de Heybech and His Table for Finding True Syzygy'. *Historia Mathematica* 19: 265–89.
Chabás, José, and Bernard R. Goldstein. 1997. 'Computational Astronomy: Five Centuries of Finding True Syzygy'. *JHA* 28: 93–105.
Chabás, José, and Bernard R. Goldstein. 2000. *Astronomy in the Iberian Peninsula: Abraham Zacut and the Transition from Manuscript to Print*. Philadelphia: American Philosophical Society (= Transactions of the American Philosophical Society 90.2).
Chabás, José, and Bernard R. Goldstein. 2003a. 'John Vimond and the Alfonsine Trepidation Model'. *JHA* 34: 163–70.
Chabás, José, and Bernard R. Goldstein. 2003b. *The Alfonsine Tables of Toledo*. Dordrecht: Kluwer.
Chabás, José, and Bernard R. Goldstein. 2004. 'Early Alfonsine Astronomy in Paris: The Tables of John Vimond (1320)'. *Suhayl* 4: 207–94.
Chabás, José, and Bernard R. Goldstein. 2009a. 'John of Murs's Tables of 1321'. *JHA* 40: 297–320.
Chabás, José, and Bernard R. Goldstein. 2009b. *The Astronomical Tables of Giovanni Bianchini*. Leiden: Brill.
Chabás, José, and Bernard R. Goldstein. 2012a. *A Survey of European Astronomical Tables in the Late Middle Ages*. Leiden: Brill.
Chabás, José, and Bernard R. Goldstein. 2012b. 'John of Murs Revisited: The *Kalendarium Solis et Lune* for 1321'. *JHA* 43: 411–37.
Chabás, José, and Bernard R. Goldstein. 2013. 'Computing Planetary Positions: User-Friendliness and the Alfonsine Corpus'. *JHA* 44: 257–76, 479–80.
Chabás, José, and Bernard R. Goldstein. 2015. 'Ibn al-Kammād's *Muqtabis* Zij and the Astronomical Tradition of Indian Origin in the Iberian Peninsula'. *AHES* 69: 577–650.
Chabás, José, and Bernard R. Goldstein. 2016. 'The Moon in the Oxford Tables of 1348'. *JHA* 47: 159–67.
Chabás, José, and Antoni Roca. 1985. *El 'Lunari' de Bernat de Granollachs: alguns aspectes de la història de l'astronomia a la Catalunya del Quatre-cents*. Barcelona: Funcadió Salvador Vives I Casajuana.
Chabás, José, and Antoni Roca. 1998. 'Early Printing of Astronomy: The *Lunari* of Bernat de Granollachs'. *Centaurus* 40: 124–34.
Champion, Matthew. 2017. *The Fullness of Time: Temporalities of the Fifteenth-Century Low Countries*. Chicago: University of Chicago Press.
Chantraine, Heinrich. 1976. 'Zum römischen Kalender'. *Hermes* 104: 115–18.
Chaparro Gómez, César, ed. 2012. *Isidoro de Sevilla: Etimologías; Libro VI: De las Sagradas Escrituras*. Paris: Les Belles Lettres.
Charles-Edwards, Thomas M. 2000. *Early Christian Ireland*. Cambridge: Cambridge University Press.
Chibnall, Marjorie. 1975. 'Pliny's *Natural History* and the Middle Ages'. In *Empire and Aftermath: Silver Latin II*, ed. Thomas A. Dorey, 57–78. London: Routledge & Kegan Paul.
Chlench, Kathin. 2006a. 'Die deutschen Fassungen der Schriften des Johannes von Gmunden'. In *Johannes von Gmunden (ca. 1384–1442)*, ed. Rudolf Simek and Kathrin Chlench, 127–36. Vienna: Fassbaender.
Chlench, Kathin. 2006b. 'Johannes von Gmunden—Handschriftenverzeichnis'. In *Johannes von Gmunden (ca. 1384–1442)*, ed. Rudolf Simek and Kathrin Chlench, 195–223. Vienna: Fassbaender.

Chlench, Kathin. 2007. *Johannes von Gmunden deutsch: Der Wiener Codex 3055; Deutsche Texte des 'Corpus astronomicum' aus dem Umkreis von Johannes von Gmunden*. Vienna: Fassbaender.

Cobban, Alan B. 1988. *The Medieval English Universities: Oxford and Cambridge to c. 1500*. Berkeley: University of California Press.

Cohen, J. M., trans. 1955. *The Histories of Gargantua and Pantagruel by François Rabelais*. Harmondsworth: Penguin.

Colgrave, Bertram, ed. 1927. *The Life of Bishop Wilfrid by Eddius Stephanus*. Cambridge: Cambridge University Press.

Colgrave, Bertram, and R. A. B. Mynors, eds. 1969. *Bede's Ecclesiastical History of the English People*. Oxford: Clarendon Press.

Comes, Mercè. 1996. 'The Accession and Recession Theory in Al-Andalus and the North of Africa'. In *From Baghdad to Barcelona*, ed. Josep Casulleras and Julio Samsó, vol. 1, 349–64. Barcelona: Instituto 'Millás Vallicrosa' de Historia de la Ciencia Árabe.

Comes, Mercè. 2001. 'Ibn al-Hā'im's Trepidation Model'. *Suhayl* 2: 291–408.

Comes, Mercè. 2002. 'Some New Maghribī Sources Dealing with Trepidation'. In *Science and Technology in the Islamic World*, ed. S. M. Razaullah Ansari, 121–41. Turnhout: Brepols.

Contreni, John J. 2005. 'Bede's Scientific Works in the Carolingian Age'. In *Bède le Vénérable*, ed. Stéphane Lebecq, Michel Perrin, and Olivier Szerwiniack, 247–59. Lille: Ceges—Université Charles de Gaulle.

Cordoliani, Alfred. 1943. 'Les traités du comput du haut Moyen Âge (526-1003)'. *Archivum Latinitatis Medii Aevi* 17: 51–72.

Cordoliani, Alfred. 1945–6. 'Les computistes insulaires et les écrits pseudo-alexandrins'. *Bibliothèque de l'École des chartes* 106: 5–34.

Cordoliani, Alfred. 1949a. 'Abbon de Fleury, Hériger de Lobbes et Gerland de Besançon sur l'ère de l'incarnation de Denys le Petit'. *Revue d'histoire ecclésiastique* 44: 463–87.

Cordoliani, Alfred. 1949b. 'Un manuscrit de comput et d'astronomie des XIIe–XIVe siècles: le ms. 467 de l'Université de Glasgow'. *Scriptorium* 3: 69–79.

Cordoliani, Alfred. 1952. 'Les manuscrits de comput ecclésiastique de la Bibliothèque Capitulaire de Tolède'. *Revista de Archivos, Bibliotecas y Museos* 58: 323–52.

Cordoliani, Alfred. 1954. 'Inventaire des manuscrits de comput ecclésiastique conservés dans les bibliothèques de Madrid (1e série)'. *Hispania Sacra* 7: 111–43.

Cordoliani, Alfred. 1955. 'L'évolution du comput ecclésiastique à Saint Gall du VIIIe au XIe siècle'. *Zeitschrift für schweizerische Kirchengeschichte* 49: 288–323.

Cordoliani, Alfred. 1958. 'Les manuscrits de la bibliothèque de Berne provenant de l'abbaye de Fleury au XIe siècle: le comput d'Abbon'. *Zeitschrift für schweizerische Kirchengeschichte* 52: 135–50.

Cordoliani, Alfred. 1960–1. 'Contribution à la littérature du comput ecclésiastique au Moyen Âge'. Pts. 1 and 2. *Studi medievali*, 3rd ser., 1: 107–37; 2: 169–208.

Cordoliani, Alfred. 1963. 'Le computiste Hermann de Reichenau'. *Miscellanea storica ligure* 3: 165–90.

Corning, Caitlin. 2006. *The Celtic and Roman Traditions: Conflict and Consensus in the Early Medieval Church*. New York: Palgrave Macmillan.

Coyne, George V., Michael A. Hoskin, and Olaf Pedersen, eds. 1983. *Gregorian Reform of the Calendar: Proceedings of the Vatican Conference to Commemorate Its 400th Anniversary, 1582–1982*. Vatican City: Specola Vaticana.

Crowther et al. 2015 = Kathleen Crowther, Ashley Nicole McCray, Leila McNeill, Amy Rodgers, and Blair Stein. 'The Book Everybody Read: Vernacular Translations of Sacrobosco's *Sphere* in the Sixteenth Century'. *JHA* 46: 4–28.

Cucina, Carla. 2011. 'Traces of Runic Lore in Italy: The Wooden Calendar 'Book' in Bologna and Its Medieval Connections'. *Classiconorroena* 29: 95–181.
Cucina, Carla. 2013. *Libri runici del computo: il calendario di Bologna e i suoi analoghi europei*. Macerata: eum.
Cullen, Olive M. 2007. 'A Question of Time or a Question of Theology: A Study of the Easter Controversy in the Insular Church'. Ph.D. diss., St Patrick's College, Maynooth.
Curtze, Maximilian. 1902. *Urkunden zur Geschichte der Mathematik im Mittelalter und der Renaissance*. Leipzig. Reprint, New York: Johnson, 1968.
Dailey, E. T. 2015. 'To Choose One Easter from Three: Oswiu's Decision and the Northumbrian Synod of AD 664'. *Peritia* 26: 47–64.
Dales, Richard C. 1989. 'The Computistical Works Ascribed to Robert Grosseteste'. *Isis* 80: 74–9.
Dall'Olmo, Umberto. 1975. 'L'eclisse di sole del 2 Agosto 1133'. *Coelum* 45: 233–43.
Dall'Olmo, Umberto. 1979. '*Eclypsis naturalis* ed *Eclypsis prodigialis* nelle cronache medioevali'. *Organon* 15: 153–66.
Davis, John, and Michael Lowne. 2015. 'An Early English Astrolabe at Gonville & Caius College, Cambridge, and Walter of Elveden's *Kalendarium*'. *JHA* 46: 257–90.
Decaluwé, Michiel. 2009. *A Successful Defeat: Eugene IV's Struggle with the Council of Basel for Ultimate Authority in the Church 1431–1449*. Brussels: Belgisch Historisch Instituut te Rome.
Decaluwé, Michiel, Thomas M. Izbicki, and Gerald Christianson, eds. 2017. *A Companion to the Council of Basel*. Leiden: Brill.
Declercq, Georges. 2000. *Anno Domini: The Origins of the Christian Era*. Turnhout: Brepols.
Declercq, Georges. 2002. 'Dionysius Exiguus and the Introduction of the Christian Era'. *Sacris Erudiri* 41: 165–246.
Degrassi, Attilio. 1963. *Inscriptiones Italiae*. Vol. 13, *Fasti et elogia*, fasc. 2, *Fasti anni Numani et Iuliani, accedunt ferialia, menologia rustica, parapegmata*. Rome: Istituto poligrafico dello stato.
Delatour, Jérôme. 1999. 'Noël le 15 décembre: la réception du calendrier grégorien en France (1582)'. *Bibliothèque de l'École des chartes* 157: 369–416. Reprinted in Hubert 2000.
Delbrugge, Laura, ed. 1999. *Andrés de Li: Reportorio de los tiempos*. London: Tamesis.
Delorme, Ferdinand M. 1939. 'De auctore Compoti sub nomine Rogeri Baconis recenter editi'. *Antonianum* 14: 313–22.
De Meis, Salvo. 2006. 'From Regiomontanus to Kepler: Accuracy and Applications of Astronomical Ephemerides'. In *Astronomy as a Model for the Sciences in Early Modern Times*, ed. Menso Folkerts and Andreas Kühne, 399–421. Augsburg: Rauner.
De Meis, Salvo. 2009. '*Almanach Dantis*: analisi dell'almanacco di Profacius'. *Giornale di Astronomia* 35, no. 2: 26–33.
Déprez, Eugène. 1899. 'Une tentative de reforme du calendrier sous Clement VI: Jean des Murs et la Chronique de Jean de Venette'. *Mélanges d'archéologie et d'histoire* 19: 131–43.
Déprez, Eugène, ed. 1925. *Clément VI (1342–1352): lettres closes, patentes et curiales*. Vol. 1.2. Paris: Boccard.
Depuydt, Leo. 1996. 'The Egyptian and Athenian Dates of Meton's Observation of the Summer Solstice (–431)'. *Ancient Society* 27: 27–45.
Depuydt, Leo. 2002. 'History of the *ḥeleq*'. In *Under One Sky*, ed. John M. Steele and Annette Imhausen, 79–107. Münster: Ugarit-Verlag.
Díaz-Fajardo, Montse. 2001. *La teoría de la trepidación en un astrónomo marroquí del siglo XV: estudio y edición crítica del 'Kitāb al-adwār fī tasyīr al-anwār' (parte primera) de Abū ʿAbd Allāh al-Baqqār*. Barcelona: Instituto 'Millás Vallicrosa' de Historia de la Ciencia Árabe.

Dickey, Bruce George. 1982. 'Adelard of Bath: An Examination Based on heretofore Unexamined Manuscripts'. Ph.D. diss., University of Toronto.

Dieten, Jan Louis van, trans. 1979. *Nikephoros Gregoras: Rhomäische Geschichte/Historia Rhomaïke*. Vol. 2.1. Stuttgart: Hiersemann.

Divjak, Johannes, and Wolfgang Wischmeyer. 2014. *Das Kalenderhandbuch von 354: Der Chronograph des Filocalus*. 2 vols. Vienna: Holzhausen.

Dobcheva, Ivana. 2013. 'The Umbrella of Carolingian Computus'. In *La compilación del saber en la Edad Media*, ed. Maria José Muñoz, Patricia Cañizares, and Cristina Martín, 211–29. Porto: Fédération Internationale des Institut d'Études Médiévales.

Dobrzycki, Jerzy. 1965/2010. 'The Theory of Precession in Medieval Astronomy'. Translated by Aleksandra Niemirycz. Chapter 1 in Dobrzycki 2010.

Dobrzycki, Jerzy. 1971. 'John Werner's Theory of the Motion of the Eighth Sphere'. In *Actes du XIIe Congrès international d'histoire des sciences, Paris 1968*, vol. IIIA, 43–5. Paris: Blanchard.

Dobrzycki, Jerzy. 1975. 'New Sources for the Prehistory of the Calendar Reform'. In *Proceedings of the XIVth International Congress of the History of Science*, vol. 2, 35–6. Tokyo: Science Council of Japan. Reprinted as chapter 2 in Dobrzycki 2010.

Dobrzycki, Jerzy. 1983. 'Astronomical Aspects of the Calendar Reform'. In Coyne, Hoskin, and Pedersen 1983: 117–27. Reprinted as Chapter 3 in Dobrzycki 2010.

Dobrzycki, Jerzy. 1987. 'The *Tabulae Resolutae*'. In *De astronomia Alphonsi regis*, ed. Mercè Comes, Roser Puig, and Julio Samsó, 71–7. Barcelona: Instituto 'Millás Vallicrosa' de Historia de la Ciencia Árabe. Reprinted as chapter 6 in Dobrzycki 2010.

Dobrzycki, Jerzy. 2010. *Selected Papers on Medieval and Renaissance Astronomy*. Ed. Jarosław Włodarczyk and Richard L. Kremer. Warsaw: Instytut Historii Nauki PAN.

Dowd, Matthew F. 2003. 'Astronomy and Computus at Oxford University in the Early Thirteenth Century: The Works of Robert Grosseteste'. Ph.D. diss., University of Notre Dame.

Drecker, Julius. 1931. 'Hermannus Contractus über das Astrolab'. *Isis* 16: 200–19.

Duhem, Pierre, ed. 1909. *Un fragment inédit de l'Opus tertium de Roger Bacon: précédé d'une étude sur ce fragment*. Quaracchi: Ex Typographia Collegii S. Bonaventurae.

Duhem, Pierre. 1911. 'Un document relatif à la réforme du calendrier'. In *Hommage à Louis Olivier*, 97–103. Paris: Maretheux.

Duhem, Pierre. 1913–59. *Le système du monde: histoire des doctrines cosmologiques de Platon à Copernic*. 10 vols. Paris: Hermann.

Duke, Dennis W. 2008. 'Four Lost Episodes in Ancient Solar Theory'. *JHA* 39: 283–96.

Duncan, David Ewing. 1998. *The Calendar: The 5000-Year Struggle to Align the Clock and the Heavens—and What Happened to the Missing Ten Days*. London: Fourth Estate.

Dutka, Jacques. 1988. 'On the Gregorian Revision of the Julian Calendar'. *The Mathematical Intelligencer* 10: 56–64.

Easton, Stewart C. 1952. *Roger Bacon and His Search for a Universal Science*. Oxford: Blackwell.

Eastwood, Bruce S. 1986. 'Plinian Astronomy in the Middle Ages and Renaissance'. In *Science in the Early Roman Empire*, ed. Roger French and Frank Greenaway, 197–251. London: Croom Helm.

Eastwood, Bruce S. 1987. 'Plinian Astronomical Diagrams in the Early Middle Ages'. In *Mathematics and Its Application to Science and Natural Philosophy in the Middle Ages*, ed. Edward Grant and John E. Murdoch, 141–72. Cambridge: Cambridge University Press.

Eastwood, Bruce S. 1994a. 'Manuscripts of Macrobius, *Commentarii in Somnium Scipionis*, before 1500'. *Manuscripta* 38: 138–55.

Eastwood, Bruce S. 1994b. 'The Astronomy of Macrobius in Carolingian Europe: Dungal's Letter of 811 to Charles the Great'. *Early Medieval Europe* 3: 117–34.

Eastwood, Bruce S. 1997. 'Astronomy in Christian Latin Europe, c. 500–c. 1150'. *JHA* 28: 235–58.

Eastwood, Bruce S. 1999. 'Calcidius's Commentary on Plato's *Timaeus* in Latin Astronomy of the Ninth to Eleventh Centuries'. In *Between Demonstration and Imagination*, ed. Lodi Nauta and Arjo Vanderjagt, 171–209. Leiden: Brill.

Eastwood, Bruce S. 2000. 'Astronomical Images and Planetary Theory in Carolingian Studies of Martianus Capella'. *JHA* 31: 1–28.

Eastwood, Bruce S. 2001. 'Johannes Scottus Eriugena, Sun-Centred Planets, and Carolingian Astronomy'. *JHA* 32: 281–324.

Eastwood, Bruce S. 2002. 'Invention and Reform in Latin Planetary Astronomy'. In *Latin Culture in the Eleventh Century*, ed. Michael W. Herren, C. J. McDonough, and Ross G. Arthur, vol. 1, 264–97. Turnhout: Brepols.

Eastwood, Bruce S. 2007. *Ordering the Heavens: Roman Astronomy and Cosmology in the Carolingian Renaissance*. Leiden: Brill.

Eastwood, Bruce S. 2011. 'The Power of Diagrams: The Place of the Anonymous Commentary in the Development of Carolingian Astronomy and Cosmology'. In *Carolingian Scholarship and Martianus Capella*, ed. Mariken Teeuwen and Sinéad O'Sullivan, 193–220. Turnhout: Brepols.

Eastwood, Bruce S., and Gerd Graßhoff. 2004. *Planetary Diagrams for Roman Astronomy in Medieval Europe, ca. 800–1500*. Philadelphia, PA: American Philosophical Society (= Transactions of the American Philosophical Society 94.3).

Edwards, Glenn M. 1985. 'The Two Redactions of Michael Scot's "Liber introductorius"'. *Traditio* 41: 329–40.

Eisermann, Falk. 2004. *Verzeichnis der typographischen Einblattdrucke des 15. Jahrhunderts im Heiligen Römischen Reich Deutscher Nation (VE 15)*. 3 vols. Wiesbaden: Reichert.

Eisner, Sigmund, ed. 1980. *The Kalendarium of Nicholas of Lynn*. Athens: University of Georgia Press.

Engelen, Eva-Maria. 1993. *Zeit, Zahl und Bild: Studien zur Verbindung von Philosophie und Wissenschaft bei Abbo von Fleury*. Berlin: de Gruyter.

Evans, James. 1998. *The History & Practice of Ancient Astronomy*. New York: Oxford University Press.

Evans, James, and J. Lennart Berggren, trans. 2006. *Geminos's 'Introduction to the Phenomena'*. Princeton: Princeton University Press.

Farnham, Willard. 1923. 'The Dayes of the Mone'. *Studies in Philology* 20: 70–82.

Favaro, Antonio. 1904. 'Nuove ricerche sul matematico Leonardo Cremonese'. *Bibliotheca Mathematica*, 3rd ser., 5: 326–41.

Federici Vescovini, Graziella, ed. 1992. *Pietro d'Abano: Trattati di astronomia*. Padua: Editoriale Programma.

Feeney, Denis. 2007. *Caesar's Calendar: Ancient Time and the Beginnings of History*. Berkeley: University of California Press.

Fernández, Tomás. 2011. 'Un fragment inédit attribué à Anatole d'Alexandrie'. *Byzantion Nea Hellás* 30: 189–202.

Finke, Heinrich, ed. 1896–1928. *Acta Concilii Constanciensis*. 4 vols. Münster: Regenberg.

Flint, Valerie I. J. 1982. 'Honorius Augustodunensis Imago Mundi'. *Archives d'histoire doctrinale et littéraire du Moyen Âge* 49: 7–153.

Folkerts, Menso. 2001. 'Early Texts on Hindu-Arabic Calculation'. *Science in Context* 14: 13–38.

Fontaine, Jacques, ed. 1960. *Isidore de Seville: Traité de la nature*. Bordeaux: Féret et Fils.

Fournet, Jean-Luc, and Anne Tihon. 2014. *Conformément aux observations d'Hipparque: le Papyrus Fouad inv. 267 A*. Louvain-la-Neuve: Institut Orientaliste.

Fournet, Jean-Luc, and Anne Tihon. 2016. 'A New Fragment of P. Fouad inv. 267A: The PSI inv. 2006'. *JHA* 47: 355–8.

Freisen, Joseph. 1893. *Geschichte des kanonischen Eherechts bis zum Verfall der Glossenliteratur*. 2nd edn. Paderborn. Reprint, Aalen: Scientia Verlag, 1963.

French, Roger. 1996. 'Foretelling the Future: Arabic Astrology and English Medicine in the Late Twelfth Century'. *Isis* 87: 453–80.

Friedberg, Emil. 1879–81. *Corpus Juris Canonici*. 2 vols. Leipzig: Tauchnitz, 1879–81.

Friedman, John B. 1985. 'Richard de Thorpe's Astronomical Kalendar and the Luxury Book Trade at York'. *Studies in the Age of the Chaucer* 7: 137–60.

Fryde, Edmund. 2000. *The Early Palaeologan Renaissance (1261–c. 1360)*. Leiden: Brill.

Gaab, Hans. 2011. *Der Altdorfer Mathematik- und Physikdozent Abdias Trew (1597–1669): Astronom, Astrologe, Kalendermacher und Theologe*. Frankfurt/Main: Deutsch.

Gack-Scheiding, Christine, ed. 1995. *Johannes de Muris: Epistola super reformatione antiqui kalendarii; Ein Beitrag zur Kalenderreform im 14. Jahrhundert*. Hannover: Hahn.

Galle, Karl. 2006. '16th-Century Astronomy and Its Professional Contexts: The Case of Copernicus and Calendar Reform'. In *Astronomy as a Model for the Sciences in Early Modern Times*, ed. Menso Folkerts and Andreas Kühne, 143–61. Augsburg: Rauner.

García Avilés, Alejandro. 2001. *El tiempo y los astros: arte, ciencia y religión en la Alta Edad Media*. Murcia: Universidad de Murcia.

Gautier Dalché, Patrick. 2009. *La Géographie de Ptolémée en Occident (IVe–XVIe siècle)*. Turnhout: Brepols.

Gautier Dalché, Patrick. 2013. 'Le "Tuyau" de Gerbert, ou la légende savante de l'astronomie: origines, thèmes, échos contemporains (avec un appendice critique)'. *Micrologus* 21: 243–76.

Geerlings, Wilhelm, ed. 2002. *Der Kalender: Aspekte einer Geschichte*. Paderborn: Schöningh.

Geldner, Ferdinand, ed. 1975. *Der Türkenkalender: 'eyn Manung der Christenheit widder die Durken', Mainz 1454; Das älteste vollständig erhaltene gedruckte Buch (Rar. 1 der Bayerischen Staatsbibliothek)*. Wiesbaden: Reichert.

Gerl, Armin. 1989. *Trigonometrisch-astronomisches Rechnen kurz vor Copernicus: Der Briefwechsel Regiomontanus-Bianchini*. Stuttgart: Steiner.

Gerlach, Karl. 1998. *The Antenicene Pascha: A Rhetorical History*. Leuven: Peeters.

Germann, Nadja. 2006. *De temporum ratione: Quadrivium und Gotteserkenntnis am Beispiel Abbos von Fleury und Hermanns von Reichenau*. Leiden: Brill.

Germann, Nadja. 2008a. 'À la recherche de la structure du temps: Abbon de Fleury et le comput'. In *Abbon, un abbé de l'an mil*, ed. Annie Dufour and Gillette Labory, 153–76. Turnhout: Brepols.

Germann, Nadja. 2008b. 'Zwischen *veritas naturae* und *fides historiae*: Zeit und Dauer bei Abbo von Fleury'. In *Das Sein der Dauer*, ed. Andreas Speer and David Wirmer, 171–95. Berlin: de Gruyter.

Gibson, Strickland, ed. 1931. *Statuta antiqua Universitatis Oxoniensis*. Oxford: Clarendon Press.

Gilomen, Hans-Jörg. 2017. 'Conciliar Bureaucracy'. In Decaluwé, Izbicki, and Christianson 2017: 167–228.

Gingerich, Owen. 1987. 'The Alfonsine Tables in the Age of Printing'. In *De astronomia Alphonsi regis*, ed. Mercè Comes, Roser Puig, and Julio Samsó, 89–95. Barcelona: Instituto 'Millás Vallicrosa' de Historia de la Ciencia Árabe.

Gingerich, Owen. 2002. 'The Trouble with Ptolemy'. *Isis* 93: 70–4.

Gingerich, Owen, and Barbara Welther. 1977. 'The Accuracy of the Toledan Tables'. In *ΠΡΙΣΜΑΤΑ: Naturwissenschaftsgeschichtliche Studien*, edited Y. Maeyama and W. G. Saltzer, 151–63. Wiesbaden: Steiner.

Ginzel, Friedrich Karl. 1906–14. *Handbuch der mathematischen und technischen Chronologie: Das Zeitrechnungswesen der Völker*. 3 vols. Leipzig: Hinrichs.

Girgensohn, Dieter. 2007. 'Von der konzilien Theorie des späteren Mittelalters zur Praxis: Pisa 1409'. In *Die Konzilien von Pisa (1409), Konstanz (1414–1418) und Basel (1431–1449)*, ed. Heribert Müller and Johannes Helmrath, 61–94. Ostfildern: Thorbecke.

Glorieux, Palémon. 1951. 'L'activité littéraire de Gerson à Lyon'. *Recherches de théologie ancienne et médiévale* 18: 238–307.

Glorieux, Palémon. 1960. *Jean Gerson: Oeuvres complètes*. Vol 2. Paris: Desclée.

Goldstein, Bernard R. 1964–5. 'On the Theory of Trepidation According to Thābit b. Qurra and al-Zarqāllu and Its Implications for Homocentric Planetary Theory'. *Centaurus* 10: 232–47.

Goldstein, Bernard R. 1966. 'A Note on the Metonic Cycle'. *Isis* 57: 115–16.

Goldstein, Bernard R. 1980. 'Solar and Lunar Velocities in the Alfonsine Tables'. *Historia Mathematica* 7: 134–40.

Goldstein, Bernard R. 1992. 'Levi ben Gerson's Contributions to Astronomy'. In *Studies on Gersonides*, ed. Gad Freudenthal, 3–19. Leiden: Brill.

Goldstein, Bernard R. 1994. 'Historical Perspectives on Copernicus's Account of Precession'. *JHA* 24: 189–97.

Goldstein, Bernard R. 2003. 'Ancient and Medieval Values for the Mean Synodic Month'. *JHA* 34: 65–74.

Goldstein, Bernard R. 2011. 'Solomon Franco on the Zero Point for Trepidation'. *Suhayl* 10: 77–83.

Goldstein, Bernard R., and Alan C. Bowen. 1991. 'The Introduction of Dated Observations and Precise Measurement in Greek Astronomy'. *AHES* 43: 93–132.

Goldstein, Bernard R., and José Chabás. 2008. 'Transmission of Computational Methods within the Alfonsine Corpus: The Case of the Tables of Nicholaus de Heybech'. *JHA* 39: 345–55.

Goldstein, Bernard R., José Chabás, and José Luis Mancha. 1994. 'Planetary and Lunar Velocities in the Castilian Alfonsine Tables'. *Proceedings of the American Philosophical Society* 138: 61–95.

Goldstein, Bernard R., and David Pingree. 1990. *Levi ben Gerson's Prognostication for the Conjunction of 1345*. Philadelphia: American Philosophical Society (= *Transactions of the American Philosophical Society* 80.6).

Görke, Winfried. 2011. *Datum und Kalender: Von der Antike bis zur Gegenwart*. Heidelberg: Springer.

Grabmann, Martin. 1936. *Mittelalterliches Geistesleben: Abhandlungen zur Geschichte der Scholastik und Mystik*. Vol. 2. Munich: Hueber.

Graff, Eric. 2010. 'The Recension of Two Sirmond Texts: *Disputatio Morini* and *De Divisionibus Temporum*'. In Warntjes and Ó Cróinín 2010: 112–42.

Graf-Stuhlhofer, Franz. 1996. *Humanismus zwischen Hof und Universität: Georg Tannstetter (Collimitius) und sein wissenschaftliches Umfeld im Wien des frühen 16. Jahrhunderts*. Vienna: WUV-Universitätsverlag.

Grafton, Anthony. 1993. *Joseph Scaliger: A Study in the History of Classical Scholarship*. Vol. 2, *Historical Chronology*. Oxford: Clarendon Press.

Granada, Miguel A., and Dario Tessicini. 2005. 'Copernicus and Fracastoro: The Dedicatory Letters to Pope Paul III, the History of Astronomy, and the Quest for Patronage'. *Studies in History and Philosophy of Science* 36: 431–76.

Grant, Edward, ed. 1966. *Nicole Oresme: 'De proportionibus proportionum' and 'Ad pauca respicientes'.* Madison, Wisc.: University of Wisconsin Press.

Grant, Edward, ed. 1971. *Nicole Oresme and the Kinematics of Circular Motion: Tractatus de commensurabilitate vel incommensurabilitate motuum celi.* Madison: University of Wisconsin Press.

Grant, Edward. 1984. 'Science and the Medieval University'. In *Rebirth, Reform and Resilience*, ed. James M. Kittelson and Pamela J. Transue, 68–102. Columbus: Ohio State University Press.

Grant, Edward. 1994. *Planets, Stars, and Orbs: The Medieval Cosmos, 1200–1687.* Cambridge: Cambridge University Press.

Green, Jonathan. 2012. *Printing and Prophecy: Prognostication and Media Change 1450–1550.* Ann Arbor, Mich.: University of Michigan Press.

Guenée, Bernard. 1991. *Between Church and State: The Lives of Four French Prelates in the Late Middle Ages.* Translated by Arthur Goldhammer. Chicago: University of Chicago Press.

Guilland, Rodolphe. 1926. *Essai sur Nicéphore Grégoras: l'homme et l'oeuvre.* Paris: Geuthner.

Guilland, Rodolphe, ed. 1927. *Correspondance de Nicéphore Grégoras: texte édité et traduit.* Paris: Les Belles Lettres.

Gumbert, J. P. 2016. *Bat Books: A Catalogue of Folded Manuscripts Containing Almanacs or Other Texts.* Turnhout: Brepols.

Gumbert-Hepp, Marijke, ed. 1987. *Computus Magistri Jacobi: een schoolboek voor tijdrekenkunde uit 1436.* Hilversum:Verloren.

Gushee, Lawrence. 1969. 'New Sources for the Biography of Johannes de Muris'. *Journal of the American Musicological Society* 22: 3–26.

Gushee, Lawrence. 1998. 'Jehan des Murs and His Milieu'. In *Musik—und die Geschichte der Philosophie und Naturwissenschaften im Mittelalter*, ed. Frank Hentschel, 339–71. Leiden: Brill.

Hackett, Jeremiah. 2016. 'From *Sapientes antiqui* at Lincoln to the New *Sapientes moderni* at Paris *c.* 1260–1280: Roger Bacon's Two Circles of Scholars'. In *Robert Grosseteste and the Pursuit of Religious and Scientific Learning in the Middle Ages*, ed. Jack P. Cunningham and Mark Hocknull, 119–42. Heidelberg: Springer.

Hagen, J. G. 1914. 'Die Gregorianische Kalenderreform'. *Stimmen aus Maria-Laach* 87: 41–51.

Hagenmeyer, Heinrich, ed. 1913. *Fulcheri Carnotensis Historia Hierosolymitana (1095–1127).* Heidelberg: Winter.

Haller et al. 1896–1936. *Concilium Basiliense: Studien und Quellen zur Geschichte des Concils von Basel.* 8 vols. Basel: Reich/Helbing & Lichtenhahn.

Hallyn, Fernand. 2008. 'Paul de Middelbourg, astrologue et astronome'. In *Esculape et Dionysos*, ed. Jean Dupèbe et al., 367–74. Geneva: Droz.

Hamann, Günther, ed. 1980. *Regiomontanus-Studien.* Sitzungsberichte der Österreichischen Akademie der Wissenschaften, phil.-hist. Kl., 364. Vienna: Verlag der Österreichschen Akademie der Wissenschaften.

Hamel, Jürgen. 1999. 'Die Kalenderreform Papst Gregors XIII. von 1582 und ihre Durchsetzung'. In *Geburt der Zeit*, ed. Hans Ottomeyer, Sven Lüken, and Micha Röhring, 292–301. Wolfratshausen: Minerva.

Hamel, Jürgen. 2002. 'Die Rolle Michael Mästlins in der Polemik um die Kalenderreform von Papst Gregor XIII'. In *Zwischen Copernicus und Kepler*, ed. Gerhard Betsch and Jürgen Hamel, 33–63. Frankfurt/Main: Deutsch.

Hamel, Jürgen. 2014. *Studien zur 'Sphaera' des Johannes de Sacrobosco*. Leipzig: Akademische Verlagsanstalt.

Hamel, Jürgen. 2017. 'Landgraf Wilhelm IV und Michael Mästlin in der Polemik um die Gregorianische Kalenderreform'. In *Le temps des astronomes*, ed. Edouard Mehl and Nicolas Roudet, 231–68. Paris: Les Belles Lettres.

Hameter, Wolfgang, Meta Niederkorn-Bruck, and Martin Scheutz, eds. 2005. *Ideologisierte Zeit: Kalender und Zeitvorstellungen im Abendland von der Antike bis zur Neuzeit*. Innsbruck: StudienVerlag.

Hannah, Robert. 2005. *Greek and Roman Calendars: Constructions of Time in the Classical World*. London: Duckworth.

Harper, Richard I. 1966. 'The *Kalendarium Regine* of Guillaume de St.-Cloud'. Ph.D. diss., Emory University.

Harrison, Kenneth. 1979. 'Luni-Solar Cycles: Their Accuracy and Some Types of Usage'. In *Saints, Scholars and Heroes*, ed. Margot H. King and Wesley M. Stevens, vol. 2, 65–78. Collegeville, Minn.: Hill Monastic Manuscript Library.

Harrison, Kenneth. 1982. 'Episodes in the History of Easter Cycles in Ireland'. In *Ireland in Early Medieval Europe*, ed. Dorothy Whitelock, Rosamond McKitterick, and David Dumville, 307–19. Cambridge: Cambridge University Press.

Harrison, Kenneth. 1984. 'A Letter from Rome to the Irish Clergy, AD 640'. *Peritia* 3: 222–9.

Hart, Cyril. 1970. 'The Ramsey Computus'. *English Historical Review* 85: 29–44.

Hart, Cyril. 2003. *Learning and Culture in Late Anglo-Saxon England and the Influence of Ramsey Abbey on the Major English Monastic Schools: A Survey of the Development of Mathematical, Medical, and Scientific Studies in England before the Norman Conquest*. Vol. 2.1. Lewiston, NY: Edwin Mellen Press.

Haskins, Charles Homer. 1924. *Studies in the History of Mediaeval Science*. Cambridge, Mass.: Harvard University Press.

Hayton, Darin. 2015. *The Crown and the Cosmos: Astrology and the Politics of Maximilian I*. Pittsburgh: University of Pittsburgh Press.

Heilbron, John L. 1999. *The Sun in the Church: Cathedrals as Solar Observatories*. Cambridge, Mass.: Harvard University Press.

Heilen, Stephan. 2015. 'Paul of Middelburg's *Prognosticum* for the Years 1484 to 1504'. In *From Māshā'allāh to Kepler*, ed. Charles Burnett and Dorian Gieseler Greenbaum, 231–78. University of Wales, Trinity Saint David: Sophia Centre Press.

Heitz, Paul, ed. 1905. *Hundert Kalender-Inkunabeln*. Strasbourg: Heitz and Muendel.

Helfand, Jessica. 2002. *Reinventing the Wheel: Volvelles, Equatoria, Planispheres, Fact-Finders, Gestational Charts*. New York: Princeton Architectural Press.

Helmrath, Johannes. 1987. *Das Basler Konzil, 1431–1449: Forschungsstand und Probleme*. Cologne: Böhlau.

Herold, Werner, ed. 2011. *Reinher von Paderborn: Computus Emendatus; Die verbesserte Osterfestberechnung von 1171*. Paderborn: Bonifatius.

Hertz, Martin, ed. 1955. *Prisciani Grammatici Caesariensis Institutionem Grammaticarum Libri XVIII*. Vol. 1, *Libros I–XII continentes*. Leipzig. Reprint, Hildesheim: Olms, 1961.

Herzog, Markwart, ed. 2002. *Der Streit um die Zeit: Zeitmessung—Kalenderreform—Gegenzeit—Endzeit*. Stuttgart: Kohlhammer.

Holford-Strevens, Leofranc. 2008. 'Paschal Lunar Calendars up to Bede'. *Peritia* 20: 165–208.

Holford-Strevens, Leofranc. 2010. 'Marital Discord in Northumbria: Lent and Easter, His and Hers'. In Warntjes and Ó Cróinín 2010: 143–58.

Holford-Strevens, Leofranc. 2011. 'Church Politics and the Computus: From Milan to the Ends of the Earth'. In Warntjes and Ó Cróinín 2011: 1–20.

Honecker, Martin. 1940. 'Die Entstehung der Kalenderreformschrift des Nikolaus von Cues'. *Historisches Jahrbuch* 60: 581–92.

Hourani, George F. 1972. 'The Medieval Translations from Arabic to Latin Made in Spain'. *The Muslim World* 62: 97–114.

Hubert, Marie-Clotilde, ed. 2000. *Construire le temps: normes et usages chronologiques du Moyen Âge à l'époque contemporaine*. Paris: Champion.

Hübner, Wolfgang. 2000. 'Une glose à la *Massa compoti* d'Alexandre de Villedieu contenant des vers sur le zodiaque'. In *Science antique, science médiévale (autour d'Avranches 235)*, ed. Louis Callebat and Olivier Desbordes, 373–89. Hildesheim: Olms.

Hugonnard-Roche, Henri. 1992. 'Problèmes méthodologiques dans l'astronomie au début du XIVe siècle'. In *Studies on Gersonides*, ed. Gad Freudenthal, 55–70. Leiden: Brill.

Hunt, Richard William. 1936. 'English Learning in the Late Twelfth Century'. *Transactions of the Royal Historical Society*, 4th ser., 19: 19–42.

Hunt, Richard William. 1984. *The Schools and the Cloister: The Life and Writings of Alexander Nequam (1157–1217)*. Edited and Revised by Margaret Gibson. Oxford: Clarendon Press.

Hunt, Tony. 2013. *Writing the Future: Prognostic Texts of Medieval England*. Paris: Classiques Garnier.

Husson, Matthieu. 2011. 'L'astronomie alphonsine dans *l'Expositio intentionis regis Alfonsii circa tabulas ejus* de Jean de Murs'. *Archives d'histoire doctrinale et littéraire du Moyen Âge* 78: 229–45.

Husson, Matthieu. 2012. 'Ways to Read a Table: Reading and Interpolation Techniques in Canons of Early Fourteenth-Century Double-Argument Tables'. *JHA* 43: 299–319.

Husson, Matthieu. 2016. 'Exploring the Temporality of Complex Computational Practice: Two Eclipse Notes by John of Murs in the ms Escorial O II 10'. *Centaurus* 58: 46–65.

Irtenkauf, Wolfgang. 1957. 'Der Computus ecclesiasticus in der Einstimmigkeit des Mittelalters'. *Archiv für Musikwissenschaft* 14: 1–15.

Jacquemard, Catherine. 2015. 'La réinvention de l'astrolabe au Moyen Âge: l'exemple du *De statu mundi*'. In *La technologie gréco-romaine: transmission, restitution et médiation*, ed. Philippe Fleury, Catherine Jacquemard, and Sophie Madeleine, 191–211. Caen: Presses universitaires de Caen.

Jacquemard, Catherine, Olivier Desbordes, and Alain Hairie. 2007. 'Du quadrant *vetustior* à *l'horologium viatorum* d'Hermann de Reichenau: étude du manuscrit Vaticano BAV Ott. lat. 1631, f. 16–17v'. *Kentron* 23: 79–125.

Jones, Alexander. 1991. 'Hipparchus's Computations of Solar Longitudes'. *JHA* 22: 101–25.

Jones, Alexander. 2000. 'Calendrica I: New Callippic Dates'. *ZPE* 129: 141–58.

Jones, Alexander. 2005. '"In order that we should not ourselves appear to be adjusting our estimates…to make them fit some predetermined amount"'. In *Wrong for the Right Reason*, ed. Jed Z. Buchwald and Allan Franklin, 17–39. Dordrecht: Springer.

Jones, Alexander. 2016. 'Unruly Sun: Solar Tables and Calculations in the Papyrus Fouad 267 A'. *JHA* 47: 76–99.

Jones, Charles W. 1934. 'The Victorian and Dionysiac Paschal Tables in the West'. *Speculum* 9: 408–21.

Jones, Charles W. 1937. 'The "Lost" Sirmond Manuscript of Bede's "Computus"'. *English Historical Review* 52: 204–19.

Jones, Charles W. 1939. *Bedae Pseudepigrapha: Scientific Writings Falsely Attributed to Bede*. Ithaca, NY: Cornell University Press.

Jones, Charles W. 1943a. 'A Legend of St Pachomius'. *Speculum* 18: 198–210.

Jones, Charles W. 1943b. *Bedae Opera de Temporibus*. Cambridge, Mass.: Mediaeval Academy of America.

Jullien, Marie-Hélène. 2010. *Clavis Scriptorum Latinorum Medii Aevi: Auctores Galliae 735–987*. Vol. 3. Turnhout: Brepols.

Juste, David. 2004. 'Neither Observation nor Astronomical Tables: An Alternative Way of Computing the Planetary Longitudes in the Early Western Middle Ages'. In *Studies in the History of the Exact Sciences in Honour of David Pingree*, ed. Charles Burnett, Jan P. Hogendijk, Kim Plofker, and Michio Yano, 181–222. Leiden: Brill.

Juste, David. 2007. *Les 'Alchandreana' primitifs: étude sur les plus anciens traités astrologiques latins d'origine arabe (Xe siècle)*. Leiden: Brill.

Juste, David. 2016. 'Hermann der Lahme und das Astrolab im Spiegel der neuesten Forschung'. In *Hermann der Lahme: Reichenauer Mönch und Universalgelehrter des 11. Jahrhunderts*, ed. Felix Heinzer and Thomas Zotz, 273–84. Stuttgart: Kohlhammer.

Kaeppeli, Thomas. 1980. *Scriptores Ordinis Praedicatorum Medii Aevi*. Vol. 3, *I–S*. Rome: Ad S. Sabinae.

Kaltenbrunner, Ferdinand. 1876. 'Die Vorgeschichte der gregorianischen Kalenderreform'. *Sitzungsberichte der philosophisch-historischen Classe der kaiserlichen Akademie der Wissenschaften* [Vienna] 82: 289–414.

Kaltenbrunner, Ferdinand. 1877. 'Die Polemik über die Gregorianische Kalenderreform'. *Sitzungsberichte der philosophisch-historischen Classe der kaiserlichen Akademie der Wissenschaften* [Vienna] 87: 485–586.

Kaltenbrunner, Ferdinand. 1878. 'Ein Kalender aus dem Anfang des 13. Jahrhunderts'. *Neues Archiv* 3: 387–95.

Kaltenbrunner, Ferdinand. 1880a. 'Beiträge zur Geschichte der Gregorianischen Kalenderreform. I. Die Commission unter Gregor XIII. nach Handschriften der Vaticanischen Bibliothek'. *Sitzungsberichte der philosophisch-historischen Classe der kaiserlichen Akademie der Wissenschaften* [Vienna] 97: 7–54.

Kaltenbrunner, Ferdinand. 1880b. 'Der Augsburger Kalenderstreit'. *Mittheilungen des Instituts für Österreichische Geschichtsforschung* 1: 497–540.

Karasawa, Kazutomo, ed. 2015. *The Old English Metrical Calendar (Menologium)*. Cambridge: D. S. Brewer.

Kaster, Robert A. 2010. *Studies on the Text of Macrobius' Saturnalia*. Oxford: Oxford University Press.

Kaster, Robert A., ed. and trans. 2011. *Macrobius: Saturnalia; Books 1–2*. Cambridge, Mass.: Harvard University Press.

Kay, Richard. 2003. 'Unwintering January (Dante, *Paradiso*, 27.142–143)'. *MLN* 118: 237–44.

Keil, Gundolf, ed. 1983. *Vom Einfluss der Gestirne auf die Gesundheit und den Charakter des Menschen: Kommentar zur Faksimile-Ausgabe des Manuskriptes C 54 der Zentralbibliothek Zürich (Nürnberger Kodex Schürstab)*. Luzern: Faksimile-Verlag.

Kendall, Calvin B., and Faith Wallis, trans. 2010. *Bede: On the Nature of Things and On Times*. Liverpool: Liverpool University Press.

Kersken, Wolfgang. 1975. *Genner beschnaid: Die Kalendergedichte und der Neumondkalender des Oswald von Wolkenstein; Überlieferung—Text—Deutung*. Göppingen: Kümmerle.

Kersken, Wolfgang. 1978. 'Die Kalendergedichte und der Neumondkalender Oswalds von Wolkenstein'. In *Gesammelte Vorträge der 600-Jahrfeier Oswalds von Wolkenstein, Seis am Schlern 1977*, ed. Hans-Dieter Mück and Ulrich Müller, 425–53. Göppingen: Kümmerle.

King, David A., and Julio Samsó (with a contribution by Bernard R. Goldstein). 2001. 'Astronomical Handbooks and Tables from the Islamic World (750–1900): An Interim Report'. *Suhayl* 2: 9–105.

King, Henry. 1978. *Geared to the Stars: The Evolution of Planetariums, Orreries, and Astronomical Clocks.* Toronto: University of Toronto Press.

King, James C., and Petrus W. Tax, eds. 1996. *Notker der Deutsche: Die kleineren Schriften.* Tübingen: Niemeyer.

Kleine, Sabine. 1995. 'Die "Mainauer Naturlehre"'. *Sudhoffs Archiv* 79: 101–14.

Klug, Rudolf. 1912. *Der Astronom Johannes von Gmunden und sein Kalender.* Sonderabdruck aus dem Programm des k. k. Staatsgymnasiums Linz 1912. Linz: Pringruber.

Klug, Rudolf. 1943. *Beiträge zur Geschichte der Universität Wien I: Johannes von Gmunden, der Begründer der Himmelskunde auf deutschem Boden; nach seinen Schriften und den Archivalien der Wiener Universität.* Vienna: Hölder-Pichler-Tempsky.

Knorr, Wilbur R. 1997. 'Sacrobosco's *Quadrans*: Date and Sources'. *JHA* 28: 187–222.

Köll, Thomas. 1983. '"Mer ist zewissen von dem schalt iar…": Der Beitrag des Stamser Mönchs Vitus de Augusta zur Kalenderreform des Jahres 1582'. *Innsbrucker Historische Studien* 6: 45–63.

Koller, Edith. 2014. *Strittige Zeiten: Kalenderreformen im Alten Reich 1582–1700.* Berlin: de Gruyter.

Korpiola, Mia. 2008. '*Tempus nuptiarum*: Ecclesiastical Influences on Wedding Times in Medieval Sweden'. In *Sacri canones servandi sunt*, ed. Pavel Krafl, 633–43. Prague: Historický ústav AV ČR.

Krafft, Fritz. 1994. 'Des Nicolaus Copernicus Bemühungen um die Bestimmung der Länge des tropischen Jahres: Zur Chronologie copernicanischer Astronomie'. In *Cosmographica et Geographica*, ed. Bernhard Fritscher and Gerhard Brey, vol. 1, 255–96. Munich: Institut für Geschichte der Naturwissenschaften.

Kremer, Richard L. 2003a. 'Thoughts on John of Saxony's Method for Finding Times of True Syzygy'. *Historia Mathematica* 30: 263–77.

Kremer, Richard L. 2003b. 'Wenzel Faber's Table for Finding True Syzygy'. *Centaurus* 45: 305–29.

Kremer, Richard L. 2007. '"Abbreviating" the Alfonsine Tables in Cracow: The *Tabulae Aureae* of Petrus Gaszowiec (1448)'. *JHA* 38: 283–304.

Kremer, Richard L. 2008. 'John of Murs, Wenzel Faber and the Computation of True Syzygy in the Fourteenth and Fifteenth Centuries'. In *Mathematics Celestial and Terrestrial*, ed. Joseph W. Dauben et al., 147–60. Stuttgart: Wissenschaftliche Verlagsgesellschaft.

Kremer, Richard L. 2011. 'Experimenting with Paper Instruments in Fifteenth- and Sixteenth-Century Astronomy: Computing Syzygies with Isotemporal Lines and Salt Dishes'. *JHA* 42: 223–58.

Kremer, Richard L. 2012. 'Marcus Schinnagel's Winged Polyptich of 1489: Astronomical Computation in Liturgical Format'. *JHA* 43: 321–45.

Kremer, Richard L. 2013. 'Hans Sporer's Xylographic Practices: A Census of Regiomontanus's Blockbook Calendar'. *Bibliothek und Wissenschaft* 46: 161–87.

Kremer, Richard L. 2017. 'Incunable Almanacs and *Practica* as Practical Knowledge Produced in Trading Zones'. In *The Structures of Practical Knowledge*, ed. Matteo Valleriani, 333–69. Dordrecht: Springer.

Krusch, Bruno. 1880. *Studien zur christlich-mittelalterlichen Chronologie: Der 84-jährige Osterzyklus und seine Quellen.* Leipzig: Veit & Comp.

Krusch, Bruno. 1884. 'Die Einführung des griechischen Paschalritus im Abendlande'. *Neues Archiv* 9: 99–169.

Krusch, Bruno. 1910. 'Das älteste fränkische Lehrbuch der dionysianischen Zeitrechnung'. In *Mélanges offerts à M. Émile Chatelain par ses élèves et ses amis*, 232–42. Paris: Champion.

Krusch, Bruno. 1938. *Studien zur christlich-mittelalterlichen Chronologie: Die Entstehung unserer heutigen Zeitrechnung*. Berlin: de Gruyter.

Kruse, Norbert. 2003. 'Eine neue Schrift Notkers des Deutschen: Der althochdeutsche Computus'. *Sprachwissenschaft* 28: 123–55.

Kuczynski, Michael P. 1987. 'A New Manuscript of Nicholas of Lynn's "Kalendarium": MS Chapel Hill 522, fols. 159r–202r'. *Traditio* 43: 299–319.

Kuithan, Rolf, and Joachim Wollasch. 1984. 'Der Kalender des Chronisten Bernold'. *Deutsches Archiv für Erforschung des Mittelalters* 40: 478–531.

Kunitzsch, Paul. 1992. 'Gerard's Translation of Astronomical Texts, Especially the *Almagest*'. In *Gerardo da Cremona*, ed. Pierluigi Pizzamiglio, 71–84. Cremona: Biblioteca Statale.

Kurtik, G. E. 1990. 'Precession Theory in Medieval Indian and Early Islamic Astronomy'. In *Interaction between Indian and Central Asian Science and Technology in Mediaeval Times*, ed. W. H. Abdi et al., vol. 1, 94–110. New Delhi: Indian National Science Academy.

Kuzenkov, Pavel. 2010. 'Correction of the Easter *Computus*: Heresy or Necessity? Fourteenth Century Byzantine Forerunners of the Gregorian Reform'. In *Orthodoxy and Heresy in Byzantium*, ed. Antonio Rigo and Pavel Ermilov, 147–58. Rome: Università degli Studi di Roma 'Tor Vergata'.

Labajos Alonso, José, ed. 2010. *Escritos académicos de Pedro de Osma*. Salamanca: Universidad Pontificia de Salamanca.

Laird, Edgar. 2012. 'Robert Grosseteste, Ptolemy, and Christian Knowledge'. In *Robert Grosseteste and His Intellectual Milieu: New Editions and Studies*, ed. John Flood, James R. Ginther, and Joseph W. Goering, 131–52. Toronto: Pontifical Institute of Mediaeval Studies.

Lamont, Roscoe. 1919. 'The Roman Calendar and Its Reformation by Julius Caesar'. *Popular Astronomy* 27: 583–95.

Lange, Ludwig. 1928. *'Paradoxe' Osterdaten im Gregorianischen Kalender und ihre Bedeutung für die moderne Kalenderreform*. Munich: Verlag der Bayerischen Akademie der Wissenschaften.

Lapidge, Michael, and Peter S. Baker. 1997. 'More Acrostic Verse by Abbo of Fleury'. *Journal of Medieval Latin* 7: 1–27.

Lattis, James M. 1994. *Between Copernicus and Galileo: Christoph Clavius and the Collapse of Ptolemaic Cosmology*. Chicago: University of Chicago Press.

Lawrence-Mathers, Anne. 2010. 'Domesticating the Calendar: The Hours and the Almanac in Tudor England'. In *Women and Writing, c. 1340–c. 1650*, edited Anne Lawrence-Mathers and Phillipa Hardman, 34–61. York: York Medieval Press.

Lazarus, Paul. 1912. *Das Basler Konzil: Seine Berufung und Leistung, seine Gliederung und seine Behördenorganisation*. Berlin: Ebering.

Lebigue, Jean-Baptiste. 2014. 'Jean de Berry à l'heure de l'Union: les *Très Riches Heures* et la réforme du calendrier à la fin du Grand Schisme'. In *Église et État, Église ou État?*, ed. Christine Barralis, Jean-Patrice Boudet, Fabrice Delivré, and Jean-Philippe Genet, 367–89. Paris: Publications de la Sorbonne.

Le Goff, Jacques, Jean Lefort, and Perrine Mane, eds. 2002. *Les calendriers: leurs enjeux dans l'espace et dans le temps*. Paris: Somogy éditions d'art.

Lehoux, Daryn. 2007. *Astronomy, Weather, and Calendars in the Ancient World: Parapegmata and Related Texts in Classical and Near Eastern Societies*. Cambridge: Cambridge University Press.

Lejbowicz, Max. 1992. 'Computus: le nombre et le temps altimédiévaux'. In *Le temps, sa mesure et sa perception au Moyen Âge*, ed. Bernard Ribémont, 151–95. Caen: Paradigme.

Lejbowicz, Max. 1997. 'Les disciplines du *quadrivium*: l'astronomie'. In *L'enseignement des disciplines à la Faculté des arts (Paris et Oxford, XIIIe–XVe siècles)*, ed. Olga Weijers and Louis Holtz, 195–216. Turnhout: Brepols.

Lejbowicz, Max. 2004. 'Présentation de Jean de Murs, "observateur et calculateur sagace et laborieux"'. In *Méthodes et statut des sciences à la fin du Moyen-Âge*, ed. Christophe Grellard, 159–80. Villeneuve d'Ascq: Presses Universitaires du Septentrion.

Lejbowicz, Max. 2006. 'Des tables pascales aux tables astronomiques et retour: formation et réception du comput patristique'. *Methodos* 6, doi:10.4000/methodos.538.

Lejbowicz, Max. 2010. 'Les Pâques baptismales d'Augustin d'Hippone, une étape contournée dans l'unification des partiques computistes latines'. In Warntjes and Ó Cróinín 2010: 1–39.

Lempire, Jean. 2007. 'Le calcul de la date de pâques dans les traités de S. Maxime le confesseur et de Georges, moine et prêtre'. *Byzantion* 77: 267–304.

Leonardi, Claudio. 1959–60. 'I codici di Marziano Capella'. 3 pts. *Aevum* 33: 443–89; 34: 1–99, 411–524.

Leonhard, Clement. 2006. *The Jewish Pesach and the Origins of the Christian Easter: Open Questions in Current Research*. Berlin: de Gruyter.

Lerner, Michel-Pierre, Alain-Philippe Segonds, and Jean-Pierre Verdet, eds. 2015. *Nicolas Copernic: De revolutionibus orbium coelestium/Des révolutions des orbes célestes*. 3 vols. Paris: Les Belles Lettres.

Leso, Tommaso. 2013. 'Columbanus in Europe: The Evidence from the *Epistulae*'. *Early Medieval Europe* 21: 358–89.

Leurquin, Régine. 1985. 'La *Tribiblos astronomique* de Théodore Méliténiote (*Vat. gr.* 792)'. *Janus* 72: 257–82.

Lichtenberg, Heiner. 2003. 'Das anpassbar zyklische, solilunare Zeitzählungssystem des gregorianischen Kalenders: Ein wissenschaftliches Meisterwerk der späten Renaissance'. *Mathematische Semesterberichte* 50: 45–76.

Lind, L. R., ed. 1958. *Ecclesiale by Alexander of Villa Dei*. Lawrence: University of Kansas Press.

Lindberg, David C. 1978. 'The Transmission of Greek and Arabic Learning to the West'. In *Science in the Middle Ages*, ed. David C. Lindberg, 52–90. Chicago: University of Chicago Press.

Lindberg, David C., ed. 1983. *Roger Bacon's Philosophy of Nature*. Oxford: Clarendon Press.

Lindhagen, Arvid. 1914. 'Länge und Alter des Mondes in den Kalendarien des Mittelalters', *Arkiv för Matematik, Astronomi och Fysik* 9, no. 36: 1–27.

Lindhagen, Arvid. 1916. 'Die Neumondtafel des Robertus Lincolniensis'. *Arkiv för Matematik, Astronomi och Fysik* 11, no. 2: 1–41.

Lindhagen, Arvid. 1917. 'Der xylographische Kalender des Johannes von Gmunden'. *Arkiv för Matematik, Astronomi och Fysik* 13, no. 1: 1–13.

Lithberg, Nils. 1953. *Computus med särskild hänsyn till runstaven och den borgerliga kalendern*. Stockholm: Norstedt.

Little, Andrew G. 1914. 'Roger Bacon's Works: With References to the MSS and Printed Editions'. In *Roger Bacon*, ed. Andrew G. Little, 375–425. Oxford: Clarendon Press.

Lluís Martos, Josep. 2014. 'La *editio princeps* del *Repertorio de los tiempos* de Andrés de Li: el proyecto editorial y la recuperación del incunable'. In *Textos, edición y público lector en los albores de la imprenta*, ed. Marta Haro Cortés and José Luis Canet, 155–86. Valencia: Universitat de València.

Logemann, Silke. 1996. *Die Franziskaner im mittelalterlichen Lüneburg*. Werl: Dietrich-Coelde-Verlag.

Lohr, Alfred, ed. 2013. *Der Computus Gerlandi: Edition, Übersetzung und Erläuterungen.* Stuttgart: Steiner.

Lohrmann, Dietrich. 1993. 'Alcuins Korrespondenz mit Karl dem Grossen über Kalender und Astronomie'. In *Science in Western and Eastern Civilzation in Carolingian Times*, ed. Paul Leo Butzer and Dietrich Lohrmann, 79–114. Basel: Birkhäuser.

Lollis, Cesare de. 1892. *Autografi di Cristoforo Colombo con prefazione e trascrizione diplomatica.* Rome: Ministero della Pubblica Istruzione.

Ludwig, Corinna. 2010. 'Die Karriere eines Bestsellers: Untersuchungen zur Entstehung und Rezeption der *Sphaera* des Johannes de Sacrobosco'. *Concilium medii aevi* 13: 153–85.

Lützelschwab, Ralf. 2007. *Flectat cardinales ad velle suum? Clement VI. und sein Kardinalskolleg; Ein Beitrag zur kurialen Politik in der Mitte des 14. Jahrhunderts.* Munich: Oldenbourg.

Mac Carthy, Bartholomew. 1901. *The Annals of Ulster.* Vol. 4. Dublin: The Royal Irish Academy.

Maeyama, Yasukatsu. 1998. 'Determination of the Sun's Orbit: Hipparchus, Ptolemy, al-Battānī, Copernicus, Tycho Brahe'. *AHES* 53: 1–49.

Maiello, Francesco. 1989. 'Tempo, potere, e cosmologica: la riforma gregoriana del calendario'. *Dimensioni e problemi della ricerca storica* 1: 102–37.

Maiello, Francesco. 1994. *Storia del calendario: la misurazione del tempo, 1450–1800.* Turin: Einaudi.

Malitz, Jürgen. 1987. 'Die Kalenderreform Caesars: Ein Beitrag zur Geschichte seiner Spätzeit'. *Ancient Society* 18: 103–31.

Malpangotto, Michela. 2008. *Regiomontano e il rinnovamento del sapere matematico e astronomico nel quattrocento.* Bari: Cacucci.

Mancha, José Luis. 1992a. 'Astronomical Use of Pinhole Images in William of Saint-Cloud's *Almanach Planetarum* (1292)'. *AHES* 43: 275–98.

Mancha, José Luis. 1992b. 'The Latin Translation of Levi ben Gerson's *Astronomy*'. In *Studies on Gersonides*, ed. Gad Freudenthal, 21–46. Leiden: Brill.

Mancha, José Luis. 1997. 'Levi ben Gerson's Astronomical Work: Chronology and Christian Context'. *Science in Context* 10: 471–93.

Mancha, José Luis. 1998. 'The Provençal Version of Levi ben Gerson's Tables for Eclipses'. *AIHS* 48: 269–353.

Mancha, José Luis. 2002–3. 'A Note on Copernicus' "Correction" of Ptolemy's Mean Synodic Month'. *Suhayl* 3: 221–9.

Mancha, José Luis. 2004. 'Al-Biṭrūjī's Theory of the Motions of the Fixed Stars'. *AHES* 58: 143–82.

Mancha, José Luis. 2012. 'The Theory of Access and Recess in Levi ben Gerson's *Astronomy*, and Its Sources'. *Aleph* 12: 37–64.

Marcos Rodríguez, Florencio. 1979–80. 'Pedro de Osma y los primeros incunables españoles'. *Revista española de teología* 39/40: 331–41.

Mariano, Ivan. 2017. 'The Council and Negotations with the Greeks'. In Decaluwé, Izbicki, and Christianson 2017: 310–39.

Markowski, Mieczysław. 1973. 'Marcin Biem of Olkusz'. In *The Cracow Circle of Nicholas Copernicus*, edited by Józef Gierowski, 7–21. Kraków: Jagellonian University Press.

Martín Baños, Pedro. 2015. 'Impresos y manuscritos desconocidos de Antonio de Nebrija: la miscelánea V.B.5622 de la Biblioteca Real de Bélgica'. *Minerva* 28: 241–60.

Martínez Gázquez, José, ed. 2011. '*Dicunt experimentatores/Dicunt qui experimento locuti sunt*: l'expérience dans les indications du calendrier de Cordoue (10e siècle)'. In '*Expertus sum': l'expérience par les sens dans la philosophie naturelle médiévale*, ed. Thomas Bénatouïl and Isabelle Draelants, 239–55. Florence: SISMEL.

Martínez Gázquez, José, and Julio Samsó. 1981. 'Una nueva traducción latina del calendario de Cordoba (siglo XIII)'. In *Textos y estudios sobre astronomía española en el siglo XIII*, ed. Juan Vernet, 9–78. Barcelona: Universidad Autónoma de Barcelona.

Marzi, Demetrio. 1896. *La questione della riforma del calendario nel Quinto Concilio Lateranense (1512–1517)*. Florence: Carnesecchi.

Marzi, Demetrio. 1897. 'Giovanni Maria Tolosani, Alessandro Piccolomini e Luigi Giglio'. *Miscellanea storica della Valdelsa* 5: 202–9.

Marzi, Demetrio. 1906. 'Nuovi studii e ricerche intorno alla questione del calendario durante i secoli XV e XVI'. In *Atti del congresso internazionale di scienze storiche (Roma, 1–9 Aprile 1903)*, vol. 3, 637–50. Rome: R. Accademia dei Lincei.

Masolini, Serena. 2016. 'Petrus de Rivo (ca. 1420–1499): Portrait(s) of a Louvain Master'. Ph.D. diss., KU Leuven.

Massa, Eugenio. 1955. *Ruggero Bacone: etica e poetica nella storia dell' 'Opus maius'*. Rome: Edizioni di Storia e Letteratura.

Matus, Zachary. 2012. 'Reconsidering Roger Bacon's Apocalypticism in Light of his Alchemical and Scientific Thought'. *Harvard Theological Review* 105: 189–222.

Mayr, Joachim. 1932. 'Das Kunstwerk des Lilius'. *Astronomische Nachrichten* 247: 429–44.

Mayr, Joachim. 1955. 'Der Computus ecclesiasticus'. *Zeitschrift für katholische Theologie* 77: 301–30.

Mayr-Harting, Henry. 2007. *Church and Cosmos in Early Ottonian Germany: The View from Cologne*. Oxford: Oxford University Press.

Mc Carthy, Daniel P. 2011. 'The Study and Use of Numbers in Early Irish Monasteries'. In *Glendalough*, ed. Charles Doherty, Linda Doran, and Mary Kelly, 219–33. Dublin: Four Courts Press.

Mc Carthy, Daniel P., and Aidan Breen. 2003. *The Ante-Nicene Christian Pasch—'De ratione paschali': The Paschal Tract of Anatolius, Bishop of Laodicea*. Dublin: Four Courts Press.

McCluskey, Stephen C. 1998. *Astronomies and Cultures in Early Medieval Europe*. Cambridge: Cambridge University Press.

McCluskey, Stephen C. 2003. 'Changing Contexts and Criteria for the Justification of Computistical Knowledge and Practice'. *JHA* 34: 201–17.

McCluskey, Stephen C. 2004. 'Astronomy, Time, and Churches in the Early Middle Ages'. In *Villard's Legacy*, ed. Marie-Therese Zenner, 197–210. Aldershot: Ashgate.

McCluskey, Stephen C. 2006. 'The Medieval Liturgical Calendar, Sacred Space, and the Orientations of Churches'. In *Proceedings of the Conference 'Time and Astronomy in Past Cultures': Toruń, March 30 – April 1, 2005*, ed. Arkadiusz Sołtysiak, 139–48. Warsaw: Institute of Archaeology.

McCluskey, Stephen C. 2011. 'Martianus and the Traditions of Early Medieval Astronomies'. In *Carolingian Scholarship and Martianus Capella*, ed. Mariken Teeuwen and Sinéad O'Sullivan, 221–44. Turnhout: Brepols.

McEvoy, James. 1983. 'The Chronology of Robert Grosseteste's Writings on Nature and Natural Philosophy'. *Speculum* 58: 614–55.

McGuire, Brian Patrick. 2005. *Jean Gerson and the Last Medieval Reformation*. University Park: Pennsylvania State University Press.

McGurk, Patrick. 1974. 'Computus Helperici: Its Transmission in England in the Eleventh and Twelfth Centuries'. *Medium Aevum* 43: 1–5.

McNutt, Jennifer Powell. 2006. 'Hesitant Steps: Acceptance of the Gregorian Calendar in Eighteenth-Century Geneva'. *Church History* 75: 544–64.

Means, Laurel. 1992a. 'Electionary, Lunary, Destinary, and Questionary: Toward Defining Categories of Middle English Prognostic Material'. *Studies in Philology* 89: 367–403.

Means, Laurel. 1992b. 'Ffor as moche as yche man may not haue Þe astrolabe: Popular Middle English Variations on the Computus'. *Speculum* 67: 595–623.

Means, Laurel. 1993. *Medieval Lunar Astrology: A Collection of Representative Middle English Texts*. Lewiston, NY: Edwin Mellen Press.

Meeus, Jean. 2002. *More Mathematical Astronomy Morsels*. Richmond, Va.: Willmann-Bell.

Mehl, Édouard. 2017. 'La révolution copernicienne de la chronologie'. In *Le temps des astronomes*, ed. Edouard Mehl and Nicolas Roudet, 269–97. Paris: Les Belles Lettres.

Mentz, Arthur. 1906. *Beiträge zur Osterfestberechnung bei den Byzantinern*. Königsberg: Leupold.

Mercier, Raymond. 1976–7. 'Studies in the Medieval Conception of Precession'. 2 pts. *AIHS* 26: 197–220; 27: 33–71.

Mercier, Raymond. 1987. 'Astronomical Tables in the Twelfth Century'. In *Adelard of Bath*, ed. Charles Burnett, 87–118. London: The Warburg Institute.

Mercier, Raymond. 1996. 'Accession and Recession: Reconstruction of the Parameters'. In *From Baghdad to Barcelona*, ed. Josep Casulleras and Julio Samsó, vol. 1, 399–47. Barcelona: Instituto 'Millás Vallicrosa' de Historia de la Ciencia Árabe.

Mercier, Raymond. 2000. 'From Tantra to Zīj'. In *Sic Itur ad Astra*, ed. Menso Folkerts and Richard Lorch, 451–60. Wiesbaden: Harrassowitz.

Mercier, Raymond. 2004. 'The Lost Zīj of al-Ṣūfī in the Twelfth-Century Tables for London and Pisa'. Chapter 8 in idem, *Studies on the Transmission of Medieval Mathematical Astronomy*. Aldershot: Ashgate.

Mercier, Raymond. 2014. 'Astronomical Tables of Abraham Bar Ḥiyya'. In *Time, Astronomy, and Calendars in the Jewish Tradition*, ed. Sacha Stern and Charles Burnett, 155–207. Leiden: Brill.

Methuen, Charlotte. 2001. 'Time Human or Time Divine? Theological Aspects in the Opposition to Gregorian Calendar Reform'. *Reformation and Renaissance Review* 3: 36–50.

Mett, Rudolf. 1989. *Regiomontanus in Italien*. Vienna: Verlag der Österreichischen Akademie der Wissenschaften.

Metzger, Marcel, ed. 1987. *Les constitutions apostoliques*. Vol. 3. Paris: Éditions du Cerf.

Meyer, Paul. 1897. 'Traités en vers provençaux sur l'astrologie et la géomancie'. *Romania* 26: 225–75.

Meyvaert, Paul. 2002. 'Discovering the Calendar (*Annalis Libellus*) Attached to Bede's Own Copy of *De Temporum Ratione*'. *Analecta Bollandiana* 120: 5–64.

Mezzi, Egidio, and Francesco Vizza. 2010. *Luigi Lilio: medico astronomo e matematico di Cirò; ideatore della riforma del calendario gregoriano*. Reggio Calabria: Laruffa Editore.

Michels, Agnes Kirsopp. 1967. *The Calendar of the Roman Republic*. Princeton: Princeton University Press.

Milanković, Milutin. 1924. 'Das Ende des julianischen Kalenders und der Kalender der orientalischen Kirchen'. *Astronomische Nachrichten* 220: 379–84 (no. 5279).

Milano, Alberto, ed. 2000. *Immagini del tempo: 500 anni di lunari e calendari da muro dalla Raccolta Bertarelli*. Basano del Grappa: Tassotti.

Millás Vallicrosa, José María. 1931. *Assaig d'história de les idees fisiques i matemátiques a la Catalunya medieval*. Vol. 1. Barcelona: Institució Patxot.

Millás Vallicrosa, José María. 1936. 'Una obra astronómica desconocida de Johannes Avendaut Hispanus'. *Osiris* 1: 451–75. Reprinted as chapter 10 in Millás Vallicrosa 1949.

Millás Vallicrosa, José María. 1943–50. *Estudios sobre Azarquiel.* Madrid: CSIC.
Millás Vallicrosa, José María. 1945. 'El "Liber de motu octave sphere" de Ṯābit ibn Qurra'. *Al-Andalus* 10: 89–108.
Millás Vallicrosa, José María. 1946. 'Un tratado de almanaque probablemente de R. Abraham Ibn Era'. In *Studies and Essays in the History of Science and Learning*, ed. M. F. Ashley Montagu, 421–32. New York: Schuman.
Millás Vallicrosa, José María, ed. 1947. *El libro de los fundamentos de las Tablas astronómicas de R. Abraham Ibn 'Ezra.* Madrid: CSIC.
Millás Vallicrosa, José María. 1949. *Estudios sobre historia de la ciencia española.* Barcelona: CSIC.
Millás Vallicrosa, José María. 1951. 'La transmisión del Almanaque desde Oriente a Occidente'. *AIHS* 4: 874–8.
Millás Vallicrosa, José María. 1963. 'Translations of Oriental Scientific Works to the End of the Thirteenth Century'. In *The Evolution of Science*, ed. Guy S. Métraux and François Crouzet, 128–67. New York: New Americans Library.
Millet, Hélène, and Monique Maillard-Luypaert. 2015. *Le schisme et la pourpre: le cardinal Pierre d'Ailly, homme de science et de foi.* Paris: Éditions du Cerf.
Milz, Joseph. 2001. 'Gerhard Mercator und die Gregorianische Kalenderreform'. *Duisburger Forschungen* 47: 65–77.
Moesgaard, Kristian Peder. 1974. 'Thābit ibn Qurra between Ptolemy and Copernicus: An Analysis of Thābit's Solar Theory'. *AHES* 12: 199–216.
Mohler, Ludwig. 1942. *Kardinal Bessarion als Theologe, Humanist und Staatsmann: Funde und Forschungen.* Vol. 3. Paderborn. Reprint, Aalen: Scientia Verlag, 1967.
Molanus, Jean. 1861. *Les quatorze livres sur l'histoire de la ville de Louvain.* Edited by P. F. X. de Ram. Vol. 1. Brussels: Hayez.
Montgomery, Scott L. 2000. *Science in Translation: Movements of Knowledge through Cultures and Time.* Chicago: University of Chicago Press.
Mooney, Linne R. 1983. 'A Medieval Latin Mnemonic for Finding the Date of Easter'. *Notes and Queries* 30: 391–2.
Mooney, Linne R. 1984. 'A Middle English Verse Compendium of Astrological Medicine'. *Medical History* 28: 406–19.
Mooney, Linne R. 1997. 'English Almanacks from Script to Print'. In *Texts and Their Contexts*, ed. John Scattergood and Julia Boffey, 11–25. Dublin: Four Courts Press.
Mooney, Linne R., ed. 1998. *The Kalendarium of John Somer.* Athens, Ga.: University of Georgia Press.
Morelon, Régis, ed. 1987. *Thābit ibn Qurra: Oeuvres d'Astronomie.* Paris: Les Belles Lettres.
Morelon, Régis. 1996. 'Eastern Arabic Astronomy between the Eighth and the Eleventh Centuries'. In *Encyclopedia of the History of Arabic Science*, ed. Roshdi Rashed, vol. 1, 20–57. London: Routledge.
Moreton, Jennifer. 1994. 'John of Sacrobosco and the Calendar'. *Viator* 25: 229–44.
Moreton, Jennifer. 1995a. 'Before Grosseteste: Roger of Hereford and Calendar Reform in Eleventh- and Twelfth-Century England'. *Isis* 86: 562–86.
Moreton, Jennifer. 1995b. 'Robert Grosseteste and the Calendar'. In *Robert Grosseteste*, ed. James McEvoy, 77–88. Turnhout: Brepols.
Moreton, Jennifer. 1998. 'Doubts about the Calendar: Bede and the Eclipse of 664'. *Isis* 89: 50–65.
Moreton, Jennifer. 2003. 'On Not Editing Grosseteste'. In *Editing Robert Grosseteste*, ed. Evelyn A. Mackie and Joseph Goering, 167–84. Toronto: University of Toronto Press.
Mosimann, Martin. 1994. *Die 'Mainauer Naturlehre' im Kontext der Wissenschaftsgeschichte.* Tübingen: Francke.

Mosshammer, Alden A. 2008. *The Easter Computus and the Origins of the Christian Era.* Oxford: Oxford University Press.

Mosshammer, Alden A. 2013. 'The *Praefatio (Prologus) Sancti Cyrilli de Paschate* and the 437-year (not 418!) Paschal List Attributed to Theophilus'. *Vigiliae Christianae* 67: 49–78.

Moyer, Gordon. 1983. 'Aloisius Lilius and the "Compendium novae rationis restituendi kalendarium"'. In Coyne, Hoskin, and Pedersen 1983: 171–88.

Mozaffari, S. Mohammad. 2016. 'A Medieval Bright Star Table: The Non-Ptolemaic Star Tables in the *Īlkhānī Zīj*'. *JHA* 47: 294–316.

Müller, Kathrin. 2008. *Visuelle Weltaneignung: Astronomische und kosmologische Diagramme in Handschriften des Mittelalters.* Göttingen: Vandenhoeck & Ruprecht.

Müller, Markus. 2009. *Beherrschte Zeit: Lebensorientierung und Zukunftsgestaltung durch Kalenderprognostik zwischen Antike und Neuzeit.* Kassel: Kassel University Press.

Müller, Tom. 2010. *'ut reiecto paschali errore veritati insistamus': Nikolaus von Kues und seine Konzilsschrift 'De reparatione kalendarii'.* Münster: Aschendorff.

Müller, Ute. 1971. 'Deutsche Mondwahrsagetexte aus dem Spätmittelalter'. Ph.D. diss., Free University of Berlin.

Mundy, John. 1943. 'John of Gmunden'. *Isis* 34: 196–205.

Murdoch, John E. 1984. *Album of Science: Antiquity and the Middle Ages* New York: Charles Scribner's Sons.

Mütz, Karl. 1996. 'Der Kalender für Graf Eberhard im Bart und der Kalender von Regiomontanus: Zwei herausragende Werke ihrer Zeit'. *Zeitschrift für Württembergische Landesgeschichte* 55: 65–91.

Mütz, Karl. 1999. 'Der Kalender des Magister Hainrich Solder: Ein Reformwerk 180 Jahre vor Papst Gregor XIII'. *Rottenburger Jahrbuch für Kirchengeschichte* 18: 167–85.

Mütz, Karl, ed. 2003. *Computus Chirometralis: Spätmittelalterliches Lehrbuch für Kalenderrechnung.* Leinfelden-Echterdingen: DRW-Verlag.

Mütz, Karl. 2008. 'Das Kalenderwerk (1404) im Tübinger Hausbuch, Handschrift Md 2 der Universitätsbibliothek Tübingen'. *Zeitschrift für Württembergische Landesgeschichte* 67: 27–43.

Muzerelle, Denis. 2000. *Manuscrits datés des bibliothèques de France.* Vol. 1, *Cambrai.* Paris: CNRS Éditions.

Naudin, Claude, ed. 2001. *De temps en temps: histoires de calendriers.* Paris: Tallandier.

Nenci, Elio, ed. 1998. *Bernardino Baldi: le vite de' matematici.* Milan: Angeli.

Neugebauer, Otto. 1957. *The Exact Sciences in Antiquity.* 2nd edn. Providence, RI: Brown University Press.

Neugebauer, Otto. 1962a. 'Thâbit ben Qurra "On the Solar Year" and "On the Motion of the Eighth Sphere"'. *Proceedings of the American Philosophical Society* 106: 264–99.

Neugebauer, Otto. 1962b. *The Astronomical Tables of Al-Khwārizmī.* Copenhagen: Munksgaard.

Neugebauer, Otto. 1989. 'From Assyriology to Renaissance Art'. *Proceedings of the American Philosophical Society* 133: 391–403.

Nicolaïdis, Efthymios. 1995. 'Quelques notes au sujet des manuscrits byzantins sur les tables astronomiques persanes: la *Παράδοσις* et la *Σύνταξις*'. *Θησαυρίσματα–Thesaurismata* 25: 39–48.

Nilsson, Martin P. 1920. *Primitive Time-Reckoning: A Study in the Origins and First Development of the Art of Counting Time among the Primitive and Early Culture Peoples.* Lund: Gleerup.

Nobis, H. M. 1983. 'The Reaction of Astronomers to the Gregorian Calendar'. In Coyne, Hoskin, and Pedersen 1983: 243–54.

North, John D. 1976. *Richard of Wallingford*. 3 vols. Oxford: Clarendon Press.

North, John D. 1977. 'The Alfonsine Tables in England'. In ΠΡΙΣΜΑΤΑ: *Naturwissenschaftsgeschichtliche Studien*, edited Y. Maeyama and W. G. Saltzer, 269–301. Wiesbaden: Steiner.

North, John D. 1980. 'Astrology and the Fortunes of Churches'. *Centaurus* 24: 181–211.

North, John D. 1983. 'The Western Calendar—"Intolerabilis, Horribilis, et Derisibilis": Four Centuries of Discontent'. In Coyne, Hoskin, and Pedersen 1983: 75–113. Reprinted with addenda in North, *The Universal Frame* (London: The Hambledon Press, 1989), 39–77.

North, John D. 1988. *Chaucer's Universe*. Oxford: Oxford University Press.

North, John D. 1992a. 'Astronomy and Mathematics'. In *The History of the University of Oxford*, vol. 2, *Late Medieval Oxford*, ed. J. I. Catto and Ralph Evans, 103–74. Oxford: Clarendon Press.

North, John D. 1992b. 'Natural Philosophy in Late Medieval Oxford'. In *The History of the University of Oxford*, vol. 2, *Late Medieval Oxford*, ed. J. I. Catto and Ralph Evans, 65–102. Oxford: Clarendon Press.

North, John D. 1996. 'Just Whose Were the Alfonsine Tables?' In *From Baghdad to Barcelona*, ed. Josep Casulleras and Julio Samsó, vol. 1, 453–75. Barcelona: Instituto 'Millás Vallicrosa' de Historia de la Ciencia Árabe.

North, John D. 1999. 'Roger Bacon and the Saracens'. In *Filosofia e scienza classica, arabo-latina medievale e l'età moderna*, ed. Graziella Federici Vescovini, 129–60. Louvain-la-Neuve: Fédération Internationale des Instituts d'Études Médiévales.

North, William L. 2007. 'History, Tradition, and Authority in Bede's Interpretation of the Council of Whitby, 664'. *Anglo-Saxon* 1: 1–53.

Nothaft, C. P. E. 2012. *Dating the Passion: The Life of Jesus and the Emergence of Scientific Chronology (200–1600)*. Leiden: Brill.

Nothaft, C. P. E. 2013a. 'Reforming the Calendar at the University of Salamanca ca. 1468: Pedro Martínez de Osma and His *Disputatio de anno*...' *eHumanista* 23: 522–56.

Nothaft, C. P. E. 2013b. Review of Carabias Torres 2012. *Aestimatio* 10: 190–204.

Nothaft, C. P. E. 2013c. 'Robert of Leicester's Treatise on the Hebrew *Computus* and the Study of Jewish Knowledge in Medieval England'. *Jewish Historical Studies* 45: 63–78.

Nothaft, C. P. E. 2013d. 'The Chronological Treatise *Autores Kalendarii* of 1317, Attributed to John of Murs: Text and Introduction'. *Cahiers de l'Institut du Moyen-Âge Grec et Latin* 82: 1–89.

Nothaft, C. P. E. 2013e. 'The Meaning of *Judaeus* and the Myth of Jewish Male Menses in a Late Medieval Astronomical School Text'. *European Journal of Jewish Studies* 7: 73–91.

Nothaft, C. P. E. 2014a. 'A Tool for Many Purposes: Hermann Zoest and the Christian Appropriation of the Jewish Calendar'. *Journal of Jewish Studies* 65: 148–68.

Nothaft, C. P. E. 2014b. 'Climate, Astrology and the Age of the World in Thirteenth-Century Thought: Giles of Lessines and Roger Bacon on the Precession of the Solar Apogee'. *Journal of the Warburg and Courtauld Institutes* 77: 35–60.

Nothaft, C. P. E. 2014c. 'John of Pulchro Rivo and John of Saxony: A *Mise au point*'. *JHA* 45: 227–45.

Nothaft, C. P. E. 2014d. *Medieval Latin Christian Texts on the Jewish Calendar: A Study with Five Editions and Translations*. Leiden: Brill.

Nothaft, C. P. E. 2014e. '*Me pudet audire Iudeum talia scire*: A Late Medieval Latin School Text on the Jewish Calendar'. In *Time, Astronomy, and Calendars in the Jewish Tradition*, ed. Sacha Stern and Charles Burnett, 327–65. Leiden: Brill.

Nothaft, C. P. E. 2014f. 'Origen, Climate Change, and the Erosion of Mountains in Giles of Lessines's Discussion of the Eternity of the World (c. 1260)'. *The Mediaeval Journal* 4, no. 1: 43–69.

Nothaft, C. P. E. 2014g. 'The Reception and Application of Arabic Science in Twelfth-Century Computistics: New Evidence from Bavaria'. *JHA* 45: 35–60.

Nothaft, C. P. E. 2015a. 'Bede's *horologium*: Observational Astronomy and the Problem of the Equinoxes in Early Medieval Europe (c.700–1100)'. *English Historical Review* 130: 1079–1101.

Nothaft, C. P. E. 2015b. 'Critical Analysis of the Alfonsine Tables in the Fourteenth Century: The Parisian *Expositio tabularum Alfonsii* of 1347'. *JHA* 46: 76–99.

Nothaft, C. P. E. 2015c. 'Johannes Keck, das Konzil von Basel und der vergessene Osterstreit des Jahres 1444'. *Deutsches Archiv für Erforschung des Mittelalters* 71: 105–47.

Nothaft, C. P. E. 2015d. 'John of Murs and the Treatise *Autores kalendarii* (1317): A Problem of Authorship'. *Sudhoffs Archiv* 99: 209–29.

Nothaft, C. P. E. 2015e. 'Medieval Astronomy in Catalonia and the South of France: The "Improved" Lunar *Kalendarium* of Friar Raymond (Ramon) Bancal (ca. 1311) and Its Predecessors'. *Llull* 38, no. 81: 101–25.

Nothaft, C. P. E. 2015f. 'Roman vs. Arabic Computistics in Twelfth-Century England: A Newly Discovered Source (*Collatio Compoti Romani et Arabici*)'. *Early Science and Medicine* 20: 187–208.

Nothaft, C. P. E. 2015g. 'Science at the Papal Palace: Clement VI and the Calendar Reform Project of 1344/45'. *Viator* 46, no. 2: 277–302.

Nothaft, C. P. E. 2015h. 'The Astronomical Data in the *Très Riches Heures* and Their Fourteenth-Century Source'. *JHA* 46: 113–29.

Nothaft, C. P. E. 2015i. 'Thomas Strzempiński, Hermann Zoest, and the Initial Stages of the Calendar Reform Project Attempted at the Council of Basel (1434–1437)'. *Cahiers de l'Institut du Moyen-Âge Grec et Latin* 84: 164–303.

Nothaft, C. P. E. 2016a. 'Duking it Out in the Arena of Time: Chronology and the Christian–Jewish Encounter (1100–1600)'. *Medieval Encounters* 22: 213–35.

Nothaft, C. P. E. 2016b. '*Vanitas vanitatum et super omnia vanitas*: The Astronomer Heinrich Selder and a Newly Discovered Fourteenth-Century Critique of Astrology'. *Erudition and the Republic of Letters* 1: 261–304.

Nothaft, C. P. E. 2016c. 'Walter Odington's *De etate mundi* and the Pursuit of a Scientific Chronology in Medieval England'. *Journal of the History of Ideas* 77: 183–201.

Nothaft, C. P. E. 2017a. 'A Reluctant Innovator: Graeco-Arabic Astronomy in the *Computus* of Magister Cunestabulus (1175)'. *Early Science and Medicine* 22: 24–54.

Nothaft, C. P. E. 2017b. 'Astronomy and Calendar Reform at the Curia of Pope Clement VI: A New Source'. *Annals of Science* 74: 1–24.

Nothaft, C. P. E. 2017c. 'Criticism of Trepidation Models and Advocacy of Linear Precession in Latin Medieval Astronomy'. *AHES* 71: 211–44.

Nothaft, C. P. E. 2017d. 'Strategic Skepticism: A Reappraisal of Nicholas of Cusa's Calendar Reform Treatise'. In *Le temps des astronomes*, ed. Edouard Mehl and Nicolas Roudet, 65–102. Paris: Les Belles Lettres.

Nothaft, C. P. E. 2017e. *Walcher of Malvern: 'De lunationibus' and 'De Dracone'; Study, Edition, Translation, and Commentary*. Turnhout: Brepols.

Nothaft, C. P. E. (forthcoming-1). 'John Holbroke, the Tables of Cambridge, and the "True Length of the Year": A Forgotten Episode in Fifteenth Century Astronomy'. *AHES*.

Nothaft, C. P. E. (forthcoming-2). 'Pierre d'Ailly and Calendrical Astronomy: The *Tractatus de vero cyclo lunari*'. In *Pierre d'Ailly, un esprit universel à l'aube du XVe siècle*, ed. Hélène Millet et al. Paris.

Nothaft, C. P. E. (forthcoming-3). 'Roger of Hereford and the Transformation of the *Computus* in Twelfth-Century England'. In *Travelling Wisdom*, ed. Christian Etheridge and Michele Campopiano. Turnhout: Brepols.

Nothaft, C. P. E. (forthcoming-4). 'Two Medieval Pioneers of Technical Chronology'. In *Early Modern Chronologies*, ed. Michał Choptiany. Leiden: Brill.

Oakley, Francis. 1964. *The Political Thought of Pierre d'Ailly: The Voluntarist Tradition*. New Haven: Yale University Press.

Oakley, Francis. 2003. *The Conciliarist Tradition: Constitutionalism in the Catholic Church 1300–1870*. Oxford: Oxford University Press.

Oakley, Francis. 2006. 'Gerson as Conciliarist'. In *A Companion to Jean Gerson*, ed. Brian Patrick McGuire, 179–204. Leiden: Brill.

O'Boyle, Cornelius. 2005. 'Astrology and Medicine in Later Medieval England: The Calendars of John Somer and Nicholas of Lynn'. *Sudhoffs Archiv* 89: 1–22.

Obrist, Barbara. 2004. *La cosmologie médiévale: textes et images*. Vol. 1, *Les fondements antiques*. Florence: SISMEL.

Obrist, Barbara. 2006, ed. *Abbon de Fleury: philosophie, sciences et comput autour de l'an mil.* 2nd edn. Paris: Centre d'histoire des sciences et des philosophies arabes et médiévales.

Obrist, Barbara. 2008. 'Abbon de Fleury: cosmologie, comput, philosophie'. In *Abbon, un abbé de l'an mil*, edited by Annie Dufour and Gillette Labory, 177–203. Turnhout: Brepols.

O'Connell, D. J. K. 1975. 'Copernicus and Calendar Reform'. In *Colloquia Copernicana III* [= Studia Copernicana 13], ed. Owen Gingerich and Jerzy Dobrzycki, 189–202. Wrocław: Wydawnictwo Polskiej Akademij Nauk.

Ó Cróinín, Dáibhí. 1982a. 'A Seventh-Century Irish Computus from the Circle of Cummianus'. *Proceedings of the Royal Irish Academy* 82C: 405–30. Reprinted as chapter 9 in Ó Cróinín 2003.

Ó Cróinín, Dáibhí. 1982b. 'Mo Sinu Maccu Min and the Computus at Bangor'. *Peritia* 1: 281–95. Reprinted as chapter 3 in Ó Cróinín 2003.

Ó Cróinín, Dáibhí. 1983. 'The Irish Provenance of Bede's Computus'. *Peritia* 2: 229–48. Reprinted as chapter 13 in Ó Cróinín 2003.

Ó Cróinín, Dáibhí. 1985. 'New Heresy for Old: Pelagianism in Ireland the Papal Letter of 640'. *Speculum* 60: 505–16. Reprinted as chapter 8 in Ó Cróinín 2003.

Ó Cróinín, Dáibhí. 1997. 'The Computistical Works of Columbanus'. In *Columbanus: Studies on the Latin Writings*, ed. Michael Lapidge, 264–70. Woodbridge: Boydell Press.

Ó Cróinín, Dáibhí. 2003. *Early Irish History and Chronology*. Dublin: Four Courts Press.

Ó Cróinín, Dáibhí. 2008. 'Dionysius Exiguus in the Classroom'. In *Mathematics Celestial and Terrestrial*, ed. Joseph W. Dauben et al., 253–74. Stuttgart: Wissenschaftliche Verlagsgesellschaft.

Ohashi, Masako. 1999. 'The Impact of the Paschal Controversy: Computus, Exegesis and Church History in Early Britain and Ireland'. Ph.D. diss., Nanzan University.

Ohashi, Masako. 2015. 'The "Real" Addressee(s) of Bede's Letter to Wicthed'. In *Early Medieval Ireland and Europe*, ed. Pádraic Moran and Immo Warntjes, 119–35. Turnhout: Brepols.

O'Grady, Standish Hayes. 1926. *Catalogue of the Irish Manuscripts in the British Museum.* Vol. 1. London: British Museum.

O'Malley, Charles Donald. 1964. *Andreas Wesalius of Brussels, 1514–1564.* Berkeley: University of California Press.

Oost, Stewart Irvin. 1953. 'The Roman Calendar in the Year of Pydna (168 B.C.)'. *Classical Philology* 48: 217–30.

Palmer, James T. 2011a. 'Calculating Time and the End of Time in the Carolingian World, c.740–820'. *English Historical Review* 126: 1307–31.

Palmer, James T. 2011b. 'Computus after the Paschal Controversy of AD 740'. In Warntjes and Ó Cróinín 2011: 213–41.

Palmer, James T. 2014. 'The Ends and Futures of Bede's *De temporum ratione*'. In *Bede and the Future*, ed. Peter Darby and Faith Wallis, 139–60. Farnham: Ashgate.

Pannekoek, Anton. 1955. 'Ptolemy's Precession'. *Vistas in Astronomy* 1: 60–6.

Panti, Cecilia. 2001. *Moti, virtù e motori celesti nella cosmologia di Roberto Grossatesta: studio ed edizione dei trattati 'De sphera', 'De cometis', 'De motu supercelestium'.* Florence: SISMEL.

Panti, Cecilia. 2003. 'Robert Grosseteste's Early Cosmology'. In *Editing Robert Grosseteste*, ed. Evelyn A. Mackie and Joseph Goering, 135–65. Toronto: University of Toronto Press.

Paravicini Bagliani, Agostino. 1973. 'Un matematico nella corte papale del secolo XIII: Campano da Novara (†1296)'. *Rivista di storia della Chiesa in Italia* 27: 98–129.

Paravicini Bagliani, Agostino. 1991. *Medicina e scienze della natura alla corte dei papi nel Duecento.* Spoleto: Centro Italiano di Studi Sull'Alto Medioevo.

Pascoe, Louis B. 2005. *Church and Reform: Bishops, Theologians, and Canon Lawyers in the Thought of Pierre d'Ailly (1351–1420).* Leiden: Brill.

Pasquale, Maria Grazia di. 2003. 'Versi computistici: proposte per una nuova edizione'. In *Poetry of the Early Medieval Europe*, ed. Edoardo d'Angelo and Francesco Stella, 171–81. Florence: SISMEL.

Pedersen, Fritz S. 1978. 'Petrus de Dacia: Tractatus Instrumenti Eclipsium'. *Cahiers de l'Institut du Moyen-Âge Grec et Latin* 25: 1–102.

Pedersen, Fritz S. 1979. 'Petrus de S. Audomaro: Tractatus de Semissis'. *Cahiers de l'Institut du Moyen-Âge Grec et Latin* 33: 1–113.

Pedersen, Fritz S., ed. 1983–4. *Petri Philomenae de Dacia et Petri de S. Audomaro Opera quadrivialia.* 2 vols. Copenhagen: Gad (= Corpus Philosophorum Danicorum Medii Aevi 10).

Pedersen, Fritz S. 1986. 'Scriptum Johannis de Sicilia super canones Azarchelis de tabulis Toletanis'. Pt. 2. *Cahiers de l'Institut du Moyen-Âge Grec et Latin* 52: 1–268.

Pedersen, Fritz S. 1998. 'The Toulouse Tables: A List of Manuscripts'. *Cahiers de l'Institut du Moyen-Âge Grec et Latin* 68: 3–12.

Pedersen, Fritz S. 2002. *The Toledan Tables: A Review of the Manuscripts and the Textual Versions with an Edition.* 4 vols. Copenhagen: Reitzel.

Pedersen, Fritz S. 2014. 'William of Saint-Cloud: *Almanach Planetarum*; an Edition of the Canons, a Few Samples from the Tables, and a Foray into the Numbers'. *Cahiers de l'Institut du Moyen-Âge Grec et Latin* 83: 1–133.

Pedersen, Olaf. 1967. 'The Life and Work of Peter Nightingale'. *Vistas in Astronomy* 9: 3–10.

Pedersen, Olaf. 1975. 'The *Corpus Astronomicum* and the Traditions of Mediaeval Latin Astronomy: A Tentative Interpretation'. In *Colloquia Copernicana III* [= Studia Copernicana 13], ed. Owen Gingerich and Jerzy Dobrzycki, 57–96. Wrocław: Wydawnictwo Polskiej Akademij Nauk.

Pedersen, Olaf. 1976. 'Petrus Philomena de Dacia: A Problem of Identity; with a Survey of the Manuscripts'. *Cahiers de l'Institut du Moyen-Âge Grec et Latin* 19: 1–54.

Pedersen, Olaf. 1978. 'Astronomy'. In *Science in the Middle Ages*, ed. David C. Lindberg, 303–37. Chicago: University of Chicago Press.

Pedersen, Olaf. 1981. 'The Origins of the "Theorica Planetarum"'. *JHA* 12: 113–23.

Pedersen, Olaf. 1983. 'The Ecclesiastical Calendar and the Life of the Church'. In Coyne, Hoskin, and Pedersen 1983: 17–74.

Pedersen, Olaf. 1985. 'In Quest of Sacrobosco'. *JHA* 16: 175–221.

Pedersen, Olaf. 1986. 'Some Astronomical Topics in Pliny'. In *Science in the Early Roman Empire*, ed. Roger French and Frank Greenaway, 162–96. London: Croom Helm.

Pelteret, David A. E. 2011. 'The Issue of Apostolic Authority at the Synod of Whitby'. In Warntjes and Ó Cróinín 2011: 150–72.

Pereira, Michela. 1976. 'Ricerche intorno al *Tractatus novus de astronomia* di Raimondo Lullo'. *Medioevo* 2: 169–226.

Perez de Tudela, Juan, ed. 1991. *Imago mundi del Cardenal Pedro d'Ailly y Juan Gerson: estudio critico*. Madrid: Testimonio Compañia Editorial.

Peri, Vittorio. 1967. *Due date, un'unica Pasqua: le origini della moderna disparità liturgica in una trattativa ecumenica tra Roma e Constantinopoli (1582–84)*. Milan: Vita e Pensiero.

Perovic, Sanja. 2012. *The Calendar in Revolutionary France: Perceptions of Time in Literature, Culture, Politics*. Cambridge: Cambridge University Press.

Pesic, Peter. 2009. 'How the Sun Stood Still: Old English Interpretations of Joshua and the Leap Year'. *Culture and Cosmos* 13, no. 2: 33–41.

Petersen, Viggo M., and Olaf Schmidt. 1967. 'The Determination of the Longitude of the Apogee of the Orbit of the Sun According to Hipparchus and Ptolemy'. *Centaurus* 12: 73–96.

Pezzella, Salvatore, ed. 1982. *Astronomia ed astrologia nel medioevo da un manoscritto inedito (sec. XIII) della città di Firenze*. Florence: La Giuntina.

Pfaff, Alfred. 1944. *Aus alten Kalendern*. Augsburg: Riger & Kranzfelder.

Philip, Alexander. 1921. *The Calendar: Its History, Structure, and Improvement*. Cambridge: Cambridge University Press.

Pillonel-Wyrsch, Roland-Pierre. 2004. *Le calcul de la date de Pâques au Moyen Âge: analyse et commentaires sur 'De Temporum Ratione' de Bède*. Fribourg: Academic Press.

Pingree, David. 1990. 'The Preceptum Canonis Ptolomei'. In *Rencontres de cultures dans la philosophie médiévale*, ed. Jacqueline Hamesse and Marta Fattori, 355–75. Louvain-la-Neuve: Université catholique de Louvain.

Pingree, David. 1997. *Preceptum Canonis Ptolomei*. Louvain-la-Neuve: Academia-Bruylant.

Pingree, David. 2000. 'Avranches 235 dans la tradition manuscrite du *Preceptum Canonis Ptolomei*'. In *Science antique, science médiévale (autour d'Avranches 235)*, ed. Louis Callebat and Olivier Desbordes, 163–69. Hildesheim: Olms.

Pitra, Jean-Baptiste-François. 1852. *Spicilegium Solesmense complectens sanctorum patrum scriptorumque ecclesiasticorum anecdota hactenus opera*. Vol. 1. Paris: Didot.

Plassard, Joël. 1975. 'Projets de réforme du calendrier à Paris au début du XIVe siècle: textes édités et commentés'. *École nationale des chartes: positions des thèses*: 175–81.

Porres de Mateo, Beatriz. 2002. 'Astronomy between Prague and Vienna in the 15th Century: The Case of John Šindel and John of Gmunden'. In *Tycho Brahe and Prague*, ed. John Robert Christianson et al., 248–55. Frankfurt/Main: Deutsch.

Porres de Mateo, Beatriz. 2006. 'Die astronomischen Tafeln des Johannes von Gmunden: Seine Lehre und Forschung an und außerhalb der Universität Wien'. In *Johannes von*

Gmunden (ca. 1384–1442), ed. Rudolf Simek and Kathrin Chlench, 105–25. Vienna: Fassbaender.

Porres de Mateo, Beatriz, and José Chabás. 1998. 'Los cánones de las *Tabulae Resolutae* para Salamanca: origen y transmisión'. *Cronos* 1: 51–83.

Porres de Mateo, Beatriz, and José Chabás. 2001. 'John of Murs's *Tabulae permanentes* for Finding True Syzygies'. *JHA* 32: 63–72.

Poulle, Emmanuel. 1964. 'Le quadrant nouveau médiévale'. *Journal des savants*, 148–67, 182–214.

Poulle, Emmanuel. 1980a. 'Jean de Murs et les Tables Alphonsines'. *Archives d'histoire doctrinale et littéraire du Moyen Âge* 47: 241–71.

Poulle, Emmanuel. 1980b. *Les instruments de la théorie des planètes selon Ptolémée: équatoires et horologerie planétaire du XIIIe au XVIe siècle*. 2 vols. Geneva: Droz.

Poulle, Emmanuel, ed. 1984. *Les Tables Alphonsines avec les canons de Jean de Saxe*. Paris: Éditions du Centre National de la Recherche Scientifique.

Poulle, Emmanuel. 1987. 'Les *Tables Alphonsines* sont-elles d'Alphonse X?' In *De astronomia Alphonsi regis*, ed. Mercè Comes, Roser Puig, and Julio Samsó, 51–69. Barcelona: Instituto 'Millás Vallicrosa' de Historia de la Ciencia Árabe.

Poulle, Emmanuel. 1988. 'The Alfonsine Tables and Alfonso X of Castille'. *JHA* 19: 97–113.

Poulle, Emmanuel. 1991. 'L'astronomie latine au XIVe siècle'. In *Gersonide en son temps*, ed. Gilbert Dahan, 253–64. Leuven: Peeters.

Poulle, Emmanuel. 1994. 'Un témoin de l'astronomie latine du XIIIe siècle: les tables de Toulouse'. In *Comprendre et maîtriser la nature au Moyen-Âge*, 55–81. Geneva: Droz.

Poulle, Emmanuel. 1998. 'L'astronomie latine au XIIe siècle'. In *Cieli e terre nei secoli XI–XII*, 75–88. Milan: Vita e pensiero.

Poulle, Emmanuel. 2004a. 'Quand le soleil a rendez-vous avec la lune'. *Micrologus* 12: 9–26.

Poulle, Emmanuel. 2004b. 'Y a-t-il du nouveau sur les tables alphonsines?' *AIHS* 34: 141–55.

Poulle, Emmanuel. 2005. 'Les astronomes parisiens au XIVe siècle et l'astronomie alphonsine'. In *Histoire littéraire de la France*, vol. 43.1, 1–54. Paris: Boccard.

Poulle, Emmanuel. 2011. 'Mesures et astronomie au Moyen Âge'. *Micrologus* 19: 73–87.

Poulle, Emmanuel, and Daniel Savoie. 1998. 'La survie de l'astronomie alphonsine'. *JHA* 29: 201–7.

Power, Amanda. 2008. '"In the last days at the end of the world": Roger Bacon and the Reform of Christendom'. *Canterbury Studies in Franciscan History* 1: 135–51.

Power, Amanda. 2011. 'Seeking Remedies for Great Danger: Contemporary Appraisals of Roger Bacon's Expertise'. In *Knowledge, Discipline, and Power in the Middle Ages*, ed. Joseph Canning, Edmund King, and Martial Staub, 63–78. Leiden: Brill.

Power, Amanda. 2013a. *Roger Bacon and the Defence of Christendom*. Cambridge: Cambridge University Press.

Power, Amanda. 2013b. 'The Cosmographical Imagination of Roger Bacon'. In *Mapping Medieval Geographies: Geographical Encounters in the Latin West and Beyond, 300–1600*, ed. Keith D. Lilley, 83–99. Cambridge: Cambridge University Press.

Price, Derek J. 1955. *The Equatorie of the Planetis: Edited from Peterhouse Ms. 75.I.* Cambridge: University Press.

Procter, Evelyn S. 1945. 'The Scientific Works of the Court of Alfonso X of Castille: The King and His Collaborators'. *Modern Language Review* 40: 12–29.

Proverbio, Edoardo. 1983. 'Copernicus and the Determination of the Length of the Tropical Year'. In Coyne, Hoskin, and Pedersen 1983: 129–34.

Ragep, F. Jamil, ed. 1993. *Naṣīr al-Dīn al-Ṭūsī's 'Memoir on Astronomy' (al-Tadhkira fī ʿilm al-hayʾa)*. 2 vols. New York: Springer.

Ragep, F. Jamil. 1996. 'Al-Battani, Cosmology, and the Early History of Trepidation in Islam'. In *From Baghdad to Barcelona*, ed. Josep Casulleras and Julio Samsó, vol. 1, 267–98. Barcelona: Instituto 'Millás Vallicrosa' de Historia de la Ciencia Árabe.

Ragep, F. Jamil. 2014. 'New Light on Shams: The Islamic Side of *ΣÀΜΨ ΠΟΥΧΆΡΗΣ*'. In *Politics, Patronage and the Transmission of Knowledge in 13th–15th Century Tabriz*, ed. Judith Pfeiffer, 231–47. Leiden: Brill.

Ramírez de Verger, Antonio, trans. 1990. *Imago mundi del Cardenal Pedro d'Ailly y Juan Gerson: introducción y traducción*. Madrid: Testimonio Compañia Editorial.

Ramírez-Weaver, Eric M. 2017. *A Saving Sciece: Capturing the Heavens in Carolingian Manuscripts*. University Park, Pa.: Pennsylvania State University Press.

Rawlins, Dennis. 2002. 'Aristarchos and the "Babylonian" Month'. In *Under One Sky: Astronomy and Mathematics in the Ancient Near East*, ed. John M. Steele and Annette Imhausen, 295–6. Münster: Ugarit-Verlag.

Reeve, Michael D. 2012. 'Excerpts from Pliny's *Natural History*'. In *Ways of Approaching Knowledge in Late Antiquity and the Early Middle Ages*, ed. Paulo Farmhouse Alberto and David Paniagua, 245–63. Nordhausen: Bautz.

Rehm, Albert. 1941. *Parapegmastudien*. Munich: Verlag der Bayerischen Akademie der Wissenschaften.

Rehm, Albert. 1949. 'Parapegma'. In *Paulys Real-Encyclopädie der classischen Altertumswissenschaft: Neue Bearbeitung*. Vol. 18.2, ed. Georg Wissowa et al., cols. 1295–366. Waldsee (Württ.): Druckenmüller.

Reich, Karin. 1990. 'Problems of Calendar Reform from Regiomontanus to the Present'. Supplement 5 (pp. 345–62) to Ernst Zinner, *Regiomontanus: His Life and Work*. Translated by Ezra Brown. Amsterdam: North-Holland.

Reich, Karin, and Eberhard Knobloch. 2004. 'Melanchthons Vorreden zu Sacroboscos "Sphaera" und zum "Computus ecclesiasticus"'. *Beiträge zur Astronomiegeschichte* 7 = Acta Historica Astronomiae 23: 14–44.

Reinhardt, Klaus, ed. 1977. *Pedro de Osma y su comentario al símbolo 'Quicumque'*. Madrid: Joyas Bibliográficas.

Ribordy, Olivier. 2017. 'Mesure du temps et réforme du calendrier: Pierre d'Ailly face aux enjeux chronologiques'. In *Le temps des astronomes*, ed. Edouard Mehl and Nicolas Roudet, 21–63. Paris: Les Belles Lettres.

Richards, E. G. 1998. *Mapping Time: The Calendar and Its History*. Oxford: Oxford University Press.

Richter, Michael. 1999. *Ireland and Her Neighbours in the Seventh Century*. Dublin: Four Courts Press.

Rico y Sinobas, Manuel, ed. 1863–7. *Libros del Saber de astronomía del Rey D. Alfonso X de Castilla*. 5 vols. Madrid: Aguado.

Rigo, Antonio. 1991. 'Bessarione, Giovanni Regiomontano e i loro studi su Tolomeo a Venezia e Roma (1462–1464)'. *Studi Veneziani*, n.s., 21: 49–110.

Rigo, Antonio. 1994. 'Gli interessi astronomici del cardinal Bessarione'. In *Bessarione e l'Umanesimo: catalogo della mostra*, ed. Gianfranco Fiaccadori, 105–17. Naples: Vivarium.

Robinson, Maureen. 2007. 'The Heritage of Medieval Errors in the Latin Manuscripts of Johannes Hispalensis (John of Seville)'. *Al-Qanṭara* 28: 41–71.

Roegel, Denis. 2004. 'The Missing New Moon of A.D. 16399 and Other Anomalies of the Gregorian Calendar'. [Intern report] A04-R-436—roegel04a, 2004, 15 p. <inria-00099868>.

Rollo-Koster, Joëlle, and Thomas M. Izbicki, eds. 2009. *A Companion to the Great Western Schism (1378–1417)*. Leiden: Brill.

Rose, Paul Lawrence. 1975. *The Italian Renaissance of Mathematics: Studies on Humanists and Mathematicians from Petrarch to Galileo*. Geneva: Droz.

Rosen, Edward. 1958. 'Galileo's Misstatements about Copernicus'. *Isis* 49: 319–30.

Rosen, Edward. 1974. 'Rheticus as Editor of Sacrobosco'. In *For Dirk Struik*, ed. R. S. Cohen, J. J. Stachel, and M. W. Wartofsky, 245–48. Dordrecht: Reidel.

Rosen, Edward. 1977. 'When did Copernicus Write the "Revolutions"?' *Sudhoffs Archiv* 61: 144–55.

Roth, Norman. 1990. 'Jewish Collaborators in Alfonso's Scientific Work'. In *Emperor of Culture*, ed. Robert I. Burns, 59–71. Philadelphia: University of Pennsylvania Press.

Rühl, Franz. 1897. *Chronologie des Mittelalters und der Neuzeit*. Berlin: Reuther & Reichard.

Rüpke, Jörg. 2006. *Zeit und Fest: Eine Kulturgeschichte des Kalenders*. Munich: Beck.

Rüpke, Jörg. 2011. *The Roman Calendar from Numa to Constantine: Time, History and the 'Fasti'*. Oxford: Wiley-Blackwell.

Russell, Josiah C. 1932. 'Hereford and Arabic Science in England about 1175–1200'. *Isis* 18: 14–25.

Russo, François. 1983. 'Contemporary Discussions on the Reform of the Calendar'. In Coyne, Hoskin, and Pedersen 1983: 287–97.

Salembier, Louis. 1932. *Le Cardinal Pierre d'Ailly: Chancelier de l'Université de Paris, Évêque du Puy et de Cambrai 1350–1420*. Tourcoing: Frère.

Sambin, Paolo. 1976–77. 'Il dottorato padovano in medicina di Paolo da Middelburgo (1480)'. *Quaderni per la storia dell'Università di Padova* 9–10: 252–6.

Samsó, Julio. 1984. 'Algunas notas sobre el modelo solar y la teoría de la precesión de los equinoccios en la obra astronómica de Alfonso X'. *Dynamis* 4: 81–114.

Samsó, Julio. 1987a. 'Alfonso X and Arabic Astronomy'. In *De astronomia Alphonsi regis*, ed. Mercè Comes, Roser Puig, and Julio Samsó, 23–38. Barcelona: Instituto 'Millás Vallicrosa' de Historia de la Ciencia Árabe.

Samsó, Julio. 1987b. 'On the Solar Model and the Precession of the Equinoxes: The Alphonsine *Zīj* and Its Arabic Sources'. In *History of Oriental Astronomy*, ed. G. Swarup, A. K. Bag, and K. S. Shukla, 175–83. Cambridge: Cambridge University Press.

Samsó, Julio. 1987c. 'Sobre el modelo de Azarquiel para determinar la oblicuidad de la ecliptica'. In *Homenaje al Prof. Darío Cabanelas Rodríguez*, vol. 2, 367–77. Granada: Universidad de Granada.

Samsó, Julio. 1992. *Las ciencias de los antiguos en Al-Andalus*. Madrid: Mapfre.

Samsó, Julio. 1994a. 'Las ciencias exactas y físico-naturales'. In *La época del gótico en la cultura española (c. 1220–c. 1480)*, ed. José Ángel Garciá de Cortázar, 555–93. Madrid: Espasa Calpe.

Samsó, Julio. 1994b. 'Trepidation in al-Andalus in the 11th Century'. Chapter 8 in idem, *Islamic Astronomy and Medieval Spain*. Aldershot: Variorum.

Samsó, Julio. 1998. 'An Outline of the History of Maghribī Zijes from the End of the Thirteenth Century'. *JHA* 29: 93–102.

Samsó, Julio. 2000. 'Maslama al-Majīṭī and the Star Table in the Treatise *De mensura astrolabii*'. In *Sic Itur ad Astra*, ed. Menso Folkerts and Richard Lorch, 506–22. Wiesbaden: Harrassowitz.

Samsó, Julio. 2001. 'Astronomical Observations in the Maghrib in the Fourteenth and Fifteenth Centuries'. *Science in Context* 14: 165–78.

Samsó, Julio. 2004a. 'Els inicis de la introducció de la ciència àrab a Europa a través de Catalunya'. In *La ciència en la història dels Països Catalans*, ed. Joan Vernet and Ramon Parés, vol. 1, 115–59. Valencia: Institut d'Estudis Catalans.

Samsó, Julio. 2004b. 'El procés de la transmissió científica al nord-est de la península Ibèrica al segle XII: els textos llatins'. In *La ciència en la història dels Països Catalans*, ed. Joan Vernet and Ramon Parés, vol. 1, 269–96. Valencia: Institut d'Estudis Catalans.

Samsó, Julio. 2012. '"Dixit Abraham Iudeus": algunas observaciones sobre los textos astronómicos latinos de Abraham Ibn Ezra'. *Iberia Judaica* 4: 171–200.

Samsó, Julio, and Francisco Castelló. 1988. 'An Hypothesis on the Epoch of Ptolemy's Star Catalogue According to the Authors of the Alfonsine Tables'. *JHA* 19: 115–20.

Samsó, Julio, and Eduardo Millás. 1994. 'Ibn al-Bannā', Ibn Isḥāq and Ibn al-Zarqālluh's Solar Theory'. Chapter 10 in idem, *Islamic Astronomy and Medieval Spain*. Aldershot: Variorum.

Samuel, Alan E. 1972. *Greek and Roman Chronology: Calendars and Years in Classical Antiquity*. Munich: Beck.

Sarton, George. 1957. *Six Wings: Men of Science in the Renaissance*. Bloomington, Ind.: Indiana University Press.

Saulnier, Stéphane. 2012. *Calendrical Variations in Second Temple Judaism: New Perspectives on the 'Date of the Last Supper' Debate*. Leiden: Brill.

Scarabelli, Luciano, ed. 1866. *Comedia di Dante degli Allagherii col commento di Jacopo della Lana*. 3 vols. Bologna: Tipografia Regia.

Schabel, Chris. 1996. 'John of Murs and Firmin of Beauval's Letter and Treatise on Calendar Reform for Clement VI'. *Cahiers de l'Institut du Moyen-Âge Grec et Latin* 66: 187–215.

Schabel, Chris. 1998. '*Ad correctionem calendarii*... The Background to Clement VI's Initiative?' *Cahiers de l'Institut du Moyen-Âge Grec et Latin* 68: 13–34.

Schäfer, Karl Heinrich. 1914. *Die Ausgaben der apostolischen Kammer unter Benedikt XII., Klemens VI. und Innocenz VI (1335–1362)*. Paderborn: Schöningh.

Scheutz, Martin. 2005. '"Den neuen bäpstischen Calender anlangende würdet derselb [...] durchaus nit gehalten": Der Gregorianische Kalender als politischer und konfessioneller Streitfall'. In Hameter, Niederkorn-Bruck, and Scheutz 2005: 116–43.

Schiavo, Armando. 1993. *La meridiana di S. Maria degli Angeli*. Rome: Libreria dello stato.

Schissel, Otmar, and Maria Ellend. 1943/49. 'Berechnung des Sonnen-, Mond- und Schaltjahreszirkels in der griechisch-christlichen Chronologie'. *Byzantinische Zeitschrift* 42: 150–7.

Schmeidler, Felix, ed. 1972. *Joannis Regiomontani Opera Collectanea: Faksimiledrucke von neun Schriften Regiomontans und einer von ihm gedruckten Schrift seines Lehrers Purbach*. Osnabrück: Zeller.

Schmid, Joseph. 1882–84. 'Zur Geschichte der Gregorianischen Kalenderreform'. 3 pts. *Historisches Jahrbuch* 3: 388–415, 543–95; 5: 52–87.

Schmid, Joseph. 1905. *Die Osterfestfrage auf dem ersten allgemeinen Konzil von Nicäa*. Vienna: Mayer.

Schmid, Joseph. 1907. *Die Osterfestberechnung in der abendländischen Kirche vom I. Allgemeinen Konzil zu Nicäa bis zum Ende des VIII. Jahrhunderts*. Freiburg im Breisgau: Herder.

Schröer, Alois. 1951. 'Das Tridentinum und Münster'. In *Das Weltkonzil von Trient: Sein Werden und Wirken*, ed. Georg Schreiber, vol. 2, 295–370. Freiburg: Herder.

Schubring, Ernst. 1883. 'Zur Erinnerung an die Gregorianische Kalenderreform (Oktober 1582)'. *Zeitschrift für Naturwissenschaften* 56: 516–38.

Schulz, Fritz. 1945. 'Bracton as a Computist'. *Traditio* 3: 265–305.

Schwartz, Eduard. 1905. *Christliche und jüdische Ostertafeln*. Berlin: Weidmann.

Shank, Michael H. 2012. 'The Geometrical Diagrams in Regiomontanus's Edition of His Own *Disputationes* (c. 1475): Background, Production, and Diffusion'. *JHA* 43: 27–55.

Shank, Michael H. 2013. 'Schools and Universities in Medieval Latin Science'. In *The Cambridge History of Science*, vol. 2, *Medieval Science*, ed. David C. Lindberg and Michael H. Shank, 207–39. Cambridge: Cambridge University Press.

Shank, Michael H. 2017. 'Regiomontanus and Astronomical Controversy in the Background of Copernicus'. In *Before Copernicus: The Cultures and Contexts of Scientific Learning in the Fifteenth Century*, ed. Rivka Feldhay and F. Jamil Ragep, 79–109. Montreal and Kingston: McGill-Queen's University Press.

Shaw, Matthew. 2011. *Time and the French Revolution: The Republican Calendar, 1789–Year XIV*. Woodbridge: Boydell Press.

Shore, Lys Ann. 1989. 'A Case Study in Medieval Nonliterary Translation: Scientific Texts from Latin to French'. In *Medieval Translators and Their Craft*, ed. Jeanette Beer, 297–327. Kalamazoo, Mich.: Medieval Institute Publications, Western Michigan University.

Short, Ian, ed. 1984. *Philippe de Thaon: Comput (MS BL Cotton Nero A. V)*. London: Anglo-Norman Text Society, 1984.

Sickel, Theodor. 1861. 'Die Lunarbuchstaben in den Kalendarien des Mittelalters'. *Sitzungsberichte der philosophisch-historischen Classe der kaiserlichen Akademie der Wissenschaften* [Vienna] 38: 153–201.

Silagiová, Zuzana. 2007. '*Est duplex cyclus*: Veršovaný komputus ze 14. století'. In *Donum magistrae*, ed. Zuzana Silagiová, Hana Šedinová, and Petr Kitzler, 154–78. Prague: Filosofia.

Silagiová, Zuzana. 2013. 'Die Sprache der Komputi'. *Archivum Latinitatis Medii Aevi* 71: 303–13.

Silverstein, Theodore. 1955. 'Liber Hermetis Mercurii Triplicis De VI Rerum Principiis'. *Archives d'histoire doctrinale et littéraire du Moyen Age* 22: 217–302.

Simek, Rudolf. 1987. 'Die mittelhochdeutschen Übertragungen von Johannes von Sacroboscos *Liber de sphaera*: Zur Funktion der astronomischen Abbildungen in den Handschriften und Frühdrucken'. *Codices manuscripti* 13: 57–76.

Simon, Eckehard. 1988. *The 'Türkenkalender' (1454) Attributed to Gutenberg and the Strasbourg Lunation Tracts*. Cambridge, Mass.: The Medieval Academy of America.

Skemer, Don C. 2010. '*Armis Gunfe*: Remembering Egyptian Days'. *Traditio* 65: 75–106.

Small, Robin. 1991. 'Incommensurability and Recurrence: From Oresme to Simmel'. *Journal of the History of Ideas* 52: 121–37.

Smith, David Eugene, ed. 1928. *Le comput manuel de Magister Anianus*. Geneva. Reprint, Geneva: Slatkine, 1977.

Smolinsky, Heribert. 2000. *Deutungen der Zeit im Streit der Konfessionen: Kontroverstheologie, Apokalyptik und Astrologie im 16. Jahrhundert*. Heidelberg: Winter.

Smoller, Laura Ackerman. 1994. *History, Prophecy, and the Stars: The Christian Astrology of Pierre d'Ailly, 1350–1420*. Princeton: Princeton University Press.

Smyth, Marina. 2017. 'Once in Four: The Leap Year in Early Medieval Thought'. In *Late Antique Calendrical Thought and Its Reception in the Early Middle Ages*, ed. Immo Warntjes and Dáibhí Ó Cróinín, 229–64. Turnhout: Brepols.

Solan, Olivier de. 1999. 'La réforme du calendrier dans une question quodlibétique d'Henri de Runen (1444)'. *Bibliothèque de l'École des chartes* 157: 171–220. Reprinted in Hubert 2000.

Solan, Olivier de. 2016. *La réforme du calendrier aux conciles de Constance et de Bâle*. Paris: CNRS Éditions.

Southern, Richard W. 1984. 'From Schools to University'. In *The History of the University of Oxford*, vol. 1, *The Early Oxford Schools*, ed. J. I. Catto, 1–36. Oxford: Clarendon Press.
Southern, Richard W. 1986. *Robert Grosseteste: The Growth of an English Mind in Medieval Europe*. Oxford: Clarendon Press.
Springsfeld, Kerstin. 2002. *Alkuins Einfluß auf die Komputistik zur Zeit Karls des Großen*. Stuttgart: Steiner.
Stahl, Alan M., ed. 2009. *The Book of Michael of Rhodes: A Fifteenth-Century Maritime Manuscript*. Vol. 2, *Transcription and Translation*. Cambridge, Mass.: MIT Press.
Starostine, Dmitri. 2006. '. . . *in die festivitatis*: Gift-Giving, Power and the Calendar in the Carolingian Kingdoms'. *Early Medieval Europe* 14: 465–86.
Stebbins, Frederick A. 1958. 'A Roman Sun-Dial'. *Journal of the Royal Astronomical Society of Canada* 52: 250–4.
Steel, Duncan. 2000. *Marking Time: The Epic Quest to Invent the Perfect Calendar*. New York: Wiley & Sons.
Steele, John M. 2000. *Observations and Predictions of Eclipse Times by Early Astronomers*. Dordrecht: Kluwer.
Steele, Robert, ed. 1922. *The Earliest Arithmetics in English*. London: Oxford University Press.
Steele, Robert, ed. 1926. *Compotus Fratris Rogeri*. Oxford: Clarendon Press (= Opera hactenus inedita Rogeri Baconi 6).
Stegemann, Viktor, ed. 1955. *Nikolaus von Cues: Die Kalenderverbesserung/ De correctione kalendarii*. Heidelberg: Kerle.
Steinmetz, Dirk. 2011. *Die Gregorianische Kalenderreform von 1582: Korrektur der christlichen Zeitrechnung in der Frühen Neuzeit*. Oftersheim: Steinmetz.
Stern, Sacha. 2001. *Calendar and Community: A History of the Jewish Calendar, Second Century bce–Tenth Century ce*. Oxford: Oxford University Press.
Stern, Sacha. 2012. *Calendars in Antiquity: Empires, States, and Societies*. Oxford: Oxford University Press.
Stevens, Wesley M. 1995. 'Bede's Scientific Achievement: The Jarrow Lecture 1985 (revised 1995)'. Chapter 2 in idem, *Cycles of Time and Scientific Learning in Medieval Europe*. Aldershot: Variorum.
Stieber, Joachim W. 1978. *Pope Eugenius IV, the Council of Basel, and the Secular and Ecclesiastical Authorities in the Empire: The Conflict over Supreme Power and Authority in the Church*. Leiden: Brill.
Stieve, Felix. 1879. 'Review of Kaltenbrunner 1876 and 1877'. *Historische Zeitschrift* 42: 127–36.
Stieve, Felix. 1880. *Der Kalenderstreit des 16. Jahrhunderts in Deutschland*. Munich: Verlag der k. Akademie.
Strecker, Karl, ed. 1923. *Poetae latini aevi Carolini*. Vol. 4.2–3. Berlin: Weidmann.
Streveler, Paul A., and Katherine H. Tachau, eds. 1995. *Seeing the Future Clearly: Questions on Future Contingents by Robert Holcot*. Toronto: Pontifical Institute of Mediaeval Studies.
Strobel, August. 1984. *Texte zur Geschichte des frühchristlichen Osterkalenders*. Münster: Aschendorff.
Struik, Dirk Jan. 1925. 'Paulus van Middelburg (1445–1533)'. *Mededeelingen van het Nederlandsch Historisch Instituut te Rome* 5: 79–118.
Stump, Phillip H. 2009. 'The Council of Constance (1414–18) and the End of the Schism'. In Rollo-Koster and Izbicki 2009: 395–442.
Stussi, Alfredo, ed. 1967. *Zibaldone da Canal: manoscritto mercantile del sec. XIV*. Venice: Comitato per la pubblicazione delle fonti relative alla storia di Venezia.

Sudmann, Stefan. 2005. *Das Basler Konzil: Synodale Praxis zwischen Routine und Revolution.* Frankfurt/Main: Lang.

Šuráň, Josef. 1997. 'The Calendar of the Future: A World Calendar with Leap Week'. *Vistas in Astronomy* 41: 493–506.

Suter, Heinrich, ed. 1914. *Die astronomischen Tafeln des Muḥammed ibn Mūsā al-Khwārizmī in der Bearbeitung des Maslama ibn Aḥmed al-Madjrīṭī und der latein. Übersetzung des Athelhard von Bath.* Copenhagen: Høst & Søn.

Svenberg, Emanuel. 1936. *De Latinska Lunaria: text och studier.* Göteborg: Elanders.

Svenberg, Emanuel. 1963. *Lunaria et Zodiologica Latina.* Göteborg: Elanders.

Swerdlow, Noel M. 1973. 'The Derivation and First Draft of Copernicus's Planetary Theory: A Translation of the Commentariolus with Commentary'. *Proceedings of the American Philosophical Society* 117: 423–512.

Swerdlow, Noel M. 1974. 'The Origin of the Gregorian Civil Calendar'. *JHA* 5: 48–9.

Swerdlow, Noel M. 1975. 'On Copernicus' Theory of Precession'. In *The Copernican Achievement*, ed. Robert S. Westman, 49–98. Berkeley: University of California Press.

Swerdlow, Noel M. 1979–80. 'Hipparchus' Determination of the Length of the Tropical Year and the Rate of Precession'. *AHES* 21: 291–309.

Swerdlow, Noel M. 1980. 'Long-Period Motions of the Earth in De revolutionibus'. *Centaurus* 24: 212–45.

Swerdlow, Noel M. 1986. 'The Length of the Year in the Original Proposal for the Gregorian Calendar'. *JHA* 17: 109–18.

Swerdlow, Noel M. 1990. 'Regiomontanus on the Critical Problems of Astronomy'. In *Nature, Experiment, and the Sciences*, ed. Trevor H. Levere and William R. Shea, 165–95. Dordrecht: Kluwer.

Swerdlow, Noel M. 2004. 'Alfonsine Tables of Toledo and Later Alfonsine Tables'. *JHA* 35: 479–84.

Swerdlow, Noel M., and Otto Neugebauer. 1984. *Mathematical Astronomy in Copernicus's De Revolutionibus.* 2 vols. New York: Springer.

Synodica 1981 = *Synodica.* Vol. 5, *Congrès pour l'examen de la question d'une celebration commune de Pâques par tous les Chrétiens le même dimanche, Chambésy-Genève, 28 juin–3 juillet 1977.* Chambésy-Geneva: Éditions du Centre Orthodox.

Taavitsainen, Irma. 1988. *Middle English Lunaries: A Study of the Genre.* Helsinki: Société Néophilologique.

Talbot, C. H. 1961. 'A Mediaeval Physician's *Vade Mecum*'. *Journal of the History of Medicine and Allied Sciences* 16: 213–33.

Tanturli, Giuliano, ed. 1997. *Philippi Villani De origine civitatis Florentie et de eiusdem famosis civibus.* Padova: Antenore.

Taub, Liba. 2003. *Ancient Meteorology.* London: Routledge.

Thacker, Alan T. 2014. 'Why did Heresy Matter to Bede? Present and Future Contexts'. In *Bede and the Future*, ed. Peter Darby and Faith Wallis, 47–66. Farnham: Ashgate.

Thilo, Georg, ed. 1878. *Servii grammatici qui feruntur in Vergilii Bucolica et Georgica commentarii.* Leipzig: Teubner.

Thomas, Antoine. 1921. 'Pierre Vidal, frère prècheur, astronome'. In *Histoire littéraire de la France*, vol. 35, 624–7. Paris. Reprint, Nendeln/Liechtenstein: Kraus, 1971.

Thomson, Ron B. 1985. 'Two Astronomical Tractates of Abbo of Fleury'. In *The Light of Nature*, ed. John D. North and J. J. Roche, 113–33. Dordrecht: Nijhoff.

Thomson, Ron B. 1988. 'Further Astronomical Material of Abbo of Fleury'. *Mediaeval Studies* 50: 671–3.

Thomson, S. Harrison. 1940. *The Writings of Robert Grosseteste, Bishop of Lincoln 1235–1253*. Cambridge: Cambridge University Press.

Thorndike, Lynn. 1923–58. *A History of Magic and Experimental Science*. 8 vols. New York: Columbia University Press.

Thorndike, Lynn. 1947. 'Blastus the Franciscan and His Works on Computus'. *Isis* 37: 46–7.

Thorndike, Lynn. 1948. 'Astronomical Observations at Paris from 1312 to 1315'. *Isis* 38: 200–5.

Thorndike, Lynn. 1949. *'The 'Sphere' of Sacrobosco and Its Commentators*. Chicago: University of Chicago Press.

Thorndike, Lynn. 1950a. *Latin Treatises on Comets Between 1238 and 1368 A.D.* Chicago: University of Chicago Press.

Thorndike, Lynn. 1950b. 'Pre-Copernican Astronomical Activity'. *Proceedings of the American Philosophical Society* 94: 321–6.

Thorndike, Lynn. 1951. 'Predictions of Eclipses in the Fourteenth Century'. *Isis* 42: 301–2.

Thorndike, Lynn. 1952. 'A Record of Eclipses for the Years 1478 to 1506'. *Isis* 43: 252–6.

Thorndike, Lynn. 1954. 'Computus'. *Speculum* 29: 223–38.

Thorndike, Lynn. 1955. 'Unde Versus'. *Traditio* 11: 163–93.

Thorndike, Lynn. 1957a. 'A New Work by Robert Holkot (Corpus Christi College, Oxford, MS 138)'. *AIHS* 10: 227–35.

Thorndike, Lynn. 1957b. 'Eclipses in the Fourteenth and Fifteenth Centuries'. *Isis* 48: 51–7.

Thorndike, Lynn. 1957c. 'Notes on Some Astronomical, Astrological, and Mathematical Manuscripts of the Bibliothèque Nationale, Paris'. *Journal of the Warburg and Courtauld Institutes* 20: 112–72.

Thorndike, Lynn. 1959. 'Notes upon Some Medieval Astronomical, Astrological and Mathematical Manuscripts at Florence, Milan, Bologna and Venice'. *Isis* 50: 33–50.

Thorndike, Lynn. 1961. 'Manuscripts of Michael Scot's *Liber Introductorius*'. In *Didascaliae*, ed. Sesto Prete, 425–47. New York: Rosenthal.

Thorndike, Lynn. 1965. *Michael Scot*. London: Nelson.

Thorndike, Lynn, and Pearl Kibre. 1963. *A Catalogue of Incipits of Mediaeval Scientific Writings in Latin*. Rev. edn. London: Mediaeval Academy of America.

Thurston, Hugh. 2002. 'Greek Mathematical Astronomy Reconsidered'. *Isis* 93: 58–69.

Tihon, Anne. 1981. 'L'astronomie byzantine (du Ve au XVe siècle)'. *Byzantion* 51: 603–24.

Tihon, Anne. 1987. 'Les tables astronomiques persanes à Constantinople dans la première moitié du XIVe siècle'. *Byzantion* 57: 471–87.

Tihon, Anne. 1999. 'Il *Trattato sulla data della Pasqua* di Barlaam comparato con quello di Niceforo Gregoras'. In *Barlaam Calabro*, ed. Antonis Fyrigos, 109–18. Rome: Gangemi.

Tihon, Anne. 2004. 'Le calcul de la date de Pâques de Stéphanos-Héraclius'. In *Philomathestatos*, ed. B. Janssens, B. Rossen and P. Van Deun, 625–46. Leuven: Peeters.

Tihon, Anne. 2011. 'Barlaam de Seminara: *Traité sur la date de Pâques*'. *Byzantion* 81: 362–411.

Tihon, Anne. 2013. 'Science in the Byzantine Empire'. In *The Cambridge History of Science*, vol. 2, *Medieval Science*, ed. David C. Lindberg and Michael H. Shank, 190–206. Cambridge: Cambridge University Press.

Tönsmeyer, Josef. 1930. 'Hermann Zoestius von Marienfeld: Ein Vertreter der konziliaren Theorie am Konzil zu Basel'. *Westfälische Zeitschrift* 87: 114–91.

Toomer, G. J. 1968. 'A Survey of the Toledan Tables'. *Osiris* 15: 5–174.

Toomer, G. J. 1969. 'The Solar Theory of az-Zarqāl: A History of Errors'. *Centaurus* 14: 306–36.

Toomer, G. J. 1973. 'Prophatius Judaeus and the Toledan Tables'. *Isis* 64: 351–5.

Toomer, G. J. 1987. 'The Solar Theory of az-Zarqāl: An Epilogue'. In *From Deferent to Equant*, ed. David A. King and George Saliba, 513–19. New York: New York Academy of Sciences.

Toomer, G. J., trans. 1998. *Ptolemy's Almagest*. Rev. edn. Princeton: Princeton University Press.

Traube, Ludwig. 1893. 'Computus Helperici'. *Neues Archiv* 18: 73–105. Reprinted in idem, *Vorlesungen und Abhandlungen*, ed. Franz Boll, 3 vols. (Munich: Beck, 1909–20), 3: 128–56.

Tschackert, Paul. 1877. *Peter von Ailli (Petrus de Alliaco): Zur Geschichte des grossen abendländischen Schisma und der Reformconcilien von Pisa und Constanz*. Gotha: Perthes.

Van de Vyver, André. 1935. 'Les oeuvres inédites d'Abbon de Fleury'. *Revue Bénédictine* 47: 125–69.

Van de Vyver, André. 1936. 'Les plus anciennes traductions latines médiévales (X^e^–XIe siècles) de traités d'astronomie et d'astrologie'. *Osiris* 1: 658–91.

Van de Vyver, André. 1947. 'Hucbald de Saint-Armand, écolâtre, et l'invention de Nombre d'or'. In *Mélanges Auguste Pelzer*, 61–79. Louvain: Éditions de l'Institut Supérieur de Philosophie.

Van Isacker, Philippe, and Ursmer Berlière, eds. 1924. *Lettres de Clément VI (1342–1352)*. Vol. 1, *1342–1346*. Rome: Institut Historique Belge.

Vennebusch, Joachim. 1978. 'Bemerkung zum "Tractatus de stellis" des Robert Holkot'. *Bulletin de philosophie médiévale* 20: 75.

Verbist, Peter. 2002a. 'Een elfde-eeuwse dialoog tussen een leermeester en een leerling: het *Liber Decennalis* van Sigebert van Gembloux'. In *In de voetsporen van Jacob van Maerlant*, edited Raoul Bauer, Marjan de Smet, Brigitte Meijns, and Paul Trio, 101–13. Leuven: Universitaire Pers Leuven.

Verbist, Peter. 2006. 'Abbon de Fleury et l'ère chrétienne vers l'an mil: Un esprit critique vis-à-vis d'une tradition erronée'. In Obrist 2006: 61–93.

Verbist, Peter. 2010. *Duelling with the Past: Medieval Authors and the Problem of the Christian Era, c. 990–1135*. Turnhout: Brepols.

Vercauteren, Fernand. 1951. 'Les médecins dans les principautés de la Belgique et du nord de la France, du VIIIe au XIIIe siècle'. *Le Moyen Âge* 57: 61–92.

Vernet, Juan, and Julio Samsó. 1996. 'The Development of Arabic Science in Andalusia'. In *Encyclopedia of the History of Arabic Science*, ed. Roshdi Rashed, vol. 1, 243–75. London: Routledge.

Vetter, Ferdinand. 1906. 'Neues zu Justinger: Kunrat Justinger als Schüler und Fortsetzer Königshofens und die ältesten Geschichtsschreiber Berns und des Laupenstreites'. *Jahrbuch für Schweizerische Geschichte* 31: 109–206.

Viladrich, Mercè. 1996. 'The *Mumtaḥan* Tradition in Al-Andalus: Analysis of Data from the *Calendar of Cordova* Related to the Entrance of the Sun in the Zodiacal Signs'. In *From Baghdad to Barcelona*, ed. Josep Casulleras and Julio Samsó, vol. 1, 253–65. Barcelona: Instituto 'Millás Vallicrosa' de Historia de la Ciencia Árabe.

Voskoboynikov, Oleg. 2014. 'Le *Liber particularis* de Michel Scot'. *Archives d'histoire doctrinale et littéraire du Moyen Âge* 81: 249–384.

Waerden, B. L. van der. 1984. 'Greek Astronomical Calendars I: The *Parapegma* of Euctemon'. *AHES* 29: 101–14.

Walker, G. S. M., ed. 1957. *Sancti Columbani Opera*. Dublin: Dublin Institute for Advanced Studies.

Wallis, Faith. 1985. 'MS Oxford, St. John's 17: A Medieval Manuscript in Its Context'. Ph.D. diss., University of Toronto.

Wallis, Faith. 1990a. 'Images of Order in the Medieval Computus'. In *Ideas of Order in the Middle Ages*, ed. Warren Ginsberg, 45–67. Binghamton: State University of New York Press.

Wallis, Faith. 1990b. 'The Church, the World and the Time: Prolegomena to a History of the Medieval *Computus*'. In *Normes et pouvoir à la fin du Moyen Âge*, ed. Marie-Claude Déprez-Masson, 15–29. Montreal: Ceres.

Wallis, Faith. 1999. *Bede: The Reckoning of Time*. Liverpool: Liverpool University Press.

Wallis, Faith. 2009. 'Michael of Rhodes and Time Reckoning: Calendar, Almanac, Prognostication'. In *The Book of Michael of Rhodes: A Fifteenth-Century Maritime Manuscript*, vol. 3, Studies, ed. Pamela O. Long, 281–319. Cambridge, Mass.: MIT Press.

Wallis, Faith. 2010. 'Bede and Science'. In *The Cambridge Companion to Bede*, ed. Scott DeGregorio, 113–26. Cambridge: Cambridge University Press.

Walsh, Maura, and Dáibhí Ó Cróinín. 1988. *Cummian's Letter 'De controversia paschali' and the 'De ratione conputandi'*. Toronto: Pontifical Institute of Mediaeval Studies.

Warntjes, Immo. 2005–6. 'A Newly Discovered Irish Computus: *Computus Einsidlensis*'. *Peritia* 19/20: 61–4.

Warntjes, Immo. 2009. 'The Earliest Occurrence of Old English *Gerīm* and Its Anglo-Irish Computistical Context'. *Anglia* 127: 91–105.

Warntjes, Immo. 2010a. 'A Newly Discovered Prologue of AD 699 to the Easter Table of Victorius of Aquitaine in an Unknown Sirmond Manuscript'. *Peritia* 21: 255–84.

Warntjes, Immo. 2010b. 'The *Argumenta* of Dionysius Exiguus and their Early Recensions'. In Warntjes and Ó Cróinín 2010: 40–111.

Warntjes, Immo. 2010c. *The Munich Computus: Text and Translation; Irish Computistics between Isidore of Seville and the Venerable Bede and Its Reception in Carolingian Times*. Stuttgart: Steiner.

Warntjes, Immo. 2011a. 'Irische Komputistik zwischen Isidor von Sevilla und Beda Venerabilis: Ursprung, karolingische Rezeption und generelle Forschungsperspektiven'. *Viator* 42 (Multilingual): 1–31.

Warntjes, Immo. 2011b. 'The *Computus Cottonianus* of AD 689: A Computistical Formulary Written for Willibrord's Frisian Mission'. In Warntjes and Ó Cróinín 2011: 173–212.

Warntjes, Immo. 2012. 'Köln als naturwissenschaftliches Zentrum in der Karolingerzeit: Die frühmittelalterliche Kölner Schule und der Beginn der fränkischen Komputistik'. In *Mittelalterliche Handschriften der Kölner Dombibliothek*, ed. Heinz Finger and Harald Horst, 41–96. Cologne: Erzbischöfliche Diözesan- und Dombibliothek.

Warntjes, Immo. 2013a. 'Die Verwendung der Volkssprache in frühmittelalterlichen Klosterschulen'. In *Wissenspaläste*, ed. Gesine Mierke and Christoph Fasbender, 153–83. Würzburg: Königshausen & Neumann.

Warntjes, Immo. 2013b. 'Seventh-Century Ireland: The Cradle of Medieval Science?' In *Music and the Stars*, ed. Mary Kelly and Charles Doherty, 44–72. Dublin: Four Courts Press.

Warntjes, Immo. 2013–14. 'An Irish Eclipse Prediction of AD 754: The Earliest in the Latin West'. *Peritia* 24/25: 108–15.

Warntjes, Immo. 2015. 'Victorius vs. Dionysius: The Irish Easter Controversy of AD 689'. In *Early Medieval Ireland and Europe*, ed. Pádraic Moran and Immo Warntjes, 33–97. Turnhout: Brepols.

Warntjes, Immo. 2016a. 'Computus as Scientific Thought in Ireland and the Early Medieval West'. In *The Irish in Early Medieval Europe*, ed. Roy Flechner and Sven Meeder, 158–78. London: Palgrave Macmillan.

Warntjes, Immo. 2016b. 'Hermann der Lahme und die Zeitrechnung: Bedeutung seiner Computistica und Forschungsperspektiven'. In *Hermann der Lahme: Reichenauer Mönch und Universalgelehrter des 11. Jahrhunderts*, ed. Felix Heinzer and Thomas Zotz, 285–321. Stuttgart: Kohlhammer.

Warntjes, Immo, and Dáibhí Ó Cróinín, eds. 2010. *Computus and Its Cultural Context in the Latin West, AD 300–1200*. Turnhout: Brepols.

Warntjes, Immo, and Dáibhí Ó Cróinín, eds. 2011. *The Easter Controversy of Late Antiquity and the Early Middle Ages: Its Manuscripts, Texts, and Tables*. Turnhout: Brepols.

Warrior, Valerie M. 1991. 'Notes on Intercalation'. *Latomus* 50: 80–7.

Warrior, Valerie M. 1992. 'Intercalation and the Action of M. Acilius Glabrio (cos. 191 BC)'. In *Studies in Latin Literature and Roman History*, vol. 6, ed. Carl Deroux, 119–44. Brussels: Latomus.

Wartenberg, Ilana. 2014. 'The Hebrew Calendrical Bookshelf in the Early Twelfth Century: The Cases of Abraham Bar Ḥiyya and Jacob Bar Samson'. In *Time, Astronomy, and Calendars in the Jewish Tradition*, ed. Sacha Stern and Charles Burnett, 97–111. Leiden: Brill.

Wattenbach, Wilhelm. 1884. 'Über Hermann von Marienfeld aus Münster'. *Sitzungsberichte der königlich preussischen Akademie der Wissenschaften zu Berlin*, phil.-hist. Kl., 9: 93–109. Reprinted in idem, *Kleine Abhandlungen zur mittelalterlichen Geschichte* (Leipzig: Zentralantiquariat der DDR, 1970), 83–99.

Wegener, Alfred. 1905. 'Die astronomischen Werke Alfons X'. *Bibliotheca Mathematica*, 3rd ser., 6: 129–85.

Weijers, Olga. 1994–2012. *Le travail intellectuel à la Faculté des arts de Paris: textes et maîtres (ca. 1200–1500)*. 9 vols. Turnhout: Brepols.

Weisheipl, James A. 1964. 'Curriculum of the Faculty of Arts at Oxford in the Early Fourteenth Century'. *Mediaeval Studies* 26: 143–85.

Weisheipl, James A. 1984. 'Science in the Thirteenth Century'. In *The History of the University of Oxford*, vol. 1, *The Early Oxford Schools*, ed. J. I. Catto, 435–69. Oxford: Clarendon Press.

Weißer, Christoph. 1982. *Studien zum mittelalterlichen Krankheitslunar: Ein Beitrag zur Geschichte laienastrologischer Fachprosa*. Pattensen: Wellm.

Welborn, Mary Catherine. 1932. 'Calendar Reform in the Thirteenth Century'. Ph.D. diss., University of Chicago.

Welker, Lorenz. 2005. 'Ottaviano Petrucci and the Political-Cultural Elite of his Time: The 1513 Print of Paul of Middelburg's *Summa Paulina de Recta Paschae Celebratione*'. In *Venezia 1501: Petrucci e la stampa musicale*, edited by Giulio Cattin and Patrizia Dalla Vecchia, 107–15. Venice: Fondazione Levi.

Westgard, Joshua A. 2010. 'Bede and the Continent in the Carolingian Age and Beyond'. In *The Cambridge Companion to Bede*, ed. Scott DeGregorio, 201–15. Cambridge: Cambridge University Press.

Wiesenbach, Joachim, ed. 1986. *Sigebert von Gembloux: Liber decennalis*. Weimar: Böhlau.

Wiesenbach, Joachim. 1991. 'Wilhelm von Hirsau: Astrolab und Astronomie im 11. Jahrhundert'. In *Hirsau: St. Peter und Paul 1091–1991*, vol. 2, ed. Klaus Schreiner, 109–56. Stuttgart: Theiss.

Wijk, Walter Émile van. 1932. *De Gregoriaansche kalender: een technisch-tijdrekenkundige studie*. Maastricht: Stols.

Wijk, Walter Émile van. 1936. *Le Nombre d'Or: étude de chronologie technique suivie du texte de la Massa compoti d'Alexandre de Villedieu*. The Hague: Nijhoff.

Wijk, Walter Émile van, ed. 1951. *Le comput emendé de Reinherus de Paderborn (1171)*. Amsterdam: North-Holland.

Wildhagen, Karl. 1961. 'Das Kalendarium der Handschrift Vitellius E XVIII (Brit. Mus.): Ein Beitrag zur Chronologie und Hagiologie Altenglands'. In *Texte und Forschungen zur Englischen Kulturgeschichte*, ed. Heinrich Boehmer et al., 68–118. Halle: Niemeyer.

Willetts, P. J. 1965. 'A Reconstructed Astronomical MS from Christchurch Library Canterbury'. *The British Museum Quarterly* 30: 22–30.

Wilson, Curtis A. 1975. 'Rheticus, Ravetz, and the "Necessity" of Copernicus' Innovation'. In *The Copernican Achievement*, ed. Robert S. Westman, 16–39. Berkeley: University of California Press.

Wingen-Trennhaus, Angelika. 1991. 'Regiomontanus als Frühdrucker in Nürnberg'. *Mitteilungen des Vereins für Geschichte der Stadt Nürnberg* 78: 17–87.

Wollasch, Joachim. 1983. 'Der Kalender Bernolds von St. Blasien'. In *St. Blasien*, ed. Heinrich Heidegger and Hugo Ott, 33–45. Zurich: Schnell & Steiner.

Wordsworth, Christopher. 1904. *The Ancient Kalendar of the University of Oxford: From Documents of the Fourteenth to the Seventeenth Century*. Oxford: Clarendon Press.

Wormald, Francis. 1934. *English Kalendars before ad 1100*. London. Reprint, Woodbridge: Boydell Press, 1988.

Wormald, Francis. 1939–46. *English Benedictine Kalendars after ad 1100*. 2 vols. London: Harrison & Sons.

Wright, Thomas. 1842–46. *Biographia Britannica Literaria; or Biography of Literary Characters of Great Britain and Ireland, Arranged in Chronological Order*. 2 vols. London: Parker.

Wright, Thomas. 1847. 'Inedited Notices Relating to Some Early Writers on the Compotus'. *Transactions of the Royal Society of Literature*, 2nd ser., 2: 68–75.

Yarza Urquiola, Valeriano, and Francisco Javier Andrés Santos, eds. 2013. *Isidoro de Sevilla: Etimologías; Libro V: De legibus—De temporibus*. Paris: Les Belles Lettres.

Zahora, Tomas. 2014. *Nature, Virtue, and the Boundaries of Encyclopaedic Knowledge: The Tropological Universe of Alexander Neckam (1157–1217)*. Turnhout: Brepols.

Zélinsky, André Nicolaévitch. 1982. 'Le calendrier chrétien avant la réforme grégorienne'. *Studi medievali*, 3rd ser., 23: 529–97.

Zemanek, Heinz. 1987. *Kalender und Chronologie: Bekanntes & Unbekanntes aus der Kalenderwissenschaft*. 4th edn. Munich: Oldenbourg.

Zepeda, Henry. 2015. 'Euclidization in the *Almagestum parvum*'. *Early Science and Medicine* 20: 48–76.

Ziggelaar, August. 1983. 'The Papal Bull of 1582 Promulgating a Reform of the Calendar'. In Coyne, Hoskin, and Pedersen 1983: 201–39.

Zinner, Ernst. 1925. *Verzeichnis der astronomischen Handschriften des deutschen Kulturgebietes*. Munich: Beck.

Zinner, Ernst. 1936. 'Petrus de Dacia, ein mittelalterlicher dänischer Astronom'. *Archeion* 18: 318–29.

Zinner, Ernst, ed. 1937. *Der deutsche Kalender des Johannes Regiomontan, Nürnberg, um 1474: Faksimiledruck nach dem Exemplar der Preußischen Staatsbibliothek*. Leipzig: Harrassowitz.

Zinner, Ernst. 1938. 'Die wissenschaftlichen Bestrebungen Regiomontans'. *Beiträge zur Inkunabelkunde*, n.s., 2: 89–103.

Zinner, Ernst. 1939. 'Die ältesten Räderuhren und modernen Sonnenuhren: Forschungen über den Ursprung der modernen Wissenschaft'. *Bericht der Naturforschenden Gesellschaft Bamberg* 28: 1–148.

Zinner, Ernst. 1951. 'Magister Alard von Diest und die Pariser Beobachtungen von 1312–15'. *Isis* 42: 38–40.

Zinner, Ernst. 1952. 'Der deutsche Volkskalender des 15. Jahrhunderts'. *Bericht der Naturforschenden Gesellschaft Bamberg* 33: 46–50.

Zinner, Ernst. 1964. *Geschichte und Bibliographie der astronomischen Literatur in Deutschland zur Zeit der Renaissance*. 2nd edn. Stuttgart: Hiersemann.

Zinner, Ernst. 1968. *Leben und Wirken des Joh. Müller von Königsberg, genannt Regiomontanus*. 2nd edn. Osnabrück: Zeller.

Zinner, Ernst. 1988. *Entstehung und Ausbreitung der Copernicanischen Lehre*. 2nd edn. Munich: Beck.

Zironda, Renato, ed. 1985. *Summario de la luna con alchune rasone de astrologie extracte da varii auctori*. Venice: Galla.

Zuidhoek, Jan. 2017. 'The Initial Year of *De ratione paschali* and the Relevance of Its Paschal Dates'. In *Late Antique Calendrical Thought and Its Reception in the Early Middle Ages*, ed. Immo Warntjes and Dáibhí Ó Cróinín, 71–93. Turnhout: Brepols.

Zurbonsen, Friedrich. 1884. *Hermannus Zoestius und seine historisch-politischen Schriften: Nach handschriftlichen Quellen des fünfzehnten Jahrhunderts*. Warendorf: Schnell.

Zurbonsen, Friedrich. 1899. 'Hermann Zoestius von Marienfeld und seine Schriften'. *Westdeutsche Zeitschrift* 18: 146–73.

Index

MANUSCRIPT INDEX

GENERAL INDEX

The manufacturer's authorised representative in the EU for product safety is Oxford University Press España S.A. of El Parque Empresarial San Fernando de Henares, Avenida de Castilla, 2 - 28830 Madrid (www.oup.es/en or product.safety@oup.com). OUP España S.A. also acts as importer into Spain of products made by the manufacturer.
Printed and bound by CPI Group (UK) Ltd, Croydon, CR0 4YY
14/09/2026
02216308-0014